AGRICULTURE AS A MIMIC OF NATURAL ECOSYSTEMS

Current Plant Science and Biotechnology in Agriculture

VOLUME 37

Aims and Scope
The book series is intended for readers ranging from advanced students to senior research scientists and corporate directors interested in acquiring in-depth, state-of-the-art knowledge about research findings and techniques related to all aspects of agricultural biotechnology. Although the previous volumes in the series dealt with plant science and biotechnology, the aim is now to also include volumes dealing with animals science, food science and microbiology. While the subject matter will relate more particularly to agricultural applications, timely topics in basic science and biotechnology will also be explored. Some volumes will report progress in rapidly advancing disciplines through proceedings of symposia and workshops while others will detail fundamental information of an enduring nature that will be referenced repeatedly.

The titles published in this series are listed at the end of this volume.

Agriculture as a Mimic of Natural Ecosystems

Edited by

E. C. LEFROY

Centre for Legumes in Mediterranean Agriculture,
University of Western Australia,
Nedlands, Australia

R. J. Hobbs

CSIRO Wildlife and Ecology
Perth, Australia

M. H. O'CONNOR

Muresk Institute of Agriculture,
Curtin University,
Northam, Australia

and

J. S. PATE

Botany Department,
University of Western Australia,
Perth, Australia

KLUWER ACADEMIC PUBLISHERS
DORDRECHT / BOSTON / LONDON

Library of Congress Cataloging-in-Publication Data

ISBN 978-90-481-5319-0

Published by Kluwer Academic Publishers,
P.O. Box 17, 3300 AA Dordrecht, The Netherlands.

Sold and distributed in North, Central and South America
by Kluwer Academic Publishers,
101 Philip Drive, Nowell, MA 02061, U.S.A.

In all other countries, sold and distributed
by Kluwer Academic Publishers,
P.O. Box 322, 3300 AH Dordrecht, The Netherlands.

Printed on acid-free paper

Printed in the Netherlands.

Agriculture as a Mimic of Natural Ecosystems

E. C. Lefroy, R. J. Hobbs, M. H. O'Connor and J. S. Pate

Table of Contents

Part 3. Case studies of multi-species systems

Participants at the workshop 'Agriculture as a Mimic of Natural Ecosystems' held at 'Munthoola', Williams, Western Australia, September 2–6, 1997. From left to right, Richard Hobbs, Pauline Grierson, Brian Keating, Tom Hatton, Kirsten Verburg, Ian Fillery (partly obscured), Mark Adams, John Williams, Tina Bell (front), Meine van Noordwijk (rear), Jack Ewel, Roslyn Prinsely, Dan Wildy (rear), Murray Unkovich (partly obscured), Deborah Neher (front), Frank Dunin (rear), John Passioura, Steve Morton (partly obscured), Wes Jackson, Michael O'Connor (front), David Arthur (rear), Bert Main, Chin Ong (front), Ted Lefroy, Brian Trenbath (front), Bob Nulsen (rear), John Bartle (partly obscured), David Pannell (front), Rae Fry (front), Todd Dawson, Torbjorn Rydberg (rear), Richard Joffre (front), Rod Lefroy (rear), and John Pate (absent; Richard Stirzaker, Joan Olsen, Warren Mason and Andrew Campbell).

Introduction

E. C. LEFROY[1], R. J. HOBBS[2], M. H. O'CONNOR[3] and J. S. PATE[4]
[1]*Centre for Legumes in Mediterranean Agriculture, The University of Western Australia, Nedlands WA 6907, Australia;* [2]*CSIRO Wildlife and Ecology, Private Bag, PO Wembley, WA 6014, Australia;* [3] *Muresk Institute of Agriculture, Curtin University of Technology, Bentley WA 6102, Australia;* [4]*Botany Department, The University of Western Australia, Nedlands WA 6907, Australia*

1. In search of sustainable agriculture

In 1792, a French scientific expedition under the command of Captain Joseph D'Entrecasteaux dropped anchor in an uncharted bay on the south western coast of Australia to carry out repairs and replenish supplies of fresh water. So impressed was D'Entrecasteaux with the skill of the seventeen year old midshipman at the masthead and the manoeuvrability of his ship in negotiating the outlying reefs he named the adjacent cape LeGrande after the sailor and the port Esperance after his ship (Jacob and Vellios, 1987). He was far less impressed with the hinterland. Indeed his account and those of other French explorers of the time describing the infertility of the soils, the lack of fresh water and in particular the prevalence of saline and brackish water is credited with deterring the French from any idea of settlement (Marchant, 1982). Charles Darwin was similarly unimpressed when the *Beagle* made landfall on this coast in 1836:

> The general bright green colour of the brushwood and other plants viewed from a distance seems to bespeak fertility; a single walk will however dispel any such illusion (quoted in Armstrong, 1985).

The British however were not to be as easily deterred as the French and by the 1850s the first European farmers had settled in the region. Initially challenged by major nutrient imbalances, trace element deficiencies and the semi-arid climate, they eventually succeeded in growing wool and later cereal crops but problems soon appeared (Western Australian Government, 1917). Soil erosion, salinity, and soil acidity induced by nitrate leaching have since seriously compromised the agricultural productivity of this region (Hobbs and Saunders, 1993) just as in many other parts of southern Australia and marginal lands the world over.

In 1992, a group of Esperance farmers disenchanted with existing approaches to sustainable agriculture began to search widely for solutions. They came across the long term research programme of The Land Institute in Kansas, USA where high seed-yielding mixtures of perennial plants are being developed as grain producing mimics of natural prairie grasslands (Jackson, 1985) and invited its director Wes Jackson to address a local conference on sustainable agriculture.

As Wes toured the grain growing areas of south western Australia talking to farmer groups (Jackson, 1993), it occurred to one of us (ECL) that the approach he was advocating deserved a much wider audience, one that included agronomists, ecologists and others concerned with the global phenomenon of agriculture undermining its resource base. At this point the concept of the workshop that resulted in this volume was born. Jackson's idea of designing agriculture as a structural and functional mimic of natural ecosystems offered a simple, integrated principle for working towards sustainable agriculture and one that might well overcome the inherent weakness of approaching problems in isolation when it is obvious to all that they involved inextricably linked elements.

But the concept also raised many questions in our minds. How can we reconcile the fundamental differences between a natural system featuring little if any export of materials and energy, its productivity shackled by low and uncertain resource availability, with agriculture where harvest and export are the primary goals? How can we confidently relate the structure and function of a mimic system when so little is known about the underlying processes that confer persistence and resilience on the natural ecosystems on which a mimic is based? Does the mimic concept provide testable hypotheses and if so, how widely can their outcomes be applied? What can serve as a model in disturbed environments where the relevance of 'original vegetation' is either meaningless or fallacious? Which characteristics of natural ecosystems are essential and how can we positively identify redundant components? Could the application of the mimic concept lead to flexible and adaptive systems which allow farmers to respond to changing economic and environmental pressures? And finally would this idea travel beyond the relatively fertile Great Plains of Kansas to environments less suited to agriculture?

From 2 to 6 September 1997, the contributors to this volume met at 'Munthoola' farm near Williams, 160 km south east of Perth, Western Australia to discuss these issues. The aim of the workshop 'Agriculture as a Mimic of Natural Ecosystems' was to explore the concept that by mimicking the structure and function of natural ecosystems we might be able to improve the sustainability of agriculture across a range of environments.

While the original impetus for the workshop was to address the declining agricultural resource base in southern Australia, we were keen to broaden the geographical scope of this meeting and evaluate the relevance of the topic to issues and environments on a global basis. Agriculturalists and ecologists working across a broad range of environments, systems and disciplines were invited accordingly, with particular emphasis on those with experience in comparative studies of natural and agricultural ecosystems. We also invited researchers who had studied multispecies agricultural systems, given that the increased complexity which they embody appears to be a logical and inevitable characteristic of agricultural systems that attempt to imitate the structure and function of natural ecosystems. This brought to our group expertise in natural and agricultural systems in sub-Saharan Africa, the Mediterranean basin, the humid tropics of Indonesia and Central America, the Great Plains of North America and the Indian sub-continent, as well as temperate and semi-arid Australia.

2. Outline of the book

This volume is divided into five sections. The first section is devoted to two chapters by Wes Jackson, in recognition of his role as the inspiration for the workshop. The first of these, co-authored with his daughter Laura Jackson, is a retrospective look at 20 years of research at The Land Institute. It summarises the conceptual and experimental development of a permanent, low-input grain producing system composed of complementary mixtures of perennial grasses, legumes and sunflowers. In the second, the ecosystem concept is moved down to the scale of the genome, to examine the implications of transferring genetic material across large evolutionary distances.

The second section, consisting of three chapters, deals with conceptual issues involved in using natural ecosystems as models for land use systems. The third section is made up of six case studies of multispecies landuse systems from around the world. The fourth section contains seven chapters that deal with the application of the mimic concept to southern Australia. In the final section, two chapters discuss the broader implications of the mimic concept for farmers and the landscape, while the concluding chapter provides a summary of the preceding discussion and an attempt to identify the emerging themes.

3. Acknowledgments

The workshop 'Agriculture as a Mimic of Natural Ecosystems' was made possible by grants from the Rural Industries Research and Development Corporation, the Meat Research Corporation, the Land and Water Resources Research and Development Corporation, Agriculture Western Australia and the Co-operative Research Centre for Legumes in Mediterranean Agriculture at The University of Western Australia. The organisers also wish to acknowledge the enthusiasm of all the participants, particularly those who were motivated to complete aspects of their studies and initiate new research which is reported in this volume. We also acknowledge Mardon Hundley and friends for bringing Wes Jackson to Esperance in 1992 and John Hamblin, Phil Price, Phil Cocks, John Williams and the departments of Plant Sciences and Botany at The University of Western Australia for their support. The Henderson family at 'Munthoola', Chris Brown, Ken Wallace, Elizabeth Pate and the Melvin family gave generously of their time and hospitality during the workshop and field trips. Shelley James assisted with compiling the manuscript. Finally we wish to acknowledge the valuable contributions made by our external referees; Henry Nix, Rex Oram, Charlie Sing, Steve Hopper, Anna Ridley, Meka Rao, Ian Nuberg, Brian Walker, David Freudenberger, David Wardle, Des Whitfield, Peter Thorburn, James Noble, Ian Noble, David Evans, Richard George, Ray Froend, Bill Loneragan, Kathie Meney, Tony Chisholm and Sally Marsh.

4. References

Armstrong P (1985) Charles Darwin in Western Australia. University of Western Australia Press

Ewel JJ, Mazzarino MJ and Berish CW (1991) Tropical soil fertility changes under monocultures and successional communities of different structure. Ecological Applications 1: 289–302

Hobbs RJ and Saunders DA (1993) Reintegrating Fragmented Landscapes: Towards Sustainable Agriculture and Nature Conservation. Springer Verlag, New York

Jackson W (1985) New Roots for Agriculture. University of Nebraska Press, Lincoln, Nebraska

Jackson W (1993) Meeting the expectations of the land. In: Lefroy EC (ed) Visions for Agriculture, pp 1–16. Miscellaneous Publication 6/93, Department of Agriculture Western Australia

Jacob TK and Vellios J (1987) Southland – Voyages of Discovery to the Great South Land. Ministry of Education, Perth WA

Marchant LR (1982) France Australe: A Study of French Explorations and Attempts to Found a Penal Colony and Strategic Base in South-western Australia, 1503–1826. Artlook books, Perth WA

Western Australian Government (1917) Report of The Royal Commission on The Mallee Belt and Esperance Lands Together with Appendices and Reports of Evidence, Government Printer, Perth, 192 pp

Developing high seed yielding perennial polycultures as a mimic of mid-grass prairie

W. JACKSON[1, *] and L. L. JACKSON[2]
[1] *The Land Institute, Salina, Kansas, USA;* [2] *Biology Department, University of Northern Iowa, Cedar Falls, Iowa, USA (*Author for correspondence: theland@igc.apc.org)*

Key words: *Cassia marilandica, Desmanthus illinoensis, Helianthus maximilianii, Leymus racemosus*, perennial polycultures, r and K selection, *Sorghum halapense* × *S. bicolor, Tripsacum dactyloides*

Abstract. Over the past two decades, work at The Land Institute has aimed to develop sustainable farming as a functional mimic of mid-grass prairie. Ecological studies of unploughed prairie suggested that an agricultural mimic should contain representatives of four main groups; warm season perennial grasses (C_4), cool season perennial grasses (C_3), perennial legumes and perennial members of the sunflower family. Then followed a systematic search of the literature and germplasm collections for suitable species within those target groups. Biological studies were carried out on six candidate species, the C_3 grass *Leymus racemosus* [Lam.] Tsvelev, legumes wild senna (*Cassia marilandica* L.) and Illinois bundleflower (*Desmanthus illinoensis* [Michx.] MacM), the C_4 grasses *Tripsacum dactyloides* [L.] L., and the hybrid *Sorghum halapense* (L. Persoo) × *S. bicolor* (L. Moench), and the perennial sunflower *Helianthus maximilianii* Schrad. Experiments involving these species in monocultures and various polyculture combinations have set out to answer four questions; Can herbaceous perennialism and high seed yield go together? Can a polyculture of perennial seed producers outyield the same species in monoculture? Can a perennial polyculture provide most of its own fertility? Can a perennial polyculture successfully manage insects, pathogens, and weeds? To the first question, evidence is presented that the long held view that there is a necessary trade-off in resource allocation between seed production and survival does not necessarily hold. It is suggested that other insights of value to managed ecosystems in general will flow from an evolutionary ecological approach to sustainable agriculture that uses nature as model.

1. Introduction

Nearly all researchers in sustainable agriculture work with species already in the crop inventory. They design experiments to explore alternative strategies for seedbed preparation, planting, maintenance, and eventual harvest. Most strategies involve diversity over time (rotations that usually include a legume) or space (polycultures). The motivation may be an interest in lowering production costs for farmers (which could help more farmers to stay on the land), or in reducing the impact on the environment or dependency on non-renewable resources such as fossil fuels.

The time frame for most investigators is the near future. This is a practical approach designed to help the farmer and reduce environmental damage now. In some cases, the research is aimed at achieving a scientific understanding of certain traditional farming methods, partly for the purpose of adding legitimacy to the wisdom of tradition and perhaps with the hope that the added scientific respect and more finely-tuned knowledge will lead to an expansion of sustainable agriculture everywhere. Examples of such research can be found in the work of Altieri (1987), Francis (1986), Gliessman (1984), the Rodale Research Center (Drinkwater et al., 1998), and elsewhere.

Workers at The Land Institute share the values of these researchers, but our research has a somewhat different motivation, approach, and time consideration. Rather than deal with problems in agriculture here and now, we address the 'problem of agriculture', the problem that everywhere it seems agriculture undermines the basis of its existence. This problem has been with both humanity and the earth since agriculture began some 8000 to 10,000 years ago. This long backward look in time has helped us to consider corrective measures for a time into the future that may be longer than that of most agricultural researchers. Rather than the ten years or less that most workers in sustainable agriculture have in mind as they design experiments or promote certain practices, we think the first benefits of our work will be realised after perhaps 25 or more years, but that most will be in the 50 to 100 year time consideration (Jackson, 1985; Jackson et al., 1987). For forestry and forage, benefits from this philosophy could be immediate.

1.1. *The prairie as standard*

The starting point in our search for solutions is nature. The term nature is, of course, elusive. Consequently, it is an elusive or slippery standard. Nevertheless, it seems to us at The Land Institute less slippery or elusive than any other standard when sustainability is our primary criterion. The nature we look to is the native prairie, and when we compare prairie with the ordinary grain crop, such as a corn or wheat field, important differences become apparent. From the sloping fields tended by humans, valuable nutrients run toward the sea where, for all practical purposes, most of them are gone for good. The prairie, on the other hand, by sucking nutrients from parent rock material or subsoil, all the while bathing them with the chemicals produced by life, actually builds soil. Even though diversity may not yield stability overall, chemical diversity inherent in plant species diversity on the prairie confronts insects and pathogens so that epidemics, so common to agricultural monocultures, are rare on the prairie. Since no creature has an all-consuming enzyme system, diversity does yield some protection from pests.

The prairie does not require, therefore, the introduction of chemicals with which humans have had no evolutionary experience. The prairie, like nearly all of nature, runs mostly on contemporary sunlight, while our modern agricultural fields benefit from the stored sunlight of flora extinct for hundreds of millions of years. A casual examination of the ordinary differences between a prairie ecosystem and an agricultural field will cause us to see that the prairie features perennials in a polyculture, while agriculture features annuals in monoculture.

Our work at The Land Institute is devoted primarily to exploring the feasibility of an agriculture that features herbaceous perennials grown in mixtures for seed production as substitutes for annual monocultures grown on ground that can erode. These mixtures would be domestic prairies, cropping systems analogous to the vegetative structure of the prairie (Jackson, 1985).

Our overall scheme is illustrated in Figure 1. In studying this figure, it is useful to keep in mind the four basic questions we address in our experiments at The Land Institute:

1. Can herbaceous perennialism and high seed yield go together? A persistent argument has been that perennial plants must reallocate some resources to parts below-ground. It could be difficult, therefore, to breed perennials to outyield annuals which died after seed set. Because perennial species differ, however, in the relative and absolute amounts of resources devoted to seed, it seemed worthwhile to search for species with high seed yield. Once these species are discovered, breeding work can begin. The potential improvement in seed yield depends on the range of existing genetic variability in the wild. To assess this variation requires adequate sampling across the geographic range of the species and then an evaluation of the collection within a common garden. So far we have mentioned two categories of herbaceous perennials: the wild introduced and the wild native species. A third category is the domestic annual that we wish to turn into a winter-hardy perennial, whether through hybridization with a close relative or some use of biotechnology.

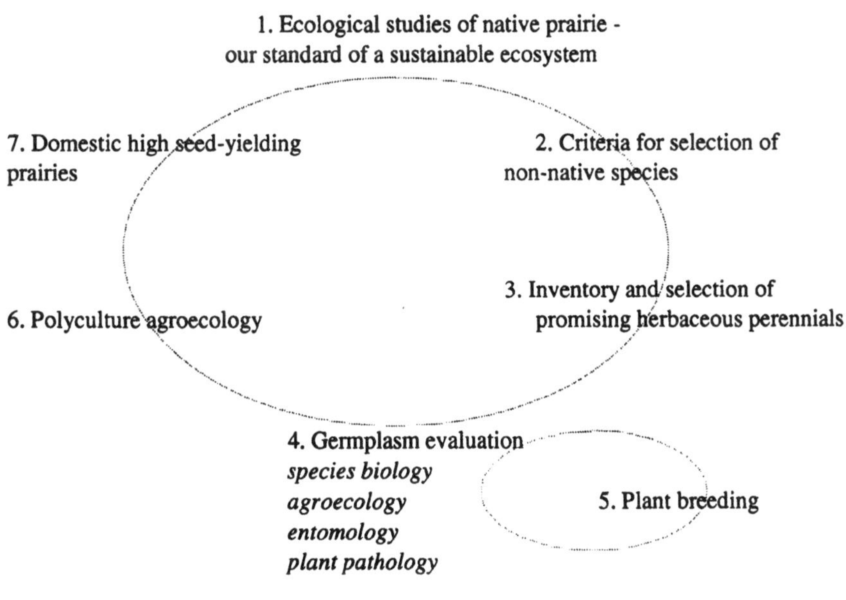

Figure 1. Research at The Land Institute.

2. Can a polyculture of perennial seed producers outyield the same species grown in monoculture? Overyielding occurs when competition between species is less intense than between members of the same species. Polycultures typically yield more than monocultures due to differences in resource use and timing of demand.
3. Can a perennial polyculture provide most of its own fertility? Can biological nitrogen-fixation and phosphate accumulation compensate for harvested nutrients? Soil nutrient pools, nutrients in the seed, and the capacity of crop plants to enrich the soil need to be documented.
4. Can a perennial mixture successfully manage insects, pathogens, and weeds? Breeding for resistant lines and studies on the effects of species diversity must converge if we are to protect a crop. Insect pests can be managed through a combination of predator attraction and preventing an insect from locating host plants. Mixtures of species, and of genotypes within species, may reduce the incidence and spread of disease. Weeds may be controlled either allelopathically or via continuous shading of the soil surface by the perennials (Jackson, 1985).

This chapter reviews our attempt to answer these questions with an emphasis on the first decade of species selection, biological studies and polyculture assembly.

2. Prairie studies

Our initial ecological studies were directed at the question of whether two species were more likely to be associated with one another than with any third species. James Peterson and Phil Haves investigated this question at the Fent prairie, eight miles north of The Land Institute, an unploughed native prairie that had never had forced grazing. Their studies indicated no preferential associations.

In a preliminary investigation, Piper (1986) asked i) how much aboveground plant life is supported each year by the prairie, and ii) what are the proportions of grasses, legumes, and composites. These are the families that dominate the native prairie and also happen to be those comprising most of our temperate agricultural species. Net biomass production at the grassland sites (500–700 g m^{-2}) was similar to that of many Midwestern crops. At their peaks, grasses composed 67% to 94% of plant matter; legumes and composites represented 16% and 11% of vegetation, respectively. Piper (1986) concluded that these results suggest that polycultures of perennial warm- and cool-season grasses and legumes are feasible.

Subsequent study evaluated the species mix of the same three families and estimated the richness, diversity, and evenness of species in this part of the mid-grass prairie and examined the phenological differences among the major families (Gernes and Piper, 1987). Their data came from clipping aboveground phytomass within twelve 0.25 m^2 quadrats on each of three prairie sites during early May, early June, and mid-August. Average phytomass per site ranged from 71.8 to 151.1 g m^{-2} in May and from 377.3 to 1077.2 g m^{-2} in August. Species diversity varied among sites and dates from 2.3 to 10.6 (unitless measure). Low diversities and high phytomass resulted from dominance by *Andropogon gerardii* on all sites. Species in the Poaceae, Euphorbiacea, Malvaceae, Labiatae, and Asteraceae were present in all three sites, but only Poaceae and Asteraceae were sampled in all sites on all dates. There were no significant differences in mean flowering and fruiting time among Poaceae, Fabaceae, and Asteraceae. The data suggest that our domestic polyculture be dominated by a C_4 grass, but associated with a C_3 grass and a few forbs.

3. Searching for high seed-yielding herbaceous perennials

The most pressing biological question at The Land Institute is whether perennialism and high seed yield can go together. Stated otherwise, how do the annual and long-term seed yields of herbaceous perennial plants compare with those of annual grain crops? The relationship between perennialism and high yield involves a question that gives emphasis to sustained production. Prairies, after all, feature perennials, but they do not feature high seed yield.

The idea of perennial grain crops has met considerable resistance. (The herbaceous perennials of which we speak are those of the polycarpic variety rather than monocarpic species which may live from year to year but die upon flowering.) The primary basis for objection comes from the theory of resource allocation in the evolution of life history strategies. A fundamental assumption of this theory is that each organism has a finite supply of resources to be allocated between competing functions of reproduction and growth. An increase in resources allocated to reproduction should reduce the ability of the organism to survive in a competitive environment. Thus, there is thought to be a strict 'trade-off' between longevity and reproduction.

This intellectually satisfying notion has its roots in thinking about animals (Cody, 1966) but was adopted soon thereafter in plants (Harper and Ogden, 1970). These early studies compared the life history characteristics (number and size of offspring and schedules of reproduction and mortality as a function of age) across species in a continuum of early- to late-successional environments. The trade-off idea was bolstered by placing annual crops in the continuum of species with different amounts of reproductive allocation. Harper (1977) and Snaydon (1984) suggested that annual crops have the highest reproductive efforts (ratio of reproductive total to dry weight) of all measured species because they are the 'weakest' due to their dependence on humans. Donald (1981) extended this logic to propose that we select for increased seed yield in crop plants by reducing the plants' competitive ability and relying more on humans to compensate for the weakness.

With further study, it became apparent that resource allocation was more difficult to measure than originally thought. First, nutrients, water or meristems may be more limiting than carbon, or their allocation may be different from that of carbon (Hickman and Pitelka, 1975; Thompson and Stewart, 1981; Abrahamson and Caswell, 1982; Watson and Casper, 1984). Different resources may be limiting at different times, for different aspects of reproduction. Clearly, biomass could not be used as a straightforward measure of reproductive effort.

Second, fruits and flowers may be green and photosynthesise, partially supporting themselves. The presence of a strong sink for photosynthate, such as a developing fruit, may elevate rates of photosynthesis locally. Therefore, the carbon content of a fruit may not reflect its cost to the plant in terms of sacrificed growth (Hole and Scott, 1983; Reekie and Bazzaz, 1987; Tissue and Nobel, 1990). This insight was important because it indicated that the idea of a fixed carbon budget ('there's only so much to go around') was not an absolute law. Several terms – reproductive effort, reproductive cost, or reproductive allocation – were used interchangeably to mean the proportion of plant dry mass devoted to reproductive structures, which were variously defined by the researchers. In most cases, no distinction was made between the *measurement* as percent dry mass and the *interpretation* of the data in terms of 'effort', 'cost', or 'allocation'.

In response to these criticisms, researchers began to measure the demographic consequences of reproduction, such as growth, risk of mortality, or reproductive output in subsequent years. Some studies employed natural experiments, in which plants with different costs of reproduction were compared (e.g. fruiting vs. non-fruiting individuals, females and males of dioecious species; Onwekwelu and Harper, 1979; Gross and Soule, 1981; Policansky, 1981; Sakai and Burris, 1985; Lovett Doust and Lovett Doust, 1987). Other studies manipulated leaf area and/or seed production experimentally (e.g. Horvitz and Schemske, 1988; Lubbers and Lechowicz, 1989; Primack and Hall, 1990; Fox and Stevens, 1991).

Many (but not all) of the demographic studies found that plants with higher reproductive output indeed experienced a demographic cost. Their results implied that, at some level, there is a direct physiological cost of reproduction. However, it is not clear to what extent induced phenotypic variation in reproduction can predict life history evolution (Reznick, 1985). None asked the questions pertinent to our work: is there genetic variation for higher reproductive output in perennials? How does this variation correlate with other features of plant longevity, such as stress tolerance or vegetative vigour? What happens to longevity when there is selection for higher reproductive output, and what is the mechanism for these correlated changes (pleiotropy)? Only one study, by Law (1979), measured negative genetic correlations between the age of first reproduction and plant longevity in *Poa annua*; no other species has been tested in this way.

At its heart, life history theory depends upon the notion of evolutionary optimisation. Species are thought to have optimised their energy allocation, such that any increase in reproductive output, no matter how small, should automatically entail reductions in some measure of plant vigour. An appropriate test of this prediction is to examine genetically based differences in seed yield within a species, to learn how those differences were created, and to compare aspects of plant vigour or longevity in the low and high yielding strains.

Jackson and Dewald (1994) studied a normal and high-yielding form of *Tripsacum dactyloides*, a wild perennial grass. The 'pistillate' condition, which is contributed by recessive alleles at a single locus, results in plants that produce five times the seed mass of normal plants due to feminisation of staminate spikelets. They found that this increase in total reproductive output, even when pollen biomass was included in the calculations, did not come at the expense of plant vigour or longevity. Seed production in pistillate plants of *T. dactyloides* was higher than in full sibs with the normal phenotype over the entire range of plant sizes, from just a few centimetres to over a metre in circumference. Seed yield of pistillate plants was also consistently higher than that of normal plants even when both types were severely defoliated. Under stressed and unstressed conditions, pistillate plants had equivalent or better growth rates, survival and carbohydrate storage, and performed equally well in the following year.

The key to these counter-intuitive results appears to lie in the physiological independence of reproductive tillers bearing seeds in the grass family. After the meristem of a vegetative tiller transforms itself into a reproductive meristem, producing floral initials where it once produced new leaves, the internodes lengthen into a tall stalk, and the grass inflorescence is elevated to 1–2 m above the ground. Once this period of internode elongation is complete, the reproductive tiller or 'seed stalk' becomes a net supplier of photosynthate to the rest of the plant. When Jackson and Dewald (1994) removed the leafy seed stalks from normal and pistillate plants, they found that vegetative growth slowed and below ground reserves were reduced relative to controls in which stalks remained intact. Removal of the leaves only from seed stalks reduced seed production in pistillate plants, but not normal plants. This result suggested that even when defoliated, the photosynthetic capability of normal seed stalks and inflorescences is entirely sufficient to fill seeds. However, pistillate plants were very sensitive to stalk defoliation. Jackson and Dewald proposed that normal individuals of *T. dactyloides* are 'overbuilt' so that some seed production is possible even under conditions of severe defoliation, while pistillate individuals may be operating closer to the maximum capacity of the seed stalk to supply photosynthate to developing seeds.

The pistillate mutation did have negative fitness consequences for seedling establishment (Jackson et al., 1992). However, these were related to average seed size and altered glume and lemma morphology, rather than to an energy related trade-off. Since the study was completed, breeding for triploid, apomictic strains of the pistillate type has continued to increase seed yields without decreasing plant size, vigour, or forage quality (C. Dewald, pers. comm.).

Early on, as we began to examine the possibility of perennialism and high yield, it was repeatedly pointed out by those familiar with the harvest of orchard crops that numerous domesticated woody perennials have not only high reproductive efforts but high yields on a per-acre basis as well. For example, Pritts and Hancock (1985) found that blueberries between three and 11 years old allocate more than 50% of their annual biomass production to fruit. In a separate study of peach trees, the proportion of dry matter allocated each year to fruit increased from less than 30% for young trees to almost 70% in mature trees (Flore and Layne, 1996). Grapes showed similarly impressive numbers. In a study covering three years of harvest of Thompson Seedless grapevines, the proportion of dry biomass partitioned to the fruit ranged from 44% to 69% of the total biomass accumulated (Williams, 1996). Table 1 highlights the yields of several domesticated fruit- and nut-producing perennials on a per-hectare basis; the average US yield for wheat is shown for comparison purposes.

Table 1. Yields of various cultivated perennials and wheat. US average yield in t ha^{-1} at 10% moisture (Marty Bender, The Land Institute).

Species	Whole[3]	Without refuse[4]
Woody domesticates[1]		
English walnuts	3.4	1.6
Hazelnuts	2.9	1.3
Apples	4.3	3.8
Pears	5.8	5.4
Oranges	6.5	3.8
Grapefruit	8.5	2.5
Herbaceous domesticates[2]		
Strawberries	3.8	3.6
Bananas (Hawaii)	4.3	2.7
Wheat[3]	na	2.5

[1] Woody trees require more than a few years to bear, so non-bearing acreage probably amounts to 5–10% of total acreage (7% for apples in Michigan; USDA National Agricultural Statistics Service, State Statistical Office, Michigan). Hence, the reported yields for total acreage are probably 90–95% of those for bearing acreage.

[2] Strawberries mature during the first year and bananas by the second year, so that only a very small portion of total acreage is non-bearing or slightly bearing. Hence, the reported yields for harvested acreage closely approximate those for total acreage.

[3] USDA (1996). Yields converted from tons per acre. Each yield is averaged over bearing acreage; the yields of strawberries and bananas, however, are for harvested acreage. Bearing acreage is defined as harvested acreage plus acreage that is only slightly bearing, not enough to be economically harvestable. See footnotes 1 and 2.

[4] USDA (1975). This source contained data on refuse content and on moisture content of the edible part. Refuse refers to cap, shell, core, or peel. Caps, shells, and cores were assumed to have the same moisture content as the edible part. Grapefruit and banana peels were assumed to have the same moisture content as orange peels.

Are woody plants somehow different from herbaceous perennials in their ability to allocate resources to reproduction? In theory, no. Among temperate deciduous species at least, both woody and herbaceous perennials must produce new leaves, vascular tissue, structural tissue, and roots each year. And more importantly, both must store some energy – annuals in seeds and perennials in dormant shoots and roots – to 'pay' for next year's emergence from dormancy.

Despite the evident productivity of many woody crops, critics insist that perennials should have more maintenance costs than do annuals, and thus be unable to afford to produce as much fruit and seed. Indeed, perennial grasses do have to pay for the metabolism of overwintering dormant shoots and roots. But annual plants also pay for the metabolism of overwintering seeds, and must invest in a whole new set of roots to explore the soil volume each spring, while perennials have a head start. In years of poor rainfall or other adverse conditions, the head start that perennials possess could more than make up for their extra metabolic costs.

The differences between annual and perennial crops are neither as great nor as fundamental as they seem. No matter whether the crop is perennial or annual, the agriculturalist's goal is to establish a plant canopy over a given area of ground each year that can quickly become physiologically independent, a net accumulator of the products of photosynthesis. Regardless of the crop's life history strategy, two things are needed to establish a plant canopy in a temperate climate: meristems and 'start-up' energy. If the meristems are distributed evenly over the area, and conditions are sufficient for nearly all meristems to produce a viable shoot, canopy establishment will require little start-up energy.

In annual crops, the meristems needed for canopy establishment are seeds, and the start-up energy is stored in the seed itself. In perennials, the meristems are dormant buds, and the start-up energy is in the plant's roots and belowground shoots. In annual crops, the proportion of last year's energy which must be stored and replanted as seed is very low because of thousands of years of selection for germination characteristics, and the cultural practice of preparing and defending (through control of competition) the seed bed. However, in wild herbaceous perennials as in wild annuals, many overwintering meristems die and much energy is wasted before a canopy can become established.

In domesticated woody perennials, we control the distribution of meristems in space and their age structure by annual pruning. Herbaceous perennials will be harder to manage. The biggest problem will be bare spots and areas that become too thick as the stand ages. We will need to select for characteristics in overwintering meristems that will promote efficient stand creation and maintenance. We will also have to investigate and develop cultural methods to create an optimum 'bud bed' functionally equivalent to our improvements in the annual crop's seed bed, and to pruning in woody crops.

Research on the perennialism and high yield question will necessarily require several years, for it amounts to an investigation of long-term demographic patterns in perennial seed production – a field that is largely unexplored. Studies thus far at The Land Institute have shown increases, decreases, or oscillations in seed yield over time. Details of these results will be presented here, but first some background on this part of our inventory phase.

3.1. *Identifying candidate species*

Our approach to the identification of high seed yielding perennials was to initiate a simultaneous three way inventory:

1. A review of the literature on seed yield in winter-hardy herbaceous perennials.
2. Establishment of an herbary consisting of 300 species of herbaceous, winter-hardy perennials.
3. A collection of 4100 accessions of over 100 species representing seven cool-season grass genera collected worldwide.

While our inventory effort continues, in recent years we have concentrated on five species plus a hybrid of our making.

3.1.1. *Literature review of seed yield in perennials*

The literature on seed yield in herbaceous perennials is dominated by results gained from species used as forage. It is of biological importance to note that in essentially every case, before a species can be suitable for forage, the problem of seed production and harvest must be solved. This means that yield, shatter resistance, and determinate seed set must be sufficiently high that a harvestable seed crop is practical. Vegetative propagation for pasture establishment is impractical. Therefore, the seed yield results for herbaceous perennials from forage crops is relevant for our work. Selection for seed in herbaceous perennials for pastures has probably been under way for millennia. A summary of reported seed yield of herbaceous perennials is shown in Table 2. Data on protein content of perennial grasses and legumes is compared with that of annual crops in Table 3.

3.1.2. *Observations from nature and herbary establishment*

At the same time we were searching the literature for yield records among perennials, we considered our own field experiences and those of colleagues. Plants that seemed to be high seed producers were planted at The Land Institute so that we might more carefully observe the problems of establishment and seed yield under cultivated conditions. Soon after, we concluded that any winter-hardy, herbaceous perennial flowering plant was a candidate for a 5 m observation row. Approximately 300 species were established in the herbary with varying degrees of success.

Table 2. Range of reported seed yields for herbaceous perennial forage species (Marty Bender, The Land Institute).

Pasture type	Seed yield (kg ha^{-1})	Species	Reference
Dryland sown grasses			
High	1011	*Sporobolus cryptandrus*	Brown (1943)
Low	337	*Elymus junceus*	Cooper (1957)
Irrigated sown grasses			
High	1939[a]	*Buchloe dactyloides*	Ahring (1964)
Low	1044	*Lolium perenne*, *Lolium multiflorum*	Wheeler (1957)
Dryland sown native grasses			
High	1011	*Sporobolus cryptandrus*	Brown (1943)
Low	225	*Elymus canadensis*, *Agropyron smithii*, *Andropogon gerardi*	Cooper (1957)
Irrigated sown native grasses			
High	1939[a]	*Buchloe dactyloides*	Ahring (1964)
Low	562	*Agropyron smithii*, *Bouteloua curtipendula*, *Eragrostis trichodes*, *Sorghastrum nutans*	Cooper (1957)
Irrigated legumes			
High	1335	*Desmanthus illinoensis*	USDA (1978)
Low	449	*Vicia villosa*	Wheeler (1957)

[a] Includes burr.

Table 3. Seed protein content of annual and perennial grasses and legumes (Marty Bender, The Land Institute).

Species	Seed yield (kg ha^{-1})	Protein (%)[a]	Protein yield (kg ha^{-1})[a]
Annuals			
Corn	6379 (USDA, 1980)	9	574
Wheat	2129 (USDA, 1980)	12	256
Soybeans	1988 (USDA, 1980)	42	835
Perennials			
Dactylis glomerata	1235 (Whyte, 1957)	27	333
Elymus canadensis	1123 (Cooper, 1957)	27	305
Bromus inermis	1058 (Wheeler, 1957)	19	201
Festuca arundinacea	1673 (USDA, 1976)	22	358
Desmanthus illinoensis	1335 (USDA, 1978)	34	454
Astragalus cicer	1123 (USDA, 1966)	40	449
Onobrychis vicifolia	563 (Thornberg, 1971)	40	449
Medicago sativa	445 (Thornberg, 1971)	37	208

[a] Earle and Jones (1962); Jones and Earle (1966); Piper et al. (1988).

The herbary was started in 1979, mostly from seeds provided by the Plant Materials Center, USDA Soil Conservation Service, near Manhattan, Kansas. By 1987 it had 265 species. Over the years, agricultural interns at The Land Institute have recorded flowering dates, plant height at flowering, and date of seed collection, and they have taken photos for reference and miscellaneous notes. In 1987, plants were rated on a four-point scale by intern Randy Kempa as follows:

1. Plant absent or declining (spring ephemerals sometimes included).
2. Plot growing well but some plants missing.
3. All plants growing vigorously.
4. Plants growing out of their plot by seeding or clonal spread.

3.1.3. *Accessions of six grass genera*
Based on the results of the literature survey, we chose six grass genera as candidates with a high likelihood of having other species with high seed potential. Two hundred species and four hybrids represented in more than 4100 total accessions from over 60 countries were obtained from the USDA Plant Introduction Center in Pullman, Washington. We planted them all, and 95% of all the accessions were established long enough for us to evaluate. Table 4 summarises the number of species, hybrids, and accessions.

We have tried to avoid being too hasty in adding a taxon to our list of species with which we would work. Neither the data from the literature nor our own observations from a single year's yields satisfied our criteria. For example, yields of buffalo grass (*Buchloe dactyloides*) at The Land Institute did not come close to the claims in the literature. *Agropyron intermedium* had high yields in the first year, but the ability of the plants to persist as perennials was not strong. So from our inventory prompted by the literature and the 4100 accessions, we added only one species to our inventory of plants for improvement, *Leymus racemosus*.

Table 4. Accessions of species and hybrids of six grass genera established at The Land Institute.

Genus	Number of species	Number of hybrids	Number of accessions
Agropyron	59	3	1207
Bromus	58		629
Elymus	26		289
Festuca	47		1255
Lolium	9	1	673
Sporobolus	13		73
Total	212		4126

3.2. *Taxa selected and reasons for their adoption*

3.2.1. *Wild senna (*Cassia marilandica*)*

This is a fairly deep-rooted legume that does not appear to form symbiotic associations with the nitrogen-fixing *Rhizobium* (Leonard, 1925). Several members of this genus are important medicinally (Duke, 1981) and as forage for livestock (Taparia et al., 1978). Erect stems sprout every spring from a woody caudex and at maturity, will vary from 0.5 to 2 m high. Flowering is from July to September. The many-flowered racemes emerge in the upper axils and in terminal panicles with the yellow flowers being pollinated by insects. The flat mature pods vary from 7 to 11 cm in length and 8 to 11 mm in width. The black ovate mature seeds measure 4 to 5 mm long and 2 to 2.5 mm wide. The chromosomes from the root tips of 10 plants in each of our 97 accessions were counted and never departed from $2n = 28$.

The species ranges from Pennsylvania to Iowa and Kansas and from southeastern Nebraska to Florida and Texas. The habitat description in two flora (Steyermark, 1963; Great Plains Flora Association, 1986) confirms our field experience concerning habitat and distribution.

In both the field and in the herbary, this species appeared to be both vigorous and high yielding. The arrangement of the pods made harvest easy. Even though flowering is spread over several weeks, the pods do not dehisce upon ripening. From the start, we doubted that it had little, if any, economic utility. Plant parts of a close relative are used as a laxative, but for such purposes, a little can go a long way. We selected it because we thought it might be our best bet to help us answer our most basic of biological questions: whether it is possible for perennialism and high seed yield to go together.

3.2.2. *Illinois bundleflower (*Desmanthus illinoensis*)*

This herbaceous perennial, like its legume relative, wild senna, also sends forth erect new stems each spring from a woody caudex. Plants may reach 2 m in height, although 1 m is more typical. When it blooms (June to August across the range), some 30 to 50 flowers will be borne in a bundle, yielding a tight-fisted cluster or bundle of brown pods. Individual pods will vary from 1.5 to 2.5 cm long and 4.5 to 6 mm wide. When mature, these flattened pods will usually contain brown seeds 3 to 5 mm long and nearly as wide. It also fixes nitrogen (Kulakow et al., in press).

Its natural distribution and preferred habitat is difficult to determine as it has been widely used in range revegetation programs and as habitat for wildlife enhancement projects. The current range extends from as far west as New Mexico and Colorado east to Texas and Alabama with the most southern locality in Florida. In the north, it can be found in Ohio, Indiana, Illinois, Minnesota, and North Dakota (Steyermark, 1963; Great Plains Flora Association, 1986). Illinois bundleflower (IBF) appears to be a high seed-yielding species, and though our primary interest is in perennial polycultures as substitutes for annual grains, we are always aware of the livestock component. The forage of IBF is preferentially chosen by grazers (Latting, 1961). Seeds contain 38% protein (Piper et al., 1988). Besides its economic potential, it can be used to answer the question as to whether perennialism and high seed yield could go together. We hope to use this species as the primary nitrogen fixer in our polycultures.

3.2.3. *Maximilian sunflower (*Helianthus maximilianii*)*
The stems of this plant rise each spring, usually in clusters but often singly, from short but thick and spreading rootstocks. There are reports from the literature of plants achieving heights of 2.5 m. This species is highly responsive to habitat in both the achievement of height and flowering. Flowering heads can be numerously arranged in the upper leaf axils or on an elongated spike or raceme-like inflorescence. One notable feature we have observed is that while last year's seeds are struggling to germinate and establish themselves as seedlings, the young shoots from the parental root stocks are experiencing rapid growth sufficient to establish a quick canopy in early spring. The species ranges from Minnesota and Manitoba to Saskatchewan south to Missouri, Oklahoma, and Texas, Maine to North Carolina westward to the Rocky Mountains and from southern Canada to Texas. It has been cultivated in gardens with the cultivar displaying more flowers and broader leaves.

It is found in numerous habitats from damp and open prairies to dry conditions. It is found in sandy waste ground and on limestone glades, ledges, and bald knobs, as well as on loess hills and deep rich soil (Great Plains Flora Association, 1986).

Sunflowers are reported to be allelopathic (Rice, 1984). We have noticed Maximilian to be highly competitive and most effective in suppressing other species when they are in the seedling stage. This characteristic, along with their relatively high yield, makes this native species most attractive. Once another species is established, however, little effect is experienced. A beneficial planting pattern may be to establish a mixture of other species in a polyculture and then follow up with Maximilian sunflower to provide weed control.

3.2.4. *Eastern gamagrass (*Tripsacum dactyloides*)*
This tall, stout perennial emerges each spring from a proaxis and achieves a height at flowering of 1 to 2 m. The terminal flowers of a primary shoot usually bear three spike-like branches. The lateral shoots bear only one. The lower parts of these branches are pistillate, and the upper two-thirds or so are staminate. The arrangement of the staminate spikelet is important; at any one locus is a pair of double-flowered florets. The pistillate part of the spikelet usually carries a single flower per locus. Seeds contain 27% protein, roughly three times that of corn and twice that of wheat. Seeds may be 1.8 times higher in methionine than corn (Bates et al., 1981). Some plants are apomictic. Two apparently distinct chromosome races predominate: a diploid of 36 and a

tetraploid of 72. It is not surprising to find an intermediate $3n = 56$ in nature.

The species covers great diversity of habitats including sand, sandstone, limestone slopes at the edge of woods and thickets, along roadsides, railroads, prairies, and even in cultivated fields. The distribution in the northern hemisphere, according to Hitchcock (1935), is from Massachusetts to Michigan, Iowa, and Nebraska, south to Florida, Oklahoma, Texas, and the West Indies. This widely distributed warm-season grass has an abundance of genetic variation even within the same ploidy levels. The basis for genetic improvement, therefore, is very great. The high protein content of the seed and high levels of methionine make this an attractive food for both humans and livestock.

An important double recessive mutant was discovered at the USDA Plant Materials Center near Manhattan, Kansas, in a plant grown from seed collected along a roadside in Ottawa County, Kansas, approximately 20 miles north of The Land Institute (Dayton, pers. comm.). Dayton's mutant is responsible for converting the male part of the flower into female, yielding from 12 to 20 times the number of seed as a normal plant, though only three to four times the yield in weight. This high seed number results from the condition mentioned above; the four ovaries per locus in what is ordinarily male result in four seeds per locus in the mutant.

This species was a candidate seed producer before the mutant was discovered and before we at The Land Institute discovered that small quantities of nitrogen were fixed in the rhizosphere of its roots. Its attractiveness as a perennial seed producer continues to improve. Forage yield and quality have been well known by stockmen for a long time.

3.2.5. *Giant wild rye (*Leymus racemosus*)*

This rhizomatous perennial of the Triticeae (which also includes wheat, rye, and barley) has culms up to 12 mm thick and, in our plots at The Land Institute, stands over 1 m tall. The flowering spikes range from 15 to 35 cm in length and 10 to 20 mm in width. The spike is densely packed with three to eight spikelets at each node with each spikelet containing four to six florets. These large spikes are borne on a strong stalk.

Burritt (1986) reports that this cool-season native of Europe and central Asia generally grows well on dry, sandy soils in Bulgaria, Rumania, Turkey, and parts of Russia (Melderis et al., 1980). Its annual growth cycle is similar to that of winter wheat. It has been introduced only a few times into the United States. Green shoots begin to emerge at The Land Institute as early as late January. Our material at The Land Institute came from the USSR and possibly explains the late January to early February emergence of green shoots. Annual above-ground biomass production rivals that of native tallgrass prairie (Gernes and Piper, 1987). Seeds have been eaten as a supplemental grain by Russian peoples, especially during drought years when annual small grains failed (Komarov, 1934).

Indigenous peoples and Viking settlers have used various species of the genus. Northern California natives ate *L. cinereus* grains. Eskimos and Aleuts ate *L. mollis* seeds (Weiner, 1980; Klebesadel, 1985). Archaeological evidence in Iceland points directly to the likelihood that Viking settlers cultivated *L. arenarius* until this century (Griffin and Rowlett, 1981). Various *Leymus* species have been grazed by sheep, cattle, and horses.

The grain of *L. arenarius* was found by Griffin and Rowlett (1981), to contain as much protein as red beans and salmon and its fatty acid content surpassed amaranth, wheat, high-protein hybrid corn, rice, and oats. They also noted that the yield of *L. arenarius* was similar to domesticated cereals cultivated in subsistence agriculture conditions. Researchers in the Soviet Union attempted to hybridise wheat and *Leymus* species from 1940 to 1975 with only moderate success (Bodrov, 1960; Tsitsin, 1978).

Nevertheless, because of its strong perennial nature and soil-holding capacity, because it is already in the human food inventory, because of the nutritional value of the various species, and because scientists have found the genus attractive enough to attempt hybrids with wheat, we have found it worthy of further exploration for incorporation into perennial polyculture.

3.2.6. Sorghum halapense × S. bicolor *hybrid*

In 1983 we developed a hybrid between Johnson grass (*S. halepense* (L.) Perso.) and commercial sorghum (*S. bicolor* (L.) Moench). *Sorghum bicolor*, an important grain and silage crop particularly in dry regions of the United States, is a perennial that is not winter-hardy. Commercial breeders have, therefore, treated it as an annual in their breeding programs. *Sorghum halepense*, commonly known as Johnson grass, is an aggressive weed, particularly in warm areas of high rainfall and humidity. It is common particularly throughout the Southeast from Florida to Texas and northward to Kansas, Nebraska, and Iowa eastward to Illinois, Indiana, Ohio, New York and Massachusetts. It is also found in the West Indies and Central and South America. It was introduced into the New World from Turkey around 1830 as a forage crop. This native of Europe and Asia normally stands less than head high, although some plants have been reported to reach 3 m. Its open, loosely spreading inflorescence is unlike the densely packed heads of its domestic relatives. The creeping rhizomes of Johnson grass have some degree of winterhardiness. We saw this physiological quality as a possible source of winter-hardiness to be combined with a tetraploid version of *S. bicolor* mentioned above. Because of the risk of rhizomatous spreading of Johnson grass, hybrids or hybrid derivatives, all plants are grown in plastic pots. The hybrid derivative seems to be much less aggressive than its Johnson grass parent.

If a relatively non-aggressive hybrid derivative could be established as a winter-hardy perennial, it would amount to the winterising of a grain crop already important in the crop inventory. Perennial sorghum could be included in polycultures and, as such, would represent the first domesticated crop to be integrated with recently wild species.

4. Ecological inventory

4.1. *Insects and pathogens*

Starting in 1985, qualitative surveys of insects and diseases were made of the prairie environment, the herbary of introduced accessions and the research plots of the more promising species. Carré (1985) reported that all sampled areas showed a diversity of both beneficial and harmful insects. Several foliar diseases were present, but few were serious with the exception of a fungal disease of Illinois bundleflower which caused significant leaf loss. The majority of the wild senna accessions showed symptoms of a virus not observed in native stands (Handley, 1986).

Subsequent insect surveys were carried out by sampling the prairie plants using sweep nets every third week from April to September and research plots from June to September. Adults were identified to the family level and the immatures to the order. Slater (1986) reported that while more families were observed in the prairie than in any single plot, the aggregate of all plots yielded more families than were present on the prairie. The ecological roles of both the adults and immatures of each family were assigned and compared across the sampled areas. Insect damage was low both in the prairie and in the research plots, with two notable exceptions. A curculionid stem cutter, *Rynchites* sp., cut the flowering stems in the *Helianthus* species and a chrysomelid, *Anomoea laticlavia*, ate the pods of Illinois bundleflower.

In a detailed survey of pathogens, Handley (1986) described eleven leaf borne diseases occurring on the six species selected for development with two considered potentially serious. She concluded that perennial plants have many of the same types of disease problems as our annual crops. Her survey also suggested possible avenues for management. The two considered serious were the leaf rust *Puccinia sorghi* in eastern gamagrass which occurred on 92% of the plants evaluated and a bacterial leaf spot affecting the lower leaves of Illinois bundleflower.

Leaf rust also infects corn and sorghum and is common in the Great Plains. To our advantage is the fact there is already great variation in susceptibility to rust. Good resistance to this pathogen is present in corn, and has remained useful over a period of many years (Don Duvick, Pioneer Seed Co., pers. comm.). Thus, it is reasonable to expect that resistant *Tripsacum* can be developed that will remain resistant over many years.

In the case of Illinois bundleflower, the leaf spot was apparently density dependent, defoliation being greater at wider spacings and as high as 40%. It was later determined that this was a splash borne disease not encountered in nature where a grassy understorey absorbs the shock of bombarding rain suggesting it would be less of a problem in our polycultures. Our observations were restricted to leaf diseases; the situation on roots and crowns is probably even more complex. Development of these species as crops will require us to understand the nature of their susceptibilities in order to develop ways to minimise disease impact. In a sustainable perennial polyculture there will be two main ways to minimise diseases: breeding for genetic resistance and employing cultural and management practices that reduce incidence or spread. Each host/pathogen interaction will require a different approach to control.

Survey work of the six species being studied as crops and native prairie plants will continue to provide insight into how diseases fit into the total picture.

4.2. *The soil environment*

A four-year study evaluated the water relationships, nutrients, and organic matter within three perennial species in monoculture – giant wild rye, eastern gamagrass (EGG), and Illinois bundleflower (IBF) (Piper, 1993). Although much of the information gathered from the study is species specific, some general conclusions may be drawn as they relate to the three major groups represented by the experimental crops (C_3 grasses, C_4 grasses and legumes).

The study species did not interact with the soil in the same way initially, with differences in patterns of soil water and nutrient change due to differences in period of establishment, resource use efficiency, growth phenology, and rooting depth. IBF and EGG stands used relatively more soil water in summer and during the establishment year. Wild rye stands were drier at shallow depths during their periods of most active growth. Seasonal water uptake patterns of the cool-season wild rye and the warm-season EGG complemented one another somewhat, but were not as striking as expected. Seasonal differences in resource uptake among crops should contribute to compatibility in mixtures.

Nitrogen depletion patterns partially reflected those of soil water. In general, phosphorus increased at the 10 cm depth in grass stands, especially in wild rye. Otherwise, there were few consistent differences among species in levels of soil carbon, phosphorus, potassium, calcium, magnesium and pH during the first two growing seasons. After three years, soil organic carbon was generally highest in EGG stands, whereas NO_3 concentration was highest for wild rye. In the fourth year, however, NO_3 was highest in wild rye at the 100 cm depth only. Additional studies of IBF and companion crops may demonstrate IBF's ability to provide nitrogen to neighbours, as has been shown in other grass-legume mixtures. Schemes for managing fertility, while minimising or eliminating synthetic inputs, in diverse perennial stands over many growing seasons remain to be devised.

Studies on the invertebrates, bacteria, and fungi need emphasis in the future, and the same holds for the soil herbivores.

5. Lessons from our biological studies

Some early experimental plantings were unsuccessful because of inappropriate seedbed preparation and germination conditions for these wild species. We lacked the advantage that comes from working with agronomic species. Nevertheless, success in establishing experiments has greatly improved over the years as the requirements of each species from planting to harvest to storage and back to planting again became better known. After several years of preliminary work the following major questions could be seriously addressed.

5.1. *Can herbaceous perennialism and high seed yield go together?*

This straightforward question does not have a straightforward answer. As shown in Table 1, domesticated herbaceous perennials such as bananas and strawberries do show high yields, and numerous woody perennials (fruit and nut trees) feature high yields as well. Of course, the ancestors of our herbaceous annual crops before agriculture yielded much less than today. An important but perhaps unanswerable question might be: how did the best of the herbaceous perennials compare with the best of the annuals in seed yield before agriculture? Questions about yield potential invariably involve yield improvement through breeding as well as improved agronomic measures.

In our experiments with the five species we have used to answer this question, seed yield over time has increased with plant size, but has also decreased with plant maturity or soil nutrient depletion. Seed yield has also simply oscillated either predictably or irregularly. No one model has helped us predict the patterns we have observed in the five perennial monocultures maintained at The Land Institute, where there was no fertiliser at any time and no irrigation beyond establishment. From our own data, wild senna yields have oscillated, but averaged 2092 ± 247 kg ha^{-1} (n = 89 plants in rows 0.75 m apart and within row spacing of 20 cm) in the second year. Illinois bundleflower produced relatively high yields during the first three years, with a peak of 1967 ± 41 kg ha^{-1} (n = 32 rows, 5 m long, 0.75 m apart). Yields of *Leymus racemosus* and Maximilian sunflower were lower at 510 ± 24 kg ha^{-1} (n = 34 rows, 5 m long, 0.5 m apart) and 304 ± 15 kg ha^{-1} (n = 32 rows, 5 m long, 0.75 m apart), respectively and both steadily declined over time. Eastern gamagrass yields increased four-fold to 884 ± 337 kg ha^{-1} (n = 6 rows, 5 m long, 1 m apart) from the first to the second year (Piper and Towne, 1988).

These baseline yields under zero input conditions for monocultures are important. Among other comparisons, they will be compared with yields achieved for these species in polyculture. This should tell us whether complementarity among compatible species can compensate for declining yields under low-input conditions.

In the early stages of our research, the theory of r and K selection was rather widely held as a reality (Cody, 1966; MacArthur and Wilson, 1967;

Pianka, 1970; Gadgil and Solbrig, 1972; Abrahamson, 1979). This theory holds that r-selected plants allocate more energy and materials to reproduction than to vegetative parts. For K-selected plants, it is the other way around. Once established, a K-selected plant is more devoted to staying alive than to reproducing. Annuals would generally be r-selected and perennials K-selected. It is an intuitively satisfying notion, but in recent years widespread questioning has developed regarding the validity of the theory (Grime, 1979; Jackson and Dewald, 1994; Reekie and Bazzaz, 1987; Reznick, 1985; Soule and Werner, 1981; Pitelka et al., 1985; Horvitz and Sckemske, 1988; section 3 above).

5.1.1. *Wild senna*

Annuals grown to produce fruit and seed usually have a high yield and high sexual reproductive effort or SRE (Singh and Stoskopf, 1971; Fischer and Turner, 1978; Johnson and Major, 1979; Wardlaw, 1980). SRE is the effort that a plant exerts in producing sexual organs (Hickman and Pitelka, 1975; Abrahamson, 1979; Abrahamson and Caswell, 1982; Thompson and Stewart, 1981). SRE can be measured in various ways, such as measuring the carbohydrates or nutrients allocated to various plant parts. One simply dries the plant to some standardised percentage of moisture, then divides the plant up into its various parts and weighs the parts (Hickman and Pitelka, 1975; Thompson and Stewart, 1981). SRE is an important measurement because it is positively correlated with high yield in our cereal crops, and plant breeders use it in their selection programs (Fischer and Turner, 1978). Another measurement is the harvest index, which involves a somewhat simpler approach. Here the seed and/or fruit weight is divided by the total aboveground weight (seed weight/above ground weight).

Burris (1984) compared the SRE for wild senna with other herbaceous perennials (Table 5) and its harvest index with those of seven annual grain crop (Table 6). The harvest index measurements show wild senna to be within or exceeding the ranges for four of the seven listed crops. Two important conclusions of Burris' study are that this herbaceous perennial has both a relatively high sexual reproductive effort and harvest index. Other research with perennial grasses has shown that seed yields can vary from year to year (Cornelius, 1950); how much and with what predictability has to be known for each species.

Berghoef and Soule (1984) measured third-year yield in 29 accessions of wild senna. Although yield had declined drastically by the third year, one accession from Chapman, Kansas averaged 856 kg ha^{-1}. The highest yielding row in this accession gave 1417 kg ha^{-1}, an accession that had averaged only 337 kg ha^{-1} in the previous year and 618 kg ha^{-1} in the first year. This accession had the highest average overall. Stiefel (1985) found total seed yield had continued to decline into the fourth year, but when plots were fertilised in that year, three- and four-year-old accessions had higher yields than in their first year by 13% and 5%, respectively. This suggested it was not senescence that caused yield decline but soil fertility.

Table 5. Mean and standard deviation of fruit weight as a percentage of total above- and below-ground biomass (dry matter basis) for wild senna and other perennials (Burris, 1984).

Species	Fruit weight (% DM)	Std. dev.	Reference
Cassia marilandica (wild senna)	27	10.2	
Hieracium floribundum	34	6.6	Abrahamson (1979)
Hesperis matronalis	41	8.3	
Houstonia caerulea	36	11.5	
Hypericum perforatum	32	5.1	
Rudbeckia hirta	39	9.2	
Rudbeckia triloba	32	1.8	
Sisyrinchium angustifolium	32	7.5	
Arisaema atrorubens	39	10.5	
Rumex crispus	46		Weaver and Cavers (1980)
Rumex obtusifolius	35		

Table 6. Ranges and means for harvest indices of wild senna and seven annual grain crops (Burris, 1984). Standard deviation 0.09, $n = 57$.

Species	Harvest index	Reference
Wild senna	0.31*	
Corn	0.32–0 47	DeLoughery and Crookston (1979)
	0.39	
	0.37	Alessi and Power (1976)
	0.26	Fischer and Turner (1978)
	0.29	
Wheat		
High yielding	0.40	Johnson and Major (1979)
Average	0.34	
Rice		
High yielding	0.47–0 .57	
Average	0.23–0 .37	
Barley		Singh and Stoskopf (1971)
Winter	0.45	
Spring	0.51	
Rye	0.27	
Oats	0.41	McDaniel and Dunphy (1978)
Soybean	0.44	Johnson and Major (1979)

Correlation between annual yields were examined over the four years. There was a strong positive correlation between yield and accession in the unfertilised plots. Apparently genotype and yield are linked, but this relationship can be masked by nutrient application. This suggested we should select for yield in an environment that has not been homogenised by inputs. Carré and Cavigelli (1985) studied the impact of plant density on yield, reporting significantly higher yields at 13.5 and 27 plants m^{-2} than at 6.5 and 3.3 plants m^{-2}.

5.1.2. *Illinois bundleflower*

Illinois bundleflower (IBF), The Land Institute's only nitrogen-fixing legume, is one of our favourite species because of its ultimate utility as a livestock food (see species description in section 3). Consequently, we have done more research with this organism than with wild senna. So far we have explored its breeding mechanism, yield, optimum density for highest yield, morphological features correlated with yield, and the amount of genetic variation present. This last question must be answered in order to predict the possibility for yield increase and competitive ability.

Breeding mechanism. Bohlen (1986) determined that IBF is self-compatible by noting that bagged inflorescences set seed. Latting (1961) had suggested the species may be wind-pollinated. Bohlen considered this improbable because pollen production is small and IBF pollen is relatively large and sticky at dehiscence. Such insects as Chrysomelidae, Halictidae, Apidae, Sphecidae, Bombidae, and Braconidae visited bundleflower in our plots that summer and are likely pollen vectors. Bohlen concluded that, to a large extent, this species is self-compatible and that our chances of obtaining seed from selfed plants and maintaining inbred lines was very good. There is also evidence of some outcrossing.

Yield studies. Gursky (1984) reported that first year yields ranged from 393 to 3412 kg ha^{-1} with a mean of 1344 across all accessions. The second year range was from 539 to 2416 kg/ha with a mean of 1164 (Bruns, 1985). Although yields were generally lower in the second year, five of the top ten yielders in 1984 were among the top ten in 1985.

Yield and density studies. Carré and Cavigelli (1985) examined the impact of plant density on yield, reporting no significant differences within the range of 3.3 to 19 plants m^{-2} in the first three years after planting. All densities yielding within the range 1382 to 1538 kg ha^{-1}. Yield declined 36% in the second year and a further 18% in the third which was thought might be partially explained by bacterial leaf spot disease causing defoliation just prior to flowering (Collins, 1986). To determine whether infection was transmitted through soil splash, Liebovitz (1987) randomly mulched two replicates with wheat straw leaving six replicates as unmulched controls. Mulching reduced defoliation, but seed yield was unaffected by the degree of defoliation in the unmulched plots.

Morphological yield indicators. In search of quick and easy yield indicators on individual plants, Rinehart (1986) measured individual stems of plants at two sites among our accessions. He measured 40 plants on a well-drained upland slope in a biculture experiment with wild senna, and 33 stems in a two-year-old plot of accessions in moist bottomland. He concluded that basal stem diameter was a fairly reliable index of both aboveground biomass and yield.

Germplasm assessment. Genetic variability is necessary for any breeding program to be successful. The objective is to produce Illinois bundleflower populations which can produce seed and forage and fix nitrogen, all in a mixed cropping system. Initial studies under the direction of our plant breeder Peter Kulakow have served to generate and define useful descriptors for future breeding work.

Mecko-Ray (1987) found that synchronous flowering was correlated with high yield. She suggested that selecting for clustering of inflorescences in determinate branches may be one method of increasing synchrony and thereby facilitating harvest. Seven of the accessions tested had 90% or more pods mature at peak harvest, indicating some synchrony of maturity. The three highest yielding accessions produced over 95% of their yield at peak harvest.

She also suggested timing and duration of flowering, plant height and branching pattern are likely to be important selection criteria when considering the compatibility of this species in a polyculture. It may be preferable to minimise overlap in flowering between species in a mixture depending on relative nutrient and water demands of other species. While a positive correlation between mean height and seed yield was found, it was suggested that shorter plants may be preferable, citing examples in Smith and Francis (1986) in which reduction in the height of a dominant crop allowed increased productivity of other crops.

Branching pattern may also be an important trait for high seed yield and compatibility in polycultures containing prostrate forms which are difficult to harvest and erect forms which allow better light infiltration and are perhaps more compatible in a mixed cropping system.

The fact that several accessions had consistently high or low seed yields across replications indicated that there were genetic differences. However, seed yield may not be the predominant goal of breeding in Illinois bundleflower, given the critical importance of nitrogen fixation to a perennial polyculture. Mecko-Ray (1987) made no observations on nodulation or nitrogen fixation in this study. Future studies of IBF need to include these factors and also to determine if there is a trade-off between nitrogen-fixation and seed yield. Future studies also need to observe uptake and utilisation of essential nutrients, as legumes tend to have high phosphorus requirements (Williams, 1983). Future evaluations should also test germplasm in both monoculture and polyculture plantings to assess differential performance between planting arrangements.

5.1.3. *Sunflower studies*

Common genomes in Helianthus *species.* In 1984, we became interested in the diploid and hexaploid perennial relatives of the high-yielding diploid annual, *Helianthus annuus*. We had concluded that Maximilian was probably the highest-yielding diploid perennial sunflower. Studies by Kostoff had led him to conclude that the hexaploid *Helianthus* species contain a genome of the high-yielding annual (cf. Heiser and Smith, 1964). We wondered if that common genome might be correlated with the high yield characteristic of the diploid annual. Adelman (1984) found that the lowest yielding perennial diploid, *H. mollis*, outyielded the two lowest-yielding hexaploid perennials. We concluded that such a genome gave those hexaploids no yield advantage over the perennial diploids lacking the annual diploid genome.

Sunflower hybrids. Calsbeek (1984) compared the yield of a race of Maximilian sunflower, three generations after its collection in the wild, with the F_1 hybrid *Helianthus grosseserratus* Martens × *H. maximilianii* Schrad. planted in a randomised block design at two densities with four replications. The Maximilian yield averaged 595 kg ha^{-1}. In spite of the extremely impressive vegetative growth of the hybrid, its seed yield was only 204 kg ha^{-1}. Bruns (1985) compared yields of a hybrid annual sunflower (Triumph 505C confectionery), an open-pollinated annual sunflower (Mammoth), and Maximilian. The mean yield for Mammoth was 931 kg/ha, the hybrid annual 2166 kg ha^{-1} and the perennial 325 kg ha^{-1}. From there on, we have concentrated on Maximilian sunflower only.

Maximilian sunflower. Riley (1984) found that seed yield averaged 456 kg ha^{-1} and did not vary significantly with density across the range of 5.5 to 13 plants m^{-2}. This lack of correlation between density and yield is supported by field observations where greater branching occurs at the lower densities and apical dominance increases at higher densities. We learned however that these results did not necessarily mean we could get away with low seeding rates as the greater apical dominance at higher densities results in seed heads in a more favourable position for harvest. Riley's results did suggest the need to gather more information about growing this potential crop as the plant density for high yielding in the first year may not be optimum later on. If so, we may have to control density by cultural practices in subsequent years or plant at less than optimal first-year density.

Kois (1985) set out to confirm whether there were yield differences between four densities, 9, 13, 22 and 67 plants m^{-2}. Again, yield did not differ significantly across the density ranges. Plant height and number of seedheads were affected most, but not branching pattern. Gernes (1986) continued this experiment by splitting the plots and thinning half back to the original spacings by hand and rototilling. In the unthinned plots, seed yield did not differ across the four densities. Plant density and number of seed heads per stem increased while plant height, branches per stem and stem height decreased. The unthinned treatment with the highest density was three times more stem-dense than its thinned pair. However, the seed yield in the thinned plots was twice that of the unthinned control. Rototilling early in the season reduced stem density but increased yield.

In the third year, 1987, only the control (never thinned) half was harvested. Again, there were no significant differences in seed production among the four density treatments. Yield was low at 33.8 kg ha^{-1}, a decline of 83.7% compared to 143.4 kg ha^{-1} in 1986 and 183.4 kg ha^{-1} in 1985 (Liebovitz, 1987). There are several possible reasons for this decline: nutrient depletion, competition, senescence, disease or some combination of these.

In search of yield indicators, Rinehart (1986) examined the relationships among basal stem diameter, aboveground phytomass, and fruit yield in Maximilian sunflower. He found that basal stem diameter predicted more of the variation in plant phytomass than in reproductive yield. Nevertheless, it seems that basal stem diameter is a fairly reliable and useful index for a crude estimate of yield.

5.1.4. *Eastern gamagrass*

Our primary question with eastern gamagrass (EGG) has been whether the pistillate form has the potential to greatly increase seed production. In 1986 we established over 4000 plants from seed, representing 122 half-sib families (opr crosses) of this species (Bohlen, 1986). Each cross or family was segregating for the pistillate mutant.

Bohlen's objective in 1986 was to establish a population containing as many genotypes of the pistillate mutant as practical. From data taken we now have a record of the transplant survivorship, a better idea of the differences in gross plant morphology in both the normal and mutant, and a comparison of the incidence and level of rust and leafspot diseases in the normal and mutant. We also have an indication of the percentage flowering and flower type. All of these categories were recorded that summer for 77 of those 122 accessions established at The Land Institute. It was found that 93% of the transplants survived, that 9% flowered the first year of establishment, and that 23% of those that did were mutants. Perhaps more importantly, we have a better idea of the kind of tasks facing those who would seek to develop perennial grasses. Accessions varied in their susceptibility to both rust and leaf spot diseases. Generally, the level of infection with leafspot was lower than with rust. On a scale of 0 to 7, the overall mean for leafspot was 2.5, which was lower than the overall mean for rust at 2.8.

Bohlen's experiments showed that survival is best when seedlings are planted in flats, overwintered outdoors, and then transplanted into a prepared seed bed. Spring-emerged seedlings can be planted directly into the field and thus avoid the intermediate step of transferring to peat pots. Direct seeding into a fall seedbed can save labour and materials, and thinning or transplanting to the necessary density can wait until spring. The most vigorous plants have the most flowers, and vegetative shoots usually do not become reproductive until they are a year or two old (Dewald and Louthan, 1979). Even though 10% of the shoots flowered the first year, perhaps because growth conditions were ideal, there was still not enough flowering on any one plant to harvest.

Handley and Boehmer (1987) studied the variability in resistance to anthracnose and leaf rust. They selected 10 of the 122 half-sib families on the basis of mean leaf rust severity in the previous year. They first divided the large field of over 4000 plants into five blocks and then selected families with the lowest and highest mean rust severity from each block. Thirty plants were chosen from within each of the ten selected families for transplanting. Each of the 300 plants was divided into six clones in early April. One planting was established at The Land Institute (three clones times 30 plants times 10 families = 900 plants) and the other at the Kerr Center for Sustainable Agriculture in east central Oklahoma near Poteau. The three replicates of each of the 300 plants were set out in a randomised complete block design at both localities. They were spaced 1 m apart with two border rows around the entire planting. The rototiller and intern hands did the weeding.

Handley and Boehmer (1987) reported differences both between and within families, indicating the likelihood of a genetic basis of plant susceptibility. The within-clone variation was low and there was a negative correlation between rust and anthracnose incidence. The mutant does not seem to be closely associated with disease susceptibility, suggesting considerable promise for the use of the mutant character to increase seed yield.

5.1.5. *Giant wild rye*

Giant wild rye (*Leymus racemosus*) was first established at The Land Institute in April 1985 in a 16-by-20 m plot consisted of 33 rows. It first flowered in May 1986. Extrapolated yield values for the 33 row segments (5 m by 0.5 m) ranged from 258 to 900 kg ha^{-1} with a mean of 510 kg ha^{-1} (Burritt, 1986). Threshing percentage was nearly constant among rows, while head number and total weight of heads per row were significantly correlated with yield. Harvest indices were calculated for two rows at 0.067 and 0.077. The latter also had the highest seed yield. Dry matter forage yields extrapolated from the total biomass yields of these two row samples were 8500 and 11,000 kg ha^{-1} respectively.

Ergot has proven to be a serious problem in *Leymus*. As deep ploughing and crop rotation are antithetical to a perennial crop on a yearly basis, other possibilities will need to be explored. Ergot-free seed is possible, but with the perennial, infestation from nearby stands of wild grasses is always possible. Stubble can be burnt at the growing season's end, but in the long run, resistant varieties will be necessary. Varietal resistance to ergot is variable since many pathogenic races exist (Jones and Clifford, 1978). Burritt (1986) concluded that as the fibrous glumes must be removed from the seed prior to human consumption, an extra step in the milling process to remove the sclerotia poses no problem if other solutions fail.

5.1.6. *Winter-hardy perennial sorghum*

Sorghum research began in 1983 when a tetraploid domestic sorghum, *Sorghum bicolor*, was crossed with Johnson grass, *S. halepense*. The F_1 was planted in 1984, with 465 plants surviving the winter. In 1985, we planted the F_2 generation. These plants displayed considerable variation in such important characteristics as flowering time, height, shattering, and panicle compactness. In 1986 Peter Kulakow acquired agronomically more favourable lines of the tetraploid *S. bicolor* and collected 27 accessions of *S. halepense* from sites in Kansas, Oklahoma, and California. Several additional techniques were employed in crossing, and a few hundred hybrid seeds were produced.

Work on this artificial hybrid is of great biological importance, as it represents a major effort to take a domesticated species that, for all practical purposes, is an annual in temperate zones and cross it with a relative that has no economic value except negatively as an aggressive weed. The biology of a perennialised domestic crop is of great importance because it would be an example of beginning with a known crop rather than developing a wild species into a domesticated form. This fusion of the domesticated and wild has philosophical implications for the 'nature as measure' paradigm.

As philosophically, biologically, and practically interesting as this process and its product might be, care is taken to prevent the spread of these hybrids and hybrid derivatives. The weedy potential of the perennial sorghum lines is not yet known, and the potential for the perennial sorghum to cross with the cultivated annual could cause weedy threats in the field. An artificial perennial sorghum could cross with the weedy populations, transferring cultivated genes to the weed populations. It is unlikely, but the competitive ability of the weed species could be enhanced by mimicking the crop or enhancing overall vigour.

5.2. *Can perennial polycultures outyield perennial monocultures?*

This question was first addressed in experiments in 1983 with a biculture of two legumes: Illinois bundleflower (IBF), a nitrogen fixer, and wild senna, a non-nitrogen fixer. Interns Mark Böhlke and Juli Neander, who established this experiment, planted four replications of five treatments: monocultures of IBF and wild senna at 30 cm within-row spacing, alternate row biculture at 30 cm spacing, and alternate plant biculture at two densities corresponding to 15 cm and 30 cm spacings. Each treatment consisted of nine 9 m rows spaced 1 m apart, with the inner 7 m of the seven middle rows used for measurement. The upland soil on which this experiment was planted was low in nitrogen and organic matter. The IBF was inoculated by pouring the *Rhizobium* into the furrows alongside the plants.

Price and Gursky (1984) recorded that there was no overyielding in the first year. Monocultures of the two species collectively had the same yield as the bicultures; relative yield totals (RYT) for the three biculture treatments ranged from 0.9 to 1.1. The second year, 1984, produced a different story.

Wild senna yielded higher than IBF in monoculture even though relative yields of wild senna were consistently lower than those of IBF in biculture. For 10 of the 11 biculture plots, the relative yield exceeded one. The differences among the three biculture treatments were not significant. Considered together, the bicultures significantly outyielded the monocultures by 29% on the average. Overall yields from the bicultures were near 1000 kg ha^{-1}.

Third-year results revealed an average relative yield for the three bicultures to be 1.94. In only one of the blocks was the relative yield less than 1.0, although in one block it exceeded 3.0. The relative yields of IBF and wild senna were almost equal. It is important to note that the absolute yields of the biculture were significantly higher than the monocultures in the second and third years, but biculture yields remained nearly constant (Braun, 1985).

From these results, some testable ideas about planting density and planting combinations were developed. Significant differences were found between the alternate-row and alternate-plant bicultures and between the two densities of alternate-plant bicultures. Also, the 30 cm spacing of the alternate-plant biculture overyielded more than the other two bicultures.

In 1985 a triculture was started from seed consisting of *Leymus racemosus*, IBF, and Maximilian sunflower. They were established in alternating rows, with each species being a neighbour of the other so that the growth and survivorship of the interplanted species could be observed and also to identify problems that would suggest alternative designs. We had problems establishing *Leymus* that first year, which made any comparison of seed yields of the three species meaningless. Maurer (1986) conducted a preliminary study to evaluate the effects of cutting Maximilian back. After cutting, the overall growth of the triculture and seed head production of Maximilian was evaluated. Seed head production and lodging decreased as the number of cuttings increased. Plants that had been cut twice continued to produce seed heads, suggesting that this plant has the potential for both forage and seed production. This is important for management purposes.

Experience gained with the first triculture was used to plan and establish a second triculture in 1987. This time two grasses were included: giant wild rye (cool-season), eastern gamagrass (warm-season), and the nitrogen-fixing legume, Illinois bundle blower. The ultimate purpose was to gain information on inter- *versus* intra-specific competition. Kendall (1987) monitored growth rates, flowering, fruiting patterns, and seed yield. The three species appeared to be compatible. IBF adjacent to the grasses produced more flowers and seeds than individual plants next to their own kind. Individuals next to eastern gamagrass had a more synchronous harvest than those neighboured by *Leymus*. A majority of later-flowering individuals were bordered by their own kind.

After a decade devoted to domesticating wild species and establishing simple mixtures, the research emphasis has shifted to exploring community assembly rules. Is it possible to design diverse ecosystems that persist? In the first series of experiments, we established a four species mix comprised of one representative of each group (C_3 grasses, C_4 grasses, legumes and sunflowers). It failed to persist, being rapidly dominated by annual weeds. By adding a minimum of four other species from the next best selections, a near weed free mix has persisted for three years after planting. Some encouragement for the development of community assembly rules comes from computer modelling of complex systems which suggests that although persistent complex mixes are statistically rare, they can be achieved through the successive addition and loss of species (Pimm, 1997).

5.3. *Can our 'domestic prairie' ecosystem sponsor its own nitrogen fertility?*

The first two studies dealing with this question contributed baseline information necessary for future experiments.

In 1985, anecdotal evidence from a physiological ecologist indicated that agricultural species tended to respond better to ammonium, whereas native prairie species responded better to a nitrate environment. We wondered whether our ecosystem should be considered agricultural, natural, or both. Gursky (1984) conducted a greenhouse experiment in which he compared seven herbaceous perennial species for response to different nitrogen sources. Each species was divided into two groups: one receiving nitrogen in nitrate form, the other as ammonium. Curly dock (*Rumex crispus*) in the buckwheat family, Maximilian sunflower (a composite), three grasses (eastern gamagrass, *Leymus*, and luna pubescent wheat grass), and *Elytrigia intermedium* var. *trichophorum* grew equally well under both conditions. The two legumes, wild senna and IBF, grew poorly in both treatments, although with these legumes, the nitrate-fed plants performed slightly better overall.

In 1987 agriculture intern Doug Dittman classified and monitored the soils of different plots at The Land Institute to observe and compare the changes within the soil over the growing season. Besides knowing our own place better, we hoped this information would help us predict which species would compete with, or complement, other species within a perennial polyculture. Dittman had a second purpose: to increase our understanding of soil nutrient and moisture use patterns within native prairie in order to establish standards against which to compare our agriculture plots. Soil moisture was monitored from May through October and soil nutrients measured in early July, August, and September. The sites included two tallgrass prairie areas at The Land Institute that had never been ploughed, three monocultures of selected species in the bottomland, and an adjacent crop of *Sorghum bicolor*.

Most of the differences among these plots were attributable to differences in soil type and plant root structure. Most interesting was the observation that the growth patterns of prairie plants more closely followed the moisture status of the B horizon than the A. This was also the case with Illinois bundleflower. Shallow-rootedness seemed to tie flowering and fruiting to the A horizon, as Dittman observed in both *Leymus* and eastern gamagrass. Overall,

Dittman's results suggest complementarity in the use of the soil resource among these species if grown together in polyculture (Dittman, 1987).

5.4. *Can a perennial polyculture adequately control pests?*

So far, experiments have only been conducted on control of weeds. Several authors have suggested that allelopathic interactions have an important role to play in biological weed control (Putnam and Duke, 1974; Lovett, 1982; Gliessman, 1983). Our Maximilian sunflower stands have remained almost weed-free. Our interests intensified in 1983 when two-year-old plantings were bare in early spring while an adjacent area was covered with wild mustard. To get a more precise assessment of this species' potential, Price (1984) studied the relationship between density of Maximilian sunflower and the growth of weeds.

Maximilian seeds were sown at six densities in a randomized block design with four replications. Each density plot consisted of two sunflower rows 1 m apart, 18.5 m long, with the inner 5 m the data source. Her goal was to measure weeds between these rows. Since sunflowers germinate more slowly than some of the weeds, plots were weeded within rows once and mowed between rows twice during the first 40 days. This assured both weed and sunflower establishment. At the end of the growing season in this establishment year, Price analysed the percentage cover using a 0.25 m^2 quadrat to compare weediness among treatments. Data were also taken on the predominant weed species. Covariance analysis showed a significant effect of sunflower density this first year. The denser the within-row plantings, the fewer the weeds. The density regression was significant (Price, 1984).

A year later, Cavigelli (1985) found that the weed biomass again declined with increasing sunflower density and that the species mix had changed. In the third year, Gernes (1986) again recorded a decline in total weed biomass with an increase in the within-row sunflower density. Over the three years of this study there were major changes in the weed flora. In the first and second years, weeds were mostly annuals. By the third year, most of the weed species were perennial, but as a percentage of total weed weight, the contribution of perennials was very small (Gernes, 1986).

One of the most interesting pilot projects we have conducted had to do with the effect of Maximilian sunflower on subsequent plantings. LaLiberte (1986) planted Illinois bundleflower, curly dock (*Rumex crispum*), wild rye (*Leymus racemosus*), and wild senna on ground where Maximilian sunflower had grown for four, three, two, one and zero years after removing all the sunflower plants, roots and all. The four new species were planted on this well-worked ground in rows perpendicular to the former Maximilian sunflower arrangement. This gave us a chance to determine if there was any residual effect due to the former presence of Maximilian. We reasoned that if persistent allelopathy or nutrient depletion were important factors, we might expect poorer germination and ultimately lower yields correlated with the length of time that the sunflowers had grown in a particular area. Neither pattern was seen. The only significant differences were for percentage emergence in curly dock and yield in Illinois bundleflower, but neither had anything to do with how long the sunflower had been growing in an area (LaLiberte, 1985).

6. The necessary marriage of agricultural science and ecology

Modern agriculture with its dependency on the greater extractive economy has increased the tendency for both researchers and farmers to understand and improve agricultural production in its own terms. The sustainable agriculture movement has the potential to help us acknowledge that ultimately, agriculture comes out of nature.

In general, scientists – including ecologists – have given priority to the parts over the whole. But the discipline of ecology has a greater potential for broadening its outlook. It is not a simple repudiation of reductionism, the prime paradigm for an emphasis on cleverness, but rather an acknowledgment that we need to go beyond, to more fully explore the laws of nature at work at the ecosystem level. These laws must surround the emergent qualities that the ecosystem displays. It is not that the whole is greater than the sum of the parts so much as the whole is different from the sum of the parts. Even though ecology is perhaps too reductionist, those who have studied bogs and alpine meadows, unploughed prairies and deciduous forests, rainforests and marshes are the best candidates for pondering and explaining what ecology does have to offer. Ecosystem studies would examine, for example, how nutrients cycle to accommodate diversity or monoculture, as well as the pathways through which sunlight flows. Ecologists are evolutionary biologists, plant population biologists, quantitative geneticists, functional morphologists, physiological ecologists, and so on. There are many more subdisciplines, depending on how one wishes to identify the categories. What holds them all together is the fact that researchers in these fields want to understand how nature's life forms work in particular ways and how they respond to the larger systems of which they are a part. Again, most of their work is reductionist, much of it necessarily so, although as we come to recognise the consequence of placing priority on part over whole, that ratio will change. Nevertheless, the knowledge these ecologists have placed on the shelves, mainly for its own sake, in the last 30 to 50 years is phenomenal. Unfortunately, much of it remains there.

What if more of this body of knowledge were applied to agriculture? And what if ecologists were to begin applying that knowledge to sustainable agriculture, using nature as the measure? The new field of agroecology, as it stands today, is not the hybrid we are talking about. Most of the workers in agroecology, whether they come out of agronomy or ecology, are still trying to understand agriculture against a standard that dates back 8000 to 10,000 years at best and, even worse, to a standard made possible by the industrial revolution and the emergence of the chemical industry. By not using nature as the standard, agroecology is apt to be watered down to an emphasis on low inputs and soil erosion, and germplasm truncation will continue at alarming rates. We have no objection to this emphasis on low input in the short run, as long as it is recognised as an intermediate step only. We simply need to prepare now to be more attentive to that body of knowledge on the ecological shelves and to bring into sustainable agriculture the large numbers of workers responsible for putting much of it there. It is imperative that we begin raising a new generation of agroecologists who represent three-way hybrids: people willing to learn from farmers (who have much to teach in the way of tradition and experience), modern agricultural researchers, and ecologists.

We cannot expect this emphasis to come from agribusiness. Success in sustainable agriculture means that we would greatly reduce our reliance on nonrenewable resources, the very stock that agribusiness relies on. Furthermore, a success story would tell how we reduced our dependence on nature as a sink for pollutants. (Nature's sinks have been treated as givens by all of us interested in externalising as many costs as possible.) This ecological emphasis will find the benefits running more to the farmer and the agricultural landscape. We cannot expect a push, therefore, to come from the business world, which supplies most of the inputs for agriculture, in anything close to the way that biotechnology got its boost in the 1980s.

To provide the necessary emphasis will require a sustained effort for several years. The 'nature as standard' idea needs high visibility. The highly respected and widely known ecologists and agriculturalists can best articulate this need and thus encourage the young researchers who will pioneer this effort by researching and writing the classic papers in this exciting new discipline.

7. Acknowledgements

An effort of numerous Land Institute interns and staff members, spanning almost two decades, is represented in this chapter. Nearly all of them are represented in the references to one or more of their papers published in The Land Institute Research Report. Our greatest debt is to them. In the early stages it was Marty Bender, our first research associate at The Land Institute, and the late Jim Peterson, followed by Walter Pickett and Dr. Judy Soule, who helped bring sense to lots of fuzzy thinking. In their wake came staff scientists Dr. Mary Handley, Dr. Peter Kulakow, and Dr. Jon Piper. Dr. David Van Tassel and Dr. Marty Bender, who returned to The Land Institute after completing his doctorate in plant ecology, now carry most of the responsibility for our research efforts.

8. References

Abrahamson WG and Caswell H (1982) On the comparative allocation of biomass, energy and nutrients in plants. Ecology 63: 982–991

Abrahamson WG (1979) Patterns of resource allocation in wild flower populations in fields and woods. American Journal of Botany 66: 72–79

Adelman P (1984) Comparison of sunflowers with different genomes. Land Report Research Supplement, The Land Institute 1: 5–6

Ahring M (1964) The Management of Buffalo grass for Seed Production in Oklahoma. Oklahoma State University Agricultural Experiment Station Technical Bulletin T-109

Alessi J and Power JF (1976) Water use by dryland corn as affected by maturity class. Agronomy Journal 68: 547–50

Altieri MA (1987) Agroecology. Westview, Boulder, CO

Baskin JM and Baskin CC (1977) Predation of Cassia marilandica seeds by *Sennius abbreviatus* (Coleoptera: Bruchidae). Bulletin of the Torrey Botany Club 104: 61–64

Bates LS, Bender M and Jackson W (1981) Eastern gamagrass, seed structure and protein quality. Cereal Chemistry 58: 138–141

Berghoef M and Soule J (1984) Third year yield in wild senna (*Senna marilandica*). Land Report Research Supplement 1: 13

Bodrov MS (1960) Hybridization between wheat and *Elymus*. In: Wide Hybridization in Plants, Akademiya Nauk USSR, pp 238–241 (Translated from Russian, 1966). National Science Foundation, Washington, DC

Bohlen P (1986a) Effect of self-pollination on *Desmanthus lllinoensis* seed production. Land Report Research Supplement 3: 9–11

Bohlen P (1986b) Multiplication and evaluation of *Tripsacum dactyloides* accessions segregating for a recessive pistillate mutant. Land Report Research Supplement 3: 30–33

Braun L (1985) Overyielding in a perennial legume biculture. Land Report Research Supplement 2: 4–7

Brown HR (1943) Growth and seed yields of native prairie plants in various habitats of the mixed prairie. Transcripts of the Kansas Academy of Science 46: 87–99

Bruns M (1985a) Illinois bundleflower: a perennial food or feed crop? Land Report 25: 6–9

Bruns M (1985b) Second year yields of Illinois bundleflower (*Desmanthus illinoensis*) accessions. Land Report Research Supplement 2: 27–28

Burris D (1984) Sexual reproductive effort of wild senna (*Senna marilandica*). Land Report Research Supplement 1: 20–22

Burritt B (1986) Yield estimation and analysis of yield components in *Leymus racemosus*. Land Report Research Supplement 3: 25–30

Calsbeek J (1984) Seed yield comparison of *Helianthus grosseseriatus* × *H. maximiliani* F_1 plants and *H. maximiliani*. Land Report Research Supplement 1: 19–20

Carré D (1985) Insect and pathogen observations. Land Report Research Supplement 2: 15–16

Carré D and Cavigelli M (1985) Effects of density on yield of Illinois bundleflower (*Desmanthus illinoensis*) and wild senna (*Senna marilandica*). Land Report Research Supplement 2: 7–9

Cavigelli M (1985) Effect of Maximilian sunflower (*Helianthus maximiliani*) on naturally occurring weeds. Land Report Research Supplement 2: 10–12

Cody ML (1966) A general theory of clutch size. Evolution 20: 174–184

Collins M (1986) *Desmanthus illinoensis* density trial, 1986. Land Report Research Supplement 3: 18–19

Cooper HW (1957) Producing and harvesting grass seed in the great plains. USDA Farmers' Bulletin No. 2112

Cornelius DR (1950) Seed production of native grasses under cultivation in eastern Kansas. Ecological Monographs 20: 1–29

DeLoughery RL and Crookston R (1979) Harvest index of corn affected by population density, maturity rating and environment. Agronomy Journal 71: 577–580

Dewald CL and Louthan VH (1979) Sequential development of shoot system components in eastern gamagrass. Rangeland Management 32: 147–151
Dittman D (1987) Soil moisture and nutrient patterns in agricultural plots and native prairie. Land Report Research Supplement 4: 14–16
Donald CM (1981) Competitive plants, communal plants, and yield in wheat crops. In: Evans LT and Peacock WJ (eds) Wheat Science – Today and Tomorrow, pp 223–247. Cambridge University Press, Cambridge
Drinkwater LE, Wagoner P and Sarrantonio M (1998) Legume-based cropping systems have reduced carbon and nitrogen losses. Nature 396: 262–265
Duke JA (1981) Handbook of legumes of worldwide economic importance. Plenum, New York
Earle FR and Jones Q (1962) Analyses of seed samples from 113 plant families. Economic Botany 16: 221–250
Fischer RA and Turner C (1978) Plant production in arid and semiarid zones. Annual Review of Plant Physiology 29: 277–317
Flore JA and Layne DR (1996) Prunus. In: Zamski E and Schaffer AA (eds) Photoassimilate Distribution in Plants and Crops: Source-Sink Relationships, pp. 825–849. Marcel Dekker, New York
Fox JF and Stevens GC (1991) Costs of reproduction in a willow: experimental responses vs. natural variation. Ecology 72: 1013–1023
Francis CA (1986) Multiple Cropping Systems. Macmillan, New York
Gadgil M and Solbrig OT (1972) The concept of r- and K-selection: Evidence from wildflowers and some theoretical considerations. American Naturalist 106: 14–31
Gernes M (1986) Relationship among cultural practices, stem density, and yield components in *Helianthus maximilianii* Schrad. Land Report Research Supplement 3: 20–22
Gernes M and Piper J (1987) Vegetation patterns in tallgrass prairie and their implications for sustainable agriculture. Land Report Research Supplement 4: 1–8
Gliessman SR (1983) Allelopathic interactions in crop-weed mixtures: Applications for weed management. Ecology 9: 991–999
Gliessman SR (1984) An agroecological approach to sustainable agriculture. In: Jackson W, Berry W and Colman B (eds) Meeting the Expectations of the Land, pp 160–171. North Point Press, Berkeley, California
Great Plains Flora Association (1986) The Great Plains Flora
Griffin LC and RM Rowlett (1981). A lost Viking cereal grain. Ethnobiology 1: 200–207
Grime JP (1979) Plant Strategies and Vegetation Processes. John Wiley and Sons, New York
Gross KL and Soule JD (1981) Differences in biomass allocation to reproductive and vegetative structures of male and female plants of a dioecious perennial herb, *Silene alba* (Miller) Kraus. American Journal of Botany 68: 801–807
Gursky M (1984) The performance and potential of *Desmanthus illinoensis* accessions. Land Report Research Supplement 1: 3–5
Handley M (1986) Survey of plant diseases in field plots at The Land Institute. Land Report Research Supplement 3: 36–38
Handley MK and Boehmer P (1987) Variability in resistance to anthracnose and leaf rust in *Tripsacum dactyloides* crosses segregation for sex ratio. Land Report Research Supplement 4: 30–35
Harper JL (1977) The population biology of plants. Academic Press, New York, NY, USA
Harper J and Ogden J (1970) The reproductive strategy of higher plants. I. The concept of strategy with special reference to Senecio vulgaris. Journal of Ecology 58: 681–698
Heiser CB and Smith DM (1964) Species crosses in *Helianthus*: II. Polyploid species. Rhodora 66: 344–358
Hickman JC and Pitelka L (1975) Dry weight indicates energy allocation in ecological strategy analysis of plants. Oecologia 21: 117–121
Hitchcock AS (1935) Manual of the Grasses of the US. USDA Miscellaneous Publication 200

Hole CC and Scott PA (1983) Effect of number and configuration of fruits, photon flux density and age on the growth and dry matter distribution of fruits of *Pisum sativum* L. Plant, Cell and Environment 6: 31–38

Horvitz CC and Schemske DW (1988) A test of the pollinator limitation hypothesis for a neotropical herb. Ecology 63: 1363–1373

Jackson L and Dewald C (1994) Predicting evolutionary consequences of greater reproductive effort in Tripsacum dactyloides. Ecology 75: 627–641

Jackson W (1985) New Roots for Agriculture. University of Nebraska Press, Lincoln

Jackson W, Berry W and Colman B (1987) Meeting the Expectations of the Land. North Point Press, Berkeley, California

Jackson LL, Dewald CL and Bohlen CC (1992) A macromutation in *Tripsacum dactyloides* (Poaceae): consequences for seed size, germination, and seedling establishment. American Journal of Botany 79: 1031–1038

Johnson DR and Major DJ (1979) Harvest index of soybean as affected by planting date and maturity rating. Agronomy Journal 71: 538–540

Jones DG and Clifford BC (1978) Cereal Diseases, 2nd ed. John Wiley & Sons, Chichester, England

Jones Q and Earle ER (1966) Chemical analyses of seed II: Oil and protein content of 759 species. Economic Botany 20: 127–155

Kendall B (1987) Growth and seed yield within a polyculture of *Leymus racemosus*, *Tripsacum dactyloides* and *Desmanthus illinoensis*. Land Report Research Supplement 4: 36–40

Klebesadel LJ (1985) Beach wild rye characteristics and uses of a native Alaskan grass of uniquely coastal distribution. Agroborealis 2: 31–38

Kois J (1985) Effect of density on seed yield and vegetative spread in first-year Maximilian sunflowers (*Helianthus maximiliani*). Land Report Research Supplement 2: 1–4

Komarov VL (1934) Flora of the USSR. Vol. II: Gramineae Botanical Institute, Academy of Sciences, USSR, Leningrad (Translated from Russian 1963). Office of Technical Services, US Dept. Commerce, Washington, DC

Kulakow PA, Benson LL and Vail JG (in press) Prospects for domesticating Illinois bundleflower. Proceedings of the First National Symposium for New Crops, Timber Press, Portland, Oregon

LaLiberte C (1985) Effect of Maximilian sunflower. Land Report Research Supplement 2: 31–32

Latting J (1961) The biology of *Desmanthus illinoensis*. Ecology 42: 487–493

Law R (1979) The cost of reproduction in annual meadow grass. American Naturalist 113: 3–16

Leonard LT (1925) Lack of nodule formation in a subfamily of the Leguminosae. Soil Science 20: 165–167

Liebovitz R (1987) Third year seed yield in plots of *Desmanthus illinoensis* and *Helianthus maximilianii*. Land Report Research Supplement 4: 17–19

Lovett JV (1982) The effects of allelochemicals on crop growth and development. In: McLoren JS (eds) Chemical manipulation of crop growth and development, pp 93–110. Butterworth Scientific, London

Lovett Doust JV and Lovett Doust L (1987) Leaf demography and clonal growth in female and male *Rumex acetosella*. Ecology 68: 2056–2058

Lubbers AE and Lechowicz MJ (1989) Effects of leaf removal on reproduction vs. belowground storage in *Trillium grandiflorum*. Ecology 70: 85–96

MacArthur RH and Wilson EO (1967) The Theory of Island Biogeography. Princeton University Press, Princeton, NJ

McDaniel ME and Dunphy DJ (1978) Differential iron chlorosis of oat cultivars. Crop Science 18: 136–138

Mecko-Ray V (1987) Evaluation of *Desmanthus illinoensis* germplasm. Land Report Research Supplement 4: 24–28

Melderis A, Humphries CJ, Tutin TG and Heathcote SA (1980) Tribe Triticeae Dumort. In:

Tutin TG, Heywood VH, Burges NA, Moore DM, Valentine DH, Walters SM and Webb DA (eds) Flora Europaea, Volume 5, pp 190–200. Cambridge University Press, Cambridge, England

Onwekwelu SS and Harper JL (1979) Sex ratio and niche differentiation in spinach (*Spinacea oleracea* L.). Nature 282: 609–611

Pianka ER (1970) On r- and K-selection. American Naturalist 104: 592–597

Pimm S (1997). In search of perennial solutions. Nature 389: 126–127

Piper J (1986) The prairie as a model for sustainable agriculture: A preliminary study. Land Report Research Supplement 3: 1–4

Piper J and Towne D (1988) Multiple year patterns of seed yield in five herbaceous perennials. Land Report Research Supplement 5: 14–18

Piper J, Henson J, Bruns M and Bender M (1988) Seed yield and quality comparison of herbaceous perennials and annual crops. In: Allen P and van Dusen D (eds) Global perspectives in agroecology and sustainable agricultural systems, pp 715–719. University of California, Santa Cruz

Piper J (1993) Soil water and nutrient change in stands of three perennial crops. Soil Science Society of America Journal 57: 497–505

Pitelka LF, Hansen JB and Ashman JW (1985) Population biology of *Clintonia borealis*. I. Ramet and patch dynamics. Journal of Ecology 73: 169–183

Policansky D (1981) Sex choice and the size advantage model in jack-in-the-pulpit (*Arisaema triphyllum*). Proceedings of the National Academy of Sciences (USA) 78: 1306–1308

Price D (1984) The effect of *Helianthus maximiliani* on naturally occurring weeds. Land Report Research Supplement 1: 1–2

Price D and Gursky M (1984) Overyielding in a perennial legume biculture. Land Report Research Supplement 1: 10–12

Primack RB and Hall P (1990) Costs of reproduction in the pink lady's slipper orchid: a four-year experimental study. American Naturalist 136: 638–656

Pritts MP and Hancock JF (1985) Lifetime biomass partitioning and yield component relationships in the Highbush Blueberry, *Vaccinium Corymbosum* L. (Ericaceae). American Journal of Botany 72: 446–452

Putnam AT and Duke WB (1974) Biological suppression of weeds: evidence for allelopathy in accessions of cucumbers. Science 185: 370–372

Reekie EG and Bazzaz FA (1987a) Reproductive effort in plants. 1. Carbon allocation to reproduction. American Naturalist 129: 876–896

Reekie EG and Bazzaz FA (1987b). Reproductive effort in plants. 2. Does carbon reflect the allocation of other resources? American Naturalist 129: 897–906

Reekie EG and Bazzaz FA (1987c). Reproductive effort in plants. 3. Effect of reproduction on vegetative activity. American Naturalist 129: 907–919

Reznick D (1985) Costs of reproduction: an evaluation of the empirical evidence. Oikos 44: 257–267

Rice E (1984) Allelopathy. Academic Press, New York

Riley K (1984) *Helianthus maximiliani* density trial. Land Report Research Supplement 1: 6–8

Rinehart D (1986) Relationships among basal stem diameter, biomass, and yield. Land Report Research Supplement 3: 11–13

Sakai AK and Burris TA (1985) Growth in male and female aspen clones: a 25 year longitudinal study. Ecology 66: 1921–1927

Singh ID and Stoskopf NC (1971) Harvest index in cereals. Agronomy Journal 63: 224–226

Slater M (1986) Survey of insects in native prairie and research plots. Land Report Research Supplement 3: 4–8

Smith M and Francis C (1986) Breeding for multiple cropping systems. In: Francis C (ed) Multiple Cropping Systems, pp. 219–247. Macmillan, New York

Snaydon RW (1984) Plant demography in an agricultural context. In: Dirzo R and Sarukhan J (eds) Perspectives on Plant Population Ecology, pp 389–407. Sinauer Associates Sunderland, Mass

Soule JD and Werner (1981) Patterns of resource allocation in plants, with special reference to *Potentilla recta* L. Bulletin of the Torrey Botanical Club 108: 311–319
Steyermark JA (1963) Flora of Missouri. Iowa State University Press, Ames
Stiefel V (1985) Yields of four- and three-year-old wild senna (*Senna marilandica*). Land Report Research Supplement 2: 29–32
Taparia AL, Talmale TW, and Sharma VV (1978) Utilization of *Cassia tora* seeds in rations of buffalo calves. Indian Journal of Animal Science 48: 804–810
Thompson K and Stewart AJA (1981) The measurement and meaning of reproductive allocation in plants. American Naturalist 117: 205–211
Thornberg A (1971) Grass and legume seed production in Montana. Agricultural Experiment Station Bulletin 333, Montana State University, Bozeman
Tissue DT and Nobel PS (1990) Carbon relations of flowering in a semelparous clonal desert perennial. Ecology 71: 273–281
Tsitsin NV (1978) Remote hybridization in plants. Nauca, Moscow, USSR
USDA (1966) Technical note 12/1/66, Soil Conservation Service, Plant Materials Center, Bridger, Montana
USDA (1975) Nutritive Value of American Foods. In: Common Units. Agricultural Handbook No. 456. Agricultural Research Service. US Government Printing Office, Washington, DC
USDA (1976) Technical Report 1975–76, Soil Conservation Service, Plant Materials Center, Coffeeville, Mississippi
USDA (1978) Annual Report. Soil Conservation Service, Plant Materials Center, Knox City, Texas
USDA (1980) Agricultural Outlook, Economics, Statistics, and Co-operative Service. June
USDA (1996) Agricultural Statistics 1995–96. US Government Printing Office, Washington, DC
Wardlaw IF (1980) Translocation and source sink relationships In: Carlson PS (ed) The biology of crop production, pp 297–339. Academic Press, New York
Watson MA and Casper BB (1984) Morphogenetic constraints on patterns of carbon distribution in plants. Annual Review of Ecology and Systematics 15: 233–258
Weaver SE and Cavers PB (1980) Reproductive effort of two perennial weed species in different habitats. Journal of Applied Ecology 17: 505–513
Weiner MA (1980) Earth medicine – Earth food. Macmillan, New York
Wheeler W (1957) Grassland Seeds. Princeton University Press, Princeton, 434 pp
Whyte RO (1957) Grasses in Agriculture. Food and Agriculture Organisation, Rome, 141 pp
Williams LE (1996) Grape. In: Zamski E and Scaffer AA (eds) Photoassimilate Distribution in Plants and Crops: Source-Sink Relationships, pp 851–869. Marcel Dekker, New York.
Williams RJ (1983) Tropical legumes. In: Melvin JG and Bray RA (eds) Genetic resources of forage crops. CSIRO, East Melbourne, Australia

From genomes to ecosystems to human communities

W. JACKSON
The Land Institute, Salina, Kansas USA; E-mail: theland@igc.apc.org

Key words: context dependency, large effect genes, small effect genes

Abstract. Genomes can be regarded as miniature ecosystems which display many of the same characteristics as that of a prairie or forest ecosystem. Human communities may likewise have characteristics common to nature's ecosystems. The forces at work which disrupt settings at the landscape level are often the same forces at work which have disrupted small human communities and now threaten to invade the nano-ecosystem of our major crops and livestock. Increasingly, the reward for the disruption of all three integrities is oligarchic power.

1. Introduction

We need coherent ideas or notions to inform our habits *without thought.* Cultural evolution is Lamarckian and serves to temper our core nature, enabling us to operate with minimum thought. As children of the enlightenment, most of us arise in the morning, unaware that we will play out, in both our formal and informal discourse, some of the thinking of Aristotle, some thinking of Francis Bacon, some thinking of Rene Descartes. These thoughts or ways of being come 'with the milk', so to speak and stand as one reason why so many of our problems seem intractable. Since ideas out of the past are deeply embedded in our thoughts including our language, to get a grip on an alternative way of thinking about the world is formidable. The assumptions which attend the dominant way of thinking have a way of reasserting themselves.

Increasingly, we hear that this or that science needs a new paradigm. What we mean by that can't be some trivial bumper sticker version. When we read *The Structure of Scientific Revolution* by Thomas Kuhn or read what philosophers of science and historians of science had to say about Kuhn's contribution, we realize that new paradigms aren't mere fads or style changes. It is even more than a change in point of view. There must be a change in consciousness which leads eventually, in the words of Thomas Kuhn himself, to a 're-evaluation of prior theory and the reinterpretation of prior fact'.

As beneficiaries of both the scientific (1500 to 1700) and industrial revolutions (1800 to present) we are beginning to have a new story which speaks to some of the old religious questions with answers quite unlike answers derived from the old story. Beginning with Copernicus and Galileo, we now have a sun-centred rather than an earth-centred system. The moral code Dante assigned to the cosmology of Ptolemy may be rich in content, insightful about human nature, but the physical structure of the earth from Rome to Hell or from Purgatory upward through the planets toward God through the angelic hierarchies is not true. Awareness of the new cosmological reality may have little to do with Dante's warning on the gate to Hell: 'abandon all hope ye who enter here'. We have no data to refute this warning.

What I am getting at is that even before 19th century geologists Lyell and Hutton paved the way for our understanding of the antiquity of the earth, discoveries were pouring in to set the stage which would eventually challenge our assumptions a century later. So to Charles Darwin's explanation of evolution through natural selection, we added a century later the unfolding of the modern molecular synthesis and a new genesis story was building. More recently we have come to accept that the eukaryotic cell itself is the product of a merger of former prokaryotes which have contributed their critical organelles and thereby made multicellularity possible. The idea of a separate biological creation, a legacy of the Genesis version, would not stand. Of course, we are quite different in some very important ways, but in the biological sense where we are as one, we are beginning to pay attention to our biological oneness and let's hope, just in time.

The fundamental dichotomy between prokaryotes and eukaryotes is now widely acknowledged. Prokaryotes lack DNA organised within a nuclear membrane. As we move up the tree of life there is cellular differentiation, but there is also an invasion of various life forms into other life forms. Little spirochetes are invading, another bacterium comes in, perhaps as an invader but leaving its mitochondria. Various cell organelles get parked inside other host cells, not to remain as refuse, but retaining at least part of their roles, becoming useful organelles for their new hosts. Some of these organelles, like mitochondria and chloroplasts, retain some DNA. Evolving here, at this nano level, is what we now call the eukaryotic cell. Prokaryotes, like the blue-green algae and bacteria are still around, but the eukaryotes represent a big new thing that would turn out to have staggering possibilities. At the point in the process of evolution in which we finally have a full-blown eukaryotic cell, eighty percent of the time there has been life on earth has passed. It doesn't matter whether we are redwood or whale, human or giraffe, dog or bull snake, carrot or broccoli, wheat, potatoes, corn. It is the same sort of code, the same amino acids. Eukaryotes all! At this level, plant and animal designations become less important than the commonalities they share.

The university freshman taking an introductory biology course with passing marks likely knows what I have just said. It is part of our general understanding of how the biological world is. The reason I bring it up is because only in the last half of the 20th century have such a large number of thinking people become aware of this biological reality. Our cultural heritage, including our language and religion, has internalized this less than it has the legacy of our earlier stories. Take the words Creation and Creature. The cognition which formerly attended them may have changed, with the new knowledge, but the legacy of the past is carried forward as a shadow over our new understanding. An older term, rarely used, will have its precision more easily preserved. An older term commonly used up to the present often has accumulated several meanings. This is the reason for the need to remind ourselves of our biological oneness as we prepare for the next important step in exploring the *dimensions* of our oneness.

2. Eukaryotic genomes: the first ecosystems

During the evolution of eukaryotic organisms, miniature ecosystems, ecosystems at the genome level have evolved. More than organelles have been left behind with their DNA. DNA from the former host of the organelle have accompanied the countless prokaryotic invasions which must have happened again and again. Hybridisation in various forms took place, at least this is what the evidence suggests. As this was happening, the architecture of the recipient genome was evolving and not in a passively receptive sort of way but by sorting out unworkable relationships from the workable ones. This point deserves emphasis, for here the reality of the architecture of the various genomes begins to provide the basis for a moral argument. At this point we begin to have a way to think about the proper role of the gene splicer, the proper role of biotechnology in general.

Imagine a Gaussian distribution. Now fold it to yield half of this statistical distribution (Figure 1). On the left in Figure 1 we see that most of the genes of that genome (the entire genetic compliment) are small-effect genes. Few are for large effects. Whether considering any eukaryotic creature, humans or corn plants, this is the general architecture, a reality with important implications. According to Strohman (1993), less than 5% of the human disease burden may be attributable to the effects of variations in single genes. This doesn't square with what we read in the press; someone has found a gene for this kind of cancer or that kind of cancer or this kind of heart disease or whatever. We hear and read only of large-effect genes. It is never mentioned that there is a very small contribution of such genes to our overall health. Small-effect genes which represent the majority provide a kind of shock absorber. When some alien gene enters the genome, be it from long or short evolutionary distance, these small-effect genes absorb the 'shock.' This absorption, however, forces a change in the genetic profile, which is to say, a change in the architecture of the genome. This has important consequences. This must be the very basis of homeostasis. It is one of those times that we say, 'Aha, this must be the way that multicellular creatures of this planet have managed to be resilient to the perturbations coming at them in a world that is fundamentally unknowable! This is Darwinian evolution at the most basic level. It had to be this way.' *The Economist* (May 22, 1998) reports that 'Evidence is accumulating that the immune system works like an

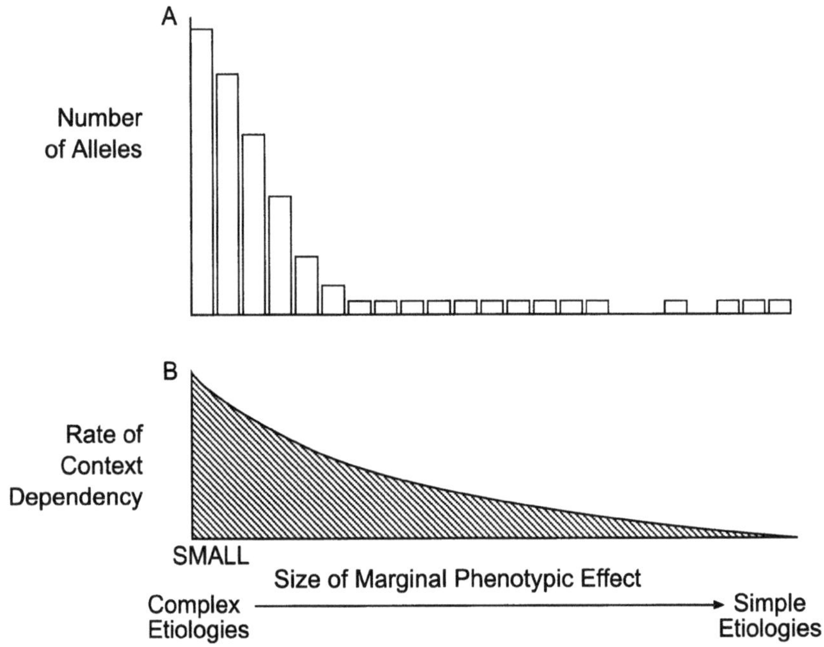

Figure 1. Characteristic genomic architecture of eukaryotes: small effect genes are many, large effect genes are few (From Sing et al., 1996).

ecosystem, in that its component cells compete with one another for survival. This could have important implications for the treatment of auto-immune diseases'.

3. Arches and spandrels

Various cathedrals are known for their arches, the structures responsible for supporting the roof. The space between the arches is called a spandrel; in churches it is usually bricked in and once bricked in, is likely to be plastered over and then painted with a fresco. The spandrel is a derivative, a place between the arches. In biological systems, that which is under genetic selection at one point may be analogous to that arch. At this point the story changes. The spandrel takes on greater significance when visitors from a town in Tuscany or Umbria or anywhere arrive in Venice and say, 'We want our church to be similar. We want to be able to paint a fresco on the plastered spandrel.' Now the spandrel itself, indeed the fresco, is an object of natural selection.

This analogy helps explain how some characteristics may be only a consequence of a derivative. The spandrel as derivative of the arch can now have an adaptive value, meaning it is subject to replication (Gould and Lewontin, 1979; Dennett, 1995).

Let's say a group from an agri-chemical company, like Monsanto arrives with an interest in having a fresco, and being bottom-line people, they set out to find the gene for fresco-hood. Now comes the basis for a moral, for if one hunts for the gene for fresco-hood and *only fresco-hood*, one had better be prepared to bring along many other traits because a change will be forced on the overall architecture of the cathedral.

For a couple of decades corn dramatically responded to an increase in nitrogen fertiliser. Breeders were selecting for 'fertiliser-response' genes. At least this was the language of the breeder and understandably so. But what was happening at a biological level was the *elimination* or neutralisation of the genes responsible for the plant's ability to discriminate in the rate of fertiliser uptake. The biological reality was assigned another name, 'fertiliser-response' genes. That era of fertiliser-response flattened, but now it is climbing again. I suspect that it is climbing because breeders are working with a different genetic ensemble responsible for that response. All genes interact. And as selection pressure is applied, as when humans seek an improved wheat variety or corn variety or improved beef cow or sheep, we are restructuring the genome of this miniature ecosystem. This can go on for a while. We have got away with this approach from the beginning of agriculture. In the industrial era we have sped this up, of course, but even with ordinary plant breeding, breeders get away with these modifications because they are forced to deal with phenotypes (the way traits are expressed) and unavoidably 'drag along' genotype (the genetic code). With gene splicing it is the other way around. Now it is a new story. It is the very 'dragging along' that has saved us, for this liability has in fact been to our advantage as it has ensured that a workable context has been retained.

As the agricultural revolution began and advanced in various places around the world, eight to ten thousand years ago, the genomes of our crops and livestock – miniature ecosystems really – were altered. With alteration came a degree of dependency on the species responsible for the alteration, *Homo sapiens*. Of course, we humans became dependent on them, too. Inter-dependence evolved as these early agriculturists altered the genetic architecture of some wild species and established, not a one-way dependency, but two-way. To be anthropomorphic for a moment, the various plants and animals were saying, 'Look what we have to offer,' and we effectively said, 'Oh yes, we would like to have that. If you provide us with a more or less assured abundant grain supply or meat supply or cream or egg supply, we'll take care of you. With you grain producers, however, we have to make some changes. Your seeds shatter. We don't want to pick them off the ground one by one, so those among you that tend not to shatter we'll pick as parents for our future crops.'

This shameful anthropomorphic version should not minimise the importance of the consequences of something so modest as the selection for shatter resistance. As wild species they were independent of us, but those first few cuts of selection for something as simple as shatter resistance must have *greatly* truncated and changed the genetic profile. Co-dependency likely arose very early in domestication. Without us the corn plant won't live, nor the wheat plant, nor nearly any other domestic crop or livestock.

The agricultural revolution as a genetic revolution heightened dependency, heightened interdependency, and domesticated us as we domesticated the crops and livestock. The nano-ecosystem of all involved species was altered. The Chicago Board of Trade genes, the fossil fuel well-head genes of the industrial era and now increasingly the computer genes – surrounding us and in us – are all ensembles of genes that would not be here if it were not for the reality of the Board of Trade, fossil fuels, computers. All of these realities have forced a change in the architecture. It is a Darwinian reality.

When Monsanto splices in genes from long evolutionary distance, such as genes from bacteria into corn, not only is it done at a speed for which there is no precedent in the long tradition of plant and animal breeding, the genes are isolated from their evolved context. Dr. Don Duvick, retired Senior Vice-President and head of research at Pioneer Hi-Bred International, Inc. says of this phenomenon that in his experience such genes eventually 'act like tar is being smeared over them'. In other words, in a few generations their expression falters. Stated otherwise, the genomic architecture changes; the background shifts. An inherent incompatibility forces the new expression to be disallowed by the nano-ecosystem. This absorption of the new trait has come at a cost – very likely a more vulnerable, more *high-tech-dependent genome*.

So here is the sequence. *Humanity* was responsible for altering the wild plants to make the major crops available several millennia ago. Those crops became dependent upon us as we became dependent on them. And now the high-tech arrangement which brings the genetic changes in the genome will create a dependency on the high-tech arrangement. In other words, the scientific and technical arrangements that a corporation or for that matter a major university are capable of are not available to the run-of-the-mill breeding program. Hence, a more specialised brand of dependency on a particular kind of economic situation is now developing.

4. Complex adaptive systems

When we consider the human from conception to death (Figure 2), we see that in the first five years, single gene diseases dominate. If we humans make it beyond age five, continue to grow and become productive adults by age fifty the multifactorial diseases are likely to take over.

In Figure 3 we see what happens leading up to a stroke or a heart attack. A fatty streak develops, a fibrous plaque, complicated lesions, haemorrhage, thrombosis, and sometime after fifty, one can experience a stroke or a heart attack. Several genetic factors contribute to these conditions; carbohydrate metabolism, lipid metabolism, blood clotting, and blood pressure regulation.

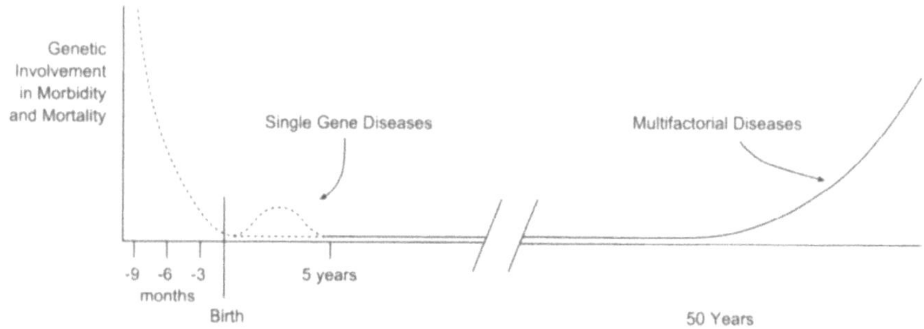

Figure 2. Single gene diseases dominate in early stages of human life while multi-factorial diseases dominate late in life (CF Sing, unpubl.).

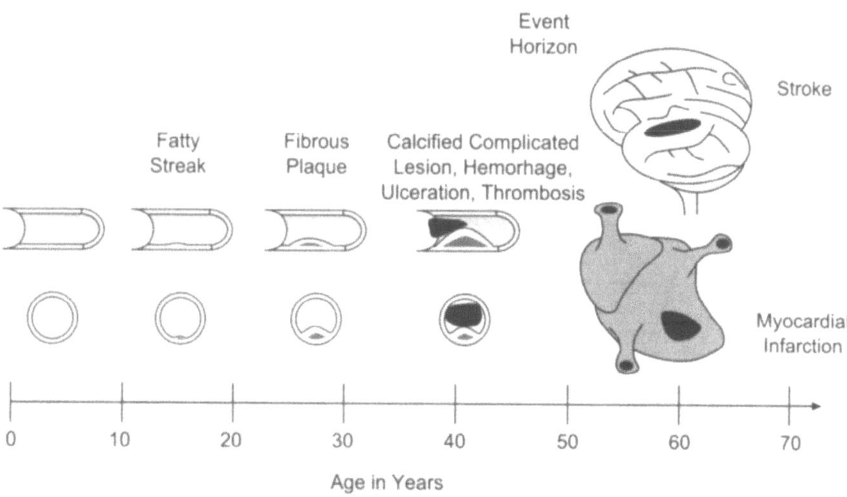

Figure 3. Genetic factors contributing to heart attack or stroke (Redrawn from McGill, 1979).

Many kinds of environments are possible. As we age these possible environments interact with a particular genotype.

When we place a grid over all of this (Figure 4), what is called a reaction surface, it appears as though it is made of rubber. This reality is not Cartesian, not Baconian, not Aristotelian, but it *is* Darwinian, deeply so. Cartesian thinking encourages us to narrow our boundary of consideration, greatly restricting the environmental variables and ignoring the product of genetic interactions. The Darwinian view encourages us to acknowledge the reality of a complex adaptive ecosystem at work and we can see it here, even at the genome level.

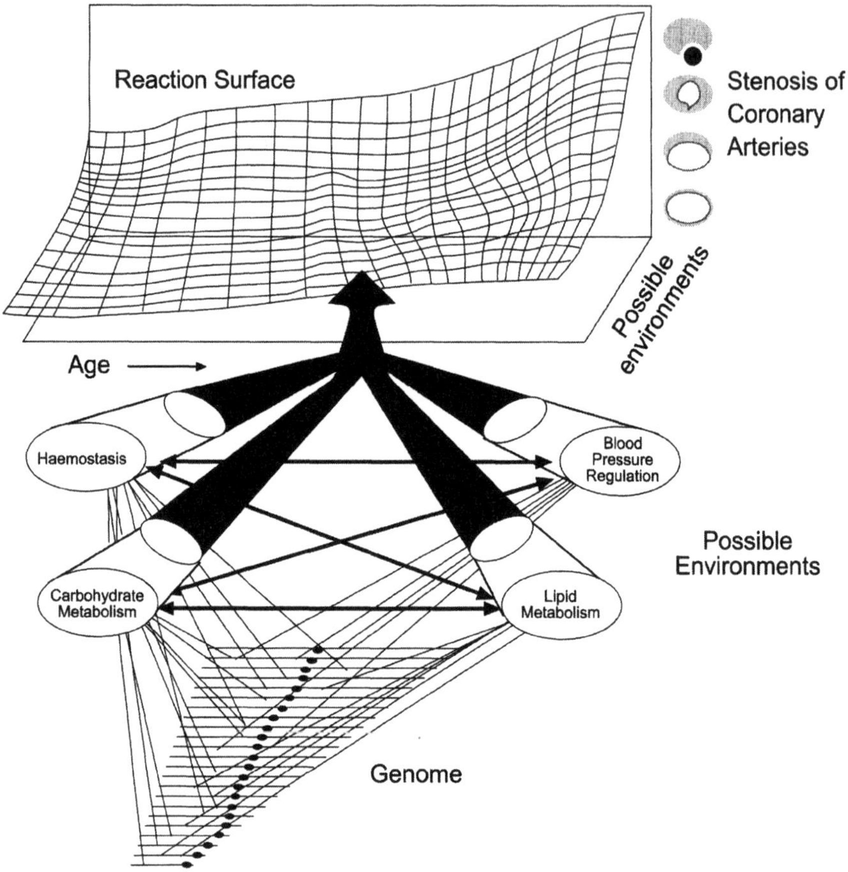

Figure 4. A reaction surface illustrating how possible environments over time interact with a particular genotype (Redrawn from Sing et al., 1996).

4.1. *Simple averages often mislead us*

The systolic blood pressure of two people in Figure 5 shows a low and high average blood pressure for two individuals. The lower line reveals little variation. The high average line reveals more variation especially at night. Conventional wisdom holds that low average blood pressure is good and high average blood pressure is bad. But here that wisdom is reversed. The person with higher variation responds to the realities of life. The low blood pressure person with little variation is not keeping his or her vascular system toned up through blood pressure changes. This is not an argument for high blood pressure but to show that with complex adaptive systems a simple index can mislead us.

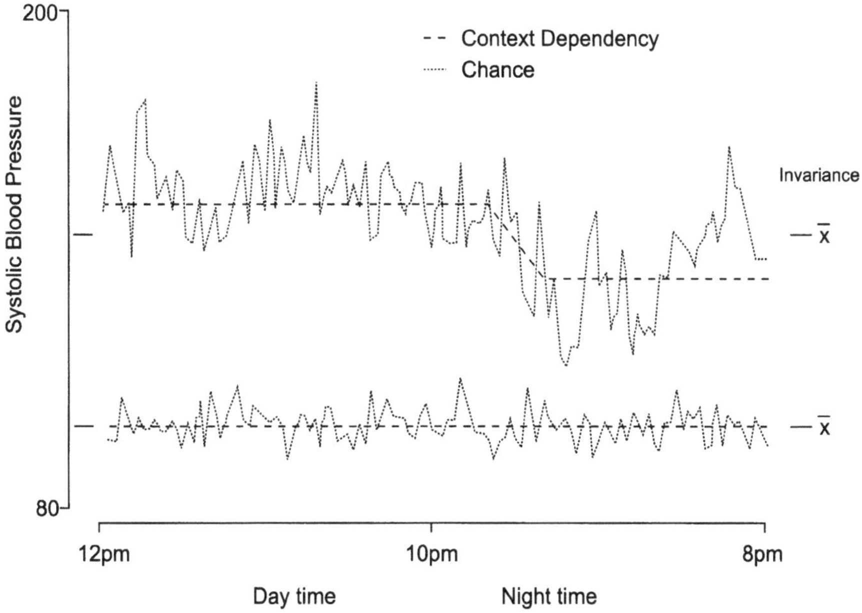

Figure 5. Systolic blood pressure of two people in which the individual with the higher average blood pressure is at higher risk of stroke or heart attack (CF Sing, unpubl.).

4.2. *A beneficial genetic quality is often dependent*

In Figure 6, big round circles represent carbohydrate and lipid metabolism in a child. Blood clotting or blood pressure regulation is a minor factor. When the same individual becomes an ageing adult, carbohydrate and lipid metabolism become relatively unimportant while blood clotting and blood pressure regulation become important.

An ecosystem experiencing succession has different requirements at different stages of maturity. The same with the changing 'human ecosystem'. Professor Richard Lewontin from Harvard put it this way: 'The appropriate metaphor for the evolution of genotype relationships is not hill climbing but walking on a trampoline' (R. Levin, pers. comm.). Until recently, students of population genetics had learned their lessons in the school which featured Sewell Wright adaptive peaks. Plant and animal populations were struggling toward those adaptive peaks. The idea of progress was implied. Instead, we're walking on this trampoline and as we do, adjustments are forced along the way.

Experience and data out the laboratories and research plots of scientists who study the genetics of populations ranging from humans to plants have forced them to turn to a more dynamic model. Classical genetics is being replaced by greater attention to complex adaptive systems.

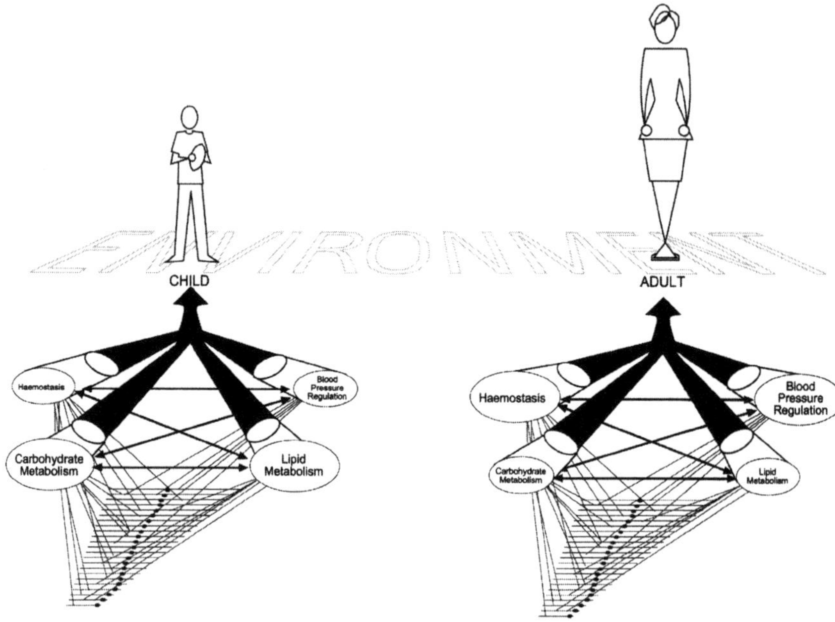

Figure 6. A beneficial genetic quality at one age may not be so at another (Redrawn from Sing et al., 1996).

4.3. *Tradition and experience as requirements across the ecological mosaic*

A prairie, like most other land-based ecosystems, features roots that hold soil in place and a prairie, again like other ecosystems, has 'genius' in it, meaning that the current arrangements are products of trial and error over the millennia. Alexander Pope in his 'Epistle to Burlington' said, 'Let Nature never be forgot, consult the genius of the place in all' (W Berry, pers. comm.). The use of the word 'consult' is inspired. Pope didn't say 'obey,' he didn't become dogmatic, he said 'consult the genius of the place'.

In the early 1930s a young graduate student at the University of Nebraska, working on a masters degree under the great ecologist Professor John Weaver, compared a wheat field to an adjacent never-ploughed native prairie near Lincoln. It turned out to be the driest year on record. But here on similar soil, similar slope, the prairie genius received and stored precipitation most efficiently and kept that prairie ecosystem intact. All the wheat plants died. The genius of that place handled the drought.

Jack Ewel (this volume) working in the tropical rain forest of Costa Rica where water is a nemesis of fertility (because water leaches nutrients downward) described a genius which operated in an opposite manner. In Costa Rica the rainforest is designed to move that moisture up and out fast. The rainforest trees, giant solar powered pumps, lift it back into that atmosphere.

Here then are two different ecosystems from the ecological mosaic. One is 'designed' to receive and hold water, the other 'designed' to receive and get rid of water. When it comes to agriculture, local adaptations in conformity with the genius of a particular part of the mosaic is called for.

Wendell Berry once said that as we come across the North American continent, cutting the forests and ploughing the prairies that 'we never have known what we were doing because we never knew what we were undoing' (Berry, 1977). This has been the case wherever agrarian settlers have followed conquerors. For example, we were unaware what we were doing to the mycorrhizal associations when we ploughed the prairie. A sample of mycorrhiza spores from never-ploughed native prairie shows high diversity. Mycorrhiza from corn and soybean ground displays no such diversity. In other words, *Homo* the homogeniser is creating simplicity at the below-ground level, as well as above, maybe *especially* at the below-ground level.

The soil, dark in colour due to the carbon fixed over the millennia as the consequence of the living world, sponsored the breadbasket of the North American continent (Figure 7). When the prairie was ploughed and planted to crops over a fifty year period, the soil organic matter dropped (Figure 8). There was an early and brief upward bump in the nutrient supply. That is the bump responsible for those huge initial yields, the yields that became so famous, the yields that attracted settlers to the prairie from east of the Appalachian Mountains all the way to Europe. That bump represents nutrient release of the roots and other living things. But notice how brief was that

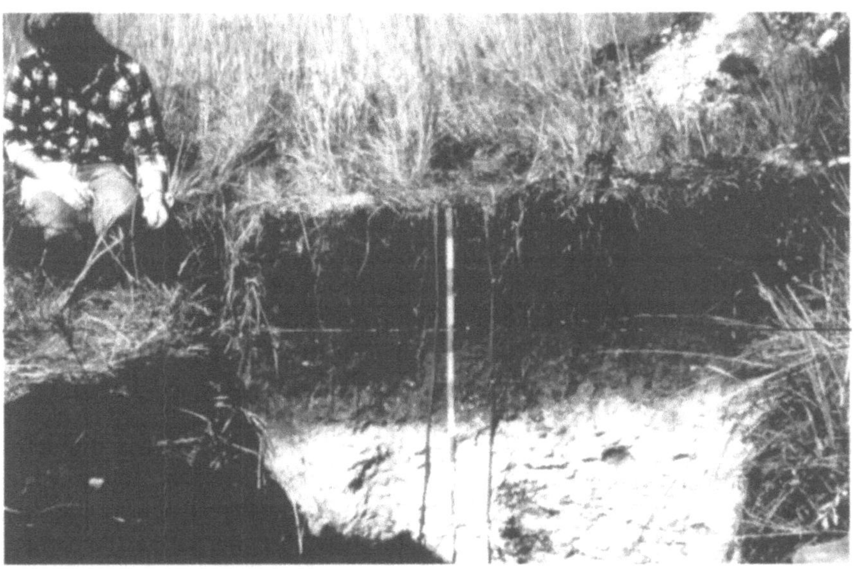

Figure 7. Central Kansas prairie soil. Note contrast between parent material and dark carbon rich soil due to prairie biota.

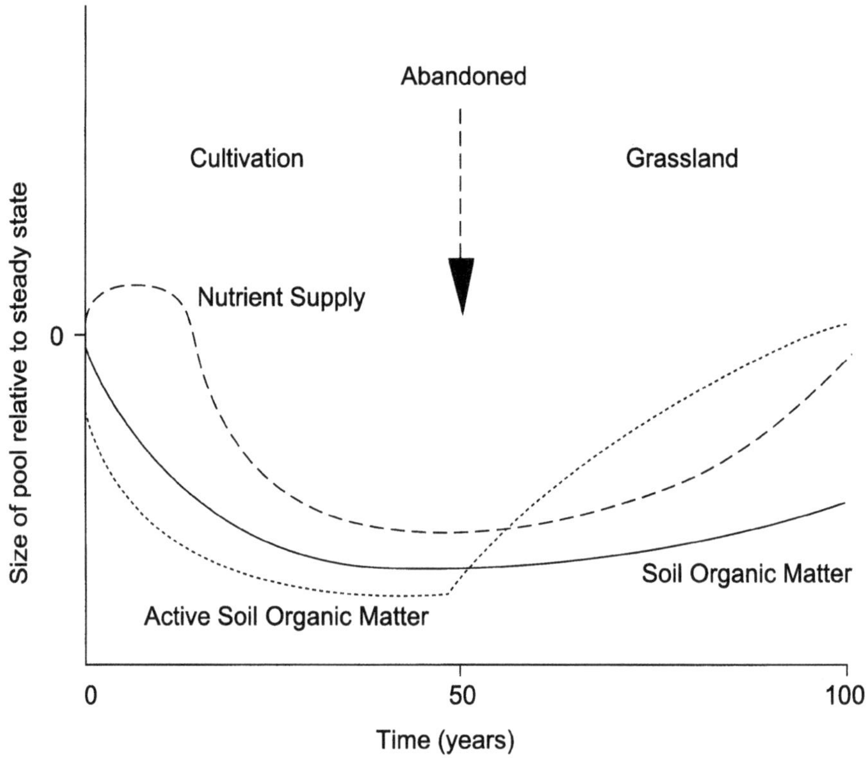

Figure 8. Exploitation and recovery: the responses of total soil organic matter, active soil organic matter and nutrient supply to 50 years of cultivation and subsequently to 50 years of recovery as perennial grassland (Redrawn from Burke et al., 1995).

surplus. All lines go down for 50 years. The good news is that after fifty years, as the prairie returns, so does active soil organic matter. Nutrient supply climbs and soil organic matter slowly comes back. Here the 'genius of the place' that was ploughed under is able to surface again, but slowly.

For millions of years in Australia the winds of the roaring forties have slammed into the south coast depositing aerosols of salt. The salt was leached downward. The vegetative structure of the eucalypts enable them to suck that moisture up on its downward trip leaving the salts stored in the deep clays. Englishmen arrive 'with vision but not with sight, visions of former places' (Berry, 1977) who clear and plant wheat, clear and plant cool season pastures for sheep. Visions of English idyllic hillsides become a reality. Unfortunately, those cool season grasses do not pull the same quantity of moisture upward with the consequence that the water table rises bringing salt to the surface. The solution? Consult the genius of the place. Plant back a vegetative structure sufficient to ensure precipitation is returned before it mobilises the fossil salt. The natural places saved by Australians, for whatever reason, become the standard for knowing what to do to preserve sheep ranching and grain production in Australia.

How we deal with the reality of the ecological mosaic is problematic. The area under the curve in Figure 9 can represent either all the land in Australia or Western Australia. Maybe it is all of the United States or Kansas. The curve will assume various shapes depending on the nature of the mosaic in a given area. At the lower end is land that will carry a wide variety of agricultural activity. These are environments with broad ecological amplitude. One might grow, for example, watermelon, sweet potatoes, cantaloupes, corn, wheat, nearly anything weather will permit. The vertical lines on the curve reveal the ecological amplitude. On the upper right we have a highly vulnerable environment. Here, one had better stick to something closer to the original vegetative structure. Failure to do so means that there will be a great loss or contamination of ecological capital. So the decision before the farmer and the culture comes down to deciding where a piece of land falls along that line. It is a way of confronting the reality of the ecological mosaic.

Figure 10 is a photograph taken in 1888 in eastern Kansas, probably showing the ruts of the Santa Fe trail. The man in the background is my great grandfather who came from England and settled with his parents in Western Pennsylvania. He arrived in Kansas in 1854. Like his ancestors, he came 'with

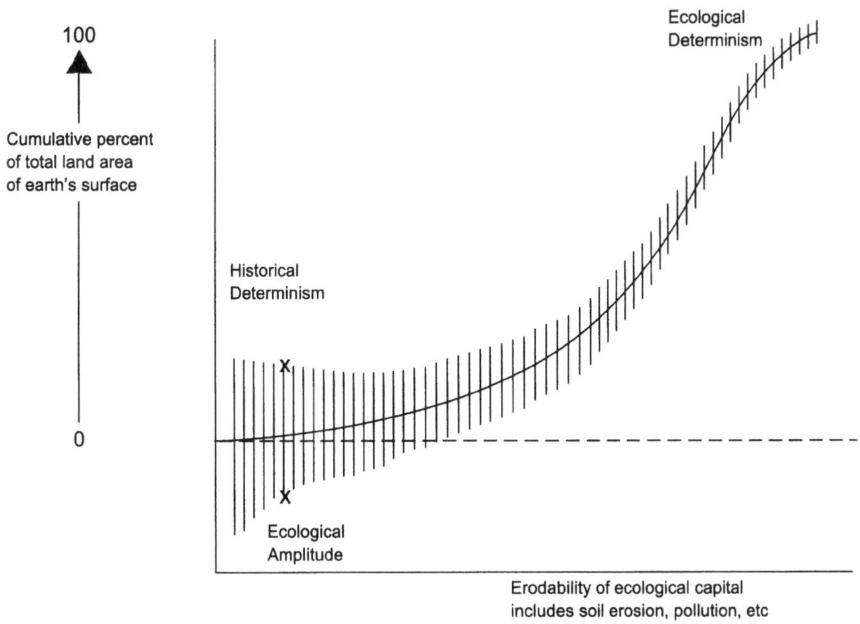

Figure 9. A generalised representation illustrating the reality of the ecological mosaic. The area under the curve can represent all the land of a specific region, accumulating to 100% to the extreme right. The curve will assume various shapes depending on the nature of the mosaic in a given area.

Figure 10. Eastern Kansas prairie in 1888.

vision but not with sight'. Family lore has it that he and his sons had to go twelve miles for wood. On that same spot now stands a sugar maple grove. Opposite these New England sugar maples is a remnant of native prairie, preserved because it features the ruts of the Santa Fe trail. Without those ruts I suspect that this would also have become woodland.

Here is a landscape we would represent close to the lower left of Figure 9, for it is an area with wide ecological amplitude. This brings up a complicated consideration. The trees my great grandfather planted, species more at home in a New England deciduous forest than in Kansas, did not reflect a consultation of the 'genius of the place'. But the trees are there and descendants of my great grandfather have harvested maple syrup. Where would this particular landscape of the mosaic fall on the graph? The ecological amplitude, the potential in terms of what will grow here without serious soil loss may be answered with forest as safely as with prairie.

What we are able to disrupt, without destroying options for future generations depends on the *forgiveness* of a particular area or place. We have to ask a similar question about a nano-ecosystem called a genome.

The late American painter, Norman Rockwell painted *The County Agent* in 1948 (Figure 11). In that painting we see the dutiful 4-H girl and we imagine her record book is in good shape. The county agent is measuring the girth of the Guernsey heifer. Brother is attentive and waiting to talk about his poultry project. Sis has her sewing project. Grandpa is in the background with the horse. Dad is in the near shadows to the left. Even the dog seems pleased with

Figure 11. Norman Rockwell's 'The County Agent'.

the demonstration. The chickens are enjoying the visit. The cat seems mildly interested. The original of this painting hangs at the University of Nebraska, the land grant university for that state.

It all seems so idyllic. But there is something wrong. With a little study we see that expertise and youth are central. Tradition and experience are peripheral. Norman Rockwell was only painting what was, but he also was painting what got writ large. It would be a safe bet that neither the daughter nor the son in the painting ended up as farmers. The removal of people from the land has taken its toll. In addition, industrial agriculture, at once simple and simplifying has ignored two essential needs of the ecological mosaic, tradition and experience.

When people, land, and community are as one, all members prosper. When regarded as competing interests, all members suffer. To move the ecosystem concept upward to the community level allows us to see the dynamics of a complex adaptive system and how the removal of large numbers of people has created a disequilibrium. Early invaders on the landscape were the agricultural technologies of the early Neolithic. From that era to the recent introductions of fossil fuels and agri-chemicals our agrarian landscapes featuring small effect interventions have had to absorb one assault after another. How long it can continue depends on the particularities of each member of the mosaic.

5. Summary

The genome of multi-cellular organisms seems to be the product of past encounters from various life forms. Cellular organelles have been added, like mitochondria, and chloroplasts. Other hereditary material has been retained and incorporated into the nucleus. The genetic program is similar enough in its dynamics to an ecosystem like a prairie that it seems worth considering the genome as an ecosystem of sorts. The small effect genes represent the majority and the major effect genes are in the minority. These small effect genes serve as a shock absorber for when alien genes enter. Some are incorporated, some rejected, some included but subdued. It is a system designed for resilience when invaded. The prairie ecosystem also features resilience due to small effects interacting.

Human communitites, before the industrial revolution, promoted and managed the products of photosynthesis through cultural arrangements. Over time those photosynthetic products, those packets of energy, were moved through harmonious routings to satisfy the human endeavour. The local nature and culture that generated that contemporary energy in turn informed the local culture. A Papago Indian in the southwestern desert of the United States has one way of managing these photosynthetic products that is different from the native in the Brazilian jungle along the Amazon. The Amish have a way of managing the cycling of nutrients and the flow of energy much different from the corn and soybean farmer in the U.S. Midwest. The Amish are using more contemporary sunlight for food production than the fossil fuel intensive industrialised farmer. Fossil fuel economies require a sufficiency of capitol rather then a sufficiency of people.

When the internal combustion engine entered our fields and travelled our roads, this high energy led to the destruction of the cultural information essential for every activity from seed preparation to harvest and storage. Fossil fuel forced a disappearance of countless arrangements and social contexts we call human. The front porch and neighbourliness disappeared. Little services went away. What is now left in the little places, after the parts have been stripped away are vulnerable people dependent upon high energy and high concentration agents in distant cities. During this evolution we seemed not to realise that we were affecting the architecture of these cultural shock absorbers – genomes, landscapes, rural communities. We were dismantling what we need in an emotional and spiritual sense as well as physical sense. Much activity formerly regarded as frenetic behaviour became the norm.

Failure to adapt at the genome level for plant and animal species we call extinction. When the ice returned after each interglacial period of the Pleistocene as large-effect phenomena, prairie ecosystems retreated with species assemblies waiting to return with the next advance. Some species likely became extinct. But the system featuring the small effect arrangements represented a kind of living ecological capital ready to return more non-living ecological capital to the stock pile. When the Pleistocene was over, when the plough came, the stock pile went into decline. And so it is the same story. When the small effects of the genome, or a land based ecosystem or a human community, are persistently assaulted by large effects either wilfully or ignorantly, we lose slack.

With agriculture and civilisation we have raised the bar. When the bar is raised for the pole vaulter, we don't expect a reduction in the absorption of the landing pit. The vaulters increase in height is regarded as less important than a safe landing. But as we have raised the bar during this brief industrial revolution phase we have also lowered the shock absorbers of the pits. At the genome level, at the ecosystem level, at the level of the human community, we have created brittleness, not resilience.

6. References

Berry W (1977) The Unsettling of America: Culture and Agriculture. Sierra Club Books, 228 pp

Burke IC, Lauenroth WK and Coffin DP (1995) Soil organic matter recovery in semiarid grasslands: implications for the conservation reserve program. Ecological Applications 5: 793–801

Dennett DC (1995) Darwin's Dangerous Idea. Simon and Schuster

Gould SJ and Lewontin R (1979) The spandrels of San Marco and the Panglossian paradigm: a critique of the adaptationist programme. Proceedings of the Royal Society B205: 581–598

McGill HC (1979) Antherosclerosis: problems in endpoints for genetic analysis. In: Sing CF and Kolnick MS (eds) Genetic Analysis of Common Diseases: Applications to Predictive Factors in Coronary Disease, pp 27–49. Alan Liss, New York

Sing CF, Havilland MB and Reilly SL (1996) Genetic architecture of common multifactorial diseases. In: Chadwick D and Cardew G (eds) Variation in the Human Genome, pp 211–229. John Wiley and Sons, Chichester

Strohman RC (1993) Ancient genes, wise bodies, unhealthy people: limits of a genetic paradigm in biology and medicine. Perspectives in Biological Medicine 37: 112–145

Agroforestry Systems **45**: 1–21, 1999.

Natural systems as models for the design of sustainable systems of land use

J. J. EWEL
Institute of Pacific Islands Forestry, Pacific Southwest Research Station, USDA Forest Service, Australia; E-mail: jackewel@gte.net

Key words: agriculture, climate, fertility, humid tropics, life zones, succession

Abstract. Natural ecosystems, whose components are the results of natural selection, are sustainable; most are productive, responsive to pests, and retentive of nutrients. Thus, they are appropriate models on which to base the design of new systems of land use. Abiotic and biotic stressors are related non-linearly; the nadir of total stress being mid-way along a gradient of environmental harshness. Superimposing the stress functions on Holdridge's life zone chart yields four broad categories of environments for agriculture: climates where annual rainfall is similar to potential evapotranspiration, plus three other categories that are either too cold, too arid, or too wet. Extremely cold lands have no potential for agriculture. Lands that are arid or infertile can be used successfully, although the cost of compensating for environmental limitations increases exponentially with increasing abiotic stress. Grazing animals (which act as trophic buffers between people and environment) have proven successful in dry and infertile environments. The humid tropical lowlands epitomise environments of low abiotic stress but overwhelming biotic intricacy. Here it pays to imitate natural systems rather than struggle to impose simplicity on ecosystems that are inherently complex. The keys to success are to (i) channel productivity into outputs of nutritional and economic importance, (ii) maintain adequate diversity to compensate for losses in a system simple enough to be horticulturally manageable, (iii) manage plants and herbivores to facilitate associational resistance and not associational susceptibility, and (iv) use perennial plants to maintain soil fertility, guard against erosion, and make full use of resources.

1. Introduction

Different environments require different solutions in the quest for systems of land use that are ecologically, socially, economically, and politically sustainable. This paper is an attempt to simplify that search through development of a theoretical basis for categorising landscapes and analysis of the kinds of solutions that might be sought for each category. The humid tropical lowlands are given special emphasis because they constitute today's agricultural frontier. Significant tracts of tropical forests – our doomed warehouses of global biological diversity – can only be saved if people make a swift transition from short-term exploitation to sustainable land use: mimicry of nature is one potential solution.

Why seek new models for agriculture? More specifically, why pursue the development of agricultural systems that imitate the structure and function of natural ecosystems? There are three important reasons. First, much high-yield agriculture is not sustainable because its practice consumes non-renewable environmental capital, especially soil and ancient groundwater. A century of petroleum-driven agriculture has yielded some striking mismatches between land use and environment. Many global transfers of agricultural practices were, in fact, little more than the cultural baggage of early colonists, with more basis in history than agronomy: extensive grain monocultures to the humid tropics are one example, and Spanish sheep to the dry subtropics are another. Native ecosystems are time-proven survivors, and it is logical to learn from them and imitate their useful traits.

Second, modern agronomy carries a cultural burden imparted by our system of education: most students are provided with little knowledge of evolution and ecology, at least not at a scale that extends beyond a particular cultivar or perhaps a cultivar and the insects and diseases that attack it. Naturally occurring ecosystems are long-term products of evolution and the accommodation of organisms to environment: they change with time, as both environment and biota change, and they run on solar power, thus making them self-sustaining. Modern agriculture, in contrast, is completely dependent upon fossil energy – fuels, machinery, fertilisers, pesticides, and all the industries that support them. Nevertheless, there is no doubt that affordable supplies of fossil energy – a non-renewable resource on any time scale relevant to humans-are dwindling (e.g. see Hatfield, 1997). Reduced supplies inevitably will be accompanied by increased costs to consumers, including farmers. Nature's solar powered systems make eminent sense for the future of food production, making it all the more regrettable that most agricultural scientists are ill-equipped to take advantage of the knowledge these systems offer.

And third, most farmers (e.g. those in much of India, Africa, and China) have not benefited significantly from the technological innovations that led to soaring agricultural productivity elsewhere in this century. Furthermore, population growth forces socially marginalised people to use agriculturally marginal lands. It is on such landscapes – places where water and nutrients are in short supply, or where high rainfall and temperatures promote competition from weeds, pest outbreaks, and nutrient leaching – that new approaches are needed. Here again, the most logical approach is surely to look to natural ecosystems as models for the design of sustainable systems of land use.

But natural systems are immensely more complex than the simplified plant-animal assemblages that humans exploit for food production. To attempt to generalise about such complexity is a risky undertaking, as any generalisations are certain to be riddled with exceptions. Nevertheless, generalisation is the only prospect for cosmopolitan solutions: with two or three decades of cheap fossil fuel remaining (Hatfield, 1997), coupled with the prospect of a doubling of the world's population in that same time frame, we simply cannot afford the luxury of solving agricultural problems solely on a hectare-by-hectare basis. We do need that site-specific detail, but decision-makers also need to be able to evaluate that detail against a global context. Thus, despite the risks and inevitable exceptions, this essay begins with an examination of the relationship between environment and biotic complexity, and it puts this relationship in a theoretical context relevant to land use. Building on that theory, landscapes are then assigned to four broad categories, each having different suitability for agricultural production. Next is a detailed analysis focused on the humid tropics, where people are still groping for sustainable systems of land use. This is followed by examples of how ecological research can provide useful information for the design of sustainable agroforests. The examples are based on highlights of research in the humid tropics, involving construction, monitoring, and assessment of fast-growing, diverse successional communities that imitate the structure of naturally occurring communities.

2. Agriculture across environmental gradients

2.1. *Theory*

From a biophysical perspective, the difficulty of achieving agricultural production may be viewed as the sum of two types of stresses: abiotic and biotic. The first involves environmental factors such as water, temperature, and soil fertility; the second, competition, herbivory, and disease.

Mitscherlich's (1923) model for crop yield is a useful starting point. He proposed that, if no factors were limiting to growth, a certain maximum yield (Ym) would be obtained, and that the response to additions of the limiting factor were proportional to the difference between current and maximum yield: i.e. the more limiting the factor, the greater the response (Eq. 1).

$$\text{Mitscherlich's Function: } \frac{dY}{dX} = (Ym - Y)k \qquad (1)$$

where Y is yield at some level of the limiting factor, X; Ym is the maximum yield obtainable; and k is a constant that describes the response of yield to increments of the limiting factor.

If we assume that the difference between maximum yield and yield at a given level of a limiting factor is a measure of abiotic stress, i.e. stress that would be alleviated if the limiting factor were incremented, then the resulting curve of abiotic stress as a function of level of environmental factor X is described by Eq. 2.

$$\text{Abiotic stress: } \frac{dS_a}{dX} = -k_a S_a \qquad (2)$$

where S_a is abiotic stress, induced by absence of a limiting factor (or excess of a debilitating environmental factor); X is the amount of the limiting (or debilitating) factor; and k_a is a constant that describes the decrement in stress per unit increment in X. The further one goes to the right along this exponentially decreasing curve (Figure 1a, dashed line), the less the abiotic restriction to growth.

The corollary is that, as abiotic factors become less constraining, the opportunity for more kinds of organisms to thrive increases: add water to a desert and get more kinds of plants, microorganisms, and herbivores, for example.

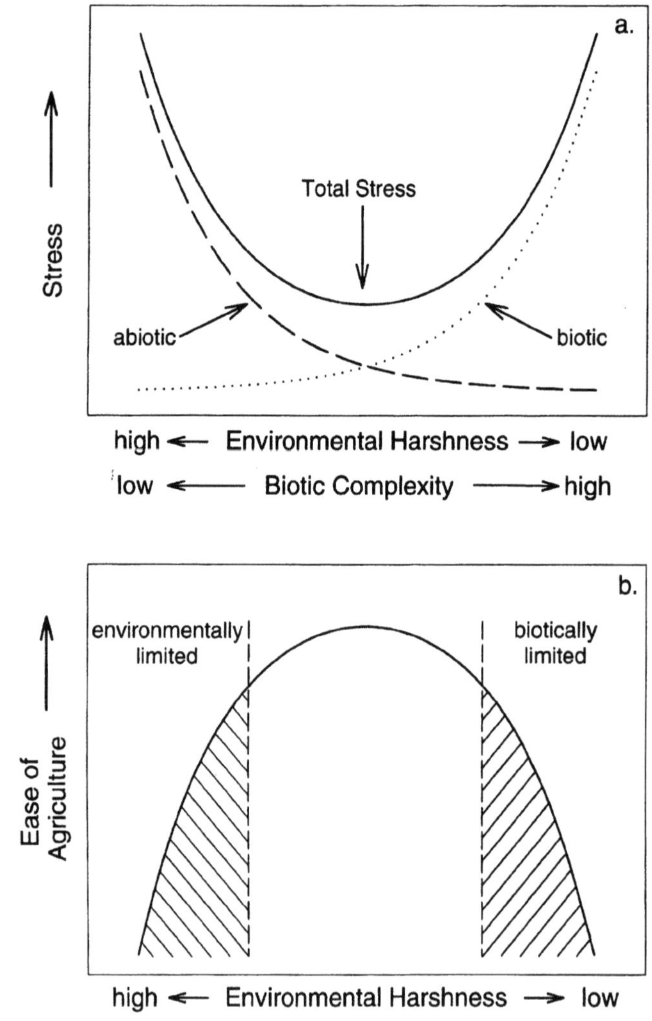

Figure 1. Stress and agriculture as a function of environment. a) Biotic and abiotic stress respond in opposite ways, and non-linearly, to environmental and biological gradients. Total stress is the sum of biotic and abiotic stresses. b) Appropriateness of different environments for agriculture. This function is the inverse of the total-stress curve in (a).

The relationship between such biotic stress and environmental factor X is described by Eq. 3.

$$\text{Biotic stress: } \frac{dS_b}{dX} = -k_b S_b \tag{3}$$

where S_b is biotic stress, induced by the competitors and consumers that thrive as limiting factor X increases, and k_b is a constant that describes the increment in biotic stress per unit increment in X. The further one goes to the right along this exponentially increasing curve (Figure 1a, dotted line), the greater the biotic impediments to agriculture. Thus, biotic impediments to agriculture (and the biotic simplicity it entails) increase as the abiotic constraints decline, and *vice versa.*

The key point is that the responses are not linear tradeoffs. Taking the sum of the responses (i.e. the sum of the dotted line and the dashed line in Figure 1a) to be the total constraint or impediment (shown as solid line in Figure 1a), we find that it is at the two endpoints that stresses are greatest. At the left-hand end of the graph, the constraints are primarily abiotic; at the right-hand end they are primarily biotic. The inverse image of this graph describes the ease with which agriculture can be accomplished in a particular environment (Figure 1b).

2.2. *Application*

It is feasible to nest global environments into categories according to their suitability for agriculture, ranging from those that are severely limited by abiotic stressors to those that are limited by biotic stressors, in accordance with the theory outlined above. One approach to broad-scale categorisation is to superimpose the concepts outlined above onto Holdridge's life zone chart (Figure 2a). Life zones are bioclimatic units determined by mean annual biotemperature and rainfall (Holdridge, 1947, 1967; Tosi, 1964). They have been used for assessment and planning at many scales, ranging from determination of local climates appropriate for specific crops (e.g. Hunter, 1959) to national descriptions of vegetation (e.g. Holdridge et al., 1971) to global assessments of the distribution of carbon and nitrogen (e.g. Post et al., 1982, 1985). There are more than 100 life zones on the earth's surface, but they can be aggregated into four clusters for an overview of appropriateness for agriculture (Figure 2b).

2.2.1. *Forgiving environments*

First, there are those life zones that have proven most propitious for agriculture, the ones clustered around the apogee of the ease-of-agriculture parabola (Figure 1b). These are (with few exceptions) the life zones in which annual rainfall does not deviate greatly from potential evapotranspiration (Figure 2); irrigation water, if needed at all, is required in only modest amounts, yet

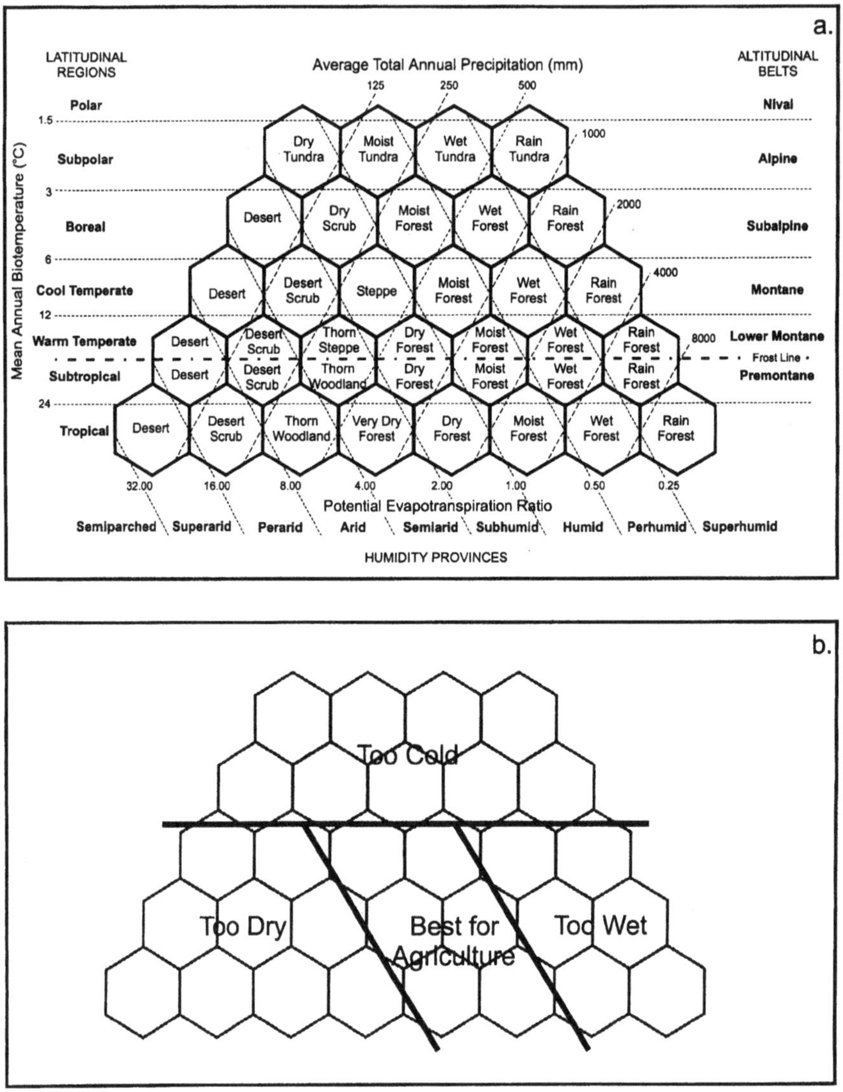

Figure 2. Categorisation of climates, and their suitability for agriculture. (a) The life zone classification scheme of Holdridge (1947, 1967). This two-dimensional figure illustrates the 38 Tropical life zones. Additional figures would be required to specifically illustrate the life zones of the six other latitudinal regions (named along the left edge of the chart); each successively more-polar region loses the lowest tier of life zones of the next most equatorial region. (b) Holdridge's life zones divided into four categories, each having different suitability for agriculture because of the stresses described by Figure 1.

rainfall is not so high that pests and nutrient leaching are overbearing problems. Not accidentally, these are the life zones of densest human settlement in many parts of the world (e.g. Tosi and Voertman, 1964). Where soils are inherently fertile (or where fertility is readily corrected, as in much of Australia), agriculture has thrived in these life zones, which encompass everything from the huge grain-producing farms of the temperate zone to tiny but prosperous mixed-crop farms of tropical and subtropical mountains. Some of the successful agroecosystems in these forgiving environments superficially resemble the natural systems they replaced. The grain monocultures of North America's Great Plains are low-diversity, short-lived versions of the native grasslands which makes the resemblance only superficial, as native grasslands are diverse and composed of perennial plants (Jackson, 1985), and orchards grown on lands that once supported native woodlands. Most, however, are products of human invention and cultural transplants.

2.2.2. *Too cold*

Of the other three clusters of life zones, all of which are more problematic, those that afford least agricultural utility are the ones that are permanently cold. From high in equatorial mountains they track a thermal decline in elevation, reaching sea level toward the poles. They lend themselves to little agricultural use other than subsistence transhumance (or, in more polar regions, seasonal hunting). The strategy in these zones is to harvest the grazers and browsers, whether domesticated or wild, which themselves harvest the meagre plant biomass produced in these inhospitable climates, just one step removed from permanent ice and snow.

2.2.3. *Too dry*

The second of the three problematic clusters consists of lands too arid to sustain agriculture without employing one, or some combination, of three tactics: (i) irrigate, (ii) schedule short-rotation crops to synchronise exactly with rains, or (iii) use grazers or browsers. Irrigation, especially in zones of high solar radiation and favourable temperatures, supports some of the highest population densities in the world. Nevertheless, it is not a panacea for agricultural development in arid zones because of the irony that irrigation is most feasible where least needed. The drier the climate, the more water required to meet the demands of evapotranspiration, the more likely the salinisation of surface soils, and the more the expenditure on infrastructure and maintenance. Inevitably, the most heavily irrigated, lowest-rainfall areas will experience or now experience salinity problems.

Timing crops to short rains is the global lottery of dryland agriculture. It is from such areas that news reports are filed annually with accounts of crop failure, sometimes accompanied by famine. One year it is Africa's Sahel, another year it is the warm-temperate end of the North American grain belt, and still another it is monsoonal South Asia. Recent change in global climate is the current scapegoat (if not the culprit), a situation that seems headed for exacerbation in the near future.

The third tactic employed in arid regions is to use large mammals to harvest, convert, and store plant biomass (Breman and de Wit, 1983; Coughenour et al., 1985). The cow (or camel, or sheep, or goat) covers a large area, harvesting and converting the primary production of natural ecosystems, its own mass rising and falling with availability of fodder. Sometimes their nomadic human shepherds live trophically downstream from the grazers and browsers, eating the animals or their metabolites (e.g. milk and blood); in other cultures the ungulates are used as draft animals in irrigated plots (Harris, 1965; Odend'hal, 1972). In both cases humans are buffered from the vagaries of rainfall by the large-biomass herbivores (Odum, 1967). In a sense, this system mimics nature, still to be seen today in East Africa's remnant herds of migrating wildebeest, gazelles, zebras, and their predators. Like so many of nature's models, this one seems doomed to extermination before we learn enough about it to facilitate even our own survival (Carr, 1964).

Southwestern Australia's wheat belt and its native vegetation, covered in detail elsewhere in this volume (Lefroy and Stirzaker, Hatton and Nulsen, Hobbs and O'Connor, and Dunin et al.), is an intriguing, if counterintuitive, example of the imminent collapse of grain agriculture in one such semiarid climate. It has one element in common with other dry regions where failure has occurred, as the introduction of grain agriculture here was as much cultural as agronomic, having arrived with farmers who had been successful wheat growers elsewhere. But it has some unique features as well, chief among them its long geological history, during which time salts rained onto the landscape as wind-born, oceanic aerosols. Leached down into the soil profile, these salts were kept at bay by the transpirational pumping of fresh groundwater into the atmosphere by deep-rooted trees and shrubs: by pulling the freshwater out from depth over spring and summer, the salts remained harmlessly combined in the clays deeper in the profile. Each winter the groundwater was recharged with rainwater percolating rapidly to depth along low-resistance pathways such as root channels (Specht, 1957; Nulsen et al., 1986). Exchanging deep-rooted plants for shallow-rooted wheat, however, has proven disastrous. No deep pumping means water accumulates at depth, dissolving the stored salts as water tables rise; as more groundwater accumulates year after year, the salts rise closer to the surface, and when those salts reach roots of annual crops or pasture grasses, farming ends.

2.2.4. *Too wet*

Finally, there is the cluster of problematic life zones that make up today's agricultural frontier: the humid lowlands of the tropics. Where nature is so lush, why is farming so difficult? The problems are basically twofold. First, there is the biological problem of the crop being subsumed by the trophic complexity of the local biota: weeds, fungi, insects, viruses, nematodes, bacteria, and rodents flourish; among these, the relatively defenceless crops suffer. Second, there is the problem of much more water coming into the system than departing via evapotranspiration. The excess moves downward through the soil or runs off across the surface, and essential elements that are not sequestered in vegetation or tightly bound in the soil are vulnerable to leaching. People have coped with environments of high biotic stress, such as tropical moist forests, in many ways, ranging from wood mining to mimicry of forest structure, and it is those solutions that are addressed in much of the remainder of this paper.

3. Humid tropical lowlands

Much use of these areas of intensive biotic interaction has been, and in some areas continues to be, primarily exploitative. This involves the harvesting of palm stems for the manufacture of rattan furniture, the tapping of trees for latex destined to become aeroplane tires, the trapping of primates for biomedical research, and the harvesting of dozens of other naturally occurring resources, some for subsistence and some for sale. Fortunes are still being made from timber exploitation, a process that will probably come to a close within a few decades as the resource is exhausted and forests are conserved for other values.

The other principal land use in warm, high-rainfall areas is shifting agriculture, which, in the context of the theory outlined above, can be regarded as moving the agricultural plot to new land in anticipation of the arrival of the stressors, i.e. before the grasses and grasshoppers invade, or before the soil becomes completely impoverished through occupation by shallow-rooted, short-lived plants. In areas of low human-population density shifting agriculture works well, but it cannot be sustained where population pressures significantly shorten rotations.

Other, potentially more sustainable, land use systems are prominent, or becoming so, in the humid tropics, especially as timber exploitation and shifting agriculture become less and less tenable. One of these is the welcome re-discovery of natural-forest silviculture, a discipline that almost disappeared after World War II, when globalisation of markets and high demand for wood led to indiscriminate harvesting; this supplanted sound forest management throughout the tropics. New life is now being breathed into silviculture, as sustainable forestry regains international credibility as an appropriate land use for the humid tropics (e.g. National Research Council, 1993; Wadsworth, 1998).

A second system, one that deviates significantly from the structure of the natural forest, is plantation agriculture – cacao (*Theobroma cacao*), coconut (*Cocos nucifera*), rubber (*Calotropis procera*), and oil palm (*Elaeis guineensis*), for example. These perennials, unlike short-lived plants, afford uninterrupted protection of soil from erosion and nutrient leaching, especially when species are aggregated into judiciously selected multiple life-form combinations (Ewel and Bigelow, 1996). Such systems yield modest amounts of biomass, but produce large revenues per unit of mass harvested.

A third system, one that has thrived for centuries in the humid tropics, is the diverse, structurally complex home garden, or agroforest, dominated by perennial plants (e.g. Falanruw, 1994; Pinedo-Vásquez and Padoch, 1996; Michon, 1997). Characterised by continuous and diverse yields, these most-complex of agroecosystems provide a broad range of products for home consumption as foods, medicines, and building materials, as well as high-value outputs for market-vanilla beans from Tabasco, durians from Sarawak, allspice from Zanzibar. Such agroforests are ecologically sustainable, but they are labour intensive and require skilful management. Furthermore, they involve a trade-off between high diversity of products and low yield. Structurally and functionally they are probably the closest mimics of natural forest yet attained.

Agroforests and shifting agriculture are often partnered land uses in the same society: the shifting agriculture field provides the main subsistence crop, such as maize, rice, cassava, sorghum, sweet potato, banana, or taro, and the agroforest provides an essential diversity of fruits, herbs, pharmaceuticals, ornamentals, and spices. Sometimes the agroforest is derived successionally from the shifting agriculture fields (Seavoy, 1973; Manner, 1981; Denevan et al., 1984), sometimes the two are intermingled across the landscape, and sometimes the shifting agriculture field is remote from the residence whereas the agroforest is literally the dooryard garden (Alvarez and Banal, 1984; Fernandes et al., 1984).

Given that, on most soils in the wet tropics, continuous grain monocultures are non-sustainable without massive subsidies of pesticides and fertilisers (cf. Sanchez and Salinas, 1981; Sanchez et al., 1982; Sanchez and Benites, 1987), what options are available? Forestry is one possibility, and plantations of export crops is another, but both of these are beyond the economic reach of most smallholders. Agroforests offer a definite possibility, but little is known of their ecological functioning. It was that lack of knowledge, coupled with concern about the modest productivity of many wooded gardens, that motivated a study involving horticultural mimicry of the fast-growing successional vegetation that follows shifting agriculture.

4. Mimicry of successional ecosystems

For five years my co-workers and I constructed plant communities designed to imitate the structure and, we hoped, the functional properties, of the successional communities that blanket the land when agricultural fields are abandoned in the humid tropical lowlands. Successional communities were chosen as models because they offered three traits of potential value to agriculture: (i) high responsiveness (resistance and resilience) to pest attack, (ii) high productivity, and (iii) high retention of soil nutrients.

The design consisted of six replications of each of four treatments:

1. natural successional vegetation (the model system) that developed without investigator intervention following slashing and burning of the existing vegetation (Ewel et al., 1981),
2. the investigators' attempt to build a mimic containing the same mix of life forms that appeared in the successional vegetation, but which consisted of species alien to the site, selected and planted by the scientists (naturally occurring colonists being weeded out),
3. a sequence of four monocultures, each consisting of a species having the same life form that dominated the successional vegetation during a particular interval: maize, *Zea mais* (when the successional vegetation was dominated by herbaceous vegetation); a second planting of maize; cassava, *Manihot esculenta* (when the successional vegetation was dominated by shrubs); and a tree, *Cordia alliodora* (when the successional vegetation was dominated by trees); and
4. a species-enriched treatment, in which vast quantities of seeds were regularly added to natural successional vegetation in an effort to assess invasibility and the functional responses of further enrichment of diversity.

The first three treatments are relevant to this discussion, and it is data from those three treatments (succession, imitation, and monocultures) that are summarised here.

The response variables assessed fell into three broad categories that together define the ecological underpinnings of sustainability: responses to pests, productivity, and soil fertility. The findings have been published in more than 20 articles and book chapters; the following discussion cites some of those results but includes some previously unpublished data as well.

4.1. *Responses to pests*

Over all, the imitation (system 2) fared only moderately worse than the model with respect to leaf tissue lost to herbivorous insects (Ewel et al., 1982; Brown and Ewel, 1987); each lost, on average, 0.2 g m^{-2} ground area d^{-1} (although the imitation did lose a significantly greater proportion of its leaf area). Some species suffered disproportionately high amounts of herbivory, but this did not lead to a community-level reduction in productivity or leaf area. This highlights an extremely important point about diverse communities – their ability to compensate for the loss of individual species. In a diverse system, the resources freed by the loss of one species are taken up by another, thereby maintaining system-wide performance.

One example of a plant-herbivore interaction illustrates a second point, and that is the impact of community structure on biotic interactions in complex systems. There is a general perception that diversity confers protection from herbivores, and indeed there is ample evidence that this is sometimes the case. Nevertheless, the opposite is sometimes true, as exemplified by the consumption of cassava by leaf-cutting ants (*Atta cephalotes*). As anticipated, cassava was heavily attacked by the ants when grown in monoculture (Blanton and Ewel, 1985; Brown and Ewel, 1987), but we did not anticipate that it would be even more heavily attacked when grown as part of the diverse mixture of species (more than 100 per plot) in the imitation. Thus, just as structural complexity can confer associational resistance to herbivores, so too can it confer associational susceptibility. In this case, the ants may have responded to the shaded environment of the imitation, to the complex structure of the imitation that may have facilitated their mobility (using other plants as trellises, for example), or to something about the imitation that made it freer of predators. The main point is that diversity cannot be counted upon to afford protection from herbivores, and sometimes it can have the opposite affect.

4.2. *Productivity*

Net primary productivity (Figure 3) was estimated by determining net biomass increments at frequent intervals by cutting subplots of the vegetation and by adding the rate of biomass death (measured as rates of litter fall, plus death of tissue that was not abscised, plus herbivory) to those increments. Two findings merit attention here.

First, the imitation was not as productive as the successional vegetation. This was probably due, in large part, to the horticultural complexity of building and maintaining such complex communities. The natural succession (the model) was always ahead of the mimic, and this was true for floristic complexity and canopy development, as well as productivity.

Second, the short-lived, herbaceous crop, maize, contributed the two extremes of productivity (Figure 3e), once being the highest (Figure 3a) and once being the lowest (Figure 3b). Is it any wonder that grain monocultures are an attractive farming venture? By the same token, it should not be any surprise that their adoption is fraught with risk.

4.3. *Soil fertility*

Changes in fertility were assessed through frequent measurements of essential elements and organic matter (Ewel et al., 1991; Raich et al., 1985), and the loss process itself was investigated through studies of leaching (Russell and Ewel, 1985) and roots (Berish, 1982; Berish and Ewel, 1988). The two main trends had been predicted beforehand: (a) the monocultures were (with one important exception, described below) much less retentive of nutrients than were the more species-rich treatments, and (b) both the natural succession and the imitation of succession were very retentive of nutrients, the model being only moderately better than the mimic.

The exception (referred to above) was the striking increase in nutrient retention by the tree monoculture over a three-year period (Figure 4). During its first year, the *Cordia* plantation was just as vulnerable to nutrient leaching as

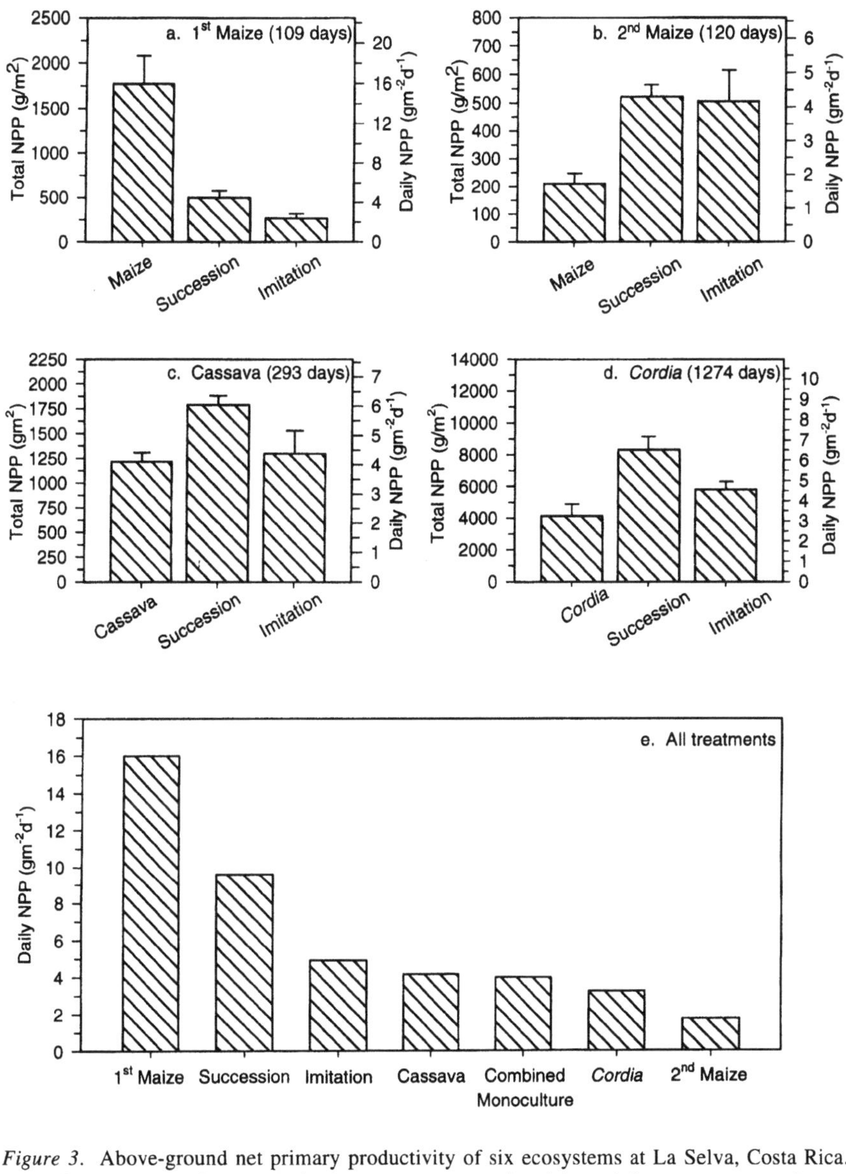

Figure 3. Above-ground net primary productivity of six ecosystems at La Selva, Costa Rica. Two of these, succession and imitation, were very species-rich, and four were monocultures. The smaller panels compare the three treatments over the same time intervals (e.g. a. compares the first maize with succession and imitation, for the 109 days when the monoculture was planted to the first maize crop), and the larger panel (e) compares productivity among all treatments, each for the time it occupied the site: five years for succession and imitation, and shorter periods for each of the four monocultures.

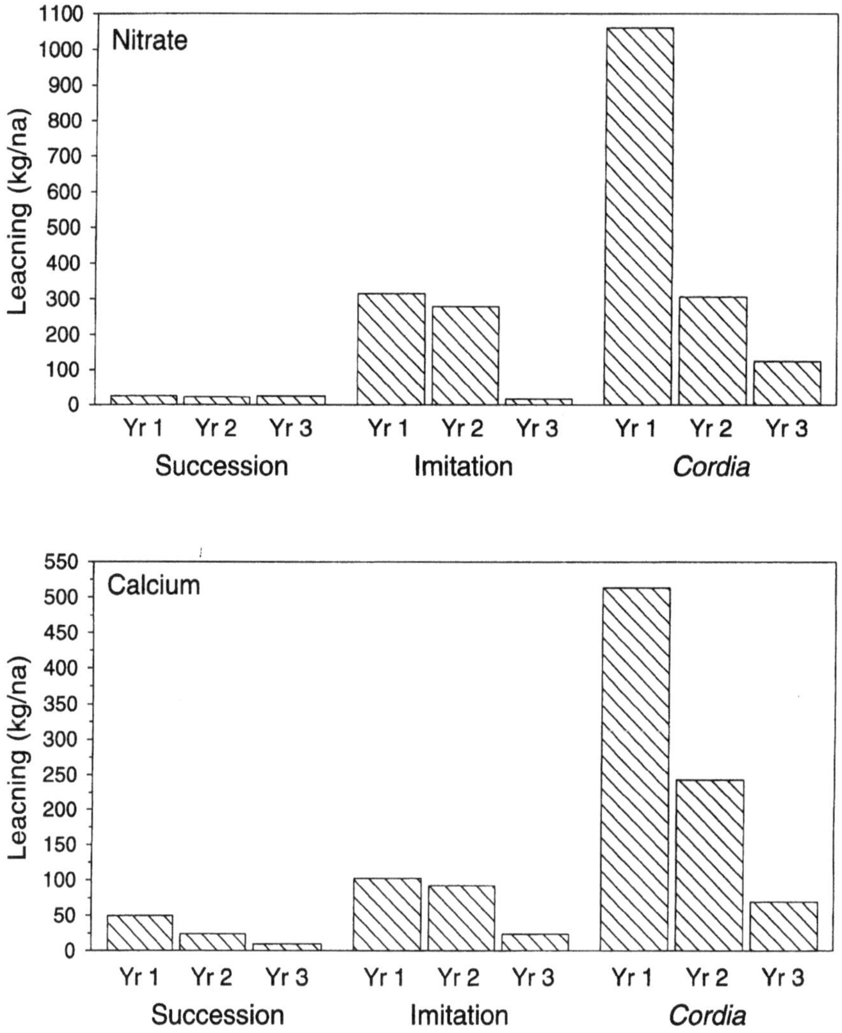

Figure 4. Annual leaching losses of nitrate (upper) and calcium (lower) from species-rich ecosystems and a monoculture of *Cordia alliodora* trees over a three-year period at La Selva, Costa Rica. Note the dramatic improvement in nutrient retention in the monoculture over the course of the three years.

the other monocultures had been; the second year, it improved substantially; and by the third year, it held nutrients almost as tightly as did the succession and its imitation. Here was a three-year-old, single-species plant community that was extremely effective at retaining nutrients – almost as effective as neighbouring communities containing more than 100 species each.

Two factors account for this remarkable capability of the tree monoculture to prevent nutrient leaching in a high-rainfall environment that loses about

1500 mm of water per annum to deep infiltration. First, it happened to be a very successful monoculture. The tree was well matched to the site; it grew quickly, and it was not heavily attacked by herbivores or disease. Had the *Cordia* trees performed poorly overall, or had patches of it succumbed to root pathogens, for example, there is no doubt that it would have proven to be every bit as susceptible to nutrient leaching as had the shorter-lived monocultures. Second, continuity of occupancy led to the development of an extensive, deep network of fine roots that effectively captured nutrients from soil solution before they could be leached (Figure 5). Longevity, as much as diversity (e.g. see Hobbs and O'Connor, this volume), is the key to nutrient retention in the humid tropics. It is this same difference in internal processes that marks the difference between the perennial native grasses of the North American prairies and their functionally deficient imitation by annual grain crops (Jackson, 1985; Soule and Piper, 1992).

5. Overview and conclusions

5.1. *Where to invest*

Blinded by past agricultural successes, society placidly ignores the agricultural crisis soon to be faced, when fossil energy becomes priced in proportion to the leverage it exerts in our global economy. No longer will we be able

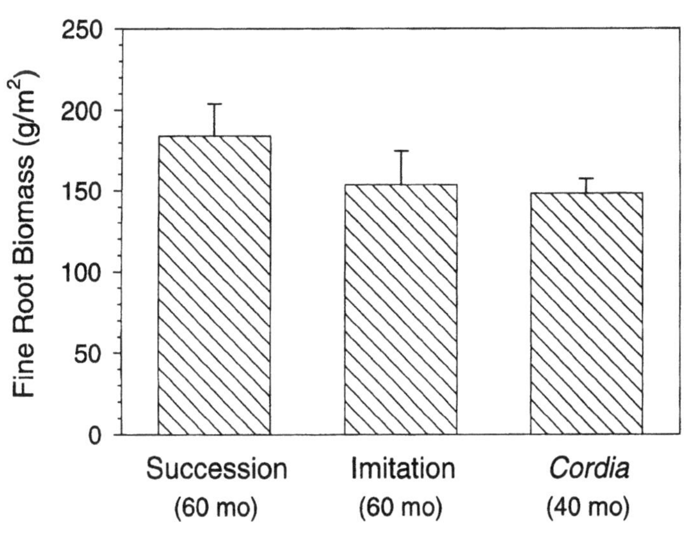

Figure 5. Comparative biomass of absorptive roots in two species-rich ecosystems (at age 5 y) and a monoculture of *Cordia alliodora* (at age 3.3 y) at La Selva, Costa Rica. It was this rapid development of a permanent network of fine roots that led to excellent nutrient retention by the tree monoculture (Figure 4).

to afford to move agricultural systems from one part of the globe to another without first giving due consideration to their ecological fit. Where can remaining reserves of fossil energy best be invested in intensive, high-technology agriculture?

Such investment makes sense in three environments. One is on reasonably fertile soils within that set of life zones where the annual water balance is neither too dry nor too wet, i.e. the very environments where agriculture has enjoyed a long history of success. To understand why this is the case, view agriculture as the human-mediated conversion of solar energy (which is dilute and abundant) to chemical energy. Investments of fossil energy merely facilitate that conversion. Greatest leverage of fossil-energy infusions is to be achieved where the conversion is least constrained by either abiotic or biotic forces because the inputs can then be focused on the conversion process itself rather than on alleviation of constraints such as water shortages or pests.

There are two other useful investments to be made, both involving displacements along the environmental gradient (Figure 1b), from either end toward the centre. Consider the example of flooded rice, or *padi*, in the humid tropics. In an environment where biotic forces might otherwise make a crop monoculture a poor competitor, flooding has the effect of increasing abiotic stress, i.e. inducing a shift from the right-hand side of the axis toward the centre. Furthermore, the *padi* system is often employed on recent alluvium and on young volcanic soils, where it concentrates water and nutrients (save nitrogen, accrued through biological fixation) from a much larger landscape. The result is high yields (and high human population densities) that, in some parts of Asia, have been sustained for centuries without significant modifications to the technology. Modest infusions of fossil-energy (in the form of high yielding varieties, nitrogen fertiliser, appropriately scaled machinery, etc.) into such systems have paid off handsomely in some instances, but in others they have led to disaster (Brown, 1970; Paddock, 1970; Wilkes and Wilkes, 1972; Harris, 1972, 1973).

Shifts in the other direction, i.e. from the left-hand, or environmentally harsh, side toward the centre of the environmental gradient (Figure 1b), are commonly achieved through intensive management of soil fertility and water supplies. A classic example from the temperate zone is the Central Valley of California, USA, which has a dry (thus low-pest, high solar radiation) climate, but where surrounding mountains provide water for irrigation. Although agricultural economists (and farmers) calculate Central Valley yields on the basis of land area cultivated, would not a more accurate assessment include the vast area of mountains that scour water out of the westerly winds sweeping inland from the ocean? It is those forested mountains that make agriculture possible in this semiarid valley.

Most other environments probably do not warrant such intensive investment. The low-temperature life zones, for example, offer little hope for agriculture, subsidised or not. The same is true for most areas in the wet tropics lowlands where, even when soil fertility constraints are alleviated through fertiliser applications, insects, diseases, and weeds overwhelm attempts at simplification. Except when value-inflated crops such as bananas and illegal drugs are grown for export, the massive inputs required to achieve biological simplification in the wet tropics are seldom warranted: heroin, cocaine, and marijuana may well become scarce in the international marketplace when petroleum is priced commensurate with its true value.

5.2. *Roles for mimicry*

It is far easier to mimic, or employ natural substitutes for, specific ecosystem processes than to try to duplicate all the complexity of nature. This strategy is likely to have potential applicability in many situations: deep-rooted, non-indigenous plants to pump groundwater in southwestern Australia (Lefroy and Stirzaker, this volume) and perennial grasses to protect soil in the Midwestern USA (Jackson, 1985) are examples. Herbivore resistance, effective display of light-capturing foliage, and symbiotic links with beneficial microorganisms can be, and have been, achieved by agronomists and crop breeders; they can also be achieved through effective mimicry – process by process – of natural ecosystems.

The other, more complex, approach is to design new land use systems using natural ecosystems which have withstood the test of evolutionary time as models. Such mimics, like their models, can be productive, pest resistant, and conservative of nutrients. But they are not without problems and risk; they can also be patchy, selectively attacked, and horticulturally complex (Ewel, 1986). Furthermore, they are sometimes accompanied by interesting side effects. For example, to achieve highest yields of many crops requires high levels of fertiliser, and those dosages tend to switch off microbially mediated input and cycling activities such as nodules, mycorrhizae, and proteoid roots. When imitating nature, therefore, it is important to reduce nutrient inputs and in the process be prepared to accept lesser – but biologically sustainable – yields. As one component within a landscape mosaic that combines intensive cultivation of short-lived crops, grazing and browsing ruminants, silviculture, and forested reservoirs for genes and water, diverse agroforests and prairie mimics can play an essential role – perhaps not a substitute for other modes of agriculture, but a complement to them.

Can these mimics of natural ecosystems provide the yields we have come to expect from modern agriculture? Not yet, and perhaps never; there are two reasons why that is the case. First, only a small fraction of large stature, perennial plants is harvested from the forest mimic; the bulk of the biomass is what sustains the protective structure of the ecosystem, and the more that is harvested the more we deviate from nature's model. Human removal of latex and large-seeded fruits that have evolved to be dispersed by animals mean that only a small fraction of the agroforest's biomass is removed i.e. it has a harvest index of 10 per cent or less. This contrasts sharply with the high harvest indices (approaching 50 per cent) of the world's big-yielding crops, mostly composed of smaller seeds (primarily grasses) and below-ground storage organs. An old, diverse agroforest might have a high gross primary productivity, but much of that gross production is consumed by respiration that sustains the huge metabolic investment in structure.

The second reason why complex, perennial systems may be destined to lower yield concerns the presumed evolutionary trade-off between reproduction and permanence. Most plant biologists would contend that allocation of photosynthate involves mutually counteractive investment alternatives, with carbon allocated to stature and perennial organs being unavailable for reproduction (i.e. for the seeds we eat) and *vice versa*. Nonetheless, there are some encouraging preliminary results to the contrary; Jackson and Dewald (1994), working with a native North American grass, showed that perennialness (the property that sustains the ecosystem in the long run) and seed set (the part that sustains the farmer in the short run) were not mutually exclusive trade-offs. This exciting finding opens the door to the prospect of perennial grass agriculture, but whether the example studied has general applicability to other species remains to be tested.

Tropical forests and temperate grasslands are not the only appropriate models by any means. For example, a search might be directed profitably toward the permanent culture of perennial plants that produce large storage organs. There are examples of such systems, though they are not widespread. For example, some Micronesians harvest only a modest portion of the mass of *Cyrtosperma* taro tubers at one time, and some indigenous peoples in South America do the same with cassava; the tuber in the ground is food in the larder. This contrasts sharply with modern production of potatoes (a short-lived perennial forced into an annual growth mode as part of its domestication), whereby the entire agroecosystem is uprooted at harvest, and the larder is a fossil-energy-cooled warehouse.

At the level of the individual species or process, mimicry of form or function makes good ecological sense, for morphologies and processes are shaped by evolution: as such, they are proven survivors in the environment that shaped them. Mimicry of the entire complexity of natural ecosystems is more problematic, but although it may be ill-suited to some situations, it is certain to be well-suited to others. The increased willingness of some modern agricultural scientists to contend with biotic complexity is especially encouraging. Witness, for example, the international boom in integrated pest management. But current efforts are insufficient, and we are not moving forward fast enough. It behoves scientists to anticipate the near-future needs of agriculture through improved understanding of how complex systems – both models and mimics – work. Agricultural ecosystems that mimic the structural and functional complexity of nature have the potential to play a crucial role for all societies. They already do so in many developing, tropical countries, and it may be those of us who have been spoiled by an excess of fossil-energy riches that are going to need them most.

6. Acknowledgements

The ideas presented in this paper were developed over a period of many years, primarily in conjunction with research supported by the U.S. National Science Foundation, most recently under award DEB 9318403. I thank the following for very helpful suggestions that improved the manuscript: Becky Brown, Ankila Hiremath, Ted Lefroy, Henry Nix, Chin Ong, and John Pate. I thank Becky Brown and Cory Berish for two years of data on productivity and nutrient concentrations in leachate, respectively, and Larry Battoe for calculation of leaching losses. Tom Cole produced the figures, and Kris Kogan compiled and formatted the manuscript.

7. References

Alvarez FN and Banal N (1984) El huerto casero: manual de agricultura orgánica. Editorial Claridad, Maunabo, Puerto Rico

Berish CW (1982) Root biomass and surface area in three successional tropical forests. Canadian Journal of Forest Research 12: 699–704

Berish CW and Ewel JJ (1988) Root development in simple and complex tropical successional ecosystems. Plant and Soil 106: 73–84

Blanton CM and Ewel JJ (1985) Leaf-cutting ant herbivory in successional and agricultural tropical ecosystems. Ecology 66: 861–869

Breman H and de Wit CT (1983) Rangeland productivity and exploitation in the Sahel. Science 221: 1341–1347

Brown BJ and Ewel JJ (1987) Herbivory in complex and simple tropical successional ecosystems. Ecology 68: 108–116

Brown LR (1970) Seeds of Change: The Green Revolution and Development in the 1970s. Praeger, New York, USA

Carr AF (1964) Ulendo: Travels of a Naturalist In and Out of Africa. Knopf, New York, USA

Coughenour MB, Ellis JE, Swift DM, Coppock DL, Galvin K, McCabe JT and Hart TC (1985) Energy extraction and use in a nomadic pastoral ecosystem. Science 230: 619–625

Denevan WM, Treacy JM, Alcorn JB, Padoch C, Denslow J and Paitan SF (1984) Indigenous agroforestry in the Peruvian Amazon: Bora Indian management of swidden fallows. Interciencia 9: 346–357

Ewel J (1986) Designing agricultural ecosystems for the humid tropics. Annual Review of Ecology and Systematics 17: 245–271

Ewel J, Benedict F, Berish C, Brown B, Gliessman S, Amador M, Bermúdez R, Martínez A, Miranda R and Price N (1982) Leaf area, light transmission, roots and leaf damage in nine tropical plant communities. Agro-Ecosystems 7: 305–326

Ewel J, Berish C, Brown B, Price N and Raich J (1981) Slash and burn impacts on a Costa Rican wet forest site. Ecology 62: 816–829

Ewel JJ and Bigelow SW (1996) Plant life-forms and tropical ecosystem functioning. Ecological Studies 122: 101–126

Ewel JJ, Mazzarino MJ and Berish CW (1991) Tropical soil fertility changes under monocultures and successional communities of different structure. Ecological Applications 1: 289–302

Falanruw MVC (1994) Food production and ecosystem management on Yap. ISLA: A Journal of Micronesian Studies 2: 5–22

Fernandes ECM, Oktingati A and Maghembe J (1984) The Chagga homegardens: a multistoried agroforestry cropping system on Mt. Kilimanjaro (Northern Tanzania). Agroforestry Systems 2: 73–86

Harris M (1965) The myth of the sacred cow. In: Leeds A and Vayda AP (eds) Man, Culture, and Animals, pp 217–228. AAAS, Washington DC
Harris M (1972) How green the revolution. Natural History 81: 28–30
Harris M (1973) The withering green revolution. Natural History 82: 20–23
Hatfield CB (1997) Oil back on the global agenda. Nature 387: 121
Holdridge LR (1947) Determination of world plant formations from simple climatic data. Science 105: 367–368
Holdridge LR (1967) Life Zone Ecology. Tropical Science Center, San José, Costa Rica (Revised edition)
Holdridge LR, Grenke WC, Hatheway WH, Liang T and Tosi JA, Jr (1971) Forest Environments in Tropical Life Zones: A Pilot Study. Pergamon Press, New York, USA
Hunter JR (1959) The Climatic Limits of Cacao, Coffee, and Rubber. Inter-American Institute of Agricultural Sciences #16, Turrialba, Costa Rica
Jackson LL and Dewald CL (1994) Predicting evolutionary consequences of greater reproductive effort in *Tripsacum dactyloides*, a perennial grass. Ecology 75: 627–641
Jackson W (1985) New Roots for Agriculture, 2nd ed. University of Nebraska Press, Lincoln, Nebraska, USA
Manner HI (1981) Ecological succession in new and old swiddens of montane Papua New Guinea. Human Ecology 9: 359–377
Michon G (1997) The agroforest alternative to *Imperata* grasslands: when smallholder agriculture and forestry reach sustainability. Agroforestry Systems 36: 105–120
Mitscherlich EA (1923) Bodenkunde für Land-und Forstwirte. Paul Parey, Berlin
National Research Council (U.S.) (1993) Sustainable Agriculture and the Environment in the Humid Tropics. National Academy of Sciences, Washington, DC, USA
Nulsen RA, Bligh KJ, Baxter IN, Solin EJ and Imrie DH (1986) The fate of rainfall in a mallee and heath vegetated catchment in southern Western Australia. Australian Journal of Ecology 11: 361–371
Odend'hal S (1972) Energetics of Indian cattle in their environment. Human Ecology 1: 3–22
Odum HT (1967) Energetics of world food production. In: The world food problem. A report of the President's Science Advisory Committee, Volume 3: Report of the Panel on the World Food Supply, pp 55–94. US Government Printing Office, Washington, DC
Paddock WC (1970) How green is the green revolution? BioScience 20: 897–902
Pinedo-Vásquez M and Padoch C (1996) Managing forest remnants and forest gardens in Peru and Indonesia. In: Schelhas J and Greenberg R (eds) Forest Patches in Tropical Landscapes, pp 327–342. Island Press, Washington, DC
Post WM, Emanuel WR, Zinke PJ and Stangenberger AG (1982) Soil carbon pools and world life zones. Nature 298: 156–159
Post WM, Pastor J, Zinke PJ and Stangenberger AG (1985) Global patterns of soil nitrogen storage. Nature 317: 613–616
Raich JW, Ewel JJ and Olivera M (1985) Soil-CO_2 efflux in simple and diverse ecosystems on a volcanic soil in Costa Rica. Turrialba 35: 33–42
Russell AE and Ewel JJ (1985) Leaching from a tropical andept during big storms: a comparison of three methods. Soil Science 139: 181–189
Sanchez PA, Bandy DE, Villachica JH and Nicholaides JJ (1982) Amazon Basin soils: management for continuous crop production. Science 216: 821–827
Sanchez PA and Benites JR (1987) Low-input cropping for acid soils of the humid tropics. Science 238: 1521–1527
Sanchez PA and Salinas JG (1981) Low-input technology for managing oxisols and ultisols in tropical America. Advances in Agronomy 34: 279–406
Seavoy RE (1973) The shading cycle in shifting cultivation. Annals of the Association of American Geographers 63: 522–528
Soule JD and Piper JK (1992) Farming in Nature's Image. Island Press, Washington, DC

Specht RL (1957) Dark island heath (Ninety Mile Plain, South Australia), iv. Soil moisture patterns produced by rainfall interception and stemflow. Australian Journal of Botany 5: 137–150

Tosi Jr. JA (1964) Climatic control of terrestrial ecosystems: a report on the Holdridge model. Economic Geography 40: 173–181

Tosi Jr. JA and Voertman RF (1964) Some environmental factors in the economic development of the tropics. Economic Geography 40: 189–205

Wadsworth FW (1998) Forest Production for Tropical America: USDA Forest Service Agricultural Handbook 710. USDA Forest Service, Government Printing Office, Washington, DC

Wilkes HG and Wilkes S (1972) The green revolution. Environment 14: 32–39

Agroforestry Systems **45**: 23–41, 1999.

How much biodiversity is enough?

A. R. MAIN
Zoology Department, The University of Western Australia, Nedlands, 6907, WA, Australia; E-mail: bymain@cyllene.uwa.edu.au

Key words: agriculture, evolution, functional mimicry, natural ecosystems, sustainability

Abstract. This paper addresses the question of how much biodiversity is enough in the context of the concept of agriculture as a mimic of nature. Following an historical review of the likely origins of ecosystems I show that the currently accepted components of biodiversity, *viz.* genetic, species and ecosystem, and their functional expression, are relevant to agriculture. Examples of adequate biodiversity are given and it is concluded that what constitutes enough biodiversity in an agricultural system is dependent upon the goal in question and will be different depending on whether the aim is, for example, to increase yield stability or deal with salinity, ground water levels, soil erosion, leaching of nutrients or weed control. The point is made that ecosystems and their composition are contingent in nature so the history of events, their frequency and intensity all need to be considered when interpreting the natural biodiversity present and thus determining what is enough in particular circumstances.

1. Introduction

Throughout this article the reader will find a mixture of hope and pessimism that the question 'how much biodiversity is enough?' can be answered. We are all products of our experiences and the approach I have taken to this question has been strongly influenced by my experiences of the precautionary principle and adaptive management. Although it was not referred to as such, my training as an aircraft navigator in the 1940s was based on the precautionary principle and adaptive management. Our task was to individually navigate our aircraft to target at night as one of perhaps hundreds in the bomber stream. The details of the target, the course to be taken and the winds and weather conditions likely to be experienced were given in a pre-flight briefing after which navigators prepared a flight plan. The goal was to arrive at target on course and at the designated time. We were taught to plan for the worst case scenario, which was that the winds and weather would not be as predicted, rendering our flight plan useless. To minimise this risk each navigator re-established the position of his aircraft every six minutes and, if necessary, altered speed and course so that the prescribed route was followed, thus ensuring arrival at the designated time. The potential for the worst case scenario to eventuate was always present and was only avoided by a continuous series of decisions. The analogy between this experience and the problem of ensuring arrival at sustainable agriculture is apparent, but a strategy of adaptive management in agriculture is only possible if the decisions taken are based on knowledge of how the world, both physical and biological, is structured and likely to work. Such knowledge allows us to view agriculture as a special case of the general expression of biodiversity in nature. So much for an explanation of my approach.

The apparent simplicity of the question, how much biodiversity is enough, is extremely deceptive because it embraces a plethora of intricate interrelationships relating to the biology of the organisms involved. The interrelationships of concern centre around agriculture as a mimic of nature. One of the dictionary meanings of mimic is to copy with minute accuracy. Another meaning is to follow the example or do after the manner of. A further term which could be used is to see agriculture as an analogue of nature when analogue is taken to be equivalence or likeness of relations or resemblance of form or function without identity of essence. It is clear from observation that it can only be the functional, not compositional, aspects of agricultural activities where resemblance to nature is possible.

In an earlier paper (Main, 1992), I discussed landscape re-integration with special reference to the region of Western Australia under the influence of Mediterranean-type climate. This region has suffered extensive ecological disturbance on a landscape scale since the settlement by Europeans less than 200 years ago. The thrust of that paper was to outline what was needed to reverse the landscape trends now apparent and restore the impaired or lost functions, e.g. soil structure, nutrient recycling and transport and the hydrologic balance. The solution envisaged was a landscape composed of natural, revegetated and farmed areas so arranged that the lost or impaired functions were restored. The primary concern was with the conditions necessary for the maintenance of species richness and emphasis was placed on the significance of physical and biological disturbances at different scales in achieving this goal (Table 1).

Agricultural and natural systems were also compared in terms of the functioning of the systems and the goals to be achieved (Table 2). Finally I discussed how the physical and biological disturbance regimes and their interactions provided boundary conditions controlling species richness.

The thrust of the 1992 paper was towards sustainable production and nature conservation essentially measured by species richness. The present paper emphasises the agricultural component of the system, principally how species richness, now renamed biodiversity, can contribute to sustainable agriculture, but more particularly how an assessment can be made that a system has enough biodiversity to satisfy long term ecological and shorter term agricultural and economic goals. It is within the mosaic of land uses envisaged in the 1992 paper that I discuss how much biodiversity is enough.

Table 1. Likely species richness in ecosystems as a result of interactions between disturbances of physical (e.g. fire, storm) and biological (e.g. predation, herbivory, disease) origin.

Biological	Intensity of disturbance Physical		
	Infrequent	Moderate	Intense
Rare or slight	Few dominant or long lived species	Species richness maintained by disturbance	Few species tolerant of stress
Moderate	Species rich: species intolerant of physical but tolerant of biological disturbances	Highest species richness	Species presence determined by tolerance of physical stress; richness maintained by biological factors
Very intense	Species poor: dominated by species tolerant of grazing or predation	Richness determined by response to moderate physical disturbance and tolerance of grazing or predation	Species poor: species presence determined by tolerance to physical and biological factors

2. Biodiversity in the context of agriculture

The involvement of biodiversity studies and concepts in agriculture is only understandable from the history of the use of the term. Concern about loss of species was first voiced when it became clear that continued clearing of wild lands for settlement was leading to fragmentation of habitat and loss of species. Realisation of the consequences of these events led to actions directed to having reservations within which species could be retained or preserved. These activities led to studies of species richness so that areas could be selected for reservation. Such endeavours were essentially driven by moral, ethical and altruistic motives and were opposed by developers on economic and commercial grounds. Essentially their opposition centred on an attitude of 'species may be lost, so what?' The contrasting approach led to a change in title from studies of species richness to biological diversity later shortened to biodiversity (see Heywood, 1996 for history). The question was then asked what commercial as distinct from intrinsic or aesthetic value was there in retaining many species. To counter this, the function of species as components of ecosystems was emphasised. So if agriculture was seen as a special case among an array of ecosystems and it was thought desirable that it should be sustainable, then it follows that the contribution of biodiversity to sustainability should be established. However, indefinitely extensive biodiversity cannot be justified in agriculture, hence the question I have been asked to address. The question of the contribution of biodiversity to sustainability

Table 2. A comparison of agricultural and natural ecosystems.

Focus	Agricultural	Natural
Goal		
Short-term	Yield	No irreplaceable losses
Mid-term	Profit to maintain system.	Minimum external input
Long-term overall	Persistence of system, economic viability	Persistence of system, self perpetuation
Basic requirements		
Plants		
Space	Appropriate husbandry	Disturbance regime
Light	Cultivation, weed control	Species exploiting stages after disturbance
Water	Management, conserving water, weed control, maintaining groundwater table	Water harvesting by some species, soil structure maintained by animals groundwater regulated by evapotranspiration
Nutrients	Applied or by husbandry or crop rotation	Naturally occurring nutrients retained by recycling but with a net loss
Animals		
Water	Rainfall, husbandry	From plants, or natural occurrence
Nutrients	Forage or appropriate husbandry	Plants of appropriate post-disturbance stage or by eating herbivores or detritivores
Shelter/space	Appropriate husbandry	Disturbance of plants at appropriate post-disturbance stage
Soil structure	Husbandry, cultivation	Above ground burrowing and soil dwelling animals plant roots, fungi, bacteria
Tolerances		
Nutrients:		
General	Low levels result in poor growth	Low levels tolerated, high levels toxic
Major nutrients	N by husbandry or applied. P applied	N fixed naturally, P naturally occurring and recycled
Trace elements	Applied	Low levels tolerated
Losses from system	N lost as gas or harvested. P lost by leaching, harvested, bound to soil fraction	N lost or gas or transported by mobile animals. P leached or transported by mobile animals
Replacement of losses	Applied or replaced by husbandry	In unfragmented systems on average redistributed by animals

Table 2. (Continued).

Focus	Agricultural	Natural
Maintenance	Management practices which retain soil, soil structure and productive state of soil	Appropriate patch dynamics determined by natural disturbance regime and size of area
Failure to persist		
Physical	Loss of soil structure changes in soil salinity, acidity, non-wetting soils, waterlogging	Loss of soil structure, changes in salinity or acidity
Environment	Changes in rainfall, erosion by wind and water	Changes in rainfall, wind and water erosion
Management	Inappropriate husbandry	Patch dynamics of inappropriate scale

cannot be answered simply because there are several ancillary questions which need attention. They are:

- How much is enough?
- Is it merely minimum biodiversity that we are seeking?
- For what purpose is the diversity to be retained?
- For how long is it to be kept?
- What spatial scales are of concern?
- Can there be a single measure by which 'enough' may be assessed?

Furthermore, there is the matter of what sort of proof is sufficient to validate a statement that a specified level of biodiversity is adequate. Ordinarily, adequacy can be inferred from descriptive, observational, or comparative studies, but the results of such studies are often open to several interpretations because of confounding influences among the possible causes, thus yielding many possible interpretations with the true cause not readily apparent. This impediment is usually overcome through experimentation. For adequate analysis in biodiversity studies, experimental design would need to be complex and the experiment of long duration. Such a demonstration has been provided by Tilman et al. (1996). The problem is to have cost effective experiments yielding timely results. So-called natural experiments may sometimes be used but they have many shortcomings because of inadequate controls. A special category of natural experiments is provided by biogeographical or palaeontological studies.

The normal approach to the study of biodiversity is to take what is present as a given and then, on the basis of biological knowledge, attribute roles or functions to the various components. There is little consideration of how the system might have evolved and generally the assumption seems to be that they are static entities. Should this be so, then the question of how much diversity is enough is meaningful because once this figure is established, with respect to the adequacy of the system as a mimic of nature, it will thereafter maintain itself. But these assumptions need to be confirmed. Additionally it is necessary to determine how it will be known that the biodiversity present is enough.

Observations relating to the diversity currently present in nature have usually dealt with vertebrates or other easily studied animals or flowering plants. Currently, biodiversity is considered to embrace the five kingdoms of organisms, Prokaryota (Bacteria) and Eukaryota (Protists, Fungi, Plants and Animals), so the usual technique is very limited in approach. However, this is recognised and justified by arguing that the readily observed groups are, or may be regarded as, surrogates for the true biodiversity present. The foregoing state of affairs is thus unlikely to be applicable to agricultural systems because it is, to all intents and purposes, impossible to know the bacteria, protists and fungi present. Moreover, it is unlikely that the plants and animals cultivated in agriculture can in any way be said to be surrogates for the inconspicuous elements such as bacteria, protists or fungi which are important in the decomposition of organic material and recycling of nutrients. Thus, the question is how is 'enough' to be established? It cannot be simple numbers of entities present, so is functionality, persistence of system, productivity or re-establishment of prior state following disturbance to be the measure? A meaningful choice may be impossible without an historical background because biology, like geology and cosmology, is essentially an historical science. The current status with respect to biodiversity is dependent on past events, i.e. biology is contingent in nature. So is it possible to consider the present as the culmination in the evolution not only of biological entities (species), but of how chance and evolution produced the present arrangements which we interpret as the functions of ecosystems? This line of enquiry is followed below.

3. The evolution of ecosystems and biodiversity

The solar system is very old. Numerical ages are expressed either as millions (10^6), or billions (10^9) of years ago, and are referred to as Ma or Ga, respectively. Crystals of zircon have been identified in Western Australia which solidified 4.3 Ga. By 3.8 Ga there is some evidence that can be interpreted as biological fractionation of isotopes of biologically useful elements, and by 3.5 Ga there is evidence of fossils interpreted as being of Prokaryotes. Examination of the subsequent sedimentary record suggests an increasing abundance of oxygen, presumably the result of photosynthesis. By 1 Ga fossils are still rare but there is a suggestion, based on the size of micro-fossils, that single celled Eukaryotes were present. However, it was not until the Cambrian explosion about 550 Ma that the Phyla of multicellular animals are recorded as fossils. By the middle of the Palaeozoic, fossils of terrestrial plants and animals were common but it was not until towards the end of the Permian (the last Era of the Palaeozoic) about 250 Ma that vertebrate herbivores are to be found. Flowering plants and birds did not occur until the Cretaceous, possibly 100 million years later. It was only at this time that ecosystems analogous to the present could have occurred. Prior to the Permian, herbivory was essentially that of insects and food chains were either based on insects or other arthropods feeding on microbially processed dead plant material (detritus). It is interesting to note that young herbivores are dependent on animal protein (milk or egg yolk) or microbial protein for survival and growth (White, 1984, 1986; Moir, 1994). Thus, nutritionally, the present remains rooted in the past and ultimately dependent on microbial synthesis.

The essentials of this long history are that in the beginning, living things were prokaryotic, lacking nuclear membranes, while organisms with nuclear membranes, the eukaryotes, are of much later origin and multicellular organisms even later still. The trend apparent in this history is the increasing structural complexity of organisms coupled with the increasing richness of the kinds of organism present, i.e. biodiversity increased over time. But it was not a simple trend of increasing diversity because the fossil record shows that at irregular but relatively long intervals there occurred extinctions of various magnitudes in the biota. These are the basis for defining the various Epochs and Eras of the geological time scale. Following each extinction event, new kinds of organism dominated the biota but characteristically, following each extinction event, biodiversity was initially reduced relative to the pre-event biota, only returning to high levels some millions of years later (Behrensmeyer et al., 1992; Shear, 1991)

The foregoing can be interpreted as indicating that once the physical environment permitted – the presence of water, dissolved nutrients, radiant energy and space in which to grow – living forms proliferated. Moreover, once biological material was present it also became a resource to be exploited, i.e. primitive food chains and ecosystems were possible. However, it was not until the Cambrian radiation of multicellular organisms and the evolution of filter feeding and other mechanisms that the exploitation of the microbial biomass present was possible. The diversity of mechanisms which evolved to harvest the microbial resource is shown by the ciliated protists, sponges, nematodes, annelids, the lophophorate metazoans (e.g. brachiopods), arthropods such as crustacea, the lamellibranch (bivalve) molluscs, the chordates (e.g. tunicates), and the detritus feeding trilobites. The origin of these morphological types ushered in the subsequent eras of high biodiversity.

Further inspection of the fossil record shows that for long periods of geological time large quantities of biomass were not consumed by other organisms thus resulting in the accumulation of plant material to yield hydrocarbon deposits such as gas, oil, and the various kinds of coal. For example, in Australia there are coal deposits of Permian, Triassic and Tertiary ages. To a large extent the presence of these deposits were dependent on the physical conditions at the site of deposition, e.g. low oxygen levels, but the evidence also suggests that the biodiversity present was insufficient to consume all primary production.

The foregoing suggests that biodiversity is the consequence of there being either physical or biological resources to be exploited. Even resources which are difficult to exploit, such as plants containing toxins, will ultimately be utilised because of the propensity of the hereditable elements of biological entities to mutate, recombine, and be resorted, thus exposing all sorts of combinations and potentials to natural selection. The consequences are that with time, all resources that are potentially biologically useful will be exploited. Thus, there is the potential for an array of herbivores, predators (including insectivores and piscivores), endo- and ectoparasites, filter feeders, and detritivores to evolve. This appears to be what happened following past extinctions. But this evolution and its subsequent development takes place in a fluctuating environment and for the reasons mentioned above, plus the possibility of migrants and colonists supplementing the endemic biota, the composition of ecosystems is unlikely to remain constant. This conclusion gives no confidence that it will ever be possible to determine how much biodiversity is enough by simply scoring the presence of species or taxonomic entities. But things may be different on scales relevant to current problems and I now proceed to examine this aspect.

4. Biodiversity

Currently biodiversity is considered to embrace the whole of the variety of life forms, plants, animals and microorganisms, the genes they contain and the ecosystems of which they form part. Thus, biodiversity is considered at the genetic, species, and ecosystem levels (Commonwealth of Australia, 1996). In this context the study of biodiversity is driven purely by a desire to know and understand nature. It has been pointed out earlier that the questioning by more pragmatically oriented workers of such motivation has led to an orientation emphasising functional roles of the biological elements in ecosystems as well as the value of diversity *per se* to ecosystem function. Essentially there is a belief that diversity is beneficial. In the context of agriculture as a mimic of nature the emphasis has been on biodiversity's influence on yield, water quality, provision of free services, flood control, and influences on productivity (Heywood, 1996). Along with these matters the question of biodiversity's contribution to sustainability, or persistence of the system, needs to be considered.

4.1. *Examples of different degrees of biodiversity*

An example of a very simple ecosystem is found at Hamelin Pool, Shark Bay (26° S, 114° E.). This is at the southern end of a rather long narrow embayment which is partly isolated from the northern portion by a weed and calcareous sand bank, the Faure Sill. The shallow hypersaline waters within Hamelin Pool are maintained by a restricted water flow over the sill. The ecosystem present is simple, dominated by algal mats, stromatolites and the bivalve *Fragum erugatum*. From time to time severe storms scour the shells from the shallow waters depositing them as beach ridges known as the Hamelin Coquina. The scouring maintains the water depth and the Holocene age of these deposits indicates that the basin has persisted for approximately 10,000 years (Geological Survey of Western Australia, 1990). The constancy in composition of the ecosystem, at least as indicated by the dominant species, is possibly related to the rigour of the environment.

Analogous simple systems are provided by the brewing and wine industries. In both cases strict physical conditions are maintained to ensure that only the desired yeast (*Saccaromyces* sp.) is present thus ensuring the quality of the product. In both cases the duration of the system is limited by the presence of a fermentable substrate. Migrants and colonists are controlled by the strict maintenance of sterile anaerobic conditions. The systems are incomplete in so far as the wastes have to be disposed of externally to the system, by storms removing the *Fragum* shells, or the mechanical disposal of the wastes from brewing or wine making. However, the examples indicate that with appropriate physical conditions and nutrient supply, simple ecosystems are possible and can be maintained.

Water supply authorities operate at the other extreme of the biodiversity scale preferring water catchments that are fully vegetated, whether by native forest, woodland or plantations. In such cases, the expectation is not that the catchment ecosystem will be unchanging, but that water collected in the reservoir will be so low in nutrients that it will not support significant blooms of algae which would impair water quality and thus render it unfit for human consumption without expensive treatment.

In the case of Hamelin Pool, 'enough' can be demonstrated by the persistence (sustainability) of the system over geological time. With the last three systems, 'enough' is demonstrated by the quality of the product – saleable wine, beer or drinking water. In the first three cases, the simple systems were maintained by the conditions of the physical environment; in the last example biological diversity present consumes most of the nutrients present so ensuring the water quality. In this example no attempt is made to quantify the diversity present. The presence of sufficient biodiversity is inferred on the basis of the quality of the end product. Until this departs from acceptable values it is assumed that the unquantified biodiversity present is enough to satisfy the function or purpose for which the vegetated catchment is retained.

5. Biodiversity and agriculture

The literature on ecosystem functioning has been reviewed by Woodward (1993), and in Heywood (1996) two aspects of their functioning are relevant to agriculture. The first is that diversity of species contributes to a constancy of biomass production while the second is that the supply of ecosystem services is related to biodiversity. Both of these characteristics have been referred to as ecosystem functions, which accords with the convention of interpreting nature as a series of mechanisms. The convention is essentially one of convenience with the prime purpose of facilitating analysis.

The foregoing historical sections indicate that nature as we experience it is the result of a series of genetic and distributional (colonising) accidents, that worked in the sense that the bearers of the new genotypes (mutants, recombinants) or colonisers or migrants made a living from the resources available in the locality. The diversity of organisms making a living we consider under the headings of genetic, species and ecosystem diversity. The consequences of organisms making a living we interpret as ecosystem structure, food chains, water and erosion control and nutrient cycling. These latter categories are often designated as ecosystem services or functions. As an anthropocentric terminology this is relatively harmless, however, it requires a sense of discipline to see that such so-called functions are the incidental consequences of the organisms present making a living. The temptation to see the perceived function as in some sense meeting design criteria must be resisted. Agriculture on the other hand is different. It is all about meeting design criteria.

5.1. *Reducing risk*

Subsistence farmers seek to avoid starvation and so adopt a risk-averse strategy, which tends to ensure that regardless of the fluctuation between seasons some crops will yield a harvest. The mimicry of nature is very close. On the other hand, industrial broad acre farming seeks to maximise yield through an emphasis on monocultures. This is a high-risk strategy if environments are unpredictable. Later articles discuss examples of reduced risk agriculture and emphasise that some of the reduced risk derives from the biodiversity present in the cropping system (Trenbath, Van Noordwijk, this volume). How to establish what is enough biodiversity to reduce risk is one of the central issues, but the risk associated with yield is not the sole problem as it is the loss of ecosystem services that contributes to many environmental problems.

Farming practices in the semi arid wheatbelt of Western Australia rely on extensive monocultures with the consequent loss of some ecosystem services, leading to rising ground water, salinisation, erosion and soil degradation. Should it be desired that agriculture avoid these problems or restore situations where they have developed, then it is necessary to see agricultural ecosystems as assemblages of organisms constructed to meet design criteria calculated to restore or maintain ecosystem services affecting soil and water quality and their distribution in the environment. But harvestable crops are necessary in order to generate income to support those in the industry and fund rehabilitation. The question is whether a diversity of crops will better ensure that incomes are more regular and support the cost of maintaining the level of biodiversity needed outside the cropping system to maintain a favourable environment within.

Woodward (1993) in his discussion of how many species are required for a functional ecosystem concludes that there is some evidence that if the individual species in an ecosystem exhibit a range of different frequency responses (i.e. where the resource requirements for optimal growth vary), then the more species, the greater the frequency richness and ecosystem stability. This implies that the greater the number of crop species the more assured one would be of achieving a harvest. This is consistent with the practices of subsistence farmers. A more assured income is also a prerequisite for a commitment to expenditure associated with maintaining landscape level ecosystem services.

From this discussion it is possible to conclude that in an agricultural ecosystem diversity is advantageous, however, it must be planned to meet specified design criteria such as yield, insurance against crop failure or supply ecosystem services which contribute to an avoidance of the harmful effects of rising water tables, salinity, soil erosion and changes in soil structure.

In the foregoing discussion only the environmental function of ecosystems has been considered within the framework of agriculture as a mimic of nature. The other two aspects of biodiversity namely genetic and species diversity can also be readily embraced within the concept. Breeding programmes for new varieties of high yield, wider environmental tolerances or disease resistance can be seen as analogous to the adaptive responses of nature to changing environments, new diseases, herbivores and predators. Similarly introduction and acclimatisation programmes can be seen as mimicking the establishment of colonists and migrants in nature.

The subsistence farmer ensures that his crops track nature by retaining his own seed or planting seed of local provenance. The industrial type farmer does not do this. Genetic diversity, new strains and crop species are nurtured elsewhere to be introduced into to the cropping regime when environmental conditions, disease or pest susceptibility warrants. These activities have never been justified in the context of mimicking nature, rather the mimicry has been inadvertent. Should the future be seen to revolve around establishing design criteria for achieving goals related to providing ecosystem services then the totality of the mimicry inherent in the whole agricultural enterprise must be appreciated.

5.2. *Nutrient cycling as an ecosystem service*

In the nutrient poor soils of the Western Australian wheatbelt one of the desirable effects of biodiversity in the agricultural system is that among a range of crops some may add to the nutrient status of the soils as in nitrogen fixation by legumes, thus avoiding the costs of applying nitrogen fertilisers in addition to phosphorus and trace elements. There is no assurance that the nitrogen so produced will be consumed. In such a case, it can be seen as an unexploited resource and may contribute to soil acidity.

In earlier discussion, unexploited resources have been seen as offering the potential for invasion. However, soluble and mobile plant nutrients are not necessarily site limited so that when they are unused they may end up far from their origin. In many cases this leads to off-site nutrient enrichment, eutrophication of water bodies, loss of biodiversity and pollution. Often the costs associated with such events are treated as externalities, with the costs of treatment to be borne by others. Such an attitude reflects a narrow concept of the extent of agricultural ecosystems, effectively limiting them to farm boundaries. Extending the ecosystem concept to catchments, local drainage basins and whole river catchments effectively means that there can be no externalities. All the communities of the region either benefit from the application of fertilisers or suffer from their excess. This seems to be a phenomenon associated with mechanised or industrial agriculture.

The areas cultivated by subsistence farmers are small in relation to non-arable land, while the opposite is the case with broad scale farming where the tilled is vastly more extensive than wild land or native vegetation. In the absence of wild land and native vegetation, there is no mechanism to remove or immobilise excess nutrients. The net result is the loss of biodiversity. When neither nitrogen nor phosphorus are limiting, single celled and filamentous algae dominate water bodies. When nitrogen is limiting and phosphorus adequate, nitrogen fixing cyanobacteria, the only organisms which can tolerate the conditions, dominate the system. Such cases are often referred to as pollution, which is true, particularly as cyanobacteria are often toxic. Yet in another sense it merely indicates a return to pre-Cambrian conditions where low oxygen levels and structurally simple organisms dominated all aquatic environments. The present cases differ because of higher nutrient levels and corresponding greater levels of biological production leading to health risks and aesthetic problems. These observations suggest that there is a case for a wider mimicking of nature at a landscape scale. The naturally or artificially vegetated water catchments that retain natural biodiversity in order to obtain pure water are examples of the systems to be mimicked. Tree lots, wind breaks and aquatic vegetation fringing watercourses are all mimics of woodland, swamp, marsh or stream fringing plant communities which may be effective in stripping excess nutrients from drainage lines, streams and waterbodies. In cases involving landscape scales the criterion of enough is satisfied when the nutrients discharged from one ecosystem are within the assimilative capacity of the receiving system without altering its biodiversity. When biodiversity has been lost then the functional replacements such as the revegetation mentioned above should be designed to satisfy the foregoing criterion, especially as nutrients end up in waterbodies, rivers and estuaries where the resulting algal blooms, loss of biodiversity and pollution from rotting biomass is legendary.

5.3. *Invasions, pests and diseases*

The emphasis in the foregoing has been that when a resource is present there will, sooner or later, be a biological entity capable of exploiting it. In this sense agriculture is providing an endless supply of novel resources for nature to exploit. Nature responds by providing what are commonly called weeds, pests, parasites, diseases or forms of these which have become resistant to control agents.

Sometimes endemic or native species become pests or weeds such as the Australian plague locust (*Chortoicetes terminifera*) which eats crops and pastures, or the native weed known as goosefoot or mint weed (*Chenopodium pumilio*) which releases toxins into the soil that inhibit seed germination and seedling growth (Cheam, 1996). More commonly, the pests or weeds are introduced as in the case of skeleton weed (*Chondrilla juncea*). This weed, for which there is currently no effective in-crop control method, reduces soil nitrogen, competes for soil moisture and reduces yield while the wiry flowering stems may, in extreme cases, prevent harvesting. Biological control agents for skeleton weed include a rust fungus, a gall mite and a gall midge but they are not universally effective (Scott and Dodd, 1996). Even escapees from domestic gardens can pose risks, as with the Mediterranean herb Paterson's curse (*Echium lycopsis*) which is now a widespread weed.

These examples demonstrate the following in relation to biodiversity; native herbivores may not remain restricted to a diet of native plants or to native ecosystems. Similarly, native plants may adapt to cultivated situations or be pre-adapted for invasion and competition by virtue of possessing allelopathic chemicals or other traits. Thus biodiversity is reduced as a prerequisite for cropping but native organisms are not necessarily habitat or ecosystem specific and may immediately invade and increase local biodiversity. In the case of skeleton weed biodiversity was inadvertently increased by the accidental introduction and then deliberately increased in the hope of achieving an effective biological control. In one sense the foregoing are simply weeds or pests in another they are modern examples of the response of nature to reduced biodiversity or the availability of unexploited or poorly exploited resources. MacArthur (1972) argued that when production is left unused, the system is open to invasion. In the present case I am viewing resources in a wider context but the principle is the same. In terms of biodiversity, nature is doing what it has always done, namely restoration following a decline. The mimicry is two sided. We may be mimicking nature, but nature sees cultivated crops as mimics of resource rich, species poor environments resulting from extinctions and reacts as it has done to these events in the past. This conclusion offers little hope that a permanent or universal solution can be found to the question of how much biodiversity is enough to prevent invasions, pests and diseases and that manipulation by the farmer will always be necessary.

6. The side effects of utilitarianism

In the foregoing I have addressed the question of how much is enough biodiversity in agriculture in terms of productivity, sustainability, erosion and soil degradation and the ability of the agricultural enterprise to resist invasion by pests or diseases. In other words, a purely utilitarian approach. Mimicry of nature driven solely by such considerations carries significant risks when short term objectives dominate and a wide ranging perspective is lacking. The short term perspective is to see agricultural biodiversity as the technical fix that will endow agriculture with indefinite sustainability. However, this approach presents two serious risks to native vegetation; new pests and weeds of agricultural land and escapees of components of the applied biodiversity, such as *Pinus* spp., non-local eucalypts and acacias in south western Australia. Such escapees need not affect the agricultural enterprise but their spread will lead to changes in the dominant species and the character and nature of the biodiversity present in non-agricultural land.

Loss of identity is not likely to affect agriculture but will, for example, have a severe impact on the tourist industry. However, this is only the immediate effect. Once the Australian identity of the biota is swamped by exotics it will become increasingly difficult for ordinary citizens to discern new invaders and loss of local biota. This will encourage an increasingly careless attitude to monitoring species' introductions and an unwillingness to spend money on control of even the most serious pests.

In a sense these potential problems associated with escapees from agricultural biodiversity are analogous to the assessment of leached nutrients as externalities. They both lead to a degradation of the environment and a decline in the community's identity with it. However, one is a relatively straightforward and essentially physical problem while the other has all the intangibles, risks and contingent states inherent in any biological problem. In part the solution is to display good anticipatory planning, care and prudence. Yet success will come only when an informed and supportive public is involved. The fear is that because of nature's demonstrated ability to adapt genetically and exploit vacant or underutilized resources by producing an endless array of diseases, weeds and pests we may be forever applying Van Valen's Red Queen hypothesis, i.e. running as fast as we can in order to stay in the same place.

7. Courses of action

Throughout the foregoing sections the emphasis has been on the difficulties associated with mimicking nature. The treatment has been at a general or non-locality specific level as summarised in Table 3. With respect to agriculture in the semi-arid landscape of Western Australia, we are aware of the dangers. The problem is to navigate in such a way that the serious unwanted conse-

Table 3. How much biodiversity is enough.

Sources	Design Goal	Measure of enough	Risk
Genetic			
within crop vars.	High yield	Reliable yield	Disease, pests
between varieties	Risk avoidance	Always some production	Reduced risk of failure
Species			
grasses	Exploit species' adaptive niche	Some harvest, effective use of soils	Diseases and pests
legumes	Crop diversity, increase soil N	Improvement in yields	Excess N alters soil chemistry
biocontrol agents	Control of diseases and pests	Reduce losses from pests and disease	Attack other than target species
Ecosystem			
crop matrix	Shelter belts to provide improved crop microclimate	Enhanced crop and pasture yield	Reduced crop and pasture yield due to competition
catchment	Shelter belts, wood lots, and enhanced remnants at sufficient density to achieve hydrologic balance	Water quality, stable water tables, little or no wind or water erosion, reduced discharge	Weed, pest and disease reservoirs created. Problems seen by communities as externalities
drainage basin	Maintain nutrient discharge within capacity of receiving system	No change in biota of receiving system	Not achieving cooperation of all communities in the drainage basin

quences – wind and water erosion, leaching of nutrients, rising water tables, salinisation, weeds, pests and diseases – are avoided. To achieve this goal it is necessary to harness the ecosystem services provided by biodiversity in the service of sustainable agriculture. The ways that this may be achieved are summarised in Table 4. At these levels, enough biodiversity is established by partitioning the problem of sustainability into its components.

Inspection of Table 4 shows that the services provided need not be apparent from casual inspection of the biota present, as in the case of nutrient recycling and leaching. The relative importance of leaching and recycling will change depending on the community of recyclers. In natural vegetation, nutrients were retained in roots (fine roots and proteoid roots), mycorrhizae, fungal mycelia, soil biota and bacteria. The retention time for nutrients is different for each of these. Nevertheless despite vegetation cover the soils of the wheatbelt were depleted of nutrients at the time of settlement. Moreover, the composition of recyclers changes following clearing and depends on the crops and husbandry but the important thing is that the presence of this component

Table 4. Use of ecosystem services to avoid or ameliorate problems in agricultural ecosystems.

Problem	Services provided by biodiversity
Erosion	
wind	Suitable vegetation to provide wind control by landscape texture, manage landscape on soil type rather than cadastral characteristics
water	Control water runoff through texture of soil cover and surface, minimal tillage, cultivate on contours
Leaching of nutrients	Controlled application of fertilisers, plant vegetation to capture nutrients, vegetate drainage lines to strip nutrients
Rising water tables	Deep rooted vegetation, increased evapotranspiration
Salinisation	Control water tables by evapotranspiration of deep rooted plants, vegetate recharge areas
Weeds, pests and disease	Ongoing adaptive response by farmer, through husbandry, biocides, biological control

cannot be established by simple inspection. Similarly an effective rooting system to control soil water is not readily observed. In each of these cases the presence of enough biodiversity is most readily established by ensuring that the undesirable outcome does not arise, i.e. no erosion, nutrients are low in discharge water, water tables do not rise, salinisation is not observed by way of salt scalds or saline water discharged.

In an ideal world it would be desirable to have enough biodiversity to prevent all the different kinds of problems from arising. In reality this is not possible, thus there may be enough to restore or maintain such functions as hydrological balance but not enough to prevent invasion. This means that intervention by the farmer will still be necessary to control weeds and pests such as through strategic grazing, but in an important ecological sense enough has been achieved.

In summary, the courses of action can be ranked in order of desirability as follows:

1. Reduce risk of crop failure
2. Restore perturbed ecosystem services such as water cycling, nutrient fixation and cycling
3. Reduce risks of invasion, pests and diseases.

It should be noted that biodiversity required increases from 1 to 3 in the sequence above.

8. Discussion

The foregoing presentation has taken the current wide ranging understanding of what is embraced by the concept of biodiversity, placed it in an historical context, then applied the understanding so gained to an agricultural situation. In part, the application derives from the concepts of landscape reintegration developed earlier (Main, 1992) but placing particular emphasis on the question of agriculture mimicking nature.

Throughout this article the immediacy of agricultural problems has been juxtaposed against the consequences of the historical evolutionary aspects of biodiversity. This has been done in the hope that agricultural solutions may not be merely the application of recipes, but reflect an appreciation of how adjusting a dynamic system need not be simply an *ad hoc* procedure. Even the most rudimentary conceptualisation or theory of how ecosystems arise will allow the intervention in modifying dynamic systems to be tempered by an anticipation of what undesirable outcomes may arise and have to be countered. After all, the stated goal is sustainable agriculture which, translated from consideration of the immediate present to a time scale in which evolutionary processes such as mutation, recombination and selection can occur, i.e. a scale in which historical ecosystem dynamics can operate (perhaps only very few generations), means that incorporation of new genotypes into ecosystems is almost inevitable. What has been proposed is a conceptual framework within which goals can be set and outcomes anticipated.

The question comes down to how wide ranging is the mimicry to be, and at what scale, local or landscape, is the mimicry to be applied? From history and current examples, I have tried to tease out guiding principles applicable at all scales. There are no clear cut answers which could form the basis of easily understood and applied recipes. The only principle to emerge is that nature, through genetic mutation, recombination, colonisation, migration and invasion, will change the genetic and species composition of the biota. Biodiversity will be in endless flux, at different rates and extent. It is within this framework that agriculture must attempt to mimic nature.

The thrust of the foregoing answers to the question of how much biodiversity is enough centres around the self similarities evident at all scales of biological processes. Table 5 summarises these similarities. The contingent nature of events is reflected by the upward arrangement of the cells as in a geological sequence. Table 5 is restricted to three scales (geological, the parade of life; ecological, dynamics of current biota; and agriculutral, emphasising the conflict between Nature and farmers' goals). Horizontally, the self similarity is reflected by comparing cells between columns.

Unlike the natural system, where assemblages recognised by us as systems are assembled by chance, conscious mimicry of nature demands that goals be set and scales and design criteria established before intervention. Knowledge of natural mechanisms relating to the maintenance and increase in biodiversity will permit better design of agricultural systems, but there is no absolute answer to the question of how much is enough due, as emphasised repeatedly above, to the contingent nature of biology. Moreover, this fact cannot be disregarded. It has profound implications for the environment and sustainable agriculture should the mimicry go seriously awry.

Table 5. Summary.

	Scale	
Geological evolutionary Parade of Life	Ecological Dynamics of current biota	Agricultural Enough biodiversity to resolve conflict between utilitarianism and self organising Nature
Self organisation, ecosystems following cycles of extinction	Self organisation of ecosystem following disturbance	Organised by farmer: self organised by Nature
Nutrient recycling, evolution of predator, prey, parasites, disease	Food chains, recycling of nutrients	Cycles of husbandry, self organising weeds and pests
Multicellular biota	Plants, animals, bacteria, fungi	Crops, livestock, pests
Pro and Eukaryotes	Soil biota	Soil biota
Self replication	Recombinant, mutant	Recombinant, mutant
Weathering, soils, nutrients in solution	Weathering soils, nutrients in solution	Clearing, erosion, loss of nutrients

9. Conclusions

1. There is no absolute answer to the question of how much biodiversity is enough because all systems are dynamic. Moreover, events and solutions in one place may not be repeated elsewhere because of the contingent nature of biology (Table 1).
2. Nature will continue to re-sort genotypes, select mutations and genotypes which are advantageous to their bearers in a changing world, i.e. Nature will continue to respond dynamically to changes in the physical and biotic environment.
3. Agriculture must develop or retain a capacity to select and introduce new genotypes, varieties and species in response to the changes arising as a result of nature's dynamics mentioned above.
4. The question of how much biodiversity is enough can only be meaningfully answered in terms of specific criteria for sustainable agriculture such as yield, the restoration of ecosystem services such as water and nutrient cycling and resistance to pests and disease.
5. Mimicry at the landscape scale should consciously follow a design, which if implemented will mimic nature with respect to hydrological balance, water and wind erosion. In terms of nutrient cycling, the condition of enough biodiversity is satisfied when nutrients discharged from one ecosystem are within the assimilative capacity of the receiving system without altering its biodiversity.
6. Mimicking nature is a two way street. Historically, nature has always

followed strategies of exploiting physical and biological resources in new and novel ways. If agriculture is to mimic nature it must understand and match the scale of nature's capacity for adaptive responses.

To return to the analogy of the navigator, it is inevitable that the weather forecast will at some point be wrong, that the flight plan will become redundant and that to stay on track it will be necessary to constantly re-establish our position. What makes agriculture even more difficult is that there is no general agreement on the position of the target.

10. References

Behrensmeyer AK, Damuth JD, Di Michelle WA, Polls R, Suess H-D and Wing SL (1992) Terrestrial Ecosystems Through Time: Evolution and Palaeoecology of Terrestrial Plants and Animals. University of Chicago Press, Chicago

Cheam A (1996) Weeds can poison crops. Western Australian Journal of Agriculture 37: 134–138

Commonwealth of Australia (1996) The National Strategy for the Conservation of Australia's Biological Diversity, Commonwealth Department of the Environment, Sport and Territories, Canberra

Geological Survey of Western Australia (1990) Geology and mineral resources of Western Australia: Western Australia geological survey. Memoir 3: 462, 602

Heywood VH (1995) Global Biodiversity Assessment. Published for the United Nations Environment Programme. Cambridge University Press

Mac Arthur RH (1972) Geographical Ecology. Harper and Rowe, New York

Main AR (1992) Landscape reintegration: problem definition. In: Hobbs RJ and Saunders DA (eds) Reintegrating Fragmented Landscapes: Towards Sustainable Production and Nature Conservation, pp 189–208. Springer-Verlag, New York

Moir RJ (1994) The 'carnivorous' herbivores. In: Chivers DJ and Langer P (eds) The Digestive System in Mammals: Food, Form and Function, pp 87–102. Cambridge University Press

Scott P R and Dodd J (1996) Skeleton weed: the current situation. Western Australian Journal of Agriculture 37: 76–81

Shear WA (1991) The early development of terrestrial ecosystems. Nature 351: 283–289

Tilman D, Wedin D and Knops J (1996) Productivity and sustainability influenced by biodiversity in grassland ecosystems. Nature 379: 718

White TCR (1984) The abundance of invertebrate herbivores in relation to the abundance of nitrogen in stressed food plants. Oecologia 63: 90–105

White TCR (1985) When is a herbivore not a herbivore? Oecologia 67: 596–597

Woodward FJ (1993) How many species are required for a functional ecosystem? In: Schulze E-D and Mooney HA (eds) Biodiversity and Ecosystem Function. Springer-Verlag, New York

Agroforestry Systems **45**: 43–55, 1999.

Moving from descriptive to predictive ecology

R. J. HOBBS[1,*] and S. R. MORTON[2]
[1]*CSIRO Wildlife & Ecology, Private Bag, P.O. Wembley, WA 6014, Australia;* [2]*CSIRO Wildlife & Ecology, P.O. Box 84, Lyneham, ACT 2602, Australia (*Author for correspondence: richard.hobbs@per.dwe.csiro.au)*

Key words: design of agricultural mimics, ecosystem function, non-equilibrium, prediction

Abstract. The science of ecology is undergoing many important shifts in emphasis and perspective which have important implications for its role in designing sustainable farming systems. In particular, a shift has occurred from the equilibrium paradigm to one which recognises the dynamic, non-equilibrium nature of ecosystems. Allied to this is the recognition that ecosystems can occur in any one of a number of alternative stable states, depending on the disturbance and management history. An increased emphasis on spatial patchiness in ecosystems has also emerged as appropriate tools have emerged to analyse spatial mosaics. These features have led to a recognition that considerable uncertainty is associated with the outcome of any particular ecosystem modification; hence predictive capacity is also low. Recent considerations of the interrelation between biodiversity and ecosystem function have also explored the questions of how many species need to be in a system to fulfil certain functions and confer resilience. We identify a set of steps that are required for the development of an agricultural system based on mimicking natural ecosystems. Central to this is identifying (1) the functions which are currently suboptimal in the agricultural system, and (2) the species which have key functional roles in the natural system, and then reaching decisions as to the array of species needed to confer system function and resilience.

1. Introduction: ecology at the crossroads

In this article we examine the ability of ecology to contribute to the design of sustainable farming systems. We examine the current state of ecology in the light of recent debate, and then highlight some of the radical changes in fundamental paradigms that have been occurring recently. We then discuss the study of the relationship between biotic diversity and ecosystem function, an area currently receiving considerable attention in ecology, and finally place these discussions in the context of developing agricultural systems.

The science of ecology has come under fire from numerous sources recently, and is currently undergoing a radical shift in emphasis. Many aspects of the science are undergoing revision. Botkin (1990) has provided a lucid and in-depth analysis of the gulf between ecological theory and empirical observations, and suggested that theory developed from mechanical and physical analogies does not fit natural systems very well. Further, natural resource management based on these theories is bound to fail since it does not take the changing nature of the environment into account. Keddy (1989) further criticised ecology for its inability to focus on important questions, rather than interesting or easily-answerable ones, and pointed out that the thrust of much ecological research is decided by a relatively small number of influential ecologists. Ecology also focuses disproportionately on certain types of organisms (particularly attractive vertebrates), and concentrates on the northern hemisphere temperate regions, which house the majority of ecologists but which have only a small proportion of the world's biota and ecological problems.

Shrader-Frechette and McCoy (1993) went further in pointing out that ecologists frequently use ambiguous, inconsistent or unclear terms, and spend a lot of time arguing over terminology and the importance of particular processes. They commented that:

> On the whole, general ecological theory has, so far, been able to provide neither the largely descriptive, scientific conclusions often necessary for conservation decisions, nor the normative basis for policy, both of which environmentalists have sought.

Peters (1991) provided further in-depth criticism of ecology and ecological methodology. The summary of his book on its cover stated that:

> Ecology suffers because it has ignored or minimised the important criterion of predictive power in assessing scientific quality. Instead, ecologists offer logical rationalisation, historical explanation and mechanistic understanding, so that the predictive failure of the science goes almost unnoticed. Given this context, ecologists fall prey to a number of minor failings that complicate and confound any assessment of the science. Even when predictions are possible, they are often vague, inaccurate, qualitative, subjective and inconsequential. Some of ecology has become scholastic problem-solving that directs concern away from the serious problems facing humanity.

Ecology, as with any science, is an evolving entity, and because it is a relatively young science dealing with a vast and complex subject matter, there is still a lot of disagreement about even the most basic of premises. The subject matter of ecology is complex and often inherently unpredictable, and results obtained in one type of system are not necessarily transferable to another. Generality and 'hard facts' are, thus, hard to come by. Matched against this, however, is the need to come up with answers to pressing environmental problems which cannot wait for theoretical arguments to be worked out. Among these is the vital task of ensuring the sustainability of agro-ecosystems, a task that all humans agree is fundamental to our survival. However, given the turbulence in ecology, how successfully can its practitioners contribute to this objective?

2. New paradigms in ecology

In line with the debate about the role of ecology described above, the science has undergone something of a quiet revolution during the past 20 years (Pickett et al., 1992; Pickett and Ostfield, 1995). Concepts that were considered firmly established in previous decades have undergone considerable revision, even reversal. Some of these re-interpretations are perhaps well known to those working in agro-ecosystems, but others are relatively recent and are still actively developing in the ecological literature. In this section, recent developments that we believe are of greatest importance to ecosystem ecology are briefly mentioned.

2.1. *The flux of nature*

Previous generations of ecologists operated largely on the assumption that the natural world was fundamentally a stable place, a collection of communities in which each species had its ordered position and in which any disturbance would result in an ordered successional progression leading through subclimax phases back to the original climax (Christensen, 1988). Ecological communities were considered to be organised, patterned collections of co-evolved species, into which incompatible species could not penetrate (Simberloff, 1982). Ecologists now speak of this era as the period of the equilibrium paradigm. In recent years we have seen this notion of organisation and stability give way to a vision of flux. Most ecologists have come to the view that the natural world is characterised more by instability than permanence, by frequent disturbance that continually pushes ecosystems in alternative directions instead of causing them to return inevitably and regularly to their original condition; more by unique specific responses than co-ordinated, predictable, tightly constrained combinations of species, as individualistic responses outweigh tendencies towards regularly occurring communities. Awareness that the natural world is an uncertain place in which disturbances are constantly causing alterations in composition of assemblages and in spatial pattern of the environment has led now to the non-equilibrium paradigm (Pickett et al., 1992; Fiedler et al., 1997). This paradigm does not hold that ecological equilibria are non-existent, but rather that they are scale-dependent and embedded in non-equilibrial conditions. Nevertheless, the non-equilibrium paradigm does imply that predictable end-points to the successional process following disturbance are rare, that multiple stable states may exist, and that some quasi-stable states can persist for long periods.

2.2. *Multiple stable states*

Disturbance inevitably sets in train some form of succession. It is apparent now that the course of the succession is difficult to predict, because the direction which the ecosystem or assemblage takes is contingent upon the particular circumstances of the disturbance and the nature of the biophysical conditions that precede and follow it. The notion of contingency brings history to the fore: history very much matters in patterns and processes of community change. As a consequence, the end-point of many successional processes is not a predictably uniform outcome; instead, several states are possible, depending on the contingent circumstances (Noble and Slatyer, 1980; Hobbs, 1994). Depending on the frequency of the disturbances, these multiple states may be stable for long periods of time, and distinct thresholds may exist which limit the transition from one state to another. The differences among outcomes of successional events in seemingly similar assemblages or ecosystems may well follow broadly interpretable patterns, but the itineraries are not easily predictable at the outset of the journey.

2.3. *Patchiness and landscape ecology*

Recognition of the importance of spatial and temporal variability, together with the increased availability of suitable tools for analysing it, has galvanised landscape ecology. Its re-emergence springs from realisation that understanding and management of the natural world depends as much on the analysis of flows of resources across ecosystems as it does on the study of quadrats. But perhaps the principal issue underpinning landscape ecology is recognition of the vital importance of patchiness (Turner and Gardner, 1991). Patchiness does not yet possess a complete or unified theory, but is a rapidly developing conceptual tool (Levin, 1989; Ostfeld et al., 1997). Patchiness focuses on the spatial matrix of ecological processes, and emphasises the fluxes of materials and organisms within and between parts of the landscape. It is a form of spatial heterogeneity in which boundaries are discernible, and in which units appear as contrasting, discrete states of physical or ecological phenomena (Ostfeld et al., 1997). An array of patches constitutes a mosaic at whatever scale is appropriate for investigation (although it is important to note that multiple scales may be important). The study of patch dynamics promises to provide a valuable framework in which to understand and to manage the landscape mosaic, although there is still much work to be done in this area.

2.4. *Prediction*

The non-equilibrium paradigm sees ecosystems as probabilistic rather than deterministic; inherently, therefore, most ecologists believe that ecosystems are characterised by uncertainty rather than by predictability. Because of the overwhelming importance of this uncertainty, ecologists have invested considerable intellectual energy in trying to comprehend environmental stochasticity – correlated variability in chance events caused by patches in a landscape experiencing a similar environment, including both physical and biotic features – and catastrophes – correlated variability of large magnitudes that occur at a low frequency. We cannot avoid the lack of predictability; consequently, there is a need to identify the bounds or conditions under which decisions can be made in the face of uncertainty. Risk analysis and adaptive management through more detailed involvement of managers in research and development, are the principal routes by which ecologists are struggling to work with unpredictability. Although admission of the extreme difficulty of prediction has initially caused ecologists to be concerned that their science is fuzzy, a focus on uncertainty and risk analysis is common to many people in the social, political and economic spheres (Graham and Wiener, 1995), and quantitative risk assessment is widely used in engineering and technology. Hence, ecology is not necessarily difficult, in this sense, for decision-makers to comprehend.

2.4.1. *Human beings and ecology*

Recognition of the inevitability of disturbance, and of its profound ecological consequences, leads inevitably to the inclusion of humans as primary agents of flux in ecosystems (Pickett et al., 1992; McDonnell and Pickett, 1993; Hobbs, 1997; Vitousek et al., 1997). Ecology is now beginning openly to extend its interest from supposedly 'natural' systems, in order to include human-dominated systems. Anthropogenic disturbance can now be incorporated into ecology in the same way as any natural disturbance, rather than being considered as distracting noise. The incorporation of human activities into ecological investigations is most obvious in the field of conservation biology, but one can predict that the ecology of agricultural systems will also undergo fresh growth in the coming years, as a result of hybrid vigour.

2.5. *Ecology and environmental management*

Pahl-Wostl (1995) has recently summarised the changing situation in ecology and its ability to provide input to environmental management (Table 1). While we do not agree entirely with all of the points raised in her original publication, we do agree with her contention that significant shifts are required in attitudes and expectations of what ecology can deliver, and how environmental risks should be assessed. Of particular note is her suggestion that risks should be measured in terms of the reduction of degrees of freedom for future action; in other words, the extent to which future options are foreclosed.

3. Biodiversity and ecosystem function

After a period of neglect, the question of how biotic diversity and ecosystem function are related is now considered one of the fundamental questions in ecology. The early neglect of this question can be traced to the fragmentation of ecology into distinct branches, most notably with a split between organism-centred population and community approaches and the material flux approach of ecosystem ecology ('things' *versus* 'stuff': Pickett et al., 1994;

Table 1. Contribution of ecological research to environmental management.

	Current attitude	Required attitude
Research	reduces uncertainties	uncovers uncertainties
	makes quantitative predictions	generates innovative, qualitative knowledge
	provides expert knowledge	engages as partner in a social dialogue
	views nature as a machine	views nature as partner
	is rigid, controlling	is flexible, adaptive
	aims at change towards preconceived goals	fosters evolution, innovative action
Risk assessment		
source	individual phenomenon or process	overall system structure and organisation
risk	undesirable events	restriction of evolutionary potential
measure	probability × damage	decrease in degrees of freedom

Source: Modified from Pahl-Wostl (1995).

Jones and Lawton, 1995). A large international program organised by SCOPE recently examined the question in detail, both from a theoretical point of view and in terms of what we know from examples from a variety of ecosystem types (Schulze and Mooney, 1993; Baskin, 1994; Mooney et al., 1996). The societal relevance of the question has also recently been explored in the context of 'ecosystem services' and 'how the diversity of life sustains us' (Baskin, 1997; Daily, 1997).

During the course of the SCOPE program, it became apparent how remarkably few data there were with which to assess the question of how biodiversity might affect ecosystem function. For instance, a review of information available for southwestern Australia provided only one study which addressed the question directly, and even then the interpretation was equivocal (Hobbs, 1992; Hobbs et al., 1995). Recently, experimental approaches to the problem have been employed, resulting in data which are widely quoted as indicating a clear link between biodiversity (equal to species number in this context) and elements of ecosystem function (Naeem et al., 1994; Tilman and Downing, 1994; Naeem et al., 1995, 1996; Tilman et al., 1996). However, the validity of the interpretation of results from these experiments has been questioned by Huston (1997), who argues that it is impossible to separate the effects of changing biodiversity from the effects of other 'hidden treatments' in the experiments. Tilman's findings have also been questioned by Aarssen (1997) who suggested that observed differences in function were related more to differences in individual species characteristics than to diversity *per se*.

Ecologists are thus currently in the middle of a flurry of activity surrounding the question of the role of biodiversity in ecosystem function, but are also apparently in a bit of a muddle at the same time. Part of the problem has been a failure to define exactly what question is being asked.

'The ecosystem function of biodiversity' is a ridiculously broad term, and both 'ecosystem function' and 'biodiversity' can be interpreted in numerous ways. Ecosystem function can refer to the primary functions of water, carbon, energy and nutrient cycling, or it can refer to the myriad of processes which go to make up these cycles, including biotic interactions. It can also be interpreted in more utilitarian way to mean 'ecosystem services' for particular human purposes, such as the supply of fresh water, disease prevention etc. Similarly, biodiversity incorporates all levels of biological organisation from genes to landscapes, although it is frequently interpreted simply as 'number of species'. Species can also be grouped in a number of different ways, and attempts are being made to define sensible groupings which have functional significance (Smith and Shugart, 1996; Woodward and Cramer, 1996). Without defining exactly what aspect of ecosystem function one is trying to relate to what element of biodiversity, it is unlikely that useful questions can be asked.

A further problem has been the lack of consideration of the impact of different kinds of species. Although it may seem slightly ridiculous to those participating in this workshop, experimental work has concentrated almost exclusively on the number of species rather than the mix of different types of species. For instance, Tilman's experimental work constructed grassland plant communities by randomly drawing species from a total species pool. Other ecological work on community assembly rules suggests that there may be readily-defined reasons why certain plant assemblages develop in response to particular environmental and biotic factors which act to 'filter' species from the regional species pool (Keddy, 1992; Weiher and Keddy, 1995). It is also clear that individual species vary greatly in terms of their functional importance (e.g. in their quantitative contribution to particular processes), and a variety of terms have been derived for species which strongly influence system structure or function: e.g. 'keystones' (Mills et al., 1993; Paine, 1995; Stone, 1995), 'drivers' (Walker, 1992), and 'ecosystem engineers' (Jones et al., 1994). Huston (1997) questions the assumption that species diversity can be divorced from the effects of species identity. Indeed, the debate needs to focus more on the importance of particular elements of biodiversity rather than the importance of biodiversity *per se*.

An allied question is the degree of functional redundancy inherent in natural communities (Walker, 1992). In practice, however, the perception of redundancy depends on the time scales and functions considered. Apparently functionally-similar species are likely to respond to environmental variation or disturbance differently and hence, may increase the resilience of the system (Main, 1992; Hobbs and Mooney, 1995; Walker, 1995). Functional redundancy thus provides 'fail safe' or 'back up' capacity.

Recent accounts of the relationship between biodiversity and ecosystem function take greater cognisance of these dual questions of the functional significance of particular biotic elements and the importance of functional redundancy in conferring system resilience (Chapin et al., 1997). Future research on these issues will not only provide a better understanding of how systems work, but will also allow assessment of which system components are functionally the most important to system integrity or persistence, and hence which components are essential both to retain in existing ecosystems and to introduce into constructed systems.

4. Relevance to designing agricultural systems

If we are to use natural ecosystems as potential models for agricultural systems, we now ask whether the above considerations have any relevance to the agricultural situation. This question has already been asked recently in a slightly different context by Jordan (1995), who noted that most resource management guidelines were to a large extent empirical and derived from the practical experience of the resource manager. Concomitant with that was the fact that ecological concepts and theory often did not translate into workable guidelines, a proposition also discussed elsewhere (Hobbs, 1998). Jordan (1995) therefore took a recent summary of important ecological ideas (Odum, 1992) and interpreted them in a resource management framework. He produced the following set of statements (slightly modified by us):

1. To develop a sustainable system, we must understand that we are part of the system.
2. To develop a sustainable system, we must analyse the system in which we are embedded. To analyse the system we need a common currency. Energy is one convenient currency with which to analyse a system. Energy flows through all ecosystems and the way it is used and stored determines the characteristics of each ecosystem. Material and information flows are also important.
3. Energy flow through natural ecosystems is not random but is controlled and self-regulated by internal feedback interactions between organisms, or between organisms and the environment.
4. Stability and sustainability of energy flow through ecosystems are enhanced by mutualistic functions. As ecosystems became large and complex, organisms within a natural community evolved towards mutually beneficial functions. Evolution has selected many mutualistic species because of the higher efficiency of the functions they perform.
5. In managed systems, the mutually beneficial functions and natural subsidies that lend stability and sustainability to natural systems are usually destroyed. For this reason, energy subsidies are usually required.
6. The stability and sustainability of a managed system can be maintained (or increased?) by replacing external energy subsidies with the mutually beneficial functions found in nature.
7. The stability and sustainability of all systems may be enhanced by maintaining species and landscape diversity.

8. The transition from non-sustainable to sustainable systems requires time and has a cost.
9. Despite the cost, there is an urgent need to make the transition.

While Jordan recognises that some of the ideas contained in these statements are still controversial and not proven in a strict scientific sense, they encapsulate most of what we wish to say in relation to agricultural systems. Statements 2–5 emphasise the importance of bridging the current gap between natural systems with internal feedbacks and mutualistic interactions, and managed systems that require heavy energy subsidies. Statement 6 implies that the way to do this is to replace lost mutualistic support functions. Important questions to be asked are: which are the most important functions, and which species or sets of species need to be replaced to achieve these? We tend to think mostly in terms of plant diversity (structural and compositional), but soil biota may be at least as important.

Statement 7 suggests that both species diversity and landscape diversity are likely to be important. From the agricultural perspective, the question develops into the vital decision as to whether species diversity must be increased (or better balanced) within individual blocks of land (i.e. increasing alpha diversity), or whether the same effect can be achieved by diversifying between blocks of land. We consider that this is one of the key issues to be addressed, and working out the necessary spatial relationships is vital if we are to make progress.

Statement 8 identifies the important question of the costs of achieving a transition to a sustainable system. An essential element of making the transition will be ensuring that these costs are met in a socially equitable manner.

5. Conclusions: has ecology anything to offer?

We have presented a picture of ecology as a science in transition, moving from one set of paradigms to another and grappling with new sets of questions. The move from description to prediction, although still problematic, is timely in relation to the theme of this book. The other major shift occurring in ecology is the recognition that classical ecologists are no longer simply concerned with the natural bits that are left in nature reserves, but have to consider areas managed for production purposes too. It is often forgotten that many of the major conceptual advances in ecology have come from simplified managed systems, and it is now clear that ecology can make a contribution to redesigning these systems to meet the new challenges they are facing. Having said that, however, it is also clear that ecology has no simple formulae or models available with which to do this. We still do not have a set of rules for the design of a system which will meet specified requirements in terms of its functioning. Nevertheless, the design of agroecosystems is a fertile area for the advancement of ecological understanding; by trying to construct something, we learn more about how it works (Jordan et al., 1987). Using natural systems as a reference point, we can start developing more effective assembly rules and design guidelines. The potential for natural ecosystems to occur in alternative states also allows for optimism that productive agricultural systems can be developed using certain elements or states of the natural system, and that we may be able to chose to mimic the state which best suits our purpose.

In the absence of clear predictive ability, we can still provide guidance which will focus efforts and maximise the likely benefits of these efforts. We suggest that the following steps are required for the development of an agricultural system which replaces external subsidies with mutualistic functions of nature:

1. Identify the system functions which are currently suboptimal in the managed system.
2. Identify the suite of species which carry out these functions in the natural ecosystem.
3. Within this suite of species, identify species with key functional roles.
4. Identify the likely range of environmental conditions and disturbances, and select an array of species needed to confer system resilience.
5. Consider how many of these species are required for the managed system, in the context of trading-off environmental risks *versus* long and short term costs and benefits. For instance, is it essential to install the full suite of species immediately, or can a phased approach be employed?
6. Consider the merits of developing mixtures of these species *versus* placing species in blocks across the landscape (i.e. increasing landscape diversity), or gradations between the two (such as alley cropping). Again, we consider this to be a fundamental issue which has important ramifications for how we frame the overall questions relating to developing mimics.
7. Provide socio-economic instruments which facilitate implementation.
8. Develop systems in an adaptive management framework, with adequate monitoring and the capacity to modify elements of the design.

Points 7 and 8 above are really central to the likely success of any attempts to develop natural systems agriculture. The task is complex and the methods by which to achieve desired outcomes will not always be obvious. Despite considerable rhetoric on the need for adaptive management of ecosystems, ecologists have only rarely attempted to work in an adaptive management framework. We now have the opportunity to do this in agroecosystems, and to put the idea into practice.

Given that natural systems are dynamic and often non-equilibrium, there is no reason to assume that sustainable agricultural systems will be static, or that there will be only one solution. Natural systems provide a multitude of examples of organisms coping with the same set of environmental and biotic constraints and stresses in vastly different ways. We rejoice in nature's diversity, and perhaps one of the best ways to mimic natural ecosystems in agriculture is to facilitate the development of a diversity of responses to today's problems.

6. References

Aarssen LW (1997) High productivity in grassland ecosystems: effected by species diversity or productive species? Oikos 80: 183–184

Baskin Y (1994) Ecologists dare to ask: how much does diversity matter? Science 264: 202–203

Baskin Y (1997) The Work of Nature: How the Diversity of Life Sustains Us. Island Press, Washington, DC

Botkin DB (1990) Discordant Harmonies. A New Ecology for the Twenty-first Century. Oxford University Press, Oxford

Chapin FS, Walker BH, Hobbs RJ, Hooper DU, Lawton JH, Sala OE and Tilman D (1997) Biotic control over the functioning of ecosystems. Science 277: 500–504

Christensen NL (1988) Succession and natural disturbance: paradigms, problems, and preservation of natural ecosystems. In: Agee JK and Johnson DR (eds) Ecosystem Management for Parks and Wilderness, pp 62–86. University of Washington Press, Seattle

Daily GC (ed) (1997) Nature's Services: Societal Dependence on Natural Ecosystems. Island Press, Washington, DC

Fiedler PL, White PS and Leidy RA (1997) The paradigm shift in ecology and its implications for conservation. In: Pickett STA, Ostfeld RS, Shachak M and Likens GE (eds) The Ecological Basis of Conservation: Heterogeneity, Ecosystems and Biodiversity, pp 83–92. Chapman and Hall, New York

Graham JD and Wiener JB (eds) (1995) Risk vs. Risk: Tradeoffs in Protecting Health and the Environment. Harvard University Press, Harvard, Massachusetts

Hobbs RJ (1992) Is biodiversity important of ecosystem functioning? Implications for research and management. In: Hobbs RJ (ed) Biodiversity in Mediterranean Ecosystems in Australia, pp 211–229. Surrey Beatty and Sons, Chipping Norton, NSW

Hobbs RJ (1994) Dynamics of vegetation mosaics: can we predict responses to global change? Écoscience 1: 346–356

Hobbs RJ (1997) Future landscapes and the future of landscape ecology. Landscape and Urban Planning 37: 1–9

Hobbs RJ (1998) Managing ecological systems and processes. In: Peterson D and Parker VT (eds) Scale Issues in Ecology. Columbia University Press, New York

Hobbs RJ, Groves RH, Hopper SD, Lambeck RJ, Lamont BB, Lavorel S, Main AR, Majer JD and Saunders DA (1995) Function of biodiversity in the mediterranean-type ecosystems of southwestern Australia. In: Davis GW and Richardson DM (eds) Mediterranean-type Ecosystems: The Function of Biodiversity, pp 233–284. Springer, New York

Hobbs RJ and Mooney HA (1995) Spatial and temporal variability in California annual grassland: results from a long-term study. Journal of Vegetation Science 6: 43–57

Huston MA (1997) Hidden treatments in ecological experiments: re-evaluating the ecosystem function of biodiversity. Oecologia 110: 449–460

Jones CG, Lawton J and Shackak M (1994) Organisms as ecosystem engineers. Oikos 69: 373–386

Jones CG and Lawton JH (eds) (1995) Linking Species and Ecosystems. Chapman and Hall, New York

Jordan CF (1995) Conservation: Replacing Quantity with Quality as a Goal for Global Management. John Wiley and Sons, Chichester

Jordan WRI, Gilpin ME and Aber JD (1987) Restoration ecology: ecological restoration as a technique for basic research. In: Jordan WR, Gilpin ME and Aber JD (eds) Restoration

Ecology: A Synthetic Approach to Ecological Research, pp 3–21. Cambridge University Press, Cambridge
Keddy PA (1989) Competition. Chapman and Hall, London
Keddy PA (1992) Assembly and response rules: two goals for predictive community ecology. Journal of Vegetation Science 3: 157–164
Levin SA (1989) Challenges in the development of a theory of community and ecosystem structure and function. In: Roughgarden J, May RM and Levin SA (eds) Perspectives in Ecological Theory, pp 242–255. Princeton University Press, Princeton
Main AR (1992) The role of diversity in ecosystem function: an overview. In: Hobbs RJ (ed) Biodiversity of Mediterranean Ecosystems in Australia, pp 77–93. Surrey Beatty and Sons, Chipping Norton, NSW
McDonnell MJ and Pickett STA (eds) (1993) Humans as Components of Ecosystems. The Ecology of Subtle Human Effects and Populated Areas. Springer Verlag, New York
Mills LS, Soulé ME and Doak DF (1993) The keystone species concept in ecology and conservation. BioScience 43: 219–224
Mooney HA, Cushman JH, Medina E, Sala OE and Schultze E-D (eds) (1996) Functional Roles of Biodiversity: A Global Perspective. John Wiley and Sons, Chichester
Naeem S, Håkansson K, Lawton JH, Crawley MJ and Thompson LJ (1996) Biodiversity and plant productivity in a model assemblage of plant species. Oikos 76: 259–264
Naeem S, Thompson LJ, Lawler SP, Lawton JH and Woodfin RM (1994) Declining biodiversity can alter the performance of ecosystems. Nature 368: 734–736
Naeem S, Thompson LJ, Lawler SP, Lawton JH and Woodfin RM (1995) Empirical evidence that declining species diversity may alter the performance of terrestrial ecosystems. Proceedings of the Royal Society of London B 347: 249–262
Noble IR and Slatyer RO (1980) The use of vital attributes to predict successional changes in plant communities subject to recurrent disturbances. Vegetatio 43: 5–21
Odum EP (1992) Great ideas in ecology for the 1990s. BioScience 42: 542–545
Ostfeld RS, Pickett STA, Shachak M and Likens GE (1997) Defining the scientific issues. In: Pickett STA, Ostfeld RS, Shachak M and Likens GE (eds) The Ecological Basis of Conservation: Heterogeneity, Ecosystems and Biodiversity, pp 3–10. Chapman and Hall, New York
Pahl-Wostl C (1995) The Dynamic Nature of Ecosystems. Chaos and Order Entwined. Wiley and Sons, Chichester
Paine RT (1995) A conversation on refining the concept of keystone species. Conservation Biology 9: 962–964
Peters RH (1991) A Critique for Ecology. Cambridge University Press, Cambridge
Pickett STA, Kolasa J and Jones CG (1994) Ecological Understanding: The Nature of Theory and the Theory of Nature. Academic Press, New York
Pickett STA and Ostfield RS (1995) The shifting paradigm in ecology. In: Knight RL and Bates SF (eds) A New Century for Natural Resources Management, pp 261–278. Island Press, Washington, DC
Pickett STA, Parker VT and Fiedler P (1992). The new paradigm in ecology: implications for conservation biology above the species level. In: Fielder P and Jain S (eds) Conservation Biology: The Theory and Practice of Nature Conservation, pp 65–88. Chapman and Hall: New York
Schulze E-D and Mooney HA (eds) (1993) Ecosystem Function of Biodiversity. Springer Verlag, Berlin
Shrader-Frechette KS and McCoy ED (1993) Method in Ecology. Strategies for Conservation. Cambridge University Press Cambridge
Simberloff D (1982) A succession of paradigms in ecology: essentialism to materialism and probabilism. In: Saarinen E (ed) Conceptual Issues in Ecology, pp 63–99. Reidel, Boston
Smith TM and Shugart HH (eds) (1996) Plant Functional Types. Cambridge University Press, Cambridge
Stone R (1995) Taking a new look at life through a functional lens. Science 269: 316–317

Tilman D and Downing JA (1994) Biodiversity and stability in grasslands. Nature 367: 363–365
Tilman D, Wedin D and Knops J (1996) Productivity and sustainability influenced by biodiversity in grassland ecosystems. Nature 379: 718–720
Turner MG and Gardner RH (eds) (1991) Quantative Methods in Landscape Ecology. The Analysis and Interpretation of Landscape Heterogeneity. Ecological Studies. Springer-Verlag, New York
Vitousek PM, Mooney HA, Lubchenco J and Melillo J (1997) Human domination of Earth's ecosystems. Science 277: 494–499
Walker BH (1992) Biodiversity and ecological redundancy. Conservation Biology 6: 18–23
Walker BH (1995) Conserving biological diversity through ecosystem resilience. Conservation Biology 9: 747–752
Weiher E and Keddy PA (1995) The assembly of experimental wetland plant communities. Oikos 73: 323–335
Woodward FI and Cramer W (1996) Plant functional types and climate change: introduction. Journal of Vegetation Science 7: 306–308

Agroforestry Systems **45**: 57–79, 1999.

The dehesa system of southern Spain and Portugal as a natural ecosystem mimic

R. JOFFRE*, S. RAMBAL and J. P. RATTE
*Centre d'Ecologie Fonctionelle et Evolutive CNRS, 1919 Route de Mende, F-34293 Montpellier Cedex 5, France (*Author for correspondence: joffre@cefe.cnrs-mop.fr)*

Key words: hydrological equilibrium, Mediterranean climate, optimality, *Quercus ilex*, *Quercus suber*, water balance

Abstract. The dehesas of the southwestern Iberian Peninsula are 'man-made' ecosystems characterised by a savannah-like physiognomy. The trees are viewed as an integrated part of the system, and as a result are planted, managed, and regularly pruned. Palynological and historical evidence of the manipulation of initial ecosystems by man to obtain a savannah-like ecosystem is presented. The ecological functions of the tree are detailed using results obtained at two complementary scales. At the local scale, strong soil structural differences and functional differences in water budget and patterns of water use are observed under and outside the tree canopy. Using the concept of ecosystem mimicry, the two coexistent components of dehesas can be compared to two distant stages of a secondary succession characterised by very different behaviours. At the regional scale, evidence of relationships between tree density and mean annual precipitation over more than 5000 km^2 suggests that the structure of these man-made agroecosystems have adjusted over the long-term and correspond to an optimal functional equilibrium based on the hydrological equilibrium hypothesis. Finally, the future of dehesas in the face of contemporary exogenous threats of economic and global environmental origin is discussed.

1. Introduction

Agroforestry systems occupy an important place in the Mediterranean Basin where they have influenced the landscape for many centuries. Despite the diversity of the ecological, economic and social conditions existing in these areas, it is possible to distinguish some of their features that are common to all such systems. Firstly, this involves the profound influence of rural history in shaping the landscape and vegetation in the Mediterranean region. The variety of land tenure and land management systems in these areas has led to the extreme diversity of agrosylvopastoral systems at the present day (Joffre et al., 1991). The second fundamental characteristic of these systems is that they have to cope with the high variability of the Mediterranean climate, in other words they reflect a capacity to adapt to natural conditions and to overcome unpredictability. In ecological terms, which are what interests us here, this is reflected in the adoption of different strategies of using the plant resources corresponding to the main growth-forms. For example, in many livestock systems, forage resources are not just composed of herbs, but include trees that are used not only as a resource in their own right but also as a regulator of hydrological stress for the underlying herbaceous stratum.

A very peculiar agroforestry system, named dehesa in Spain and montado in Portugal, dominates the landscape of the southwestern Iberian Peninsula (Joffre et al., 1988b; San Miguel, 1994; Gómez Guttierez and Pérez Fernández, 1996). Characterised by the presence of a savannah-like open tree layer, mainly dominated by Mediterranean evergreen oaks – holm oak (*Quercus ilex*) and cork oak (*Q. suber*) – and to a lesser extent by the deciduous *Q. pyrenaica* and *Q. faginea*, they occupy more than 5,800,000 ha in the western and south-western provinces of Spain, representing 52% of total utilised agrarian area within these province (Campos Palacin, 1992) and more than 500,000 ha in southern Portugal.

In this paper we present some complementary points of view dealing with the origin, the history, and the present functioning of these dehesa systems. It seems unquestionable that the dehesa is a man-made ecosystem and that its two layered structure, tree and grass, is dependent on human practices and management. Nevertheless, some characteristics, particularly tree density, seem partly controlled by edaphic and climatic resources. Many empirical studies conducted on tropical savannahs have emphasised the importance of water resources in regulating the balance between woody plants and herbaceous vegetation (see Joffre and Rambal, 1993; Scholes and Archer, 1997). From a theoretical point of view, Eagleson (1982) developed hypotheses on patterns of plant water use at various time-scales (see Hatton et al., 1997 for a comprehensive account) and demonstrated that where a marked seasonality in water availability occurs, a mixed formation of grasses and woody plants is the only stable equilibrium state. These ecological optimality hypotheses lead to the prediction of an optimum tree density that is stable with respect to perturbations and metastable with respect to climatic variability. The structure of dehesas mimics those of tropical savannahs. We hypothesise that this structure, particularly present tree density managed from a long-term perspective by the farmers, follows an ecohydrological equilibrium *sensu* Eagleson (Eagleson, 1982; Eagleson and Tellers, 1982) and could be interpreted as the result of an ecological optimality, the distribution of tree density being indicative of its control by water availability.

2. Some definitions and key-characteristics

The Spanish term dehesa, the equivalent of the Portuguese montado, depicts both a landscape and a type of land use in the Iberian Peninsula. A brief history of the word dehesa in relation to changes in land utilisation will be given in section 4. Today, the dehesa can be defined as a multi-purpose agroforestry system with scattered oak trees planted by man (Joffre et al., 1988b).

The traditional system was highly diversified in terms of livestock types (sheep, goats, Iberian pigs, and cattle). Recent changes, since 1965, have led to an increase of cattle and decrease in sheep production (Joffre et al., 1988b). Pigs graze the seasonal acorn production, between October and February, gaining about 60 kg live weight during 75 days. Ruperez (1957) reported that 9 kg of *Q. ilex* acorns corresponds to the production of 1 kg of pork meat. Today, the grazing of the high value Iberian pig is the most profitable component of the dehesa system.

The herbaceous layer is comprised of either cultivated cereals (oats, barley, wheat) or, more commonly, native vegetation dominated by annual species, which are used as grazing resources. Control of invasion of the dehesa by shrub species, matorral (mainly *Cistus ladaniferus*, *C. salviaefolius* and *C. monspeliensis*), has been traditionally composed of two complementary aspects: 1) manual uprooting in the central areas and 2) the clearing and ploughing of peripheral areas by landless peasants possessing a plough horse. In exchange for their labour, the owner would give these peasants permission to cultivate cereal crops on the newly reclaimed land and to have a share in the charcoal produced from the cleared vegetation. This was made economic, despite the low productivity, by the extremely low salaries paid to the workforce. At the end of the 1950's, the manner in which the dehesas were commercially exploited was thus broadly similar to that which occurred a century earlier. Today, this control is entirely performed through mechanical work.

The trees are viewed as an integrated part of the system, and as a result are planted, managed, and regularly pruned. Several prunings are done during the life of the oaks (Ruperez, 1957; Montoya, 1988). Cork oak is pruned quite differently to holm oak. In the first case, the main product of the tree is the cork whilst in the second, the aim is to maximise acorn production. Ruperez (1957) describes the successive pruning of holm oak: during the early stages, a form pruning is conducted in order to obtain a crown with three main vertical branches. This favours the horizontal spread of the fruiting branches. Maintenance pruning every seven to nine years allows opening of the crown and sustainability of acorn production. Production pruning provides firewood and browse during severe droughts, reducing leaf area.

The holm oak trees were selected for the production of sweet acorns as a high quality stock feed. Holm oaks have a high direct value as fodder crop, providing acorns and leafy branches. Cork is the main resource of *Q. suber*. Several other secondary resources such as firewood, charcoal and tannin are nevertheless still important in the dehesa economy.

3. A palaeoecological and archaeological point of view

The intriguing problem of the origin of this very peculiar landscape of south-western Iberian Peninsula has not been investigated over the entire area concerned. In the southern part of the dehesa distribution (province of Huelva), palynological work by Stevenson (1985a,b, 1988) and Stevenson and Moore (1988) allow reconstruction of the vegetation changes over the period 4500 BC to 1900 AD. The first anthropic action was the clearing of initial oak and pine forests around 4500 BP, coinciding with high occurrences of *Vitis* pollen. This event could represent a period of early viticulture and agricultural intensification. Using pollen diagrams, charcoal fragment analysis, radiocarbon dating of the cores and modern pollen rain to establish analogy between fossil data and dominant vegetation-types, Stevenson and Harrison (1992) were able to define and describe six major phases of land-use for this region (Figure 1): 'the simple prehistoric dehesa established in Phase I, and developed in Phase II, is followed by a millenium of deforestation (Phase III), after which a more developed dehesa succeeds in the early Iberian period, to last throughout the Roman, Visigoth, Arab and Medieval periods (Phase IV). In the early modern period, pine and oak/ash forests characterise Phase V and VI'. In phase II, the peak of oak pollen is accompanied by high occurrences in the ruderals *Rumex* and *Plantago* indicating a strong modification of the understorey with shrub clearing and invasion by herbaceous species: 'it looks like the beginning of a true oak dehesa but with more limited management than in the sophisticated classical and Medieval dehesa of Phase IV'. Phase IV (500 BC–1200 AD) sees a dehesa system, now matured into a permanent feature of the landscape maintained for about 1700 years, coinciding with the establishment of urban civilisation. The increase of pine trees during the Phase V could be the result of plantation for naval supplies. Pine plantations were established in the Marismillas region (80 km west of the Stevenson study

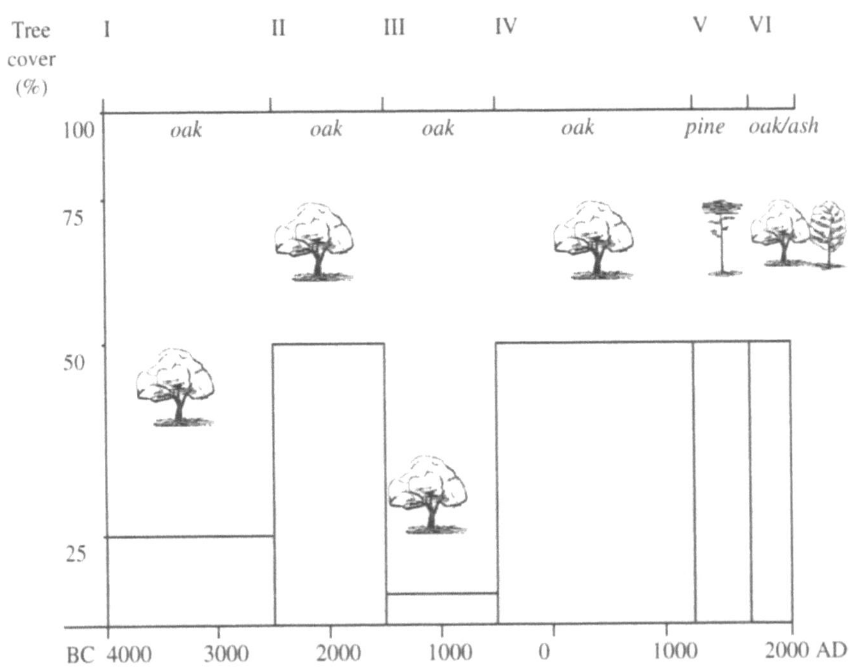

Figure 1. Palaeoecological history of some landscapes of the Huelva province, south-west Spain. Source: Stevenson and Harrison (1992).

site) from 1737, with a very rapid increase in the planted area (Granados Corona et al., 1984). Finally dehesas were re-established in the last Phase with plantations of ash (*Fraxinus* sp.) to provide browse for plough oxen.

4. An historical point of view

During the historical period, numerous written sources enable a precise meaning of the word dehesa and the formation of the corresponding landscapes. The majority of the following information on the semantic history of the word dehesa comes from Vacher (1984). According to various sources, such as the Diccionario Enciclopédico de la Agricultura, Ganaderia e Industrias Rurales (1895), the Codigos Antiguos de Espana (1885) and the Diccionario Etimologico (1977), the terms dehesa and dehesado appeared firstly in the oldest Spanish code, the Visigothic Fuero de Juzgo (VIIth century) in a law concerning the protection of fields against free ranging.

During the reconquest of southern Spain by the kings of Castile (VIIth to XVth centuries), the word dehesa specified rested grazing lands conceded by the King to military and religious orders and civil communities. This official protection was granted to avoid overgrazing by the numerous unauthorised herds of the transhumance (Mangas Navas, 1982). In the mid XIVth century, more than 3 million sheep moved each winter from Castile to southwestern Spain. The powerful breeders' organisation, *Mesta*, obliged the shepherds to keep the herds away from cereal fields, gardens, vineyards, stubble and dehesas (Klein, 1920).

The dehesas were of considerable importance for the economical autonomy of the Castile villages (Mangas Navas, 1982). These dehesas represented a generally large fraction of the territory of the villages. Royal protection was more easily conceded to civil communities than to military and religious orders (Cabo, 1978). Not only were these dehesas strictly defined, their use was restricted to specified types of livestock: the *dehesa boyal* was only grazed by plough oxen, the *dehesa yegual* by mares and foals, and the *dehesa carniceril* by animals assigned to slaughtering.

In the course of centuries, the reference to grazing, initially absent in the first definitions of the word, become more and more important. This semantic evolution corresponds to a strong modification of agricultural equilibrium from the XVIIth century. The transhumance declined, and grazing lands became seriously threatened with cereal cropping (Caxa de Leruela, 1630). This short insight does not explain the absence in all the meanings of the word dehesa of any criteria connected to the trees. Nevertheless, there are numerous references since the Roman period to the importance of the oaks. Strabo and Pline (in Bauer, 1980) emphasised the role of sweet acorns both for human food consumption and to feed domestic animals during the Roman period (see also section 3 for previous periods). Later, the importance to the farmers of exploiting the fodder resources of the trees was recognised by law (Fuero Real de Espana, 1273). The Catholic Kings ordinances, from the Xth century, favoured restoration and conservation of the trees for animal and human uses (Bauer, 1980). From the Xth to the XVIIIth centuries, various laws recommended the reforestation of dehesas with holm oak and cork oak. In 1716, the King Felipe V ordered local authorities to protect and replant the south western Iberian dehesas (Mangas Navas, 1982). It is of interest to point out that very often the ownership of the land and of the trees was distinct (Humbert, 1980). This juridical distinction between soil and tree was frequent in Mediterranean civilisations.

The presence of the trees is indicated in the first Spanish property register, the famous Catastro de la Ensenada, drawn up between 1750 and 1760. At the same time, maps of the Maritime Provinces of Spain were drawn giving an extraordinarily precise inventory of the trees and their location (Prat, 1754; Gomez Cruz, 1991) (Figure 2). This survey was primarily undertaken as an inventory of timber for naval supply at a time when naval construction was very active and important. A law was introduced in 1748 specifying that every tree growing at less than 25 leagues (i.e. 138 km) from the coast or from a navigable river was under the authority of the navy. As a consequence, cutting a tree was only permitted by licence. The abundance of the trees is next highlighted in Madoz' Diccionario Geográfico-Estadístico-Histórico covering all Spanish continental and overseas territories (Madoz, 1845; Sanchez Zurro, 1986). This dictionary pointed out the abundance of holm oak and cork oak for the Sierra Norte of Sevilla and more generally for the Sierra Morena.

5. An ecological point of view

The principal ecological constraints acting on the dehesa ecosystem can be grouped under three headings:

1. water resource;
2. nutrient availability; and
3. human management.

These constraints are highly modified by the unique vegetative structure of the dehesas, which is made up of two ecological components. The first, located outside the tree canopy, is composed of herbaceous plants, while the second, is composed of the oak tree and its associated herbaceous stratum. This difference in structure leads to predictable variation in the relative importance of the above three constraints. For the tree-grass component, the climatic constraint is generally considered to be of primary importance. This discussion focuses on what is currently known about the functional ecology of the tree grass component principally using research conducted in the dehesas of the Sierra Morena (Andalusia) as a model (Joffre and Rambal, 1988a, 1993; Joffre, 1996; Infante et al., 1997).

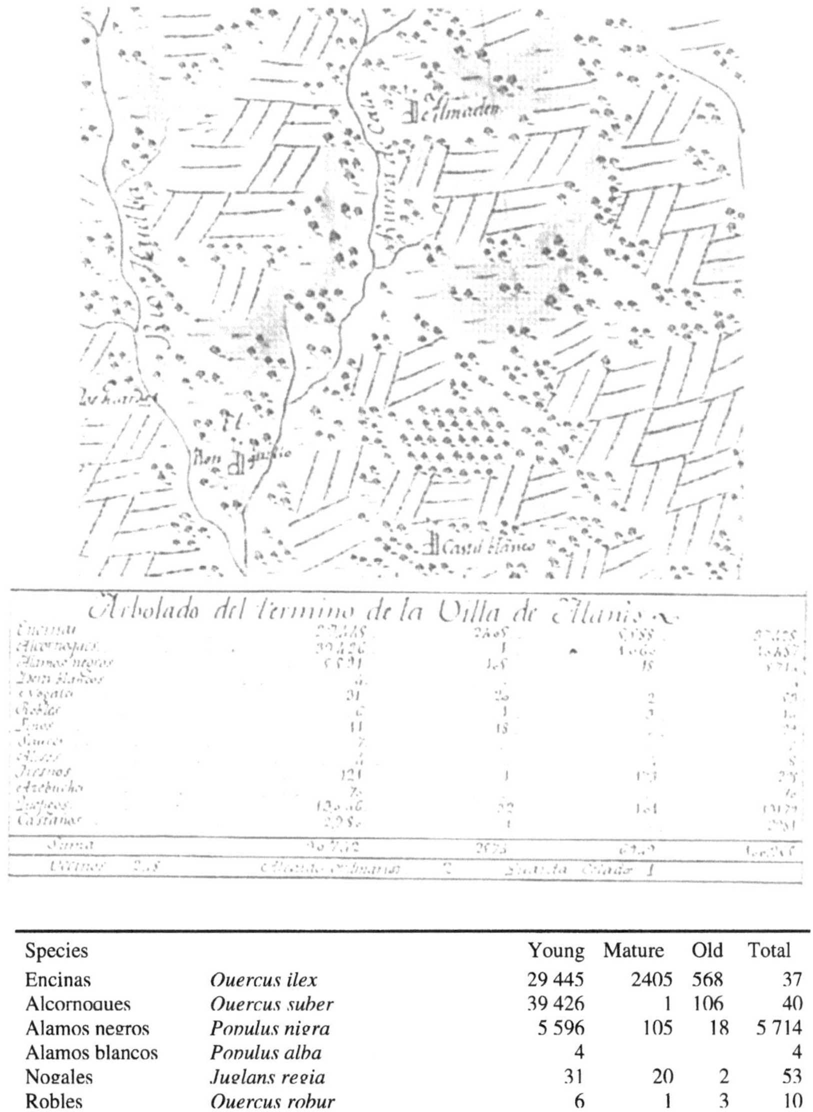

Species		Young	Mature	Old	Total
Encinas	*Quercus ilex*	29 445	2405	568	37
Alcornoques	*Quercus suber*	39 426	1	106	40
Alamos negros	*Populus nigra*	5 596	105	18	5 714
Alamos blancos	*Populus alba*	4			4
Nogales	*Juglans regia*	31	20	2	53
Robles	*Quercus robur*	6	1	3	10
Pinos	*Pinus sp*	11	18		29
Sauces	*Salix fragilis*	7			7
Alisos	*Alnus glutinosa*	4		4	8
Fresnos	*Fraxinus angustifolia*	121	1	173	295
Acebuches	*Olea europea ssp. silvestris*	70			70
Quejigos	*Quercus faginea*	13 046	32	101	13
Castaños	*Castanea sativa*	2 980	1		2981

Figure 2. Excerpt from 'Carta Geográfica o Mapa General de los pueblos, montes y sus principales arboledas de la provincia de marina de Sevilla', Antonio Prat 1754 (General Map of the villages and tree layer for the province of Sevilla) and the corresponding tree survey from the shire of Alanis, Sevilla, Spain.

5.1. *Climate variability, climatic drought and actual drought*

The high variability of precipitation and the length of the summer drought are among the most important climatic factors affecting vegetation function. Mean annual precipitation at Sevilla is 572 ± 176 mm but varies from less than 400 mm to more than 1000 mm (Figure 3). Several occurrences of three or four consecutive dry years can be observed over the period 1865–1993. The summer drought, defined as the interval between the last day of spring and the first day of autumn with rainfall events higher than 5 mm or 10 mm, has a mean value of about 120 days (Figure 4), while a summer drought of 150 days has a return period of 4.25 years (Joffre and Rambal, 1993).

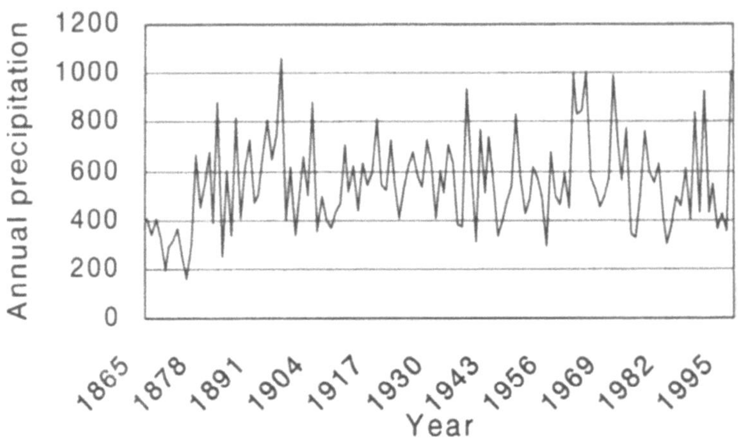

Figure 3. Precipitation records for Sevilla, Spain from 1865 to 1990.

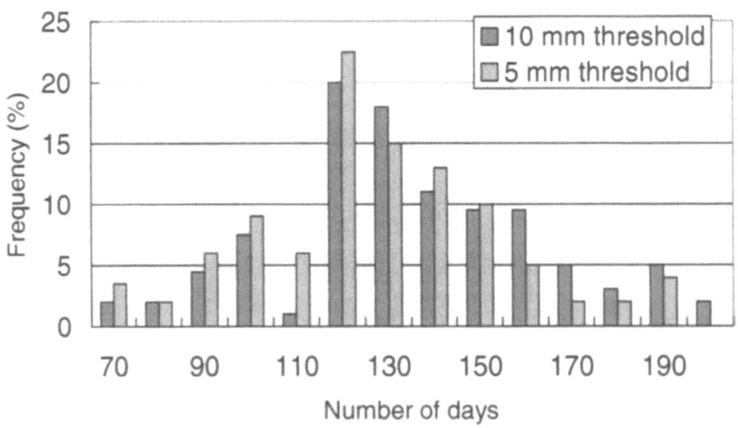

Figure 4. Duration of summer drought period (expressed in days) at Sevilla, Spain for the period 1922–1984.

Moreover, the absence of statistical relationship between annual rainfall and summer drought highlights the unpredictability of rainfall.

The water stress experienced by a plant is not only dependent on climatic conditions but depends also on edaphic conditions. Rambal (1984a) proposed a numerical approach to actual drought using both Penman potential evaporation (E_p) and maximal soil water reserve R_{max}:

$$R_{(i+1)} = \text{Min}[(1 - E_p/R_{max})\, R_{(i)} + P,\, R_{max})$$

where R is the soil water reserve (expressed in mm), E_p Penman potential evaporation (mm) and P precipitation (mm). We calculated the seasonal time-course of actual soil water on a 10 day basis over the 1975–1995 period based on a maximal soil reserve of 250 mm (Joffre and Rambal, 1993).

The soil water reserve reached this maximal value in less than half the years, and during two periods of four consecutive years there was a strong soil water limitation (Figure 5). The same result was obtained over several meteorological stations of southwestern Spain indicating the spatial homogeneity of the precipitation regime (Font Tullot, 1983). This means that the persistent woody elements of the dehesa (the trees) must be able to overcome recurrent multi-year drought periods without mortality. What are the mechanisms involved? Results obtained at two complementary scales will give us some pertinent information on: 1) how tree cover influences soil properties and water balance at the local scale and 2) whether variation in tree density reflects changes in water resources at the regional scale.

5.2. *Local scale*

Considering the dehesa as a juxtaposition of two ecological components, it is possible to compare soil characteristics and patterns of soil water storage and

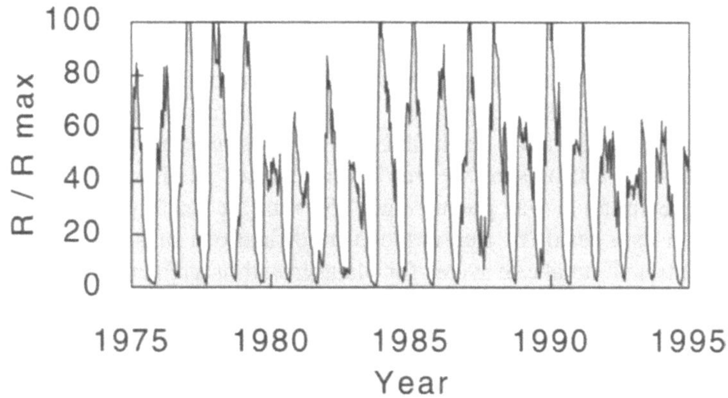

Figure 5. Time-course of simulated soil water reserve R expressed as percent of the maximum soil water reserve R_{max} for the period 1975–1995, Sevilla, Spain.

to quantify the water balance of these two components. The presence of isolated trees was shown to cause marked variations in physical properties such as dry bulk density (Joffre and Rambal, 1988a). Soils under the tree canopy were richer in nutrients and organic matter (Vacher, 1984; Escudero, 1985; Joffre, 1987). It should be noted that nutrient enrichment under the canopies was a feature of various mechanisms. Trees act as nutrient pumps, drawing nutrients from deep horizons and laterally from areas beyond the canopy, depositing them beneath the canopy through litter decomposition (Scholes and Archer, 1997). Another mechanism linked with animal behaviour seeking shade may be of importance. Transfer of nutrient under the canopies by livestock through dung deposition certainly plays a role in enhanced fertility. It has been found in three dehesas of the Sierra Morena region that the dry bulk density of the upper soil layer and moisture retention curves were significantly different and soil water content was always significantly higher under tree cover than in open areas (Joffre and Rambal, 1988a). The increase in maximum soil water storage of soils, calculated for the first 150 cm, estimated from the difference between field capacity and the driest profile observed, ranged from 40 to 110 mm under the trees. This allowed important time lags in late spring in the drying of the upper soil layers. The lowest soil water content for a given depth was reached more than one month later under tree canopies than outside. This buffering effect of the trees allows a longer growing season for understorey grasses.

Differences in water budget between the two components of the dehesas were highlighted by estimating monthly and annual time-scale water balance (Joffre and Rambal, 1993). Mean annual actual evaporation (*Ea*), estimated over three consecutive years, was 400 mm and 591 mm outside and under the trees respectively. This difference in water uptake between the two ecological components of the dehesas is in agreement with previous results on shrub and grass vegetation in the Mediterranean region. In the open areas, losses due to drainage and surface runoff were large (ranging from 65 to 100% of the *Ea*) whereas under the tree canopy losses represented only 20 to 40% of the *Ea*. Linear regression of *Ea* against annual rainfall (*P*) showed a significant correlation for the grass component but not for the tree-grass component. Under the tree canopy, when annual precipitation was less than 570 mm, water yield was negligible and all precipitation was lost by evaporation. In contrast, outside the tree canopy, water yield occurred as soon as annual precipitation was greater than 250 mm, i.e. each year. These results show that as a result of tree-induced modifications of soil properties, the availability and uptake of water for plant growth is greater for the tree-grass component.

The strong soil structural differences and the functional differences in water budgets and patterns of water use observed between the two components are coarsely similar to those found between early and late stages of succession. Using the concept of ecosystem mimicry, the two coexistent components of dehesas can, in fact, be compared to two distant stages of a secondary succession characterised by very different behaviours. Due to the long human history of these savannah-like man-made ecosystems, it can be argued that the modifications of functional properties of the tree-grass component are a result of the improvement of ecosystem function due to tree establishment and maintenance over many centuries. The situation is quite distinct from that of clearing a forest to obtain a similar landscape, where the modification of soil properties occurs more in the cleared areas and normally leads to impoverishment of soils, a decrease of organic matter and degradation of physical and chemical properties such as acidification. The trees in the dehesas cannot be considered as remnant vegetation of an hypothetical initial forest but as a structural element acting as an ecological engineer over many centuries (see section 2) in a complex man-made agroforestry system. Several studies have shown that the nutrient enrichment in tree-dominated patches of savannahs was strongly related with tree age (Bernhard-Reversat, 1982; Barth, 1980).

5.3. *Regional scale*

Based on the hypothesis that under seasonally water-limited resources, vegetation systems will develop a canopy density that produces both minimum water stress and maximum biomass, the theoretical work of Eagleson (Eagleson, 1982; Eagleson and Tellers, 1982; Eagleson and Segarra, 1985) has emphasised that a mixed formation of grasses and woody plants, i.e. savannah, is the only stable equilibrium state. These functional models permit the prediction of tree density according to the level of water resources. This concept of water resources integrates evaporative demand, amount and temporal pattern of precipitation and hydrodynamic properties of the soil. Hence, the distribution of woody plants is indicative of control by moisture availability and is a key to the conceptualisation of an ecosystem-level model of oak savannahs. In Eagleson's theory, ecosystem stability, which is characterised by the persistence of the tree layer, seems to be more fragile when climatic changes are significant, particularly when they involve a decrease of precipitation. In order to verify this hypothesis, we selected a large region of southern Spain encompassing a gradient of annual precipitation and dominated by dehesa landscapes (Figures 6 and 7). This area was predominantly grazed with occasional cropping and the dehesas were subjected to the same management system over the entire region. Consequently, the researched relationships between structure (i.e. tree density) and ecological resources would not be obscured and/or counterbalanced by local modifications of land management. Comparison of tree density and climatic resources was achieved through GIS methodology. Over this zone, a map of precipitation was overlaid with tree density mapped by remote sensing. The methods we used are briefly described below.

Figure 6. Dehesa stand with a density of 45–50 trees ha^{-1} in a 750 mm annual rainfall area, Sevilla, Spain.

Figure 7. Typical landscape with density < 10 trees ha^{-1} in a 500 mm annual rainfall area, Sevilla, Spain.

5.3.1. *Study area and regional estimation of precipitation*

A study was conducted over a 250 km × 150 km part of Sierra Morena and Meseta of southern Spain (Figure 8). Monthly and annual precipitation from 173 weather stations located in this zone were averaged over the period 1951–1991. The spatial distribution of precipitation was investigated using a geostatistical analysis. In its simplest form, this procedure involved a three-step process:

1. defining the semi-variogram, i.e. the degree of spatial correlation among the data;
2. interpolating values between measured points based on the degree of spatial correlation encountered (see Webster and Oliver, 1990 for a comprehensive account);
3. mapping the estimates.

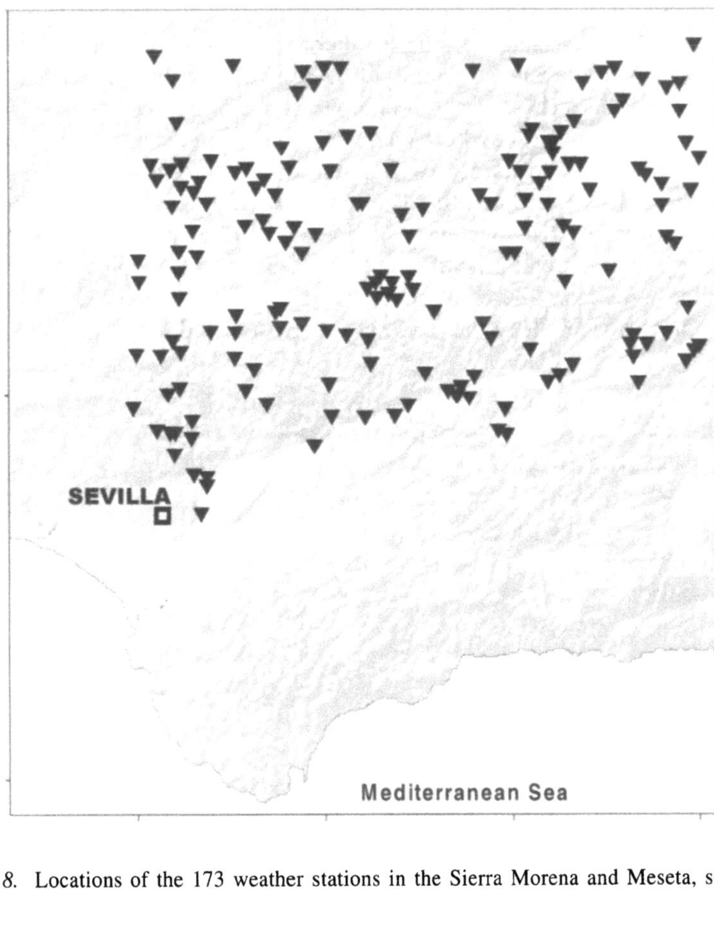

Figure 8. Locations of the 173 weather stations in the Sierra Morena and Meseta, southern Spain.

The experimental semi-variogram obtained for our set of data have been fitted to bounded spherical models. The second step uses semi-variogram parameters to interpolate values for points not measured using kriging algorithms (Trangmar et al., 1985). Maps were based on these kriged data provided by Geopack and drawn using the software program Surfer.

5.3.2. *Tree density estimation*

The estimation of tree density using the highest spatial resolution data (SPOT panchromatic = 10 m resolution) is based on the assumption that for sparse trees on a contrasting herbaceous background, spatial filters could provide a direct mapping of tree cover. Four scenes of SPOT-HRV data (scenes 31-274, 31-272, 33-273 and 35-272) were used. These scenes were selected in order to cover the range of annual precipitation in this region (Figure 9). Aerial photographs at the scale of 1/40,000 are available for part of the study area (data acquisition in October 1984). In addition to these data, several ground observations were made from 1983 to 1995. In each scene, two test areas of approximately 10 × 10 km, corresponding to 1024 × 1024 SPOT panchromatic pixels were selected. Using a supervised approach, homogeneous test sites representative of most common tree densities have been defined on the basis of visual inspection of aerial photographs and the trees directly counted on aerial photographs. The SPOT data was processed to enhance local contrast between one pixel and its nearest neighbours. According to the size of tree crowns (less than 10 m) and the distance between trees, three by three pixel windows were selected. A numerical filtering approach involving two edge-enhancement filters and a level-slicing procedure was adopted for discriminating the trees from the background (Joffre and Lacaze, 1993). The result was a binary image where the value of 0 corresponded to background pixels,

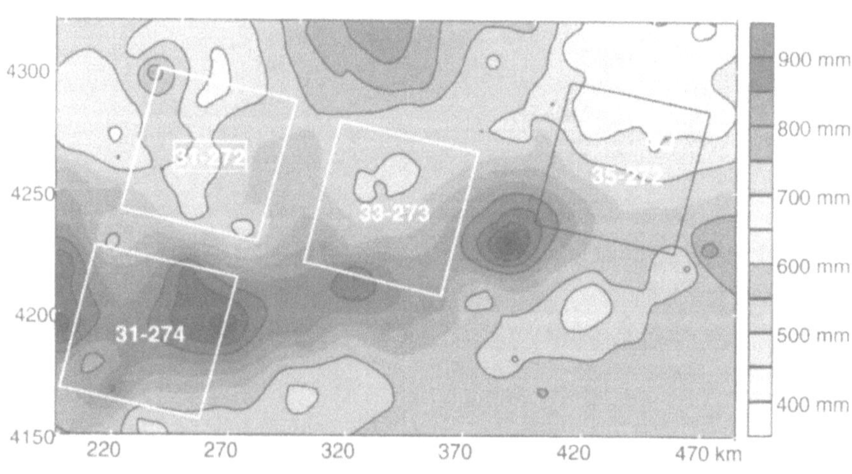

Figure 9. Locations of the SPOT scenes on a map of mean annual rainfall for part of Sevilla province, Spain.

and one for pixels belonging to class tree. So, for each test site, it was then possible to compare directly the mean value of remote estimation of trees and the number of trees counted from aerial photographs. This calibration procedure was then validated on other distinct sites and finally applied to the entire scene. Resulting maps of tree density were obtained by averaging pixels over three by three pixel windows.

5.3.3. *Looking for a functional equilibrium*

Over the studied zones, mean annual precipitation ranged from 500 mm to more than 800 mm, covering semi-arid to subhumid conditions according to the Emberger's classification of Mediterranean climate (Daget, 1977) (Figure 9). Geo-referenced climatic and tree density maps were overlaid over 10,800 km^2. Our hypothesis relating the structure of dehesas to water resources is only relevant for landscapes where slopes would not affect water balance. Consequently, a digital elevation model was used to select zones with slopes of less than 5% and 10%. These masks were applied over the density maps to obtain a coincidence tabulation report. In the same way, a mask was built from Corine Land Cover data (CEC, 1993) to exclude cultivation, plantation forest, water, bare soils and urban areas. Soils over the study zones are derived from acidic bedrocks (mainly granites and schist). These do not exhibit significant differences in physical properties, so we did not consider the soil type in our sampling strategy and excluded this layer from our co-variant analysis. The resulting database used for the coincidence tabulation analysis covered 5421 km^2 when excluding slopes higher than 5% and 5942 km^2 when excluding slopes higher than 10%.

The two thresholds of 5% and 10% of slopes for the pixels eliminated for the analysis gave very close results. The distribution frequency of tree density along classes of annual rainfall (Figure 10) showed a very clear trend. As rainfall increased, mean tree density increased up to 40 trees per hectare. The distributions, asymmetric for the lower values of rainfall, tend to greater regularity as the rainfall reached 650–700 mm. For the areas where mean annual rainfall was higher than 650 mm, the highest classes of tree density (> 50 trees ha^{-1}) represented more than 10% of the dehesas.

We have seen (section 5.2) that under the tree canopy, when annual precipitation was less than 570 mm, water yield was negligible and all precipitation was lost by evapotranspiration. These results are in accordance with studies conducted in other Mediterranean climate zones (chaparral of California, Ng and Miller, 1980; garrigue in southern France, Rambal, 1984b). It seems that under 600 mm of mean annual rainfall, the tree density corresponding to a metastable equilibrium was low (< 20 tree ha^{-1}). One can interpret this result with the previous values of around 600 mm of annual evapotranspiration for the tree-grass component. Under a mean annual precipitation of 600 mm, it is possible that competition between the two components of the dehesa, the grass outside the tree canopy and tree-grass

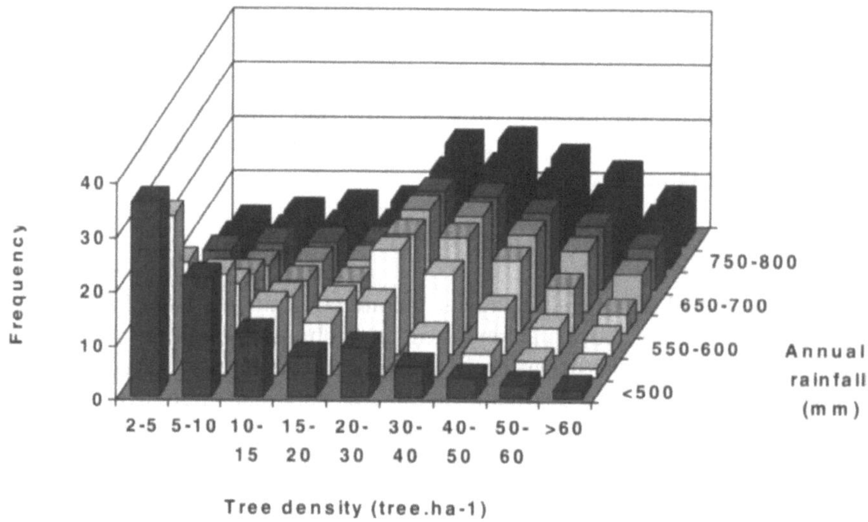

Figure 10. Frequency distribution of tree density against mean annual rainfall, Sevilla, Spain.

component, sets the boundary for the optimal equilibrium density. In these climatic conditions, the survival of the trees in the event of severe drought (three consecutive years, return frequency 20–30 years at Sevilla) is only possible if the root system of the tree can extend beyond the influence of the tree canopy (Eastham et al., 1988). An extension of the root system 0.5 m outside the projected limits of an average crown diameter of 7 m would increase the volume of soil exploited and hence the availability of water by ca. 30%.

In this zone, it appears that tree density is suboptimal as neither severe tree water stress (predawn leaf water potential lower than –2.5 Mpa) nor tree mortality was observed over a period of 15 years. That means that no tree-tree negative interaction for water use occurs. By contrast, in a closed dense woodlands of *Q. ilex* in Catalonia, a high degree of mortality was observed following a severe drought in 1994 during this period (J. Piñol and coll., pers. comm.). Better knowledge of the extension of the root system is needed to support this hypothesis about the adjustment of tree density facing severe climatic drought. Nevertheless, empirical evidence of relationships between tree density and mean annual precipitation over more than 5000 km^2 suggests that the structure of these man-made agroecosystems are adjusted over the long-term and correspond to the hydrological equilibrium hypothesis of Eagleson's theory.

6. A prospective point of view

Two issues present uncertainty for the future of the dehesas. The first is their development as an agriculturally productive system within the local socio-economic conditions and the general agricultural politics of the European Union. The second concerns the stability of this complex ecological system in the face of long-term climatic change caused by increasing atmospheric CO_2.

6.1. *Land use modifications*

Considering the evolution of the main land use types in the Sierra Norte of Sevilla, Vacher et al. (1985) showed that there was very little change in the dehesas of this region between 1750 and 1956. Between 1956 and 1977, however, dramatic changes have occurred. These have been the result of three predominant human activities:

a) a decrease in agro-pastoral management practice;
b) an increase in deforestation and clearing processes in order to extend cropping lands; and
c) reforestation.

When grazing pressure is removed, dehesas are invaded by Mediterranean matorral species (Lacaze and Joffre, 1987; Vacher, 1984), which in turn increase the fire risk. Spain and Portugal have been members of the European Union (EU) since 1986 and as a consequence subjected to the Common Agricultural Policy (CAP). Within the marginal agricultural areas, the two main regulations affecting land use changes are Regulation 2078/92 (Agricultural production methods compatible with protection of the environment and maintenance of the countryside) and Regulation 2080/92 (Community aid scheme for forestry measures in agriculture). The incentives offered to take cropping land out of production and to re-aforest marginal agricultural land could promote a strong and rapid land use change (Anon., 1992; Lawson, 1996; Ridley, 1997). An ambitious re-aforestation program concerning the region with large dehesa areas was implemented in 1995. Regulation recognises that premiums are necessary to compensate landowners for the loss of income during the non-productive period of aforested agricultural land. In some parts of the dehesa areas, abandonment of grazing activity for re-aforestation program could be interpreted as a strong decline of sustainability of the system as the promoted forest schemes do not take into account the difficulty of establishment and maintenance of the trees planted at higher density in this very variable and harsh environment.

6.2. *Stability of vegetation functional equilibrium facing changing climate*

For southern Europe, a general consensus of simulation models (Global Circulation Models) is that a doubling of atmospheric carbon dioxide could lead to a decrease of precipitation and, more importantly, an increase in the probability of extreme events such as severe drought. In Europe, Gregory and Mitchell (1995) showed that in middle latitudes (35–40° N), the average decrease in total annual rainfall is associated with a decrease in the number of rainy days. Moreover, this decrease is associated with a decrease in the frequency of low rainfall days. This means that there will be an increased frequency of dry spells despite an increase in the number of heavy rain events. Rambal and Hoff (1998) showed in a simulation exercise based on contemporary data at Puéchabon (10 km west of Montpellier) that a change in daily rainfall distribution, based on a strong increase in the frequency of occurrence of the largest rainfall class and a decrease in the amount and frequency of low rainfall, results in a slight decrease of annual rainfall (–3%) and evapotranspiration (–5.5%). More interestingly, they showed that the critical period defined for *Q. ilex* by a leaf water potential threshold fixed here at –4.2 MPa, corresponding to the water potential at which the loss of xylem conductivity is approximately 90% (Lo Gullo and Salleo, 1993), increased by 16%. In addition, results of the GCM of the Hadley Center for Climate Prediction (UK Meteorological Office) based on a 1% annual gradual increase, resulting in a doubling atmospheric CO_2 after 70 years (Murphy, 1995; Murphy and Michell, 1995), showed a decrease of about 80 mm for the annual rainfall of Sevilla in 75 years (Figure 11). This decrease would have dramatic consequences in

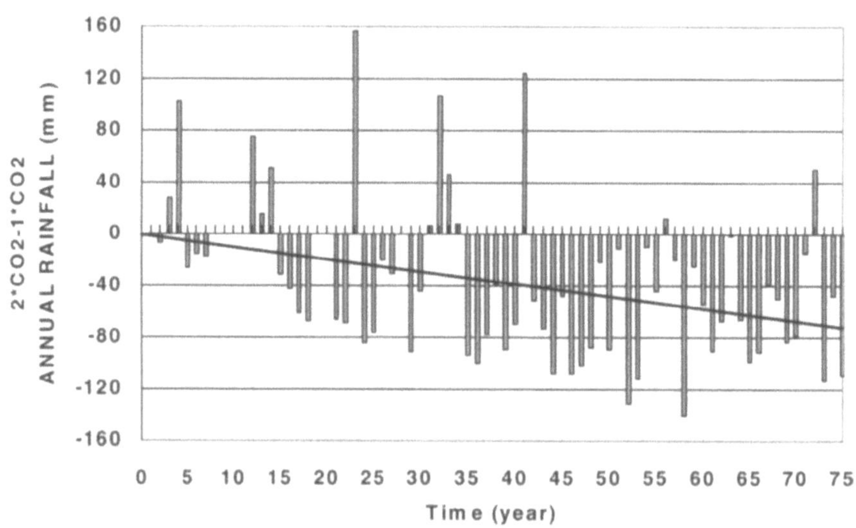

Figure 11. Changes in annual rainfall amounts for Sevilla, Spain, using the transient simulation model. The changes are with respect to the full 75-year control means.

the optimal distribution frequency of the trees for all the regions of dehesas where present annual rainfall is lower than 700 mm, as shown by Figure 10. The foreseeable changes in both the total amount and distribution of precipitation would therefore affect the structure of 88% of the studied region. A regulation mechanism could nevertheless be proposed. Elimination of individual trees which experience the most severe climatic stress would contribute to both a redistribution of water resources in favour of the remaining trees and a new functional equilibrium for the long-term persistence of the community.

7. Conclusion

The historical perspective clearly shows the influence of farmers' long-term decisions in shaping the dehesa landscapes. To quote Ridley's (1997) investigation of the sustainability of Iberian agroforestry systems, 'the concept of time scales is very marked in European agroforestry systems, illustrated by the Portuguese traditional saying "you plant vines (or crops) for yourself, olives for your children and cork oak for your grandchildren"'. It should be noted equivalent sayings also exist in Spain and Languedoc. The most intriguing and interesting characteristic of the present dehesas is the existence of a spaced tree component with an important productive function. We have not considered the composition of the grass layer as it is more directly related to particular practices (year-long continuous grazing, periodic ploughing) and could be modified without greatly altering the functioning of these system (Joffre et al., 1988, 1991). It should be noted that the present productivity of this layer seems low compared to similar systems in Mediterranean-type climate, either in Europe or in Australia. Nevertheless, various native species of high agronomic interest, legumes and perennial grasses, are remarkably adapted to this harsh environment and grow naturally in the dehesas. The potential of the local native flora is so interesting that for many years Australian scientists have collected and improved numerous pasture species from the Mediterranean Basin. A sustainable increase in total productivity would need to be based on optimisation of the multiple plant (forage, acorns, firewood, charcoal) and animal (sheep, cattle, pigs) components. Where short-term maximisation of the herbaceous layer has been the major goal, as in the case in Australian crop and pasture systems, large land degradation problems have resulted. Soil acidification and salinisation have been attributed to the high intensity of production and low water use of agricultural species compared with the native vegetation they replaced. The profitability of the dehesas has been based on the diversity of products and permitted the long-term ecological sustainability by a sub-optimisation of the resources for many centuries. This is still more evident when considering all the ecosystem services provided by the dehesas to the human population (Lefroy et al., 1993; Chapin et al., 1997), in addition to the farmers.

To gain an ecological perspective, we applied a simple monofactorial analysis to the present tree structure and function of the dehesa, based principally on water resources. Two distinct areas within the dehesa zone were defined, apparently determined by different factors. The first corresponds roughly to the areas receiving less than 650–700 mm annual rainfall, the second to areas receiving more than 650–700 mm. For the drier areas, results showed that tree density is related to rainfall amount and seems to conform to ecohydrological optimality theories. In the higher rainfall zone, tree density does not increase with rainfall. This pattern suggested that the dehesa structure is under the control of additional factors likely to be related to historic, socio-economic, and agronomic origins. Two possible candidates are:

1. limited labour for tree management (establishment, pruning); and
2. the choice by the farmers to maintain a more open structure to optimise grass production.

Diversity of production is characteristic of the dehesa in both components (forage, acorn, wood, cork, charcoal). The connection between suboptimal tree density and whole ecosystem production, i.e. the trade off between ecological and economic characteristics of these peculiar agroforestry systems, is the result of a long-term landscape management by local populations. Given the current exogenous threats of economic and global environmental origin, this diversity of production does not guarantee the persistence of the dehesas in the near future.

8. Acknowledgements

We gratefully thank Ted Lefroy for inviting Richard Joffre to the workshop 'Agriculture as a mimic of natural ecosystems' and all the participants for their stimulating discussions. We thank Dr D Viner of the Climatic Research Unit, School of Environmental Sciences, University of East Anglia, Norwich for providing the 75-year precipitation data of control and transient simulations from the Hadley Center GCM. This study was partly supported by European Union MOST (contract EV5V-CT92-0210) project.

9. References

Anonymous (1992) Council regulation (EEC) No. 2078/92 of 30 June 1992 on agricultural methods compatible with the requirements of protection of the environment and maintenance of the countryside. Official Journal of the European Communities NL 215/85

Barth RC (1980) Influence of pinyon pine trees on soil chemical and physical properties. Soil Science Society of America Journal 44: 112–114

Bauer E (1980) Los montes de España en la historia. Publicacion del Ministerio de Agricultura, Madrid, Spain, 613 pp

Bernhard-Reversat F (1982) Biogeochemical cycle of nitrogen in a semi-arid savanna. Oikos 38: 321–332

Cabo AA (1978) Antecedentes históricos de las dehesas salmantinas. In: Estudio integrado y multidisciplinario de la dehesa salmantina 1 Estudio fisiográfico descriptivo, pp 63–98. CEBA, Salamanca, Spain
Campos Palacín P (1992) Spain. In: Wibe S and Jones T (eds) Forests, market and intervention failures. Five case studies, pp 165–200. Earthscan, London, UK
Caxa de Leruela M (1630) Restauracíon de la abundancia antigua de España. Edicion a cargo de JP Le Flem 1975, Instituto de Estudios Fiscales, Ministerio de Hacienda, Madrid, Spain, 180 pp
Chapin III FS, Walker BH, Hobbs RJ, Hooper DU, Lawton JH, Sala OE and Tilman D (1997) Biotic control over the functioning of ecosystems. Science 277: 500–503
Commission of the European Communities (1993) Corine land cover. Guide Technique. Catalogue 15 CD-NA-12585-FR-C, Office for official publications of the European Communities, Luxembourg
Daget P (1977) Le bioclimat méditerranéen: analyse des formes climatiques par le système d'Emberger. Vegetatio 34: 87–103
Eagleson PS (1982) Ecological optimality in water-limited natural soil-vegetation systems. 1. Theory and hypothesis. Water Resources Research 18: 325–340
Eagleson PS and Segarra RI (1985) Water limited equilibrium of savanna vegetation systems. Water Resources Research 21: 1483–1493
Eagleson PS and Tellers TE (1982) Ecological optimality in water limited natural soil-vegetation systems. 2. Tests and application. Water Resources Research 18: 341–354
Eastham J, Rose CW, Cameron DM, Rance SJ and Talsma T (1988) The effect of tree spacing on evaporation from an agroforestry experiment. Agricultural and Forest Meteorology 42: 355–368
Escudero A (1985) Efectos de arboles aislados sobre las propriedades quimicos del suelo. Revue d'Ecologie et Biologie du Sol 22: 149–159
Font Tullot I (1983) Climatología de España y Portugal. Instituto Nacional de Meteorología, Madrid, Spain, 296 pp
Gómez Cruz M (1991) Atlas histórico-forestal de Andalucía Siglo XVIII Agencia de Medio Ambiente, Junta de Andalucía. Universidad de Granada, Granada, Spain, 71 pp
Gómez Guttierez JM and Pérez Fernández M (1996) The dehesas, silvopastoral systems in semiarid Mediterranean regions with poor soils, seasonal climate and extensive utilisation. In: Etienne M (ed) Western european silvopastoral systems, pp 55–70. INRA Editions, Paris, France
Granados Corona M, Martin Vicente A, Fernandez Ales R and Garcia Novo F (1984) Etude diachronique d'un écosystème à longue échelle. La pinéde de Marismillas (Parc National de Doñana. Mélanges de la Casa de Velazquez 20: 394–418
Gregory JM and Mitchell JFB (1995) Simulation of daily variability of surface temperature and precipitation over Europe in the current and 2 × CO_2 climates using the UKMO climate model. Quaterly Journal of the Royal Meteorological Society 121: 1451–1476
Hatton TJ, Salvucci GD and Wu HI (1997) Eagleson's optimality theory of an ecohydrological equilibrium: quo vadis? Functional Ecology 11: 665–674
Humbert A (1980) Le 'monte' dans les chaines subbétiques centrales (Espagne du Sud). Publications Université Paris-Sorbonne, Paris, France, 224 pp
Infante JM, Rambal S and Joffre R (1997) Modelling transpiration in holm-oak savannah: scaling up from the leaf to the tree scale. Agricultural and Forest Meteorology 87: 273–289
Joffre R (1987) Contraintes du milieu et réponses de la végétation herbacée dans les dehesas de la Sierra Norte (Andalousie, Espagne). Thèse Université des Sciences et Techniques du Languedoc, Montpellier, France, 186 pp
Joffre R (ed) (1996) Mediterranean oak forests: basis for apprpriate protection strategy-MOST project. Final Report. Directorate General XII, Comission of the European Communities, Brussels, Belgium, 363 pp
Joffre R and Rambal S (1988a) Soil water improvement by trees in the rangelands of southern Spain. Oecologia Plantarum 9: 405–422

Joffre R and Lacaze B (1993) Estimating tree density in oak savanna-like 'dehesa' of southern Spain from SPOT data. International Journal of Remote Sensing 14: 685–697

Joffre R, Vacher J, De Los Llanos C and Long G (1988b) The dehesa: an agrosilvopastoral system of the Mediterranean region with special reference to the Sierra Morena area of Spain Agroforestry Systems 6: 71–96

Joffre R, Hubert B and Meuret M (1991) Les systèmes agro-sylvo-pastoraux méditerranéens: enjeux et réflexions pour une gestion raisonnée. Dossier MAB 10, UNESCO, Paris, France, 96 pp

Joffre R and Rambal S (1993) How tree cover influences the water balance of Mediterranean rangelands. Ecology 74: 570–582

Klein J (1920) The Mesta: a study in Spanish economic history 1273–1836. Cambridge University Press, Cambridge, UK, 460 pp

Lacaze B and Joffre R (1987) Caractérisation de formations végétales meditérranéennes à partir de données Thematic Mapper. Une étude de cas en Andalousie (Espagne). International Journal of Remote Sensing 8: 1319–1333

Lawson G (1996) Incentives for forest establishment and management in the European Union with particular reference to Ireland, Luxembourg, Portugal and Spain. UK Institute of Chartered Foresters.

Lefroy EC, Salerian J and Hobbs RJ (1993) Integrating economic and ecological considerations: a theoretical framework. In: Hobbs RJ and Saunders DA (eds) Reintegrating fragmented landscapes – Towards sustainable production and nature conservation, pp 209–244. Springer-Verlag, New-York, USA

Lo Gullo MA and Salleo S (1993) Different vulnerabilities of *Quercus ilex* L. to freeze and summer drought-induced xylem embolism: an ecological interpretation. Plant Cell and Environment 16: 511–519

Madoz P (1845–1850) Diccionario geográfico estatddístico histórico de España y sus posesiones de ultramar. Madrid, Spain, 11688 pp

Mangas Navas JM (1982) El regimen comunal agrario de los consejos de Castilla. Ministerio de Agricultura, Serie Estudios, Madrid, Spain, 318 pp

Montoya JM (1980) Los alcornocales. Instituto Nacional de Investigaciones Agrarias, Madrid, Spain, 155 pp

Murphy JM (1995) Transient response of the Hadley Centre coupled ocean-atmosphere model to increasing carbon dioxide. Part I: control climate and flux adjustment. Journal of Climate 8: 36–56

Murphy JM and Mitchell JFB (1995) Transient response of the Hadley Centre coupled ocean-atmosphere model to increasing carbon dioxide. Part II: spatial and temporal structure of response. Journal of Climate 8: 57–80

Ng E and Miller PC (1980) Soil moisture relations in the Southern California chaparral. Ecology 61: 98–107

Prat A (1754) Carta geográfica o mapa general de los pueblos y sus principales arboledas de la provincia de Marina de Sevilla. Biblioteca Nacional, Madrid, Spain

Rambal S (1984a) Sécheresse réelle et sécheresse calculée. Bulletin de la Société Botanique de France, Actualités Botaniques 2/3/4: 295–301

Rambal S (1984b) Water balance and pattern of root evapotranspiration by a *Quercus coccifera* L. evergreen scrub. Oecologia (Berlin) 62: 18–25

Rambal S and Hoff C (1998) Mediterranean ecosystems and fire: the threats of Global Change. In: Moreno JM (ed) Large forest fires, pp 187–213. Backhurst Publishers, Leiden, The Netherlands

Ridley A (1997) Remnant vegetation travelling fellowship. Report to the Land and Water Ressources Research and Development Corporation, LWRRDC Agriculture Victoria, Victoria, Australia

Rupérez A (1957) La encina y sus tratamientos, Ediciones Selvícolas, Madrid, Spain, 151 pp

San Miguel A (1994) La dehesa española, origen, tipología, caracteristicas y gestíon. Escuela

Técnica Superior de Ingenieros de Montes. Fundacion Conde del valle de Salazar, Madrid, Spain, 96 pp
Sanchez Zurro D (1986) Edicion Facsimil del Diccionario de Pascual Madoz. Provincia de Sevilla, Ambito Ediciones, Valladolid, Spain, 420 pp
Scholes RJ and Archer SR (1997) Tree-grass interactions in savannas. Annual Review of Ecology and Systematics 28: 517–544
Stevenson AC (1985a) Studies in the vegetational history of S.W. Spain. I. Modern pollen rain in the Doñana National park, Huelva. Journal of Biogeography 12: 243–268
Stevenson AC (1985b) Studies in the vegetational history of SW Spain. II. Palynological investigations at Laguna de las Madres, SW Spain. Journal of Biogeography 12: 293–314
Stevenson AC and Harrisson RJ (1992) Ancient forests in Spain: a model for land-use and dry forest management in South-west Spain from 4000 BC to 1900 AD. Proceedings of the Prehistoric Society 58: 227–247
Stevenson AC and Moore PD (1988) Studies in the vegetational history of SW Spain. IV. Palynological investigations at El Acebrón, Huelva. Journal of Biogeography 15: 339–361
Trangmar BB, Yost RS and Uehara G (1985) Application of geostatistics to spatial studies of soil properties. Advances in Agronomy 38: 45–94
Vacher J (1984) Analyse phyto et agro-écologique des dehesas pastorales de la Sierra Norte. Thèse Université des Sciences et Techniques du Languedoc, Montpellier, France, 195 pp
Vacher J, Joffre R, Ortega F, Fernandez Ales R and Martin Vicente A. (1985) L'organisation de l'espace dans la Sierra Norte de Séville (Sierra Morena) et les problèmes actuels des dehesas. Revue Géographique des Pyrénées et du Sud-Ouest 56: 179–201
Webster R and Oliver MA (1990) Statistical methods in soil and land resource survey. Oxford University Press, Oxford, UK, 316 pp

Agroforestry Systems **45**: 81–107, 1999.

Multispecies cropping systems in India

Predictions of their productivity, stability, resilience and ecological sustainability

B. R. TRENBATH
Centre for Legumes in Mediterranean Agriculture, The University of Western Australia, Nedlands, WA 6907, Australia; E-mail: btrenbath@agric.wa.gov.au

Key words: intercropping, shifting cultivation, simulation, subsistence farming

Abstract. Several traditional Indian cropping systems are used as examples of agriculture imitating the multispecies character of natural ecosystems. Modelling of their productivity and dynamics suggests they have potential advantages in production, stability of output, resilience to perturbation, and ecological sustainability, although they are harder to manage. Extra diversity in a cropping system can increase the production of a subsistence diet through either biochemical or ecological complementation. Stability of a cropping system may be improved through the incorporation of more crop species. Within a mixed crop, compensatory growth by the stronger component will tend to increase stability of final total yield. Where a two component intercrop has a regular production advantage, the land area required to produce a person s subsistence with a certain low level of risk of failure may be much less than if the crops are grown separately. Where a crop mixture contains contrasting components, the production penalty due to a disaster may be helpfully spread over time so that resilience of the system is increased. The compensatory growth of less-damaged components makes mixtures more resilient. Multispecies systems under intensification stress may be much less resilient than unstressed ones. Unless they are well managed, they can collapse. Where high output is desired, sustainability can only be attained through an understanding of the underlying processes. Intensification can lead to increased production up to a certain level, but such an increase is usually at the expense of subsequent production

1. Introduction

Obtaining a sufficiently regular supply of food for subsistence is the primary goal of most human populations living in the rural tropics. If the Malthusian predictions are valid, it may become the major concern of mankind worldwide. Many studies have shown that the intensification and simplification of food producing systems renders them liable to instability and collapse (Cox and Atkins, 1979; Altieri, 1995). With the world's supply of arable land contracting in the face of land degradation and urban expansion, a movement towards intensification of existing farming systems, shortening of food chains, and reduction of cash cropping in its many forms, seems to be the only way in which the growing population of the earth can be fed (Ehrlich and Ehrlich, 1972; Worldwatch Institute, 1994). Even if the average productivity of systems measured over time and regions is satisfactory, there are still questions of what sort of dynamic behaviour of these systems at a local scale is still acceptable. In the period remaining before chronic shortages of food are expected to affect large areas, there are reasons to ask what lessons can agricultural scientists draw from natural ecosystems with respect to their productivity and dynamics.

Three observations seem pertinent. First, the productivity of unexploited natural vegetation has apparently been sustained over thousands of years. Second, natural vegetation contains a diversity of species. Third, of the farming systems that history suggests were potentially sustainable, many seem to have involved several or many species. From these observations follows the question: to what extent is sustained agricultural productivity dependent on species diversity? This question can be put in another way by asking what services can contemporary agriculture obtain from increased species diversity.

In cropping systems, some services can be immediately identified. One is the restoration of soil fertility provided to shifting cultivators during regrowth of natural vegetation after a period of cropping; at the same time yield-reducing weeds are suppressed. Another service provided to farmers is the supplementation of soil nitrogen by leguminous species and their symbionts; this benefits non-leguminous crops that follow them. Yet another service, seen in rotations and crop mixtures, is the protection of plants of individual crop species from their host-specific predators and disease organisms; populations of such crop enemies tend to explode when they find pure stands offering uniformly accessible food. Each of these services has its own dynamics, requiring special conditions and more or less time. Only the first two of these services will be considered here. The effects of diversity in cropping systems on pest and disease attack have been considered elsewhere (Vandermeer, 1989; Trenbath, 1993).

In this paper I will describe case studies using mathematical models based largely on Indian data first to illustrate some productivity benefits of growing two species together rather than only a single best species, and then to consider some effects of diversity of species on three aspects of the dynamic behaviour of cropping systems. These behavioural aspects are: the stability (constancy) through time of the stream of outputs from the systems; the systems' capacity to recover from perturbations (their resilience); and finally the systems' ecological sustainability. I then discuss what case studies show are the special difficulties and advantages for farmers who operate systems which mimic natural ones to the extent of involving a diversity of species, grown either together or sequentially.

2. Productivity

The production advantages gained from using a diversity of crop species within cropping systems can be classified into two main types: those gained from using them as mixtures ('intercropping'), and those gained from growing the crop species as combinations of pure stands. In the second type, the pure stands may be either separated in time as in sequential and rotational systems or separated in space as in 'shared sole' cropping, where sole crops share the land at the same time. The advantages of intercropping have been well explored in the last two decades (e.g. Willey, 1979a,b; Francis, 1986; Fukai, 1993), and those of crop sequences and rotations have long been well known (e.g. Puckridge and French, 1983; Francis, 1986), but those at a farm or landscape level from shared sole cropping have rarely been documented. The two types of advantage are illustrated in this section by a theoretical example from near Hyderabad, India, where in many low-income families, most of the food energy and protein comes from meals prepared from millet or sorghum and grain legumes.

There are risks in balancing a family's diet by exchanges of crop products with neighbours or people at the market, and so farmers will often want to grow a balanced diet on their own land. Wishing to also grow as large an area of cash crops as possible, they will probably allocate as small an area to subsistence requirements as they dare. They need to find what that necessary area is and decide how to allocate it to different species. For such farmers, the problem of subsistence is essentially how to grow a complete diet on the smallest possible land area.

If the cereal and pulse are produced from adjacent sole crops, the difference in biochemical constitution of their products usually provides the first advantage of diversity. The high energy yield per unit of land from the cereal is complemented by the higher protein yield of the pulse. In some cases, this energy/protein complementation is accompanied by a complementation of the types of essential amino acids that are present in the two dietary components. The effect of this kind of nutritional complementation at the level of amino-acids is shown in Table 1 (Bressani and Scrimshaw, 1961) where the growth rates of test rats are shown to vary with the proportions of protein in the diet from maize and cowpea. Although the total protein in the diets is the same, a mixture of protein sources provides a greater quantity of 'balanced diet'.

To illustrate an apparently new approach to the question of how to use a limited land area as efficiently as possible for providing balanced diet, we

Table 1. Effect of source of dietary protein on growth rates of rats.

%Protein from maize	%Protein from cowpea	Growth rate (relative units)
100	0	1.22
75	25	1.59
50	50	1.84
25	75	1.82
0	100	1.41

Source: Bressani and Scrimshaw (1961).

consider a theoretical case where a farm family obtains subsistence only from a dietary mixture of millet (*Pennisetum typhoides*) and groundnut (*Arachis hypogaea*). The use of groundnut here is illustrative rather than realistic since in India groundnut is usually traded as a snack food. From the data available (Table 2), and regarding millet and groundnut individually as effectively complete foods, it seems that in millet tryptophan is the limiting nutrient, and in groundnut energy content is limiting. For simplicity, supplies of vegetables and other foods from the home garden are ignored, except that supplements of vitamins and minerals from this source are assumed to prevent their supply from being limiting. On the family's land, the theoretical staples millet and groundnut are taken to grow well either as combinations of sole crops or as an intercrop containing the two crops in various row arrangements. The management of each crop is taken to be the same as in sole crop, and storage loss of products is ignored.

If the family plants the available area to sole crops of millet and groundnut, the yields that would be harvested from different allocations of space to the species can be drawn as a replacement diagram (Figure 1a). The yield of each crop depends simply on the proportion of the area planted to it. But such results can be looked at in another way. From data on energy and other contents of the two main crop products, it is possible to calculate what area of each crop has to be grown as a sole crop, if yields are average, to provide the minimum requirements of all nutrients for, say, one person for a year (e.g. a Man Equivalent-year or one ME-year) (Table 2). In Figure 1b these

Table 2. Parameter values used in construction of Figure 1.

Topic	Millet (source)	Groundnut (source)
Average yields (kg ha^{-1} y^{-1})	2226 (Gautam, 1996)	1185 (Gautam, 1996)
Portion edible (%)	100 (Platt, 1962)	70 (Platt, 1962)
Total energy yield (GJ ha^{-1} y^{-1})	32.56 (From carbohydrate and protein)[a, b]	17.06 (FAO, 1949)[b]
Protein conc. (%)	13.5 (Hulse et al., 1980)	
Tryptophan (g $100g^{-1}$ prot)	0.8 (Lenter, 1992)[c]	
Nitrogen conc. (%)		4.69 (FAO, 1970)
Tryptophan conc. (mg g^{-1} N)		65 (FAO, 1970)
Protein digestibility (%)	90 (Aykroyd and Doughty, 1970)[a]	90 (FAO, 1959)
Human energy requirement (GJ ME^{-1} y^{-1})	3.589 (FAO, 1974)	
Human tryptophan requirement (kg ME^{-1} y^{-1})	0.182 (Andrew, 1957)	

[a] Calculated from carbohydrate and protein concentrations using standard energy contents: 17.5 MJ kg^{-1} for carbohydrate, and 16.7 MJ kg^{-1} for protein.
[b] Energy digestibility of cooked foods is assumed to be 100%.
[c] Units converted using protein = 6.25 × N.
[d] Estimated on the basis that protein digestibility of wheat flour varies between 86 and 91% (Aykroyd and Doughty, 1970).

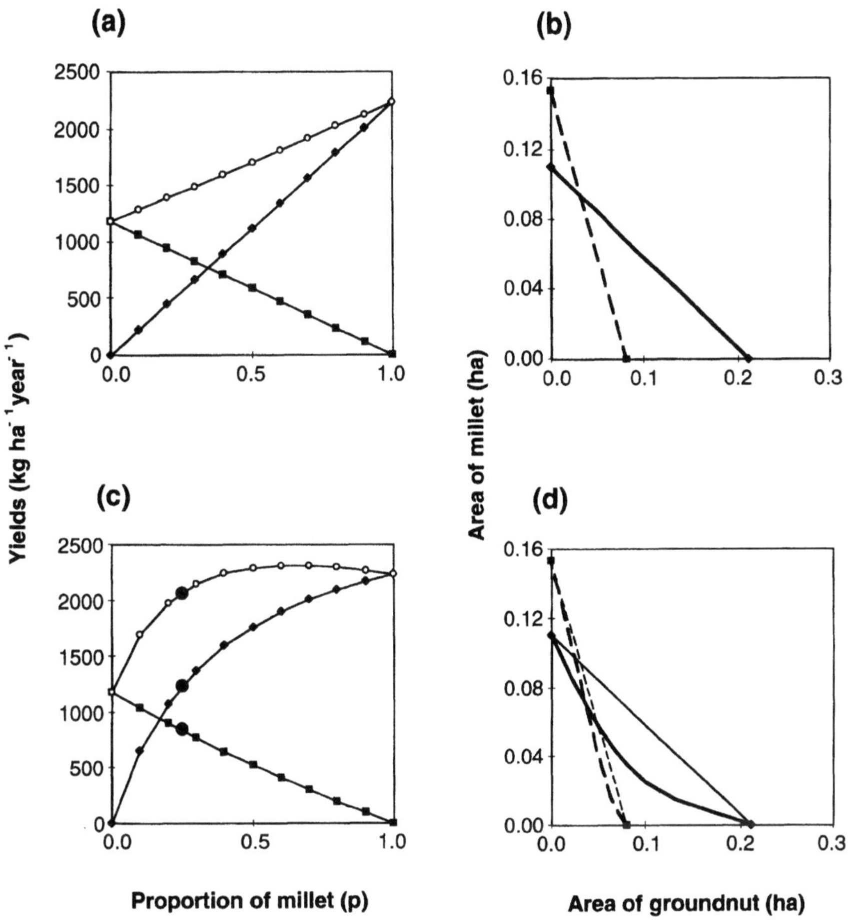

Figure 1. (a, c) Comparison of the yields of pure stands and mixtures of millet (M) and groundnut (G) based on performance in Hyderabad, India. (b, d) Comparison of the areas needed of the same pure stands and mixtures to provide a complete subsistence diet for one Man-Equivalent (1 ME). (a) Replacement series diagram showing yields of the two crops if pure stands are grown on various fractions of 1 ha of land (i.e. if the proportion of the land devoted to millet is P = 0.2, that devoted to groundnut is $1 - P = 0.8$). (b) Minimum areas of the two crops that have to be grown as pure stands to provide 1 ME's requirement of food energy (◆, ———) and tryptophan (■, – – –). The minimum area that *simultaneously* provides enough of both dietary components for 1 ME is found as the intersection of the two straight lines. (c) Replacement series diagram showing predicted yields if the two crops are grown as a *mixture* on 1 ha of land with the components planted on various proportions of it. The points are calculated using the model described in the text, tuned to exactly agree with data (●) of Reddy and Willey (1981); e.g. when P = 0.25, the LER = 1.26. (d) Minimum areas to be planted to the two crops within a mixture to provide 1 ME's requirement of food energy and tryptophan (symbols as in (b)). The minimum area that *simultaneously* provides enough of both dietary components for 1 ME is found as the intersection of the two curved lines. With pure stands in (b), the minimum total area needed is 0.124 ha, whereas, with mixtures in (d), the minimum total area needed is 0.106 ha (c.f. 0.154 ha for pure millet only and 0.210 for pure groundnut only).

areas of sole millet and sole groundnut are plotted as diamonds (continuous line) assuming only energy is limiting and squares (broken line) assuming only tryptophan is limiting. Thus, to get enough *energy* for 1 ME, a larger area of relatively low energy-yielding groundnut has to be grown (0.21 ha) than of the high energy-yielding millet (0.11 ha). Conversely, to get enough *tryptophan* for 1 ME, a larger area of low-tryptophan-yielding millet (0.15 ha) has to be grown than of groundnut (0.08 ha).

By drawing the results on axes of land area, we have the problem set up for finding a linear programming solution of the problem 'what is the combination of activities needing the least land to get a balanced diet for 1 ME?' (Trenbath, 1981). One poor solution would be to grow 0.21 ha of groundnut but there is an incorrect balance of energy and tryptophan. Another rather better solution would be to grow 0.15 ha of millet but the balance would again be poor. However, the optimal solution is indicated by the intersection of the lines: 0.093 ha of millet and 0.031 ha of groundnut, giving a total of 0.124 ha. Here, the balance of energy and tryptophan is ideal. This diet is rated good (FAO, 1964) in the sense that slightly more than 10% of the energy is in the form of protein.

The method can now be extended to intercropping where mixtures are of a replacement-series type (de Wit, 1960). Due to unequal competition for environmental resources in a mixed stand, plants of one component usually grow bigger in mixture than they do in pure stand. The opposite usually applies to the other component(s). In the simplest case, in a 1:1 mixture, this tendency is seen as an *x%* gain in per-plant yield in one component being balanced by a corresponding *x%* reduction in the other, at least in terms of biomass production (Trenbath, 1974). Because the component with the higher-yielding pure stand usually has a higher competitive ability than the other, the total mixture yield is usually higher than a value half-way between the pure stand yields. If the mixture yields are plotted against an axis of mixture composition, the line of the mixture yield of the dominant component is convex and that of the suppressed component is concave.

Even where intercrops are not strict replacement-series mixtures, agronomists have argued that, among possible outcomes of competition, the simplest case can be recognised using an index of land productivity, the Land Equivalent Ratio (LER), given by

$$\text{LER} = y_i/y_{ii} + y_j/y_{jj}$$

where y_i and y_j are the yield ha^{-1} of the mixture components i and j, and y_{ii} and y_{jj} are the corresponding yields ha^{-1} of the sole crops planted at optimum density and in the same conditions of soil and crop management as in the mixture. In the simplest case, LER = 1. This indicates that in the 1:1 mixture (by planted area), the percentage increase of yield per unit of planted area in one component over that in its pure stand is exactly matched by the same percentage decrease in the other. It also indicates that exactly the same yield obtained from any mixture on a hectare of land can be obtained on one hectare planted to pure stands in the right proportions.

Under conditions that have been much studied, competition or other types of interactions can lead to LER values well above 1. Such cases usually involve differing growth rhythms, differing soil nutrient requirements (as in legumes and non-legumes), differing rooting depths, release of auto-toxic chemicals or specific effects on pest or disease organisms (Trenbath, 1976, 1993). Thus, in many sorghum/pigeonpea intercrops, the LER is about 1.5 (Rao, 1986). This would imply that 1.5 ha of sole crops would be needed to produce the same yields as produced by the intercrop on 1 ha. With this and other combinations, occasionally LERs above 1.7 are found.

A simple mathematical model can fit most examples of the replacement series diagram for simple-case intercrops (de Wit, 1960):

$$y_i = y_{ii}p_ik_{ij}/(p_ik_{ij} + 1 - p_i)$$

where p_i is the proportion of land surface in the intercrop planted to component i, and k_{ij} is a 'crowding coefficient' expressing the competitive ability of i with respect to j. An analogous equation can model the yield of the other component. In the simplest case when LER = 1, then $k_{ij} = 1/k_{ji}$ or $k_{ij}k_{ji} = 1$. However, when LER > 1, then $k_{ij}k_{ji} > 1$ also. To model this latter case, it is convenient to estimate a raw k_{ij} from $(y_{ii}/y_{jj})q$ and then to multiply both k_{ij} and k_{ji} by a factor $1 + r$, where values of q and r are used in fitting the curves. To fit this modified de Wit model exactly to the original data points in Figure 1c, values of $q = 1.84$ and $r = 0.497$ were used.

This modified model gives an estimate of the yields of the rest of the replacement series (Figure 1c). Using the estimated yields from this replacement series graph allowed the plotting of an area-requirements graph equivalent to that given in Figure 1b for the sole crops. The sole crop lines and the intercrop curves are presented together for comparison in Figure 1d. The intersection of the curves shows the minimal area requirements for obtaining 1 ME from millet and groundnut planted in an intercrop. These requirements are 0.071 ha of millet and 0.035 ha of groundnut, giving a total of 0.106 ha. At this proportion of millet ($P = 0.67$), the LER is 1.17, which is usually interpreted as a 17% 'advantage' for intercropping. Reasoning based on LER might suggest that the family should grow the intercrop with the highest LER (1.27, from $P = 0.32$) but then the present approach can show that the nutritional balance obtained would not be good and it would take more land to obtain enough for 1 ME of both limiting components of the diet. Comparing the optimal solutions shows that on average the optimal intercrop with $P = 0.67$ is 0.124/0.106 = 1.17 times more efficient than the optimal sole crop combination with $P = 0.75$ in providing for subsistence, i.e. intercropping is 17% more efficient in producing an ideally balanced subsistence diet. This suggests what proportion of the two species should be grown. It might be noted that the appearance of the 17% figure in both calculations is quite fortuitous.

Summarising the figures from this productivity case study, to produce 1 ME of subsistence diet, the best single pure stand to grow is 0.15 ha of millet, the best combined pure-stand solution is 0.124 ha with a millet proportion of 0.75, and the best mixed crop solution is 0.106 ha with a millet proportion of 0.67. There can thus be an increase in land-use efficiency of 21% from including another species as a sole crop, and a further 17% increase from mixing species at the finer scale of an intercrop. In an intercrop like sorghum/pigeon pea, where LERs are often 1.6, the last figure could be much higher than 17%.

3. Stability

For subsistence farmers, variation of conditions from year to year leads to corresponding variation in annual yields and thus in the adequacy of food supply. The stability (or degree of constancy) of crop yield measured over seasons and years is of critical importance where availability of land and storage facilities are limited.

Where the farm's food supply depends on two contrasting species of crop, their differing responses to environmental fluctuations may well cause individual yields to be negatively correlated. A theoretical study by Van Noordwijk et al. (1994) has shown that with a negative correlation, and given a fixed area cultivated, the farmer will get a more stable food supply from sharing the area between pure stands of the two crops, than from growing the more stable-yielding species over the whole area. This result is true for a wide range of proportions of the crops. On the other hand, even if the correlation between yields of the two crops is positive due, say, to a common response to rainfall in a dry environment, the combination will theoretically still give greater yield stability, although over a narrower range of proportions, provided the correlation is not too positive. These theoretical results seem to explain well why, apart from the culinary and dietary advantages of a mixed diet, subsistence farmers often insure themselves and family against insufficient production by growing pure-stand combinations of crop species and genotypes.

It might be asked whether the stability advantage from growing a pair of crops might be increased by larger numbers of crops. The diminishing returns of extra stability from extra species are shown by a simple exercise in which a unit of land is supposed to be planted to between 1 and 10 crops occupying equal fractions of the area in each case. If, say, the potential yearly energy yield of each is 1000 MJ ha^{-1} with a standard deviation of 10 and zero correlation between crops, over 30 years the total energy yields will be less variable between years where more crops are grown (Figure 2). This is seen either in the trends of falling standard deviation of the yearly total yields or in their narrowing range. Intercorrelations between the various crops will again affect the results: weak positive correlations will weaken the effect, and

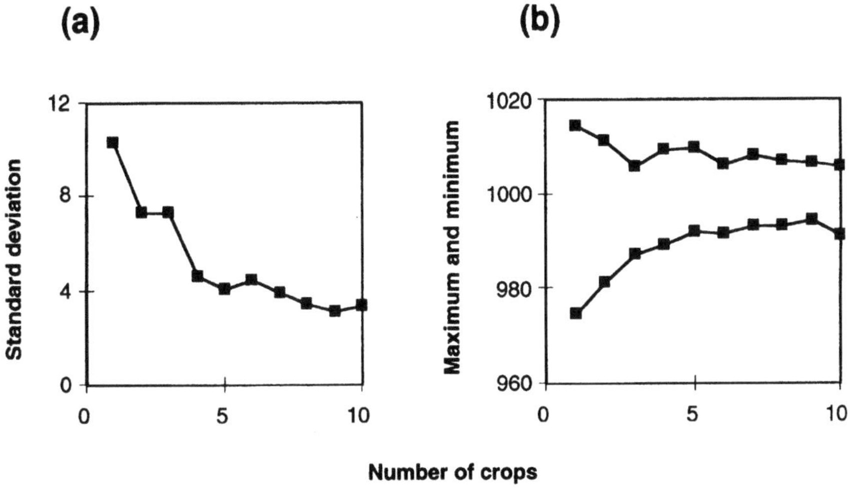

Figure 2. Predictions of (a) standard deviation of annual yield, and (b) range of annual yield where a fixed area is planted to equal areas of increasing numbers of crops. The results are based on 30 years of simulated annual random drawings from normal distributions of yields where all crops have the same mean yield and standard deviation (1000 and 10 units ha^{-1} respectively).

negative ones will strengthen it, although bringing the line in Figure 2 ato an asymptote at a lower number of crops (M. Van Noordwijk, pers. comm.).

Among the costs of dealing with more crops are, however, the added overheads of labour and knowledge required for the adequate cultivation of each one. As the number of crops increases, more and more conflicts will arise in the application of timely interventions to each so that achieved yields will fall. If total yield thus starts to fall at high crop number, the farmer will be then forced to trade off average yield against stability of yield. The number of crops she/he will grow will then be set by the acceptable level of this trade-off as determined by her/his individual circumstances (reserves of cash, storage facilities for crop products, perceived risks etc.).

Whether the chosen number of crop species is to be grown as sole crops or as intercrops is a further decision which can affect the stability of yields. This follows from the results of comparisons of the stability of two-component intercrops with that of the more stable of their components grown in sole crop. Four studies, including two in India, suggest that intercrop yields are more stable (Table 3). However, in the best replicated data set (Rao et al., 1979) with 48 comparisons involving locations, nitrogen treatments and seasons, the same stability measure was calculated for 'shared soles'. For this, it was assumed that in each case, the same land area was shared between sole crops planted on areas proportional to their yields in intercrop. The inverse of the coefficient of variation (1/CV) for the 'shared soles' was 2.22, suggesting that most of the apparent stability advantage of intercropping was simply due to the presence of two rather than one species.

Table 3. Yield stability of two-component intercrops and sole crops in cereal/legume pairs as measured by the reciprocal of the coefficient of variation. Most of the intercrops included contain components in a 1:1 proportion.

Crop	Sole cereal	Sole legume	Intercrop	Author(s)
Various combinations	2.04	1.54	2.94	Papadakis (1941)[a]
Maize/soybean	1.66	1.96	2.28	Rao et al. (1979)[b]
Sorghum/pigeon pea	2.10	2.21	2.60	Rao and Willey (1981)[c]
Maize/climbing bean	4.24	3.40	4.37	Smith and Francis (1986)[d]

[a] Six sites in Greece; [b] Seven sites in SE Asia and Australia; [c] Forty sites in India; [d] Twenty sites in Colombia. *Source:* Trenbath (1984b).

Where, as in the 1:1 maize/soybean mixtures, the greater stability of the two-species system is combined with a large advantage in average production, the risk of a farmer failing to reach a cash target is much less with an intercrop than with shared soles (Rao et al., 1979). The risk with shared soles is again less than with either sole crop, in line with the suggestion that the use of two crops rather than just one is responsible for most of the stability advantage. The situation is likely to be the same if subsistence is the target. Although an example from near Hyderabad lacks the calculation for shared soles, it compares the risk of failure to obtain a family's energy subsistence using various cropping systems. It shows that to reduce the risk to 5%, the area needed under intercropping is 0.85 ha compared with that under sole sorghum, the next best provider of food energy, which is 1.6 ha (Trenbath, 1984b; Figure 3). This result relies on data on variability of the energy yields of the individual crops. Interestingly, when the correlation between the energy yields of the crops is taken into account and an optimal solution for the farmer's problem is sought using quadratic programming (Bernard and Nix, 1973), the results are near-identical. In this environment, with these species and mixture composition, the differing reactions of the species to various types of season were therefore not great or consistent enough to alter a conclusion based on an assumption of zero correlation.

If the requirement is tightened to make the target a complete subsistence diet grown on a minimum area, the farmer will need to know the best proportion of the land planted to the species, either in sole- or inter-crops. The approach introduced in the previous section can show how variability of seasons affects the best choice of proportion of the species to be planted. Ignoring correlation, the effect of the observed variances on the optimal sole- and inter-crop proportions is seen in Figure 4. Twenty-four years optimal proportions are shown for each of the two kinds of crop. The (filled) intercrop points are much more dispersed than those (open) of the sole crops because the variation of competitive abilities creates a further component of variation. Comparing the two approaches, the mean (± SE) optimal proportion of millet for intercropping (0.66 ± 0.015) is clearly different from that

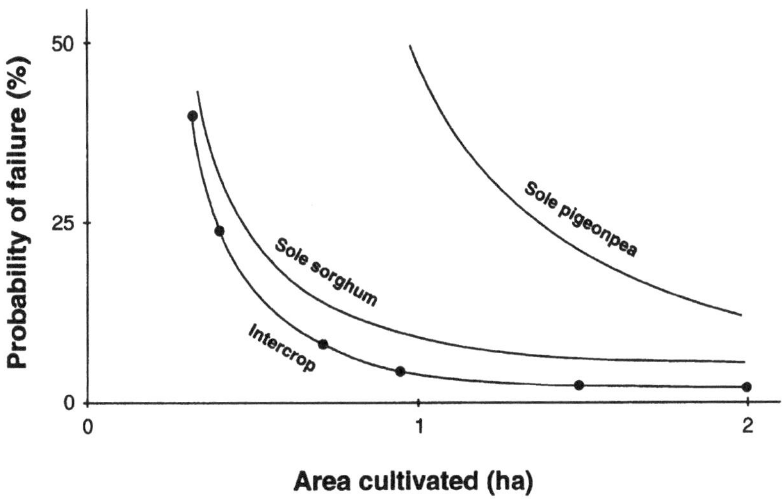

Figure 3. Probability of failure to produce enough food energy for the subsistence of a family of 3 ME (Man Equivalents) from different areas of land planted to either sole sorghum, sole pigeonpea or a 1:1 intercrop. Lines are based on tables of standard normal deviates where the means and standard deviations of the annual energy yields are calculated from Hyderabad data of M. R. Rao (pers. comm.). Filled points are optimal solutions calculated using quadratic programming based on a variance-covariance matrix of energy yields found from the same data. Source: Redrawn after Trenbath (1984b).

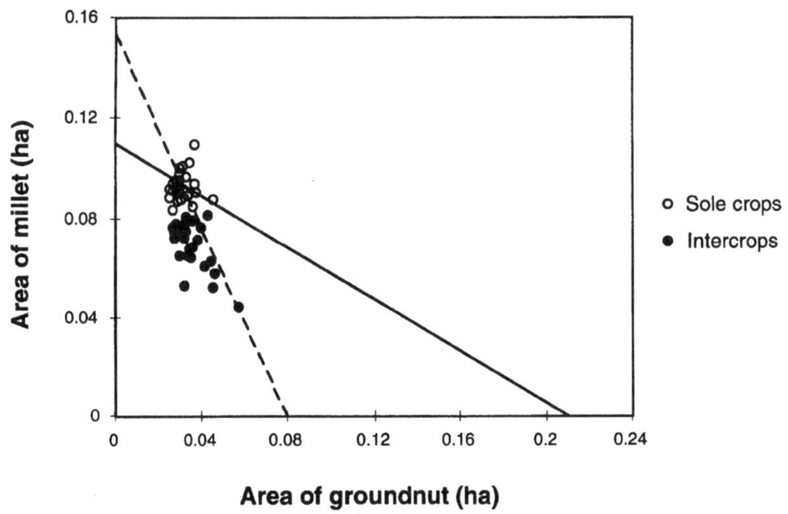

Figure 4. Comparison of minimum areas providing a complete diet for 1 ME, India, when land is planted either to combinations of sole crops and an intercrop when variability of millet and groundnut sole crop yields is considered. Mean yields at Hyderabad, India of: millet, 2226 kg ha^{-1}, and groundnut, 1185 kg ha^{-1} (Reddy and Willey, 1981). A standard deviation of 200 kg ha^{-1} is assumed for both crops and covariance is assumed zero. Twenty four of each type of point are plotted. The original pure stand lines from Figure 1b are drawn again for comparison.

for sole cropping (0.75 ± 0.006). However, neither is different from the corresponding optimum calculated using mean yields. The lower proportion of millet in the intercrop is needed to compensate for its higher competitive ability.

4. Resilience

The resilience of a cropping system is the extent of its capacity to survive perturbation or stress with minimum effect on productivity (Holling, 1973; Trenbath, 1984b). It is thus a crucial component of stability. Under certain conditions, multispecies cropping systems, either intercrops or sequences of crops, show resilience superior to that of sole crops.

When some acute stress strikes a farming system, the presence of components with differing response times to the stress can sometimes make it easier for the farmer to recover from the set-back caused by the stress. This is seen in a mixed irrigated system under test in Kerala in the early 80's where coconut palms were under-planted with grass (cut and carried for forage) and black pepper was trained up the coconut stems. In a theoretical exercise, the effects on the farmer's cash-flow of a prolonged breakdown in the irrigation system were calculated (B. R. Trenbath, P. K. R. Nair, J. Matthew and T. Varghese, unpubl. data). The death of the grass would lead to the earliest loss of income (from milk sales), while the effects on the pepper and the coconut harvests would be delayed up to one and two years ahead respectively (Figure 5). This staggering of the effects through time was expected to greatly ease the farmer's cash flow situation.

When the components of a cropping system are more similar in phenology, as in annual crop mixtures, responses tend to be simultaneous, but they may vary greatly in intensity. When a pure stand of a crop suffers damage from, say, wind or hail, chance will dictate that some plants are less damaged than others. The less damaged ones will grow more vigorously than otherwise and compensate for the poorer growth of those severely damaged. When a mixture of contrasting annual species suffers damage from the same agent, often one component suffers less so that compensation then occurs between components. The intercrop will then show more resilience.

When the damage is biological, i.e. due to pest or disease attack, damage may be totally confined to one susceptible component; in such cases compensation may allow the yield of the mixture to be well above that of the same area planted to a 1:1 combination of pure stands. Where a mixture component is eliminated by pest or frost attack early enough in the life of a planted mixed stand, the growth capacity of the remaining component(s) may be sufficient to almost completely compensate for the lost component. If the intercrop is already well grown when a component is removed, the capacity of the remaining plants to recover will be more limited. In Figure 6, the biomass of intercropped pigeonpea lags further and further behind that in a sole crop

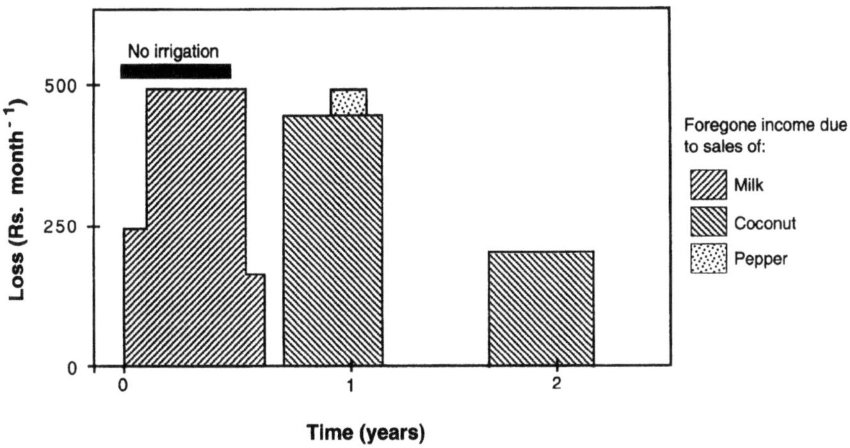

Figure 5. Distribution through time of estimated losses (Rupees month^{-1}) experienced by a one-hectare model coconut farm in the Central Plantation Crops Research Institute, Kerala, as a result of a breakdown in irrigation during the dry season. Losses are due to foregone income from milk (i.e. due to death of underplanted grass), coconuts, and black pepper trained up the coconut stems. Source: Redrawn from Trenbath (1984b).

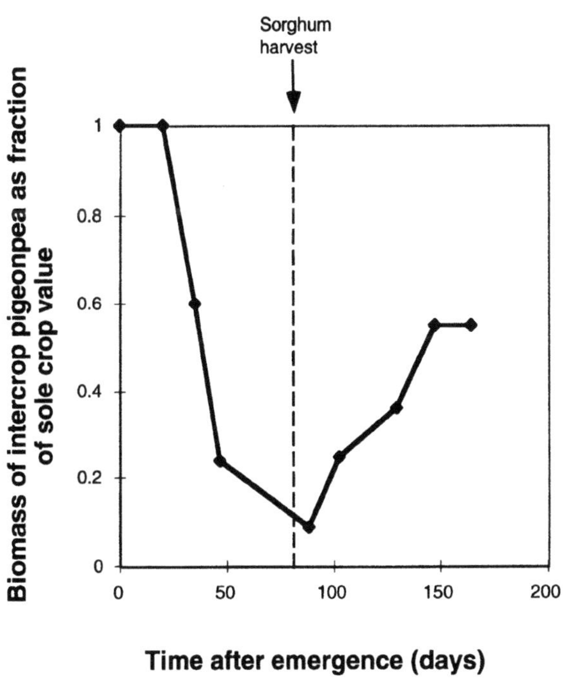

Figure 6. Recovery of pigeonpea in a sorghum/pigeonpea intercrop at Hyderabad when stress due to the presence of sorghum is released due to its removal at harvest. The time of sorghum harvest is shown as the dashed line. Recovery is indicated by the return of the intercrop pigeonpea biomass towards its sole crop value. *Source:* data from Natarajan and Willey (1981).

as it is steadily suppressed by the more competitive sorghum (Natarajan and Willey, 1981). Nevertheless, after the sorghum is removed at harvest, pigeonpea partly recovers to yield as much as 60% of the biomass that it would have given if no sorghum had been planted with it.

Where the planting of two crops is staggered in time, as in relay or sequential cropping, and the crops are exploiting a single stored resource such as water or nitrogen, a failure of the first crop will leave more of the resource for the second crop. Better growth by the second crop may then lead to the combined yields or returns being little affected by the loss of the first crop. If the first crop is a legume on whose fixed nitrogen the second crop normally depends, the loss of the first crop just before it sets seed may leave more nitrogen in the soil than usual for the second crop. Some compensation can then occur. If, however, the legume is lost before it has had time to fix much nitrogen, a following non-legume crop may perform poorly. A flexible approach by the farmer to the choice of species to grow will be an advantage but local seed supplies or contractual arrangements may limit the options.

A type of sequential cropping system in which changes over time determine farmer's interventions is the shifting cultivation system practised by many resource poor tribes people in hilly forested areas of India (Ramakrishnan, 1984; Das, 1996). This system is periodically stressed, not by the environment, but intentionally by burning and clearing of the forest, followed by a phase of cropping. If the farmer has cropped for too long or too intensively, the balance between the grass weeds of the cropping phase and the regrowing stumps of the cut trees will favour the grass, which then dominates the plot preventing re-growth of the trees. If this happens, the land becomes a low-fertility grassland where tree colonisation is usually prevented by frequent fires. Soil fertility restoration is less effective under grass (Van Noordwijk et al., 1997) and so the cropping system collapses to a much lower level of production. Survival of the high-yield system thus depends on how well the tree component withstands disturbance due to cropping.

A model developed using data from similar systems in the Philippines and Thailand produces results qualitatively similar to the main features of the Indian systems described by Ramakrishnan (1984). In Figure 7a, the model (Trenbath et al., 1989) is applied to six cycles of a non-intensive shifting cultivation system where the cropping phase of one crop per year lasts two years and the fallow phase lasts 15 years. A sample of output shows the regular recovery of tree biomass after each cropping phase and an associated partial recovery of soil fertility. Grass biomass remains negligible throughout.

A switch (Figure 7b) to a more rigorous removal of forest biomass before cropping weakens the tree re-growth and fertility restoration. This favours the grass until in the fifth cycle grass inhibits tree regrowth and in the sixth cycle it becomes dominant. Fertility remains low under the grass. Another type of intensification involving a switch to two crops per year (Figure 7c) causes a much faster loss of tree biomass and soil fertility, and an earlier invasion of grass. Intensification is thus seen to make the system more sensitive to the

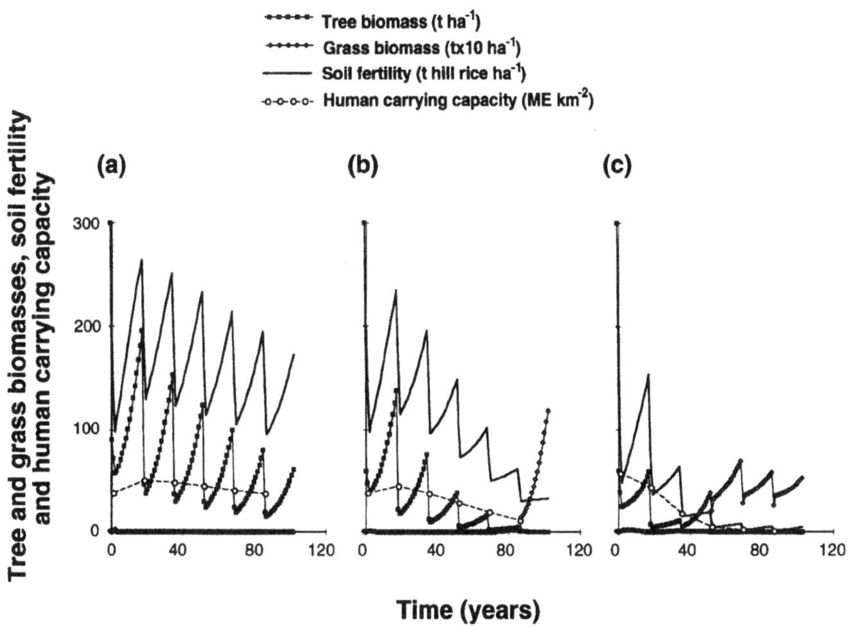

Figure 7. Simulated time courses of tree biomass, grass biomass, soil fertility (measured as yield of hill rice), and human carrying capacity (Man-Equivalents, ME) in six cycles of a shifting cultivation system in forest from data collected in the Philippines and Thailand. Each cycle contains two years of cropping followed by 15 years of forest regrowth (fallow). Simulations were according to the model and parameter values described in Trenbath et al. (1989). (a) Low intensity system with only slow fertility decline; (b) as in (a) but the fraction of the tree and grass biomass left on clearing for cropping is reduced from 0.3 to 0.2; (c) as in (b) but system further intensified by increasing the number of crops taken per year from one to two. Intensification leads to some increased carrying capacity in (c) but an earlier invasion by grass and collapse of fertility.

regular cycle of stress. The more intensive the cropping, the sooner the stress will cause the system to collapse.

Assuming that this model correctly represents the qualitative features of the re-growth phase, a 'phase-plane' diagram (Figure 8) drawn using this model illustrates the subtlety of the problem facing the farmer who is trying to intensify the system as much as possible without causing it to collapse. In the diagram, drawn for a medium soil fertility level, the streamlines and arrows show the way the system's state will move during the fallow phase from any starting point defined by a tree biomass and a grass biomass. If the farmer's cropping activities have landed the system on the high-tree side of the curved separatrix, the trees will recover and suppress the grass. The system will move toward the all-tree equilibrium. On the other hand, if the system starts a fallow phase on the low-tree side of the separatrix, the system will move toward the all-grass equilibrium. The farmer will see the problem as one of finding ways of intensifying the system as much as possible without allowing grass to take

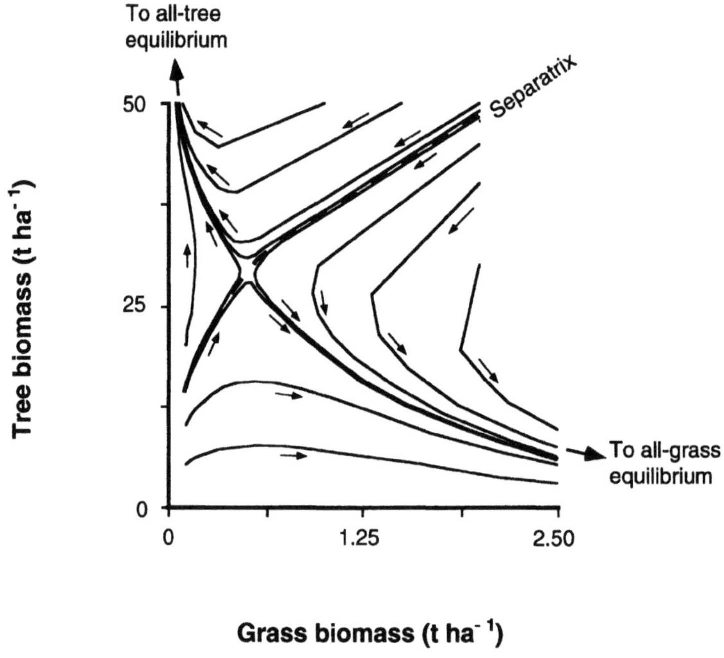

Figure 8. Phase-plane diagram of the fallow phase of a shifting cultivation system, with the dynamics in various parts of the tree-grass state space indicated by lines following trajectories of the system according to simulations carried out using the model used to give Figure 7. Thus, a system with 37 t ha^{-1} of tree biomass and 0.7 t ha^{-1} of grass will recover its tree cover, but one with the same biomass of grass but only 12 t ha^{-1} of tree will not. A curved separatrix is present dividing the state space into two domains each with its own equilibrium point. The two equilibrium points of the system are: all-tree and high fertility, and all-grass and low fertility. Neither appears on this scaled-up portion of the space. Soil fertility in this plane is equivalent to a crop yield of 2 t ha^{-1} of hill rice.

over the plot. In terms of this diagram, the problem is one of steering the system as close to the separatrix as possible without actually crossing it. Runs of the model show how complex this task is, for the separatrix is not only curved but its position moves to the right with higher fertility and to the left with lower fertility. Evidence from India and many parts of the hilly tropics is that hopes of higher yields have driven the system in many areas over the separatrix and into permanently unproductive grassland (Ramakrishnan, 1984; Allen et al., 1995; Xi, 1995). Models of these systems may help identify simple decision rules useful in improving the ways they are managed.

5. Ecological sustainability

The ecological sustainability of an agricultural system can be defined as the extent to which the ecological basis of the system allows it to operate for an indefinite period (Gliessman, 1989). Ecological sustainability thus contrasts with economic, social and political sustainability which depends on how the system relates to human values.

Many tropical cropping systems have been in operation for hundreds of years and so may be said to be ecologically sustainable, at least in their traditional forms. Recently, though, due to increasing populations to be fed, farmers have been intensifying such systems in such ways that they are no longer sustainable. Indeed, yield declines are being frequently reported to affect considerable areas (McCown and Jones, 1992; Wendt et al., 1994; Akonde et al., 1996; Subedi, 1997). Usually, farmers have too little technical or financial support to be able to find remedies. At some point, the labour efficiency (output of product per unit of labour input) of such systems will become so low that social factors will dictate that they must be abandoned or radically modified.

Systems studies are, however, often able to pinpoint the ecological factors leading to the yield declines. For instance, in the case of shifting cultivation considered in the last section, the rapid fall in human carrying capacity which accompanies the system 'flip' as it crosses the separatrix (Figure 8) is due to the loss of the trees. Trees, especially leguminous species, are responsible for restoration of soil fertility between cropping phases, and for suppression of the grass weeds that invade crops.

A cropping system tested experimentally for use on the plateau around Hyderabad is a two-year rotation of sorghum/medium-duration pigeonpea intercrop (year one) and sole sorghum followed by safflower (year two). Rego and Seeling (1996) showed that the sorghum yields could be maintained over 10 years by the nitrogen input from the pigeonpea intercrop. In contrast, under continuous sole sorghum followed by safflower, nitrogen uptake and yield of the sorghum declined (T. J. Rego, pers. comm.).

An analysis of the likely production dynamics of a very similar system using a simple model (Trenbath, 1984b) showed how yields and their sustainability may depend on the proportion of legume in the intercrop. Thus, in Figure 9, this model is used to predict time courses of 'available' soil nitrogen and human carrying capacity for a range of proportions of pigeonpea planted in the year-one intercrop. Human carrying capacity was measured in terms of food energy only, and parameter values were as in the original work, except that:

a) yield potentials have been brought in line with data of Rego and Seeling (1996; sorghum maximum yield = 4.6 t ha^{-1}) and Ali (1996; pigeonpea yield = 1.4 t ha^{-1});
b) initial available soil nitrogen was set at 30 kg ha^{-1}, and
c) a rainfall addition of 3 kg N ha^{-1} y^{-1} is included (T. J. Rego, pers. comm.).

According to these results, the commonly used 33% pigeonpea leads to a 21% increase in carrying capacity over 30 years, and an intercrop with 66% shows nearly double this rise. In contrast, a system with 0% legume in the

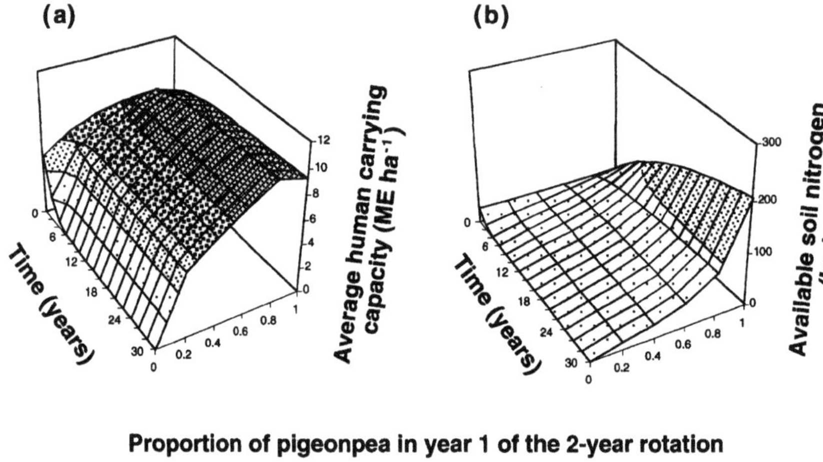

Figure 9. Changes through time in (a) the predicted human carrying capacity and (b) the predicted 'available' soil nitrogen of land under a two-course rotation of sorghum/pigeonpea intercrop followed by sorghum sole crop at Hyderabad, India. The land area is taken to be divided in two so that half is in each course of the rotation in any year; results of the two courses are averaged. The proportion of pigeonpea in the intercrop course is varied from all sorghum to all pigeonpea. *Source:* Trenbath (1984b).

intercrop (i.e. without legume at all in the system) would lose about 70% of its carrying capacity over the same period. Predicted changes of 'available' soil nitrogen parallel changes in total soil nitrogen observed by Rego and Seeling (1996) over 10 years, although their sorghum yields did not change so dramatically (TJ Rego, pers. comm.). These predictions are consistent with expectations according to residual effects measured after legumes planted as sole crops or in intercrops (Kumar Rao et al., 1996) and with observations that intercropping of non-legumes with legumes helps to maintain yields (Finlay, 1974; IARI, 1976; Kang et al., 1991). The simulation results suggest further that if the legume proportion in the intercrop nears 100% (thus making the system a sorghum-pigeonpea rotation), the lower potential yield of the sole-crop pigeonpea will lead to a penalty in terms of lost carrying capacity.

In the shifting cultivation system, even before the stage of grass invasion, if yields are compared over cycles, the cropping phase faces likely yield declines under intensification. This is widely observed in the tropics (e.g. Zinke et al., 1978). In the Indian versions of the system, the usual responses to increased population pressure are to take more crops before abandonment of the plot and to shorten the number of years in fallow (t_f) (Ramakrishnan, 1984). In terms of the model's management parameters, more crops in the cropping phase can be expressed as either more years (t_c) in this phase or as more crops y^{-1} (n). Obtained using a simplified model (Trenbath, 1984a), Figure 10 shows the simulated responses of human carrying capacity (again measured in food energy terms) and rate of soil fertility decline (measured in

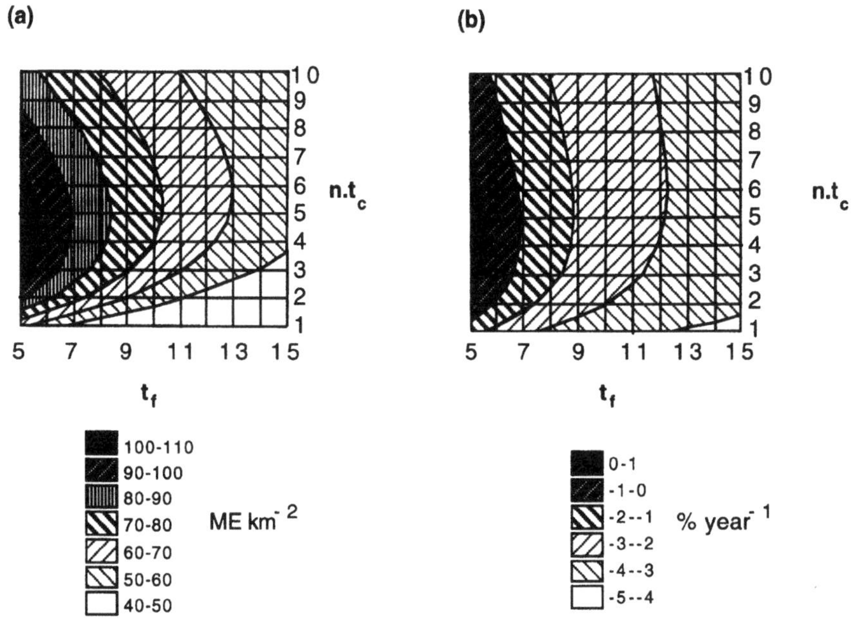

Figure 10. Predicted responses of (a) human carrying capacity and (b) soil fertility change in a shifting cultivation system to varied lengths of cropping and fallow phases (t_c and t_f years respectively). The system starts at a soil fertility equivalent to a hill rice production of 2 t ha^{-1}. One crop is taken per year (n = 1). Results obtained with a simplified version of the model used to make Figure 7; parameter values as in Trenbath (1984a) except that the area of land considered is 1 km^2. For any level of tf, both carrying capacity and rate of soil fertility loss are maximal at about t_c = 4 years. When t_c = 1 and t_f = 15, soil fertility actually rises.

average % change of production y^{-1}) to variation of the two intensification variables, $n \times t_c$ and t_f. Responses relate to the next cycle of the rotation. Over the ranges shown, the responses are similar to both forms of intensification, i.e. both rise with shortened fallow but first rise then fall when more crops are taken. This last observation agrees with the carrying capacity changes in Figure 7c where moving from one to two crops $year^{-1}$ gives an initially increased carrying capacity but thereafter a faster fall; the faster fall is associated with a earlier collapse of the system into low-fertility grassland. The common deadly negative correlation between productivity and ecological sustainability is clearly present in this system.

6. Discussion

6.1. *Which species?*

In mature landscapes, the same form of natural vegetation persists over hundreds of years and thus it is clearly 'ecologically sustainable'. Ecologists working in these natural ecosystems have shown that, through time, their materials budgets balance, biodiversity and species populations are maintained within limits, and the distribution and ratios of various kinds of vegetational texture (e.g. successional patches) remain roughly constant. Agriculturalists wishing to design their own systems to be similarly sustainable have realised their need to understand the mechanisms underlying these system properties, and have drawn many insights from the ecological literature (Hart, 1986; Altieri, 1995; Gliessman, 1989; Vandermeer et al., 1998).

The case studies presented above have shown a range of examples of multispecies cropping systems which, by suitable management, can achieve desirable dynamic properties. Not shown, however, were some background analyses to the modelling exercises which showed how the precise properties of the species are often very critical in achieving desirable system behaviour. This suggests that a part of ecological theory which ought to help agriculturalists devise systems to farmers' specifications are studies on how the functioning of ecosystems depends on the specific properties of the component species. Unfortunately, although ecologists are able to specify the general categories of species types needed for certain ecosystem functions, the development of this aspect of their science is still largely descriptive. The precise implications of the observed biodiversity for system function are often unclear. Because agriculturalists have historically been under more pressure than ecologists to make systems perform to strict specifications, they are generally better equipped than ecologists to prescribe required properties for component species.

Results of many experiments support the conclusion from modelling that to achieve target performance of a multispecies agricultural system, species have to be carefully selected (e.g. Davies et al., 1993). For example, no random choice of species to be combined with sorghum would give the remarkable land-economising result of Figure 3, where the land area needed to give a 95% certainty of achieving a target energy yield is approximately halved by the addition of pigeonpea. In this environment, only pigeonpea has the two specific traits shown to be responsible for big production advantages in this intercrop. These traits are a slow start to growth offering little competition to the fast growing early maturing sorghum, and a later maturity in which early suppression is compensated for by a much increased harvest index (Natarajan and Willey, 1980; Ranganathan, 1993). This particular combination has been well enough researched to show that not only the choice of species but the choice of genotype is critical (Rao and Willey, 1983).

Whereas in some systems, farmers make the selection of species at planting time, in others a quasi-random set of species establishes and these are subjected to selection by the farmer. Thus, in the cropping phase of some shifting cultivation systems, certain adventitious species are conserved as non-weeds to be kept for use in cooking or as medicine, whereas others are uprooted (Chacon and Gliessman, 1982; Ramakrishnan, 1984). In the forest fallow stage of similar systems, the composition of natural regrowth is also commonly manipulated to conserve useful species (Bandy et al., 1993; Kwesiga, 1995; de Jong, 1996).

6.2. *Managing species numbers and populations*

While large numbers of species may be present as fellow-travellers in subsistence systems, with many being used in some way or other, major management effort is usually applied to relatively few species, usually a single main food provider (Francis, 1986). This seems to be due to the virtues of specialisation and the difficulty of effectively managing many crop species simultaneously. As numbers of managed species increase, the numbers of pairwise combinations of possible incompatibilities increase. If each has its own requirements of timely intervention, the avoidance of clashes between requirements rapidly becomes harder. The parallel increases in time- and knowledge-overheads associated with each extra farm management task seem to explain why strongly managed components of cropping systems are limited in number (e.g. Hore et al., 1997).

Just as the number of species is an issue for the farmer, the number and spatial distribution of plants of the different species has to be decided. With regard to relative numbers of plants in intercrops, the modelled results of planting different proportions of pigeonpea in the intercrop within a two-course rotation show remarkable effects on the levels of production and its sustainability (Figure 9). If the proportion of legume is very high, the system seems sustainable for ever (ignoring disease) but the production as judged by human carrying capacity is less than the maximum. If the proportion is very low, the production falls to half the initial level in about eight years. Over a limited range of high proportions, there is a trade-off between the levels of carrying capacity and of its sustainability, while at lower proportions these are positively correlated.

Where only part of a farmer's cropping is for subsistence, subsistence production will usually be confined to the smallest area necessary (Anon, 1994). In the common Indian case where subsistence is obtained from a mix of cereal and pulse staples, for this a target ratio of two products may be aimed for. As shown in Figure 1, the minimum area needed to obtain one person's 'complete diet' as judged by a dietetical analysis is 20% less if two sole crops are grown than if just one species, millet (the better yielder), is planted over the whole area. If the two species are intercropped, the minimum area shrinks by a further 15%. The modelling of the competitive effects in an average season of intercrop growth allows a comparison of the best planting proportion of legume to be made for sole crops (0.25) and for an intercrop (0.33). Modelling of the system for a range of seasons shows that the best planting proportion in intercrop is likely to vary much more than in the sole crop system (Figure 4). The higher proportion of the legume in the best intercrop is explained by its partial suppression by the cereal. This preliminary case study shows how any calculation of best proportions will need a synthesis of nutritional and agronomic research.

Due to the added complexity of interactions among components of crop mixtures, especially when perennials are involved, it is harder to design interventions for managing such components than for managing sole crops (Roger and Rao, 1990; Rao and Coe, 1992). For instance, the presence of competition means that the timing of interventions in intercrops is more critical than in sole crops. Even under constant uniform conditions, delay of a few days in the planting of one component of a mixture can alter its final yield greatly whereas the same delay in planting a sole crop may make no difference. The situation recalls the difference between managing the interaction between grazing animals and growing pasture on one hand and managing a pasture for hay on the other. The chance of disaster with the plant-animal system is much greater. This is partly due to the presence of the same kind of alternative domains with separatrices between them as have been demonstrated in the case study of shifting cultivation (Figure 8; Noy-Meir, 1975; Walker et al., 1981).

Nevertheless, progress is being made with intercrops in the direction of defining the conditions for planting, and for deciding the timing and intensity of later interventions. As research tools, models more general than that of de Wit (1960) are now being applied to mixed crops; although they require more data to estimate their parameters, they are able to estimate the planting densities required to give particular ratios of yields (Ranganathan, 1993). A problem facing farmers growing mixtures of annuals and perennials is that the balance swings with passing time in favour of the perennial. Many empirical studies have shown that the perennial needs to be pruned to restore the balance but timings and intensities are hard to judge (Duguma et al., 1988); only recently has a science-based framework been suggested to guide pruning (Ranganathan, 1993).

Some comparisons have been made between the responses to stress of multi- and single-species stands. In mixed legume-non-legume intercrops, the response to nitrogen deficiency, caused, say, by untimely leaching rains is likely to be much less marked than in stands of pure nonlegume (Davis et al., 1986). Again, the desk study of the spreading through time of the response of a three-component agroforestry stand (Figure 5) shows that the financial impact of a breakdown in irrigation is likely to be less severe than in any pure stand. While compensatory growth can help mixed stands be resilient in the face of some environmental stresses, as intensification imposes a stress of its own on a system, a multispecies system may become more sensitive to new stresses, either different in type or of the same sort. Some key species or group of species may be lost and production may swiftly collapse. Thus, in the shifting cultivation example, a sequence of excessively long cropping phases may bring the system to a stage when the forest suddenly fails to regrow (Figure 7). Further examples of chronic stress leading to system collapse can be cited, although not specifically from India (Walker et al., 1981; Trenbath and Smith, 1981; Trenbath et al., 1989).

6.3. *Trade-offs and conclusion*

As in many other forms of economic activity, trade-offs are present which could affect some decisions. A common problem is that while multispecies systems have advantages in production and behaviour, they are more difficult to manage. Among the difficulties already quoted are their sensitivity to timing of interventions, the tendency of some to collapse under stress (Figure 7), and their heavier requirements for farmer skill and research effort. This last point is illustrated in the stochastic variant of the 'complete diet' study (Figure 4) which shows that the yield advantage from growing an ideal-proportion mixture can only be obtained at the expense of greater uncertainty about what the ideal proportion is. If the best planting proportion for a mixture has a larger variance than that for sole crops, more experimentation will be needed to estimate its value within given confidence limits.

A particularly deadly trade-off affecting most agricultural systems involves productivity and sustainability (e.g. Bellows et al., 1996). It is seen in several of the case studies presented here. Thus, intensification of the shifting cultivation system by doubling the number of crops taken before abandonment of the land makes its productivity greater in the short term, but makes it fall faster with time (Figure 7c). Furthermore, when the effects of two forms of intensification of the same system are presented in all combinations over a wide range (Figure 10), the positive correlation of productivity and rate of decline of soil fertility is seen to be near-perfect.

It might be asked whether the relatively high productivity of some mixed cropping systems will be associated with correspondingly fast soil degradation. Few long-term experiments have been suitable to test this but the answer is probably 'not necessarily'. One reason is that the heavier canopies of intercrops created by planting together components each at near its sole crop optimum density (e.g. sorghum/pigeonpea), will better protect the soil from radiation, excessively high temperatures and fast oxidation rates of organic matter, at least during the growing season. Another reason is that such heavy canopies, leaf-fall and stubbles will better protect surface soil from loss by erosion. Such effects may or may not compensate any extra loss of nutrients removed in crop products. With regards to nitrogen, greater removal should not necessarily lead to faster impoverishment. This is illustrated in Figure 9 where the relatively high yields of the 60% pigeonpea ($P = 0.6$) mixture are expected to continue for at least 30 years but with only a moderate loss of soil nitrogen.

This special relationship between legumes and non-legumes mediated by nitrogen fixation has featured in several of the Indian case studies presented in this paper. Other special relationships, sometimes mediated by pest or disease organisms are known but are less general. All such special relationships need to be taken into account when designing cropping systems in which advantages in productivity and dynamic properties are expected to rely on some degree of biodiversity.

7. Acknowledgements

For helpful information and discussion I am grateful to R. Ranganathan, B. L. Das, P. S. Ramakrishnan, T. J. Rego, D. A. Neher and C. I. Bhagat. I thank M. Van Noordwijk, M. R. Rao and an anonymous referee for helpful comments on the manuscript.

8. References

Akonde TP, Leihner DE and Steinmuller N (1996) Alley cropping on an ultisol in sub-humid Benin. Part 1. Long-term effect on maize, cassava and tree productivity. Agroforestry Systems 34: 1–12

Allen BJ, Bourke RM and Hide RL (1995) The sustainability of Papua New Guinea agricultural systems: the conceptual background. Global Environmental Change 5: 297–312

Ali (1996) Pigeonpea-based cropping systems in the semi-arid tropics. In: Ito O, Adu-Gyamfi JJ, Katayama K, Kumar Rao JVDK and Rego TJ (eds) Dynamics of roots and nitrogen in cropping systems of the semi-arid tropics), pp 41–58. Japan International Research Center for Agricultural Sciences, Ibaraki

Altieri MA (1995) Agroecology, 2nd ed. Westview, Boulder, 433 pp

Andrew JS (1957) Proteins and nutrition. Cereal Science Today 2: 242; cited from Reeves JT (1958) The relationship of yield and protein in wheat. MSc thesis, The University of Western Australia, 118 pp

Anon (1994) Caught in an export trap. International Agricultural Development 14: 9–10

Aykroyd WR and Doughty J (1970) Wheat in Human Nutrition. Food and Agriculture Organisation of the United Nations, Rome, 163 pp

Bandy DE, Garrity DP and Sanchez PA (1993) The worldwide problem of slash-and-burn agriculture. Agroforestry Today 5: 2–6

Barnard CS and Nix JS (1973) Farm planning and control. Cambridge University Press, UK, 549 pp

Bellows BC, Hildebrand PE, and Hubbell DM (1996) Sustainability of bean production systems on steep slopes in Costa Rica. Agricultural Systems 50: 391–410

Bressani R and Scrimshaw NS (1961); cited from Aykroyd WR and Doughty J (1964) Legumes in human nutrition. Food and Agriculture Organisation of the United Nations, Rome, 138 pp

Chacon JJ and Gliessman SR (1982) Use of the 'non-weed' concept in traditional tropical agroecosystems of southeastern Mexico. Agroecosystems 8: 1–11

de Wit CT (1960) On competition. Verslag Landbouwerk Onderzoek 66: 1–82

Cox GW and Atkins MD (1979) Agricultural ecology: an analysis of world food production systems. WH Freeman and Co., San Francisco, 721 pp

Das BL (1996) Slash-and-burn in India. Agroforestry Today 8: 3

Davis JHC, Wooley JN and Moreno RA (1986) Multiple cropping with legumes and starchy roots. In: Francis CA (ed) Multiple cropping systems, pp 133–160. Macmillan, New York

Duguma B, Kang BT and Okali DUU (1988) Effect of pruning intensities of three woody leguminous species grown in alley cropping with maize and cowpea on an alfisol. Agroforestry Systems 6: 19–35

Ehrlich PR and Ehrlich AH (1972) Population, resources, environment: issues in human ecology. WH Freeman, San Francisco, 509 pp

FAO (1949) Food composition tables for international use. Food and Agriculture Organisation of the United Nations, Rome, 62 pp. Cited from Aykroyd and Doughty (1964)

FAO (1959) Report of the FAO/CCTA technical meeting on legumes in agriculture and human nutrition in Africa, Bukavu, Belgian Congo, 1958. Cited from Aykroyd and Doughty (1964)

FAO (1970) Amino-acid content of foods and biological data on proteins. Food and Agriculture Organisation of the United Nations, Rome, 285 pp

FAO (1974) Handbook on human nutritional requirements. Food and Agriculture Organisation of the United Nations, Rome, 66 pp

Finlay RC (1974) Intercropping soybeans with cereals. Regional Soybean Conference, Addis-Ababa, October 14–17, 1974, pp 77–85

Francis CA (1986) Introduction: distribution and importance of multiple cropping. In: Francis CA (ed) Multiple cropping systems, pp 1–19. Macmillan, New York

Fukai S (ed) (1993) Intercropping – bases of productivity. Field Crops Research (special issue) 34: 239–467

Gautam RC (1996) Intercropping in pearlmillet: a remunerative practice of rainfed farming. Indian Farming 45: 21–31

Gliessman SR (1989) Quantifying the agroecological component of sustainable agriculture: a goal. In: Gliessman SR (ed) Agroecology – researching the ecological basis for sustainable agriculture, pp 366–370. Springer-Verlag, New York

Hart RD (1986) Ecological framework for multiple cropping research. In: Francis CA (ed) Multiple cropping systems, pp 40–56. Macmillan, New York

Holling CS (1973) Resilience and stability of ecological systgems. Annual Review of Ecology and Systematics 4: 1–23

Hore DK, Govind S and Singh IP (1997) Collecting Citrus germ plasm from the Mizoram and Tripura hills of India. Plant Genetics Resources Newsletter No. 110: 57–59

Hulse et al. (1980) Cited from Hoseney RC, Andrews DJ and Clark H (1987). Sorghum and pearl millet. In: Olson RA and Frey KJ (eds) Nutritional quality of cereal grains: genetic and agronomic improvement, pp 397–456. American Society of Agronomy, Madison, Wisconsin

IARI (1976) Beneficial effects from the introduction of legumes into crop rotations and intercropping systems. Indian Agricultural Research Institute Reporter 1: 1–5

Jong W (1996) Swidden-fallow agroforestry in Amazonia: diversity at close distance. Agroforestry Systems 34: 277–290

Kang BT, Versteeg MN, Osiname O and Gichuru M (1991) Agroforestry in Africa's humid tropics: three success stories. Agroforestry Today 3: 4

Kumar Rao JVDK, Wani SP and Lee KK (1996) Biological nitrogen fixation through grain legumes in different cropping systems of the semi-arid tropics. In: Ito O, Adu-Gyamfi JJ, Katayama K, Kumar Rao JVDK and Rego TJ (eds) Dynamics of roots and nitrogen in cropping systems of the semi-arid tropics, pp 323–334. Japan International Research Center for Agricultural Sciences, Ibaraki

Lentner C (ed) (1992) Geigy Scientific Tables. CIBA-GEIGY, Basle, Switzerland. Vol. 1, 295 pp

McCown RL and Jones RK (1992) Agriculture of semi-arid Eastern Kenya: problems and possibilities. In: Probert ME (ed) A search for strategies for sustainable dryland cropping in semi-arid Eastern Kenya, pp 8–14. Australian Centre for International Agricultural Research, Canberra

Natarajan M and Willey RW (1980) Sorghum-pigeonpea intercropping and the effects of plant population density. 1. Growth and yield. Journal of Agricultural Science, Cambridge 95: 51–58

Noy-Meir I (1975) Stability of grazing ecosystems: an application of predator-prey graphs. Journal of Ecology 63: 459–481

Papadakis JS (1941) Small grains and winter legumes grown mixed for grain production. Agronomy Journal 33: 504–511
Petterson DS and Mackintosh JB (1994) The chemical composition and nutritive value of Australian grain legumes. Grains Research and Development Corporation, Canberra, 68 pp
Platt BS (1962) Tables of representative values of foods commonly used in tropical countries. Medical Research Council, Her Majesty's Stationery Office, London, 46 pp
Puckridge DW and French RJ (1983) The annual legume pasture in cereal-ley farming systems of southern Australia: a review. Agriculture, Ecosystems and Environment 9: 229–267
Ramakrishnan PS (1984) The science behind the rotation al bush fallow agriculture system (jhum). Proceedings of the Indian Academy of Science (Plant Sciences) 93: 379–400
Ranganathan R (1993) Analysis of yield advantages in mixed cropping. Ph.D. Thesis, Landbouwuniversiteit, Wageningen, 93 pp
Rao MR (1986) Cereals in multiple cropping. In: Francis CA (ed) Multiple Cropping Systems, pp 96–132. Macmillan, New York
Rao MR and Coe R (1992) Evaluating the results of agroforestry research. Agroforestry Today 4: 4–8
Rao MR and Willey RW (1978) Current status of intercropping research and some suggested experimental approaches. Proceedings of the 2nd INPUTS Review Meeting, EastWest Center, 17 May 1978, pp 123–137. Honolulu, Hawaii
Rao MR and Willey RW (1981) Stability of performance of a pigeonpea/sorghum intercrop system. Proceedings of the International Workshop on Intercropping, 10–13 January 1979, Hyderabad, India, pp 306–317. International Crops Research Institute for the Semi-Arid Tropics, Patancheru, AP India
Rao MR and Willey RW (1983). Effects of genotype in cereal/pigeonpea intercropping on the alfisols of the semi-arid tropics of India. Experimental Agriculture 19: 67–78
Reddy MS and Willey RW (1981) A study of pearl millet/groundnut intercropping with particular emphasis on the efficiencies of leaf canopy and rooting pattern. Proceedings of the International Workshop on Intercropping, 10–13 January 1979, Hyderabad, India, pp 202–209. International Crops Research Institute for the Semi-Arid Tropics, Patancheru, AP India
Rego TJ and Seeling B (1996) Long-term effects of legume-based cropping systems on soil nitrogen status and mineralization in vertisols. In: Ito O, Adu-Gyamfi JJ, Katayama K, Kumar Rao JVDK and Rego TJ (eds) Dynamics of roots and nitrogen in cropping systems of the semi-arid tropics, pp 469–479. Japan International Research Center for Agricultural Sciences, Ibaraki
Roger JH and Rao MR (1990) Discovering the hard facts. Agroforestry Today 2: 4–7
Smith ME and Francis CA (1986) Breeding for multiple cropping systems. In: Francis CA (ed) Multiple cropping systems, pp 219–249. Macmillan, New York
Subedi KD (1997) Wheat intercropped with tori (*Brassica campestris* var. *toria*) and pea (*Pisum sativum*) in the subsistence farming system of the Nepalese hills. Journal of Agricultural Science, Cambridge 128: 283–289
Trenbath BR (1974) Biomass productivity of mixtures. Advances in Agronomy 26: 177–210
Trenbath BR (1976) Plant interactions in mixed crop communities. In: Papendick RI, Sanchez PA, Triplett GB and Bronson RD (eds) Multiple Cropping, pp 129–169. American Society of Agronomy, Special Publication, Madison
Trenbath BR (1981) Comment on a paper by Mead and Riley on the statistical aspects of intercropping. Journal of the Royal Statistical Society 144: 492–493
Trenbath BR (1984a) Decline of soil fertility and the collapse of shifting cultivation systems under intensification. In: Chadwick AC and Sutton SL (eds) Tropical Rain-forest: the Leeds Symposium, pp 279–292. Leeds Philosophical and Literary Society, Leeds
Trenbath BR (1984b) The dynamic proporties of mixed crops. In: Frontiers of Research in Agriculture, Proceedings of the Indian Statistical Institute Golden Jubilee International Conference, Calcutta, Sept/Oct 1982, pp 265–286. Indian Statistical Institute, Calcutta

Trenbath BR (1993) Intercropping for the management of pests and diseases. Field Crops Research 34: 381–405

Trenbath BR, Conway GR and Craig IA (1989) Threats to sustainability in intensified agricultural systems: analysis and implications for management. In: Gliessman SR (ed) Agroecology-researching the ecological basis for sustainable agriculture, pp 337–365. Springer-Verlag, New York

Trenbath BR and Smith ADM (1980) Basic concepts for a systems analysis of eucalypt dieback in New England. In: Old KM, Kile GA and Ohmart CP (eds) Eucalypt dieback in Forests and Woodlands, pp 234–244. Commonwealth Scientific and Industrial Organisation, Melbourne

Vandermeer J (1989) The Ecology of Intercropping. Cambridge University Press, Cambridge, UK, 237 pp

Vandermeer J, Van Noordwijk M, Ong C, Anderson J and Perfecto Y (1998) Global change and multi-species agroecosystems: concepts and issues. Agriculture, Ecosystems and Environment 67: 1–22

Van Noordwijk M, Hairiah K, Partoharjono S, Labios RV and Garrity D P (1997) Sustainable food-crop based production systems, as alternative to *Imperata* grasslands? Agroforestry Systems 36: 55–82

Walker BH, Ludwig D, Holling CS and Peterman RM (1981) Stability of semi-arid savannah grazing systems. Journal of Ecology 69: 473–498

Wendt JW, Jones RB and Itimu OA (1994) An integrated approach to soil fertility improvement in Malawi, including agroforestry. In: Crasswell ET and Simpson J (eds) Soil fertility and climatic constraints in dryland agriculture, pp 74–79. Australian Council for International Agricultural Research, Canberra

Willey RW (1979a) Intercropping – its importance and research needs. Part I. Competition and yield advantages. Field Crop Abstracts 32: 1–10

Willey RW (1979b) Intercropping – its importance and research needs. Part II. Agronomy and research approaches. Field Crop Abstracts 32: 73–85

Worldwatch Institute (1994) Report on progress toward a sustainable society. WW Norton and Co., New York. 265 pp

Xi C (1995) Slash-and-burn in China. Agroforestry Today 7: 2–3

Zinke PJ, Sabhasri S and Kunstadter P (1978) Soil fertility aspects of the Luao forest fallow system of shifting cultivation. In: Kunstadter P, Chapman EC and Sabhasri S (eds) Farmers in the forest, pp 134–159. University Press of Hawaii, Honolulu

Agroforestry Systems **45**: 109–129, 1999.

Why tree-crop interactions in agroforestry appear at odds with tree-grass interactions in tropical savannahs

C. K. ONG[1,*] and R. R. B. LEAKEY[2]
[1]*International Centre for Research in Agroforestry, P.O. Box 30677, Nairobi, Kenya;*
[2]*Institute of Terrestrial Ecology, Bush Estate, Penicuik, Midlothian, Scotland, EH26 0QB*
*(*Author for correspondence: E-mail: c.ong@cgnet.com)*

Key words: alley cropping, competition, complementarity, light interception, resource capture, water use

Abstract. This paper describes recent research findings on resource sharing between trees and crops in the semiarid tropics and attempts to reconcile this information with current knowledge of the interactions between savannah trees and understorey vegetation by examining agroforestry systems from the perspective of succession. In general, productivity of natural vegetation under savannah trees increases as rainfall decreases, while the opposite occurs in agroforestry. One explanation is that in the savannah, the beneficial effects of microclimatic improvements (e.g. lower temperatures and evaporation losses) are greater in more xeric environments. Mature savannah trees have a high proportion of woody above-ground structure compared to foliage, so that the amount of water 'saved' (largely by reduction in soil evaporation) is greater than water 'lost' through transpiration by trees. By contrast, in agroforestry practices such as alley cropping where tree density is high, any beneficial effects of the trees on microclimate are negated by reductions in soil moisture due to increasing interception losses and tree transpiration. While investment in woody structure can improve the water economy beneath agroforestry trees, it inevitably reduces the growth rate of the trees and thus increases the time required for improved understorey productivity. Therefore, agroforesters prefer trees with more direct and immediate benefits to farmers. The greatest opportunity for simultaneous agroforestry practices is therefore to fill niches within the landscape where resources are currently under-utilised by crops. In this way, agroforestry can mimic the large scale patch dynamics and successional progression of a natural ecosystem.

1. Introduction

Farmers in drylands have for centuries utilised a wide range of agroforestry practices, largely by preserving and managing a few scattered mature trees (Rocheleau et al., 1988; Shankarnarayan et al., 1987). Recent attempts to promote agroforestry through further tree planting in such environments have been disappointing (Singh et al., 1989; Nair, 1993; Cooper et al., 1996; Rao et al., 1998). The above reviews have highlighted the limitations of, for example, alley cropping in the semiarid tropics, where below-ground competition for water between trees and crops frequently outweighed the benefits of soil enrichment and microclimate improvements. These reviews offered few solutions, despite substantial evidence of improved soil nutrient, microclimatic conditions and increased vegetative production beneath large isolated trees in most savannah ecosystems (Huntley and Walker, 1982; Rhoades, 1997;

Belsky and Amundson, 1997). It might be expected that similar processes could be developed in agroecosystems.

The hydrological impact of fast growing exotic trees has caused concern to local peoples in many tropical countries, especially in semiarid areas, due to the 'Eucalyptus controversy'. In southern India, eucalypt plantations have been reported to not only extract all the rainfall that enters the soil, but also to utilise an additional 100 mm of water from each metre depth of soil that the roots penetrate (Calder et al., 1997). This is a matter of concern as recent studies in Karnataka, India, have shown that eucalypt roots can reach to a depth of 8 m in three years (Calder et al., 1997). In the Sahel, roots of mature, well-scattered trees can even reach the water-table at 30 m (Deans et al., 1995). Consequently, when planning trees in association with crops in drylands, it is essential to consider the implications of increased water use on the medium and longer-term water budgets. Special consideration should be paid to the source of water used by trees, the rate of water depletion below the crop rooting zone and the prospects for deep recharge as using water at faster than replacement rates demonstrates a serious ecological consequence of not mimicking the resource use patterns of locally adapted species.

2. Resource use in savannah ecosystems

According to the savannah literature, grasses utilise the topsoil water while tree roots have exclusive access to deeper water, creating a clear niche separation (Weltzin and Coughenour, 1990; Deans et al., 1995). However, Belsky and her colleagues (1993) observed that the shallow rooting baobab (*Adansonia digitata*) has the same beneficial effect on understorey soils and vegetation as the deep-rooted *Acacia tortilis*. One explanation could be that both species have a higher proportion of woody aboveground structure than foliage, so that more soil water is saved by reducing soil evaporation than is lost as transpiration, although interception loss would still exist. In contrast, the proportion of woody structure in most new simultaneous agroforestry systems is deliberately kept to a minimum by either frequent pruning or selection of leafy tree species (Van Noordwijk and Ong, this volume).

If the beneficial effects of savannah and parkland trees on soil properties are linked to trees with a high proportion of woody aboveground structures, then it would take a long time before the beneficial effects can be realised, since investment in woody structure slows tree growth. Therefore it is not surprising that the positive interactions between mature, deep-rooted and widely spaced *Faidherbia albida* trees (the 'albida' effect) would require 20–40 years in Ethiopia (Poschen, 1986). Such long time scales are well beyond the planning horizon of many farmers for the relatively small benefit in crop productivity and may well explain why farmers rarely plant these trees, even though they are well aware of tree species which improve soil and climate.

Instead of focusing agroforestry primarily on soil amelioration, it would perhaps be more worthwhile to focus attention on selection of trees to provide direct and more immediate benefits to farmers, with minimum loss of crop productivity. It is perhaps not surprising that farmers are already beginning to experiment with such systems. For example, in the drylands of eastern Kenya, farmers have recently developed an intensive parkland system using a fast-growing indigenous species, *Melia volkensii* (Meliaceae). This tree provides high value timber in five to eight years plus fodder during the dry season, without an apparent loss in productivity in associated crops (Stewart and Bromley, 1994). Consequently, ICRAF is currently investigating the tree-crop interactions in the *Melia* system in farmers fields.

3. Agroforestry in drylands

Early studies of agroforestry began with the general assumption that complementarity was the norm as trees are believed as a rule to be more deeply rooted than crops. However, researchers were soon discouraged by the substantial reduction in associated crop yield and they began to focus on below ground competition for resources. In this paper we trace the progress of agroforestry research in drylands by introducing them as a series of hypotheses. The first hypothesis states that where trees can exploit resources that are unavailable to crops, agroforestry can increase productivity per unit area of land through the efficient use of resources (Cannell et al., 1996); in other words through spatial complementarity. The second hypothesis is based on the assumption that competition for below ground resources is inevitable as tree and crop root systems are likely to have similar distribution in the soil profile (Jonsson et al., 1988; Singh et al., 1989; Schroth, 1995). It suggests that to reduce the negative effects of competition, tree spacing should be increased to minimise competition for water with crops (Singh et al., 1989). An extension of this idea is that to further minimise negative effects of trees on crops they should be matched to appropriate niches within the farm (Van Noordwijk and Ong, 1996). This hypothesis is also based on spatial complementarity but suggests it may be the exception rather than the rule. Third, it has been suggested that attention to differences in tree phenology could result in temporal complementarity, in which trees like *Faidherbia albida*, with its reverse phenology (Ong et al., 1992), make growth demands at a different time from that of the crops. Finally, it has been hypothesised that trees will harvest resources from their surroundings and accumulate them under the tree (Breman and Kessler, 1995). As the positive influences of woody plants on productivity are due to the redistribution of resources within the landscape it will always be difficult to develop agroforestry without risking the depletion resources at a distance from the tree. An intriguing possibility is the prospect of capitalising on this phenomenon of resource capture at a distance through strategic use of 'hydraulic lift'. This is the process in which water taken up by roots from moist zones of soil is transported through the root system and released into drier soil (Dawson, 1993, Pate and Dawson, this volume). Rainfall captured through stem flow, especially by a woody canopy, can be stored deep in the soil close to the roots and be returned to the topsoil beneath the canopy by hydraulic lift for later use with associated benefit for understorey species. Current modelling studies suggest that this strategy is only important to the tree as a means of penetrating dry soil zones to reach deep reserves of water (Van Noordwijk and Ong, this volume).

This paper examines these four hypotheses using results of recent research investigations on simultaneous agroforestry practices, primar3ily from Machakos, Kenya. It then attempts to reconcile such information with current knowledge on the ecology of savannah vegetation. Recent insights into the effects of savannah trees on understorey vegetation and soils should provide valuable clues on how to reduce the negative effects of below-ground competition in agroforestry while retaining the positive effects of trees seen in natural ecosystems. It is this opportunity for agroforests to mimic the interactions between trees and other plants in natural ecosystems that led to the recent redefinition of agroforestry (Leakey, 1996), in which different agroforestry practices are viewed as stages in the development of an agroecological succession akin to the dynamics of natural ecosystems. Over time, the increasing integration of trees into landuse systems through agroforestry can be seen as the passage towards a mature agroforest of increasing integrity. Similarly, with increasing scale, the integration of various agroforestry practices into the landscape is like the formation of a complex mosaic of patches in an ecosystem, each of which is composed of many niches. These niches are occupied by different organisms, making the system ecologically stable and biologically diverse (Leakey, 1996). In systems like this, the physiological interactions between the tree and crop components of the agroecosystem are more likely to mimic those of natural ecosystems.

3.1. *Hypothesis 1: Agroforestry systems can increase the efficient use of resources by combining species with different patterns of resource uptake*

The concept of complementary resource use is not new in ecological studies. It was proposed by de Wit (1960) and others (Trenbath and Harper, 1973; Willey, 1985; Vandermeer, 1989), that mixtures of species may have greater capacity to exploit growth resources and hence be more productive than monocultures. The principles of resource capture have been used to examine the influence of agroforestry on ecosystem function, i.e. the capture of light, water and nutrients (Ong and Black, 1994), and to better understand the ecological basis of sustainability of tropical forests (Ewel, 1986; Ewel and Bigelow, 1995). The theme of complementary resource use has been explored extensively in intercropping studies in the tropics (Trenbath, 1974; Fukai and Trenbath, 1993; Sinoquet and Cruz, 1995; Trenbath, this volume). However, these recent re-interpretations of published results indicate that increased yield from combinations of annual crops was *not* always associated with greater resource capture or utilisation. For example, the choice of fast-growing, competitive tree species, such as *Leucaena leucocephala* Lam., has often been found to be inappropriate for alley cropping despite their good coppicing and nitrogen-fixing capacity. Several workers have shown that in dryland conditions, leucaena is very competitive with intercrops such as millet, sorghum, groundnut and maize (Singh et al., 1989; Jama et al., 1995). Consequently, it has been recommended that research attention should focus on tree species with more compatible root distribution (i.e. better spatial complementarity – Van Noordwijk and Purnomosidhi, 1995).

The capture of growth resources by trees and crops can be grouped into the broad categories of neutral, complementary or competitive interactions (Figure 1). In the neutral or trade-off category, trees and crops exploit the same pool of resources so that increases in capture by one species result in a proportional decrease in capture by the associated species. If trees were able to tap resources unavailable to crops, then the overall capture would be increased as shown by the convex curve in Figure 1, i.e. complementary use of resources. In the third category, negative (competitive) interactions between the associated species could result in serious reduction in the ability of one or both species to capture growth resources (concave curve). It is important to bear in mind that in agroforestry, tree-crop interactions may change from one category to another depending on the age, size and population of the

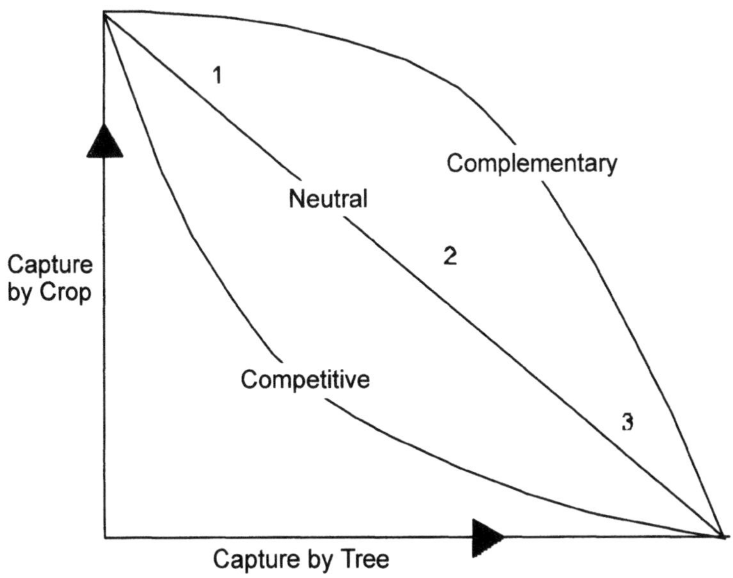

Figure 1. Resource capture by trees and crops showing competitive, complementary and neutral interactions 1. Parkland and savannah (Late stage agroecosystem), 2. Boundary planting, 3. Alley cropping (Early stage agroecosystem)

dominant species (as in a natural ecosystem), as well as the supply and accessibility of the limiting growth resources. Later, we use this diagram to illustrate the discrepancy between the findings of various agroforestry practices and that of the savannah ecosystem. We also suggest that the competitive interactions are more often associated with the early successional stages of an agroecosystem, while the complementary interactions occur during the late successional stages of mature agroforests, such as the farmed parklands of Sahelian Africa.

Such ideas on capture of water and nutrients, coupled with recent innovations in instrumentation (mini-rhizotrons, sap flow methodologies), have stimulated a resurgence in root research (Schroth, 1995; Van Noordwijk and Purnomosidhi, 1995; Khan and Ong, 1996) and attention to spatial complementarity in root distribution and the potential beneficial effects of deep rooting.

The benefit of deep rooted vegetation for maintaining ecosystem functioning is emphasised in parts of Australia, where the removal of the native perennial vegetation and its replacement by shallow-rooted annual crops and pastures has led to a profound change in the pattern of energy capture by vegetation, less efficient use of rainfall, increased vertical and lateral water movement, and hence rising water-tables and associated salinity (Hatton and Nulsen, this volume; Dunin et al., this volume). In the Sahel, removal of vegetation with deep roots has also led to increased leakiness (Cuff et al., 1993). In this situation, deep drainage on the sandy soils increased from 10–20 mm to 200–300 mm per year (Gaze et al., 1997) and a pulse of nitrogen is on its way down to the water table (Edmunds, 1991; Deans et al., 1995). In western Kenya, *Sesbania sesban* has been shown to utilise nitrogen below the rooting zone of maize (Mekonnen et al., 1997) and to make this available to crops through its leaf litter. This indicates that nutrients can be captured at depth by agroforestry trees and redistributed vertically within the soil profile. Evidence also indicates that the network of tree roots in agroforestry practices are able to reduce the leaching of nutrients from the surface soil layers, so reducing the leakiness of the nutrient cycle (Buresh and Tian, 1998). Earlier research on South African savannahs has also shown that tree roots extend into the open grassland, providing a ‘safety net’ for recycling water and nutrients, and accounting for 60% of the total below-ground biomass (Huntley and Morris, 1982). Walker and Noy-Meir (1982) also presented a conceptual model of the African savannah in which the grass roots are restricted to the upper layers which dry out first, while the tree roots have access to deeper layers, which are usually above the wilting point.

One of the earliest detailed studies of resource capture in agroforestry practices was that for a C_4 crop with a C_3 tree in semiarid India, described by Monteith et al. (1991) and Corlett et al. (1992a, 1992b). In this work in Hyderabad (annual rainfall of 700 mm), involving a *Leucaena leucocephala*/*Pennisetum americanum* (millet) alley cropping system, radiation interception and soil moisture were measured, but soil moisture results were unreliable because of highly variable soil physical properties. Total intercepted radiation during the rainy season was 40% greater in the alley crop than in sole millet, primarily because the presence of leucaena increased fractional interception during the early stages of the growing season, while the millet provided a more complete ground cover across the alleys during the later stages of the season. The sole leucaena intercepted only 6% more light than the alley crop during the rainy season. The sole leucaena and alley leucaena intercepted twice as much radiation again as millet during the following long dry season after millet harvest. The evidence from this study shows that the main advantage of alley cropping was in extending the growing period into the dry season and increasing the annual light interception. However, interception by the more efficient C_4 crop was reduced to only half that of the sole millet during the rainy season and accounted for 15% of the total intercepted by the alley system. This, therefore, falls within the lower end of the complementary curve (Figure 1). In this study from India, the alley crop produced 7 t ha^{-1} biomass compared to 4.7 t ha^{-1} of sole millet and 4 t ha^{-1} by sole trees despite the high amount of light interception, because of the low photosynthetic rate or conversion coefficient of the C_3 species. The conversion coefficient or 'radiation use efficiency' (ε) is defined here as the ratio of biomass production to intercepted light per unit area and provides a measure of the efficiency with which the captured light is used to produce new biomass.

Subsequent resource capture studies at Hyderabad examined the partitioning of water and light in an agroforestry system comprising of two C_3 species, perennial pigeon pea (*Cajanus cajan*) and groundnut (*Arachis hypogaea*), using tube solarimeters and sap flow gauges (Marshall, 1995; Ong et al., 1996). Monocrops of each species were compared with line-planted and dispersed mixtures containing identical pigeon pea populations. Monocropped groundnut and pigeon pea intercepted about 850 MJ m^{-2} each in year two, but the dispersed system intercepted twice as much light during the same period. Although ε of the understorey groundnut increased by 28%, interception by groundnut contributed only 19% of the total interception in the dispersed system. Again this falls within the lower end of the complementary curve of Figure 1.

Both Hyderabad studies showed that the less efficient C_3 overstorey (tree) component dominated the total light interception, while the increased ε of the understorey component was insufficient to compensate for the reduced light interception. These results are typical of many alley cropping studies where the tree populations were so dense that reduction in crop yield was inevitable, because the trees captured most of the resources at the expense of the crops. Although crop yields were seriously reduced, these are examples of complementary interactions, but are often misinterpreted as competitive, because without the sole tree controls is was not possible to determine whether the alley cropping did capture more resources.

Similar investigations in semiarid Kenya confirmed that, on deep alfisols, alley cropping leucaena/maize has the potential to increase light interception and to double water use as compared to sole maize stands (Howard et al., 1995), but that this is unlikely on shallow soils (McIntyre et al., 1996). These findings show that, by extending the period of growth, tropical agroforestry practices can substantially increase light interception and water use as compared to the shorter growth period of most annual cropping systems. This will only be the case if the water resource is not fully utilised by the annual crop. However, the conversion efficiency of the intercepted light in agroforestry practices will generally be lower than that reported for crops and savannah grasslands, since the photosynthetic system of trees is invariably C_3 and, thus, less efficient than that of C_4 crop plants. However, values of ε obtained from trees and forests are highly variable and greatly underestimated, since they are usually based on above-ground production. This focus on above-ground production is because quantification of tree root biomass is notoriously difficult, due to continuous decay (turnover) and consumption by soil biota.

As shown above for light, agroforestry offers substantial scope for spatial and temporal complementarity of water use, resulting from improved exploitation of soil water and off-season rainfall. However, the opportunity for significant complementarity is likely to be limited unless the species involved differ appreciably in their patterns or duration of rooting. One very successful form of agroforestry in drylands, which is characterised by complementarity, is the scattered mature trees of the Sahelian parklands, which have a discontinuous tree canopy (5–10 trees ha^{-1}). The more limited potential complementarity between high density trees and crops has been highlighted by recent work for a range of tree species at Machakos, Kenya. This showed that when rainfall was low (250 mm) maize yield was linearly and negatively related to the amount of water used by the trees (Figure 2). However, this relationship broke down when rainfall exceeded 650 mm, illustrating that the trees can then use water from the same soil profile as the maize without major negative effects on crop yield.

In water-limited environments, the quantity of biomass produced (W) depends on the quantity of water captured and the efficiency (e_w) with which this is used to produce biomass (expressed as units of biomass per unit water transpired). W is often linearly related to the quantity of water transpired, indicating that e_w is conservative (de Wit, 1958; Azam-Ali, 1983; Cooper et al., 1987). This relationship depends on the close linkage between CO_2 and water vapour fluxes due to the role of stomata in regulating the exchange of both gases. However, atmospheric saturation deficit (D) may exert a strong modifying influence on e_w, as is considered further below.

The e_w values for tropical C_4 cereals are often a little more than double those for C_3 species under equivalent conditions. For example, experiments in India under similar mean saturation deficits (2.0–2.5 kPa) provided season-long values of 3.9 and 4.6 g kg^{-1} for millet, compared to 1.5–2.0 g kg^{-1} for groundnut (Ong et al., 1987; Matthews et al., 1988; Azam-Ali et al., 1989). However, e_w is not always higher in C_4 species. Similar values of e_w have been

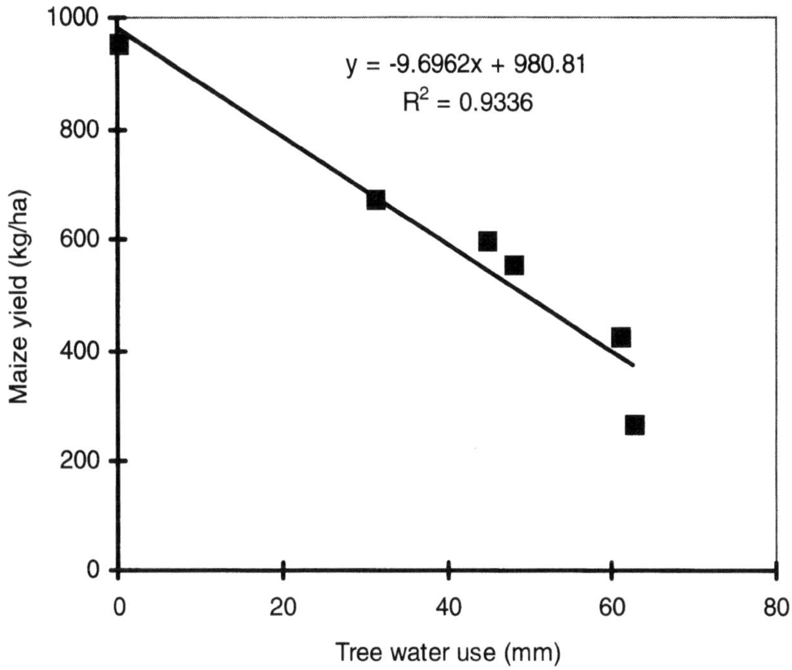

Figure 2. Regression of tree water use and maize yield in the 1996 long rains, Matchakos, Kenya.

reported for drought tolerant C_3 species, such as cowpea and cotton, as well as for relatively drought-sensitive cultivars of the C_4 species, sorghum and maize (Rees, 1986).

In theory then, the potential to improve the e_w of 'late-stages' in agroforestry succession is limited compared to 'early-stage' practices, such as intercropping, because the understorey crops are usually C_4 species and the overstorey trees are invariably C_3 species. Improvement in e_w is most likely if the understorey crop is a C_3 species. Such species are usually light saturated in the open and partial shade may have little effect on assimilation with the result that e_w is improved by concurrent reduction in transpiration. This may explain why cotton yield in the Sahel is not reduced by the heavy shading of karité (*Vitellaria paradoxa*) and néré (*Parkia biglobosa*) in parklands, while yields of millet and sorghum were reduced by 60% under the same trees (Kater et al., 1992). The same reason may explain the observation in the South and Central American savannahs that C_3 grasses are found only under trees and never grow in open grassland dominated by C_4 grasses (Medina, 1982). Surprisingly, there is no clear distinction between the C_3 and C_4 species under the trees in African savannahs, which are dominated by C_4 grasses. For example, in Zimbabwe, *Panicum maximum*, a C_4 grass, occurs as pure stands under *Combretum molle, Terminalia sericea* and *Albizzia* spp. trees but seldom occur in the open. The successful establishment of *P. maximum* under the savannah trees appears to be determined by its ability to germinate through the tree leaf litter, which prevents the survival of other C_4 grasses, abundant in the open (Kennard and Walker, 1973). In East Africa, the dominance of *P. maximum* under tree canopies was attributed to its higher water use efficiency and the higher grazing selection pressure in the open (Kinyamario et al., 1995). These examples of the parkland and savannah systems suggest that they represent the upper end of the complementary curve in Figure 1.

3.2. *Hypothesis 2: Competition for below-ground resources is inevitable and it is best managed by manipulating the interface between species in a mixture*

Early studies of spatial complementarity in agroforestry began by examining the rooting architecture of trees and crops grown as pure stands. For example, Jonsson et al. (1988) described the vertical distribution of five tree species at Morogoro, Tanzania, and concluded that their root distribution were similar to maize except for *Eucalyptus camaldulensis*, which had uniform distribution to 1 m. Thus, they concluded that there is little prospect of spatial complementarity if these trees and crops were grown in combination. Gregory (1996) summarised six other such studies of root distribution of trees, which essentially supported the conclusion of Jonsson et al. (1988). The study of roots of agroforestry systems where crops and trees are grown concurrently is much more difficult and tends to be confined to species which are easily distinguishable on the basis of visual features. Therefore species such as *Cassia* or *Senna* spp. which have a black outer coating on their roots are very convenient and frequently used (Hauser, 1993; Rao et al., 1993). These studies also showed that in semiarid environments, there was still considerable overlap between the roots of trees and crops especially in the top 0.5–1.0 m of the soil, implying that most trees will be competing with crops for both water and nutrients in the top 0.5 m of most soils.

What is the extent of spatial complementarity in water use when there is such a considerable overlap of the two rooting systems? Early results at Machakos, Kenya showed that there was no advantage in water uptake between a *Senna spectabilis*/maize or cowpea alley cropping and sole crop stands when rainfall ranged from 200–250 mm during the rainy season. This was because there was little water recharge below the crop root zone (McIntyre et al., 1997). However, recent measurements of soil moisture profiles of the same alley cropping systems indicate that when recharge occurred following 547 mm of rainfall, tree roots are still able to exploit more moisture below the rooting zone of the crops (Figure 3) even when there was a complete overlap of the root systems of trees and crops. Before the onset of the long rains, soil moisture content under the alleys was already 26 mm drier than in the pure crop treatment and the differences between the treatments increased to 31 mm after the long rains when only 235 mm of rain fell. During the dry

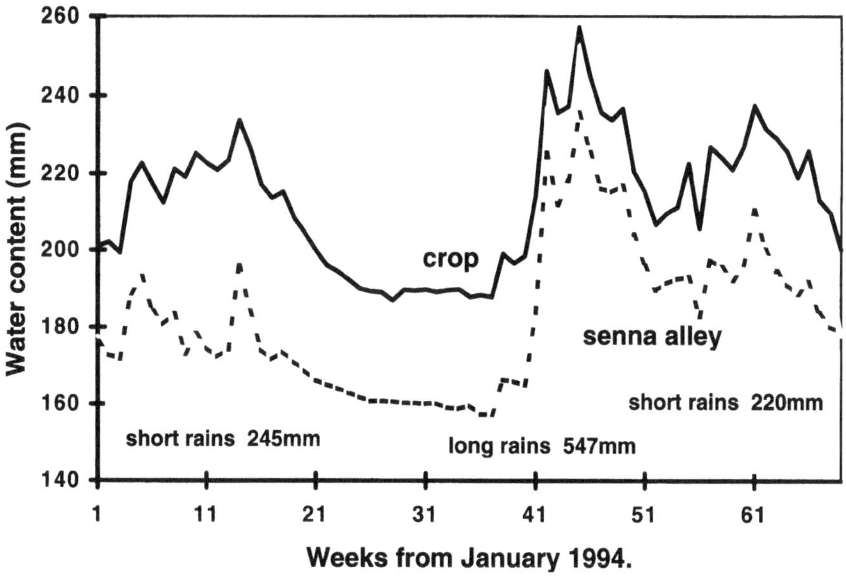

Figure 3. Weekly soil water content to 165 cm under annual crops (maize-cowpea rotation) and alley cropping with *Senna spectabilis* during three cropping seasons from January 1994 to November 1995 at Machakos, Kenya.

season, only a small amount of water was left under the alleys as soil moisture content had reached 14%, which is close to the permanent wilting point (12%) at 1.6 m depth. In this case, crop yield was reduced by 40%, indicating this is an example of the competitive interaction shown in Figure 1. In sharp contrast, soil moisture content remained at 18% below 1.0 m depth (not shown) under the pure crop treatment even after two successive low rainfall seasons. The profile was almost completely recharged in the short rains of 1995, when 547 mm of rain fell. Again, the moisture depletion rate under the alley was significantly greater than in the sole crop treatment. It is the extent of this competition for water by trees that has led to the conclusion by modellers that agroforestry is inappropriate in tropical areas with rainfall less than 800 mm (Cannell et al., 1998). It is relevant to note that the traditional and highly successful agroforestry parklands of semiarid areas are characterised by discontinuous tree canopies with only 15% cover (Breman and Kessler, 1995), while the model is based on continuous tree cover.

Where soil wetting is mainly confined to the crop root zone, competition by trees for soil moisture can be so severe that the risk of crop failure is greatly increased. For example, McIntyre et al. (1996) reported that total evapotranspiration in *Senna spectabilis*/maize or cowpea alley cropping systems in Machakos, Kenya, was either the same as or less than that of pure crops. Measurements in this and other experiments at Machakos revealed that there were still soil evaporation losses (42–48%) in both agroforestry and sole crop treatments because most of the rains occurred early in the cropping season, before: (i) there was adequate tree canopy cover, and (ii) crop roots were extensive enough to exploit the soil water.

Further explanations for the lack of spatial complementarity in alley cropping were advanced by Van Noordwijk and Purnomosidhi (1995), who observed that repeated shoot pruning of trees in alley cropping had the danger of enhancing below-ground competition by increasing the proportion of superficial roots. Van Noordwijk and Purnomosidhi imposed three pruning heights (50, 75, 100 cm) on five tree species (*Paraserianthes falcataria, Gliricidia sepium, Peltophorum dasyrachis, Senna siamea*, and *Calliandra calothyrsus*) in the sub-humid site at Lampung, Sumatra, Indonesia and they found that while pruning had no effect on shoot:root ratios or the relative importance of the tap root, lower pruning regimes promoted more horizontal root growth, providing less spatial complementarity.

Direct measurement of tree transpiration with porometry is tedious and complex. The recent availability of robust sap flow gauges offers a unique opportunity for partitioning water use between trees and crops. Heat balance and heat pulse systems have been used routinely for several years at ICRISAT and ICRAF in a range of agroforestry practices (e.g. Howard et al., 1995, 1997; Lott et al., 1996). The sap flow approach confers the major benefit of allowing daily transpiration by the components of agroforestry systems to be followed continuously and reliably. The heat balance and heat pulse techniques have been modified to measure water movement through the main roots of trees in an approach which is proving invaluable in establishing the quantity of water extracted from the crop rooting zone, and hence the below-ground competitive impact of the trees on crops. The time courses for sap flow through the lateral roots of grevillea in semiarid Kenya (Figure 4) show considerable diurnal variation and indicate that sap flow decreased with increasing distance from the trunk. The trees nevertheless extracted significant quantities of water from the crop rooting zone at distances of up to 190 cm from the trunk. Experiments in which the lateral roots were progressively severed, indicated that these three-year old trees were capable of extracting up to 80% of their water requirements from beneath the crop rooting zone, in agreement with results obtained for the same species from experiments in which sap flow through the trunk was measured while the soil around the trees was excavated to a depth of 60 cm (Howard et al., 1997). In this case the effects on crop yield were minimal, illustrating spatial complementarity.

Recent measurements of the long term (three year) effects of shoot pruning on two species, *S. spectabilis* and *G. sepium*, at Machakos, Kenya confirmed the findings in Lampung, Indonesia in showing that rooting depths of both pruned trees and crops were almost identical and restricted to the top 1 m, whereas the rooting depth of unpruned trees was much deeper and extended to 160 cm. Thus, the rooting profile of the unpruned trees had a greater spatial complementarity with crops than that of pruned trees. However, there was little difference in terms of their competitiveness with associated crops because of the higher leaf area and hence transpiration of the unpruned trees. The

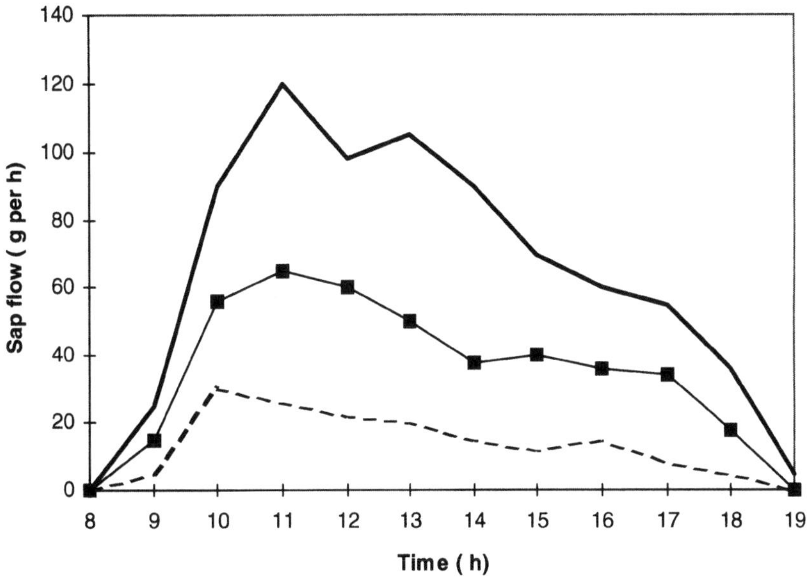

Figure 4. Diurnal trends in sapflow of lateral roots at distances of 50 cm (solid line), 115 cm (filled squares) and 150 cm (dashed line) from trunk of *Grevillea robusta* trees. *Source:* Modified from Lott et al. (1996).

evidence so far suggests that below-ground competition is inevitable in alley cropping where water is limiting (i.e. evaporation exceeds rainfall). This suggests that successful design of alley cropping in each environment hinges on managing the degree of interface between trees and crops to optimise the value of tree products while minimising the impact on crop yield (Lefroy and Stirzaker, this volume).

Recently, the opposite of hydraulic lift described by Dawson (1993) has been reported from Machakos and elsewhere, in which water is taken from the topsoil and transported into the subsoil (Burgess et al., 1998; Smith et al., 1998). This mechanism would enhance the competitiveness of deep-rooted trees and shrubs. Perhaps the selection of fast-growing, deep-rooting trees in agroforestry designs has inadvertently favoured such competitive species?

3.3. *Hypothesis 3: Tree phenology can be selectively used to create temporal complementarity*

In the semiarid tropics, water competition between trees and crops can occur both within the rainy season and through the build-up of a soil water deficit caused by tree transpiration during the dry season. Inter-seasonal water deficit can be particularly severe when there is a succession of droughts and when recharge is small and infrequent (McIntyre et al., 1996; Govindarajan et al., 1996). Thus, temporal complementarity is only possible if residual water is available after the crop harvest or if rainfall occurs when there is no scope for cropping. For example, in Hyderabad, India, 20% (152 mm) of the annual rainfall occurs outside the normal cropping season and even the traditional intercrop of sorghum/pigeonpea utilises only 41% of the annual rainfall. The remainder of the rain is lost as runoff (26%) or deep percolation (33%). In such environments, agroforestry practices increase rainfall utilisation by extending the growing season. Thus, the total rainfall utilisation of perennial pigeonpea/ groundnut systems was 84%, about twice the value reported for the most productive intercrop systems in Hyderabad (Ong et al., 1992). This is one of the largest complementary interactions reported (Ranganathan and de Wit, 1996) and falls on the middle of the complementary curve in Figure 1.

The most remarkable example of temporal complementarity in water use is the unusual phenology of the Sahelian tree, *Faidherbia albida*, which is in leaf in the dry season and sheds its leaves in the rainy season. It retains this phenology even when planted in the Deccan plateau of India, where the water-table is too deep for access by tree roots. One of the few deliberate experiments in which *F. albida* was compared with a tree with conventional leaf phenology is that reported by Ong et al. (1992) at Hyderabad. Comparison of sap flow rates of *F. albida* and a local Indian tree, *Albizia lebbek*, show that transpiration by *F. albida* begins when the understorey crops have developed a full canopy later in the rainy season. In contrast, *A. lebbeck* produces a full canopy well before the onset of the rains and sheds its leaves when the *F. albida* starts to develop its canopy. This is clearly an example of the neutral category (Figure 1).

3.4. *Hypothesis 4: Trees cause a redistribution of resources within the landscape*

Unlike the evidence presented for early-successional agroforestry practices like alley cropping, improved soil water and nutrient status have been reported under single mature tree canopies (late stage agroecosystems) in savannah areas in Africa, Central and South America and North America (Belsky and Amundson, 1997; Rhoades, 1997). There is strong evidence that nutrient redistribution by deep and extensive root systems as well as capture of resources by tree canopies are responsible for the improved soil nutrition beneath *Prosopis juliflora* (Triedmann and Klemmedson, 1973). In Kenyan savannahs, Belsky and her colleagues observed improved microclimate, greater soil biotic activity and N-mineralisation, greater infiltration rate and greater beneficial effects in more xeric (from 750 mm to 450 mm rainfall) environments (Belsky, 1994). They also demonstrated that plants grown in the open sites were more nutrient-limited than those under the tree canopy, and that artificial shade applied to plants in the open generated smaller increases than that of vegetation observed under trees. However, Belsky was unable to firmly conclude whether nutrient enrichment or microclimate changes were more important in increasing understorey productivity. It is significant to note that the positive effects of trees on understorey vegetation are limited to certain sites and species combinations, including both nitrogen-fixing and non-nitrogen fixing trees. It is also difficult to determine precisely whether the tree-grass interactions in tropical savannahs are typical of the competitive category (Figure 1). The problem is that evidence from the literature is primarily based on plot level analyses. Belsky et al. (1989) postulated that the canopy zones of savannah trees, which are mostly shallow-rooted, were enriched by nutrients brought from surrounding grasslands by both tree roots and seed-eating weaver birds, and not from deep underground. It has also been suggested that water taken up by plant roots from moist zones of soil is transported through the root system and released into drier soil (Dawson, 1993). These two examples of resource redistribution would thus be an example of the neutral category in Figure 1 because they compensate for competition for other resources.

Evidence from a series of shade cloth trials on maize and beans at Machakos indicated a small beneficial effect of shading on crop temperature and crop production when rainfall is sub-optimal for crop production (Lott, unpublished PhD thesis). There was no evidence of improved soil moisture under the shade of tree canopies. In contrast, Rhoades (1995) reported increased soil water (4–53% greater than in the open) in the crop root zone beneath *Faidherbia albida* canopies in Malawi (see also Joffre, this volume). In theory, trees can increase soil water content underneath their canopies, if the water saved by shading effects on soil evaporation and rainfall redistribution (e.g. funnelling of intercepted rainfall as stem flow), exceeds that removed by the root systems beneath tree canopies. At high tree densities, the proportion of rainfall lost as interception by tree canopies and used for tree transpiration would exceed that saved by shading and stem flow, resulting in drier soil below the tree canopy. This may be one of the most important factors for the observed difference savannah and alley cropping findings.

4. Trees and deep water use

A study of water recharge was made at Machakos, Kenya, over four consecutive drought seasons (220, 250, 120, 230 mm) in two different layers of the 1.75 m deep soil profile over a 12-month period (Figure 5). It examined patterns of soil moisture content under: (a) a sole annual crop plot and (b), a plot of crop with a single row of four year-old *Grevillea robusta* trees. As expected from root profile observations (ICRAF, 1996) the annual crops utilised water mainly from the top 75 cm of the soil profile and substantial amounts of water remained below this depth, even during the short rains 1996/7 when only 180 mm fell compared to average rainfall of 350 mm. Recharge below 75 cm did occur in the long rains of 1997, but during this period there was less evidence of increasing soil moisture below the crop

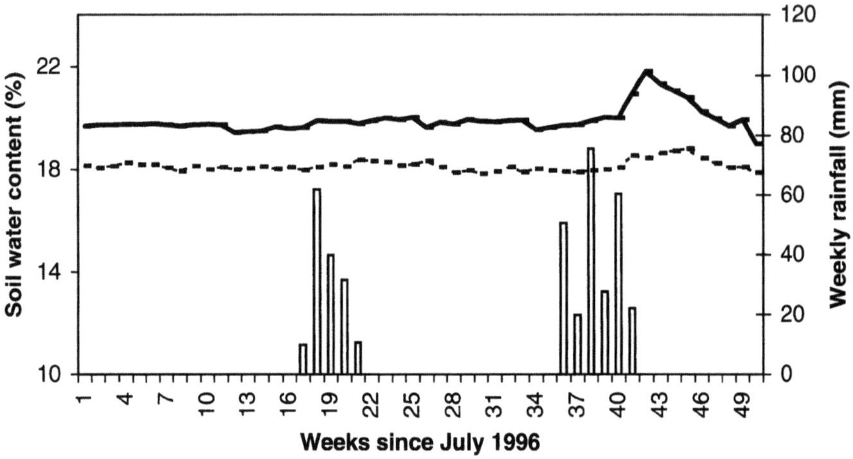

Figure 5. Soil moisture content below the crop rooting zone (75–175 cm) under an annual cropping system of maize and beans (solid line) and under *Grevillea robusta* plus crops (dashed line), Matchakos, Kenya.

rooting depth in the *G. robusta* plot. Over the twelve-month period, *G. robusta* abstracted only an extra 14.5 mm of subsoil water more than the annual crop treatment.

Our evidence suggests that although *G. robusta* is able to utilise water from below the crop-rooting zone, the annual rate of subsoil depletion is slow, unlike the high rate reported for eucalypt plantations (Calder et al., 1997). Nevertheless, the soil profile was not recharged until the fifth season. In contrast, a similar analysis for a fast growing native species, *Melia volkensii*, in the same trial provided a much higher rate (42 mm per annum) of subsoil water depletion. Clearly, this rate of water depletion is unsustainable, unless the *M. volkensii* tree population is greatly reduced. Water extraction will clearly increase as trees grow larger, which could have serious implications for water relations on sites with shallow soils and low water holding capacity, unless canopy management, such as pruning, is applied to reduce tree water use.

Where groundwater is accessible to tree roots, there is clear evidence for spatial complementarity. In the Majjia valley of Niger, measurements of stable isotopes of oxygen in plant sap, groundwater and water in the soil profile of windbreaks showed that neem trees, *Azadirachta indica*, obtained a large portion of their water from the surface layers of the soil only after rain. During the dry season, the tree roots extracted groundwater at 6 m depth (Smith et al., 1997). In contrast, at Sadore, where groundwater was at a depth of 35 m, they found that both the trees and millet obtained water from the same 2–3 m of the soil throughout the year (rainfall of 450 mm).

5. Conclusions

Our understanding of the ecological processes in alley cropping and other simultaneous agroforestry practices in the drylands has advanced considerably during the last few years. In reconciling this information with the vast ecological literature on tropical savannahs we suggest it is useful to consider agroforestry practices in terms of early and later stages of succession. From this perspective, alley cropping with fast growing, regularly pruned trees is analogous to early succession characterised by a high degree of competition while parkland agroforestry with widely spaced trees is analogous to later stage succession with opportunities for temporal and spatial complementarity. Furthermore, it would probably take a long time for the beneficial effects of microclimate and soil improvements of mature savannah trees to be realised. It would probably be more realistic to select trees which could provide direct cash benefits to farmers through their products, and to accept that in the longer term they will provide indirect environmental benefits arising from a more complex agroecosystem (Leakey, 1998). Economic gains from marketable products could compensate for any loss in crop yield (Kessler, 1992). It is, however, important to point out that current understanding of resource capture by agroforestry practices is based on well-managed small plots, often in research stations, in which about 30–45% of the rainfall is used for transpiration. Such levels of rainfall utilisation are rarely achieved in farmers' fields, or at the landscape level, in sub-Saharan Africa because of low inputs, and there are still ample opportunities for increasing water use by incorporating trees into landuse practices. For example, Rockstrom (1997) reported that only 6 to 16% of the total rainfall in a watershed in Niger is utilised by pearl millet for transpiration and the remainder is lost by soil evaporation (40%) or by deep drainage (33 to 40%). In contrast, plot level studies at Machakos by McIntyre et al. (1997) reported a rainfall utilisation of 40–45% by maize and cowpea for transpiration and the rest was lost as soil evaporation, thus limiting the opportunity for agroforestry.

The greatest opportunity for simultaneous agroforestry practices in the drylands of the tropics would seem to be the exploitation of the complementary interactions between crops and mature trees grown for their marketable products. Such trees can be fitted into the landscape by (i) exploiting the under-utilised niches within and around the farms, such as footpaths and home compounds, (ii) by integrating crops with boundary plantings, scattered trees and vegetable gardens and (iii) by integrating crops with fodder banks protected by hedges. In this way, agroforestry would allow the regeneration of parklands and lead to the successional development and patch dynamics of an agroecosystem that is akin to natural ecological succession (Leakey, 1996).

6. References

Azam-Ali SN (1983) Seasonal estimates of transpiration from a millet crop using a porometer. Agricultural and Forest Meteorology 30: 13–24

Azam-Ali SN, Simmonds LP, Nageswara Rao RC and Williams TH (1989) Population, growth and water use of groundnut maintained on stored water. 3. Dry matter, water use and light interception. Experimental Agriculture 25: 77–86

Belsky AJ (1994) Influence of trees on savannah productivity: test of shade, nutrients, and tree-grass competition. Ecology 75: 922–932

Belsky AJ and Amundson RG (1997) Influence of savanna trees and shrubs on understorey grasses and soils: new directions in research. In: Bergstrom L and Kirchmann H (eds) Carbon and Nutrient Dynamics in Natural and Agricultural Tropical Ecosystems. CAB International (in press)

Belsky AJ, Amundson RG, Duxbury JM, Riha SJ and Mwonga SM (1989) The effects of trees on their physical, chemical and biological environments in semiarid savanna in Kenya. Journal of Applied Ecology 26: 1005–1024

Belsky AJ, Mwonga SM, Amundson RG, Duxbury JM and Ali AR (1993) Comparative effects of isolated trees on their understorey environments in high- and low-rainfall savannas. Journal of Applied Ecology 30: 143–155

Breman H and Kessler JJ (1995) Woody Plants in Agroecosystems of Semiarid Regions. Springer Verlag, Berlin, Heidelberg

Buresh RJ and Tian G (1998) Soil improvement in sub-Saharan Africa. Agroforestry Systems 38: 51–76

Burgess SSO, Adams MA, Turner NC and Ong CK (1998) Redistribution of water within plant root systems. Oecologia 115: 306–311

Calder IR, Rosier PTW and Prasanna KT (1997) Eucalyptus water use greater than rainfall input-a possible explanation from southern India. Hydrology and Earth Syst Sci (in press)

Cannell MGR, van Noordwijk M and Ong CK (1996) The central agroforestry hypothesis: the trees must acquire resources that the crop would not otherwise acquire. Agroforestry Systems 34: 27–31

Cannell MGR, Mobbs DC and Lawson GJ (1998) Complementarity of light and water use in tropical agroforests II. Modelling theoretical tree production and potential crop yield in arid to humid climates. Forest Ecology and Management 102: 275–282

Cooper PJM, Gregory PJ, Tully D and Harris HG (1987) Improving water use efficiency of annual crops in rainfed farming systems of West Asia and North Africa. Experimental Agriculture 23: 113–158

Cooper PJM, Leakey RRB, Rao MR and Reynolds L (1996) Agroforestry and the mitigation of land degradation in the humid and sub-humid tropics of Africa. Experimental Agriculture 32: 235–290

Corlett JE, Black CR, Ong CK and Monteith JL (1992b) Above and below-ground interactions in a leucaena/millet alley cropping system. 2. Light interception and dry matter production. Agricultural and Forest Meteorology 60: 73–91

Corlett JE, Ong CK, Black CR and Monteith JL (1992a) Above and below-ground interactions in a leucaena/millet alley cropping system. 1. Experimental design, instrumentation and diurnal trends. Agricultural and Forest Meteorology 60: 53–72

Cuff AD, Allen SJ, Gash JHC, Lloyd CR and Wallace JS (1993) Energy and water budgets of an area of patterned woodland in the Sahel. Agricultural and Forest Meteorology 66: 65–80

Dawson TE (1993) Hydraulic lift and water use by plants: implications for water balance, performance and plant-plant interactions. Oecologia 95: 565–574

Deans JD, Lindley DK and Munro RC (1995) Deep Beneath Trees in Senegal, Annual Report 1993–1994, 12–14, Institute of Terrestrial Ecology, Bush Estate, Penicuik Scotland, UK

de Wit C (1958) Transpiration and Crop Yields. Agricultural Research Proceedings 59, Wageningen

de Wit C (1960) On competition. Verslagen van Landbouwkunige Onderzoekingen 66: 1–82

Edmunds WM (1991) Groundwater recharge in the west African Sahel. NERC News no 17, 8–10

Ewel JJ (1986) Designing agricultural ecosystems for the humid tropics. Annual Review of Ecology and Systematics 17: 245–271

Ewel JJ and Bigelow SW (1995) Plant life forms and tropical ecosystem functioning. In: Orians GH, Dirzo R and Cushman JH (ed) Biodiversity and Ecosystem Processes in Tropical Forests, pp 101–126. Springer-Verlag. Berlin, Germany

Fukai S and Trenbath BR (1993) Processes involving intercrop productivity and yields of component crops. Field Crops Research 34: 247–271

Gaze SR, Simmonds LP, Brower J and Bouma J (1997) Measurement of surface redistribution of rainfall and modelling its effect on water balance calculations for a millet field on sandy soil in Niger. Journal of Hydrology (in press)

Govinadrajan M, Rao MR, Mathuva MN and Nair PK (1996) Soil-water and root dynamics under hedgerow intercropping in semiarid Kenya. Agronomy Journal 88: 513–520

Gregory PJ (1996) Approaches to modelling the uptake of water and nutrients in agroforestry systems. Agroforestry Systems 34: 51–65

Hauser S (1993) Root distribution of *Dactyladenia* (*Acioa*) *barteri* and *Senna* (*Cassia*) *siamea* in alley cropping. Implication for field experimentation. Agroforestry Systems 24: 111–121

Howard SB, Ong CK, Black CR and Khan AAH (1997) Using sap flow gauges to quantify water uptake by tree roots from beneath the crop rooting zone in agroforestry systems. Agroforestry Systems 35: 15–29

Howard SB, Ong CK, Rao MR, Mathuva M and Black CR (1995) Partitioning of light and water in leucaena/maize agroforestry systems. In: Sinoquet H and Cruz P (eds) The Ecophysiology of Tropical Intercropping, pp 175–187. INRA Editions, Paris

Huntley BJ and Morris JW (1982) Structure of Nylsvley savanna. In: Huntley BJ and Walker BH (ed) Ecology of Tropical Savannas, pp 433–455. Springer Verlag, Berlin

Huntley BJ and Walker BH (eds) (1982) Ecology of Tropical Savannas. Springer Verlag, Berlin, 669 pp

ICRAF (1996) Annual Report 1996, pp 263, International Centre for Research in Agroforestry, Nairobi, Kenya

ICRAF (1997) ICRAF Medium Term Plan 1998–2000, International Centre for Research in Agroforestry, Nairobi, Kenya, 73 pp

Jama B, Nair PKR and Rao MR (1995) Productivity of hedgerow shrubs and maize under alley cropping and block planting systems in semiarid Kenya. Agroforestry Systems 31: 257–274

Jonsson K, Fidgeland L, Maghembe JA and Högberg P (1988) The vertical root distribution of fine roots of five tree species and maize in Morogoro, Tanzania. Agroforestry Systems 6: 63–69

Kater LJM, Kante S and Budelman A (1992) Karité (*Vitellaria paradoxa*) and néré (*Parkia biglobosa*) associated with crops in South Mali. Agroforestry Systems 18: 89–105

Kennard DG and Walker BH (1973) Relationships between tree canopy cover and *Panicum maximum* in the vicinity of Fort Victoria. Rhodesian Journal of Agricultural Research 11: 145–153

Kessler JJ (1992) The influence of karité (*Vitellaria paradoxa*) and néré (*Parkia biglobosa*) trees on sorghum production in Burkina Faso. Agroforesty Systems 17: 97–118

Khan AAH and Ong CK (1996) A low cost heat pulse method for measuring tree root water uptake. Agroforestry Today 7: 19–22

Kinyamario JI, Trlica MJ and Njoka TJ (1995) Influence of tree shade on plant water status, gas exchange, and water use efficiency of *Panicum maximum* Jacq. and *Themeda triandra* Forsk. in a Kenya savanna. African Journal of Ecology 33: 114–123

Leakey RRB (1996) Definition of agroforestry revisited. Agroforestry Today 8: 5–7

Leakey RRB (1998) Agroforestry for biodiversity in farming systems. In: Collins W and Qualset C (eds) The Importance of Biodiversity in Agroecosystems, pp 127–145. CRC Press, New York, USA

Lott JE, Khan AAH, Ong CK and Black CK (1996) Sap flow measurements of lateral roots in agroforestry systems. Tree Physiology 16: 995–1001

Marshall FM (1995) Resource partitioning and productivity of perennial pigeonpea/groundnut agroforestry systems in India. PhD thesis, University of Nottingham,UK

Matthews RB, Harris D, Nageswara Rao RC, Williams JH and Wadia KDR (1988) The physiological basis for yield differences between four genotypes of groundnut (*Arachis hypogaea*) in response to drought. I. Dry matter production and water use. Experimental Agriculture 24: 191–202

McIntyre BD,Riha SJ and Ong CK (1996) Light interception and evapotransipration in hedgerow agroforestry systems. Agricultural and Forest Meteorology 81: 31–40

McIntyre BD, Riha SJ and Ong CK (1997) Competition for water in a hedge-intercrop system. Field Crop Research 52: 151–160

Medina E (1982) Physiological ecology of neotropical savanna plants. In: Huntley BJ and Walker BH (eds) Ecology of Tropical Savannas, pp 308–335. Springer Verlag, Berlin

Mekonnen K, Buresh RJ and Jama B (1997) Root and inorganic nitrogen distribution in sesbania fallow, natural fallow and maize fields. Plant and Soil 188: 319–327

Monteith JL, Ong CK and Corlett JE (1991) Microclimatic interactions in agroforestry systems. Forest Ecology and Management 45: 31–44

Nair PKR (1993) An Introduction to Agroforestry. Kluwer Academic Publishers, Dordrecht, The Netherlands, 499 pp

Ong CK and Black CR (1994) Complementarity in resource use in intercropping and agroforestry systems. In: Monteith JL, Scott RK and Unsworth MH (eds) Resource Capture by Crops. Nottingham University Press, pp 255–278. Loughborough, UK

Ong CK, Black CR, Marshall FM and Corlett JE (1996) Principles of resource capture and utilization of light and water. In: Ong CK and Huxley PA (eds) Tree-Crop Interactions: A Physiological Approach. CAB International, pp 73–158. Wallingford, UK

Ong CK, Odongo JCW, Marshall F and Black CR (1992) Water use of agroforestry systems in semiarid India. In: Calder IR, Hall LR and Adlard PG (eds) Growth and Water Use of Plantations, Proceedings of the International Symposium on the Growth and Water Use of Forest plantations, Bangalore, 7–11 February,1991, pp 347–358. Chichester, Wiley

Ong CK, Simmonds LP and Matthews RB (1987) Responses to saturation deficit in a stand of groundnut (*Arachis hypogaea* L.). 2. Growth and development. Annals of Botany 59: 121–128

Poschen P (1986) An evaluation of the *Acacia albidia*-based agroforestry practices in the Hararghe highlands of eastern Ethiopia. Agroforestry Systems 4: 31–56

Ranganathan R and de Wit C (1996) Mixed cropping of annuals and woody plants: an analytical approach to productivity and management. In: Ong CK and Huxley PA (eds) Tree-Crop Interactions: A Physiological Approach, pp 25–50. CAB International Wallingford, UK

Rao MR, Muraya P and Huxley PA (1993) Observations of some tree root systems in agroforestry intercrop situations and their graphical representation. Experimental Agriculture 29: 183–194

Rao, MR, Nair, PK and Ong CK (1998) Biophysical interactions in tropical agroforestry systems. Agroforestry Systems 38: 3–49

Rees DJ (1986) The effects of population density and intercropping with cowpea on the water use and growth of sorghum in semiarid conditions in Botswana. Agricultural and Forest Meteorology 37: 293–308

Rhoades CC (1995) Seasonal patterns of nitrogen mineralisation and soil moisture beneath *Faidherbia albida* (syn *Acacia albida*) in central Malawi. Agroforestry Systems 29: 133–145

Rhoades CC (1997) Single-tree influences on soil properties in agroforestry: lessons from natural forest and savanna ecosystems. Agroforestry Systems 35: 71–94

Rocheleau D, Weber F and Field-Juma A (1988) Agroforestry in Dryland Africa. International Council for Research in Agroforestry, Nairobi, Kenya, 311 pp

Rockstrom, J (1997) On-farm agrohydrological analysis of the Sahelian yield crisis: rainfall partitioning, soil nutrients and water use efficiency of pearl millet. PhD thesis, University of Stockholm

Schroth G (1995) Tree root characteristics as criteria for species selection and systems design in agroforestry. Agroforestry Systems 30: 125–143

Shankarnarayan KA, Harsh LN and Kathju S (1987) Agroforestry in the arid zones of India. Agroforestry Systems 5: 69–88

Singh RP, Ong CK and Saharan N (1989) Above and below-ground interactions in alley cropping in semiarid India. Agroforestry Systems 9: 259–274

Sinoquet H and Cruz P (1995) Ecophysiology of Tropical Intercropping. INRA, Paris, 483 pp

Smith DM, Jackson NA, Roberts JM and Ong CK (1998) Reverse flow in tree roots and downward siphoning of water by *Grevillea robusta*. Functional Ecology (submitted)

Smith DM, Jarvis PG and Odongo JCW (1997) Sources of water used by trees and millet in Sahelian windbreak systems. Journal of Hydrology 198: 140–153

Stewart M and Bromley T (1994) Use of *Melia volkensii* in a semiarid agroforestry systems in Kenya. Commonwealth Forestry Review 73: 128–131

Tiedemann AR and Klemmedson JO (1973) Effect of mesquite on physical and chemical properties of the soil. Journal of Range Management 26: 27–29

Trenbath BR (1974) Biomass productivity of mixtures. Advances in Agronomy 26: 177–210

Trenbath BR and Harper JL (1973) Neighbour effects in the genus *Avena*.1. Comparison of crop species. Journal of Applied Ecology 10: 379–400

Vandermeer J (1989) The Ecology of Intercropping. Cambridge University Press, UK

Van Noordwijk M and Ong CK (1996) Lateral resource flow and capture – the key to scaling up agroforestry results. Agroforestry Today 7: 29–31

Van Noordwijk M and Purnomosidhi P (1995) Root architecture in relation to tree-soil-crop interactions and shoot pruning in agroforestry. Agroforestry Systems 30: 161–173

Walker BH and Noy-Meir I (1982) Aspects of stability and resilience of savanna ecosystems. In: Huntley BJ and Walker BH (eds) Ecology of Tropical Savannas, pp 556–590. Springer Verlag, Berlin

Weltzin JF and Coughenour MB (1990) Savanna tree influence on understorey vegetation and soil nutrients in north western Kenya. Journal of Vegetation Science 1: 325–334

Willey RW (1985) Evaluation and presentation of intercropping advantages. Experimental Agriculture 21: 119–133

Agroforestry Systems **45**: 131–158, 1999.

Can the ecosystem mimic hypotheses be applied to farms in African savannahs?

M. VAN NOORDWIJK[1,*] and C. K. ONG[2]
[1] *International Centre for Research in Agroforestry (ICRAF), P.O. Box 161, Bogor 16001, Indonesia;* [2] *ICRAF, United Nations Avenue, Gigiri, Nairobi, P.O. Box 30677, Kenya (*Author for correspondence: m.van-noordwijk@cgiar.org)*

Key words: competition, fertility, microclimate, risk, stability, water

Abstract. The first ecosystem mimic hypothesis suggests clear advantages if man-made land use systems do not deviate greatly in their resource use patterns from natural ecosystems typical of a given climatic zone. The second hypothesis claims that additional advantages will accrue if agroecosystems also maintain a substantial part of the diversity of natural systems. We test these hypotheses for the savannah zone of sub-Saharan Africa, with its low soil fertility and variable rainfall. Where annual food crops replace the natural grass understorey of savannah systems, water use will decrease and stream and groundwater flow change, unless tree density increases relative to the natural situation. Increasing tree density, however, will decrease crop yields, unless the trees meet specific criteria. Food crop production in the parkland systems may benefit from lower temperatures under tree canopies, but water use by trees providing this shade will prevent crops from benefiting. In old parkland trees that farmers have traditionally retained when opening fields for crops, water use per unit shade is less than in most fast growing trees introduced for agroforestry trials. Strong competition between plants adapted to years with different rainfall patterns may stabilise total system productivity – but this will be appreciated by a farmer only if the components are of comparable value. The best precondition for farmers to maintain diversity in their agroecosystem hinges on the availability of a broad basket of choices, without clear winners or 'best bets'.

1. Introduction

1.1. *Two mimicry hypotheses*

Only 8% of Africa's land area comprises fertile soils and the remaining 92% can be defined as marginal lands, which are either too dry, too wet or too saline (Yaker, 1993). Inherent soil fertility is low as most of the continent is geologically old and highly eroded and recent volcanic activity strictly localised. Large areas of apparently 'unsuitable' marginal lands are cleared each year for agricultural production, leading to further land degradation. Yet recent studies of some of these natural ecosystems, especially tropical grasslands, have shown that they can be highly productive (Solbrig et al., 1996). If current agricultural interventions are indeed inappropriate, would it then not be better to mimic savannah ecosystems, seen today in the East African Serengeti, where large mammals harvest and convert the primary production of the vegetation whether grass or trees. Tropical savannahs have long histories of human occupation and the strategies of plants, animals, and people in savannah areas can accordingly provide clues for designing and managing new strategies in line with present needs and circumstances. However, previous attempts to improve the management of savannahs have been disappointing (Hadley, 1985), and few have explicitly addressed the balance between trees and annual or perennial understorey vegetation, while attempting to ascribe value to both.

The first 'mimicry hypothesis' (Ewel, this volume) argues that agricultural systems should benefit from imitating the structure and function of natural ecosystems, since components of the latter result from natural selection towards sustainability and the ability to adjust to perturbations. As our second mimicry hypothesis, we will subscribe that, apart from structure and function, agroecosystems shall benefit from resembling the diversity of natural ecosystems as well.

These mimicry hypotheses, especially the second one, might well be viewed as rooted in the long standing 'romantic' perspective of imitating Mother Nature, but they have rejuvenated and been appreciably modified in the wake of environmental interests over the last three decades and are now much more open to a rigorously objective scientific analysis. The 'benefits' both hypotheses can refer to are:

- a larger total production if combinations of different plant life forms are used rather than a monoculture;
- a higher efficiency in terms of human operation since less effort will be needed to achieve the same results where one works 'along with' rather than 'against' nature;
- reduced downstream effects if total resource capture in an agroecosystem can be engineered to simulate that of the natural system which it replaces. For example, both additional and reduced water use can grossly affect groundwater flow patterns and have unexpected effects elsewhere;
- better maintenance of environmental service functions in the landscape.

The hypotheses refer to two potential classes of compatibility between agroecosystem and natural ecosystem: total resource use (in the context of this paper relating to either to water and nutrients) or diversity.

The diversity issue, as addressed in our second hypothesis, is clearly the more speculative and contentious. We will focus on the role of diversity in managing risk, i.e. essentially the old adage of 'don't put all your eggs in one basket', and relate the relevance of diversity to stability and resilience of resource capture in a fluctuating environment.

1.2. *Segregate or integrate agriculture and environmental service functions?*

The mimicry hypotheses suggest that an 'integrated' approach to land use, in which ecological functions in the landscape are affected as little as possible, is superior to a 'segregated' approach in which parts of the landscape are managed towards maximum productivity and other parts reserved for 'environmental services' (Van Noordwijk et al., 1997). These environmental services may include the reduction of disturbed patterns of 'lateral interactions' in the form of surface or groundwater flow of solutes and sediments, wind assisted gaseous exchange of water vapour, specific gasses and particles, or movement of organisms, whether linked to pests and diseases and their potential control, or to biodiversity of intrinsic value.

The above issues are not purely of academic interest. Per-capita food production continues to decline in sub-Saharan Africa. The question is not just to maintain the current productivity of agroecosystems but how to achieve a considerable increase without compromising agroecosystem resilience. To meet the human population expansion, the annual sustained growth of agricultural production in Africa will have to rise by 4% over the next 25 years, i.e. 1% above the population growth rate, rather than 1% below as is the current situation (Badiane and Delgado, 1995). Total production growth will have to come from increases in productivity per cultivated area if we are to cure 'land hunger' (Van Keulen and Breman, 1990). To achieve this it may be more economical to restrict production to the best land and pursue a 'segregated' pathway for development of land use, where negative externalities of land use or a lack of environmental services can be compensated for elsewhere.

The declining per capita food production of Sub-Saharan Africa is perceived to be largely due to the existing policy environment which does not sufficiently favour smallholder rural development, or provide sufficient financial incentives for farmers to intensify their land use (Sanchez and Leakey, 1997). Priority issues in this respect are past and ongoing depletion of soil fertility as a fundamental biophysical constraint to food security and the domestication of indigenous trees to produce high value products. One may then ask whether these priorities are compatible with the mimicry hypotheses.

The objective of attaining food security can be achieved in different ways, with increasing degrees of market integration, namely by:

- stable agronomic production of the main staple food by the vast majority of the population;
- production of staple food on a few specialised farms within a specific country/region, together with a stable source of income (derived in or outside of agricultural enterprises) for all inhabitants to allow them to buy their staple requirements;
- as above, but with a population earning sufficient income to import food on a sustained basis from specialized farms elsewhere in the world.

Trees can contribute specifically to food security in three ways:

- by providing the main staple or emergency foods in times of shortage (e.g. from planting desert date, *Balanites aegyptiaca*);

- by increasing the opportunities for growing staple food crops by enhancing soil fertility (e.g. planting of N_2 fixing woody species);
- by providing income from high value products ('money in farmers pockets').

The first and third ways are based on direct value of tree products, whereas the second one refers to an added value derived indirectly from trees in a support function role.

1.3. *Soil fertility and tree domestication related to the mimicry hypotheses*

If the nutrient resource base for agricultural production has to be increased, one may ask whether such replenishment can be accomplished in the context of systems mimicking nature, or within simple, specialized crop production systems? There are some indications of positive interactions or at least additive effects between soil organic matter (provided by trees) and inorganic fertilisers (Breman and Kessler, 1997; Palm et al., 1997), but it is not yet clear whether agroforestry has a major role to play in the soil replenishment initiatives which are currently underway (Sanchez et al., 1997). The analysis by Breman and Kessler (1995, 1997) of the added value of woody plants in the Sahel and other semi-arid regions, suggests that trees may indeed increase nutrient and water availability for annual food crops and pasture production, but only if trees:

- have the right properties (can generate stem flow of water, are deep rooted, are unpalatable and of low 'quality', and thus able to contribute more to long term accumulation of soil organic matter rather than rapid mineralisation, exhibit slow growth rate and have a long life span);
- are used at an appropriate tree density;
- are properly managed by pruning, lopping and controlled harvesting;
- are growing on appropriate sites and soil types in which the above goals can be realistically achieved.

Trees with the best set of properties, such as *Faidherbia albida*, may increase grain yield in the Sudano-savannah zone by 200–400 kg ha^{-1}, provided they have an average tree canopy cover of about 20%, can gain access to the water-table, and are not foliated during the crop growing season. A general 'grow more trees' approach is not likely to meet such conditions.

Tree domestication aims at converting wild species with proven direct or indirect value, currently collected from the wild or maintained in agroecosystems, into crops or service components of agroforestry systems of high potential value. It involves genetic selection and may involve vegetative propagation to capture and perpetuate the properties of superior individual members of tree populations. Guidelines have been developed for species prioritisation aimed at identifying what one might term 'cinderella trees', namely species overlooked by science but long appreciated by local people.

High-value trees can fit in specific niches on farms while leaving more open land to staple food crops or annual crops grown for markets (Sanchez and Leakey, 1997). The current approach used in this arena has some elements in common with the mimic hypothesis, but it is focused on domesticating the components, rather than domesticating the system. Indeed, Michon and de Foresta (1997a, 1997b) have contrasted domestication of agroforests to tree domestication, and argued that if the agroforest context is kept in mind during the tree selection process, tree improvement does not necessarily lead to germplasm which is only suitable for orchard-like tree monocultures. Elaborating on their argument, we suggest that a strategy for domesticating the system might start by asking which elements of the current system rank of lowest value and try to increase the performance of these or find ecological equivalents as replacements. However, for this to happen, it would require fundamental understanding of how both natural and partially domesticated agroecosystems function. The targets of tree domestication as currently formulated do not include traits to promote compatibility between the trees and other components of a particular agroforestry system. For example, faster growth, shorter statue and earlier fruit production are desirable for fruit trees, but these characteristics obviously make the same subjects less suitable for employment as 'parkland trees'.

1.4. *Competition, diversity and the second mimicry hypothesis*

Competition between trees and crops is often viewed as an inherently negative aspect of agroforestry systems, which one has to avoid or reduce by appropriate management protocols. Indeed, this view has led to the conclusion that one should shy away altogether from simultaneous agroforestry systems in drier climatic zones and rather focus on fallow rotations (Sanchez, 1995). As competition for water and nutrients is an inevitable feature of natural ecosystems, a purely negative view of competition couched in such terms appears inconsistent with the mimicry hypotheses, and is possibly an inappropriate basis for developing agroecosystems. But it remains debatable whether competition is so inherently negative that the above conclusion is warranted. No doubt it depends on ones perspective and on the way one evaluates the different components of the system. Cannell et al. (1996) related a biophysical advantage of agroforestry to at least some degree of complementarity of resource capture, as opposed purely to absence of competition between components. Van Noordwijk (1996) explored the degree of complementarity needed to provide a total value beyond that of the best separate monoculture performances of the system's components. If the value attached to the two competitors is equal (per unit resources captured), even a slight complementarity can be sufficient to make the combination worthwhile. If, however, one component has no discernible direct value, its presence in a system can only be justified if there is strong complementarity in resource capture and effective transfer of captured resources from it to other valuable components of the system. So competition need not be a disadvantage from the perspective of management of an agroecosystem, as long as the manager in question is a 'generalist', who places value on several, if not all, components of the system. In such a situation, the presence of strong competitors may contribute to the stability (reduced variance) of the combined system, as we will see below.

As an example of the above, farmers in semi-arid Kenya have developed a parkland agroforestry system which includes a highly competitive, fast growing native species, *Melia volkensii*, because it provides a wide range of products for household needs as well as generating income to cover the cost of children's school fees. It is a member of the mahogany family, valued principally because its high quality timber is termite resistant, coarse grained and easy to work. At the same time, its leaves and fruits provide a good fodder for goats, sheep and cattle, and the leaves can be used as a repellent for fleas and ticks. Typically, a farmer establishes five to ten trees, mainly from wildings, on a farm, and these trees can generate more income than crops from the rest of the farm over a period of five to eight years. Thus, this species is considered as 'compatible' with the agroecosystem even though it is highly competitive with crops.

1.5. *Outline of this chapter*

In the pages which follow, we will focus our discussion of the application of mimicry hypotheses to the savannah zone, as there are considerable differences of opinion about feasible and desirable pathways for development of agroecosystems in this zone (Jones et al., 1998). The parklands of the savannah zone can be compared to the Spanish 'Dehesa' system (Joffre et al., this volume), and the forest-derived landscape in NE Thailand (Sae-Lee et al., 1992). The vegetation in all these areas consists partly of trees deliberately retained after natural or fallow vegetation has been cleared, or of trees deliberately planted. The proportions of such spontaneous and planted trees can vary noticeably between regions but, in all cases, trees spared during land clearing are maintained essentially because of their potential useful products or functions, and/or compatibility with crops. In such cases, tree species diversity can be considerable and intra-specific variation may still resemble that of a natural tree population (Boyle, pers. comm.). Furthermore, if these trees are propagated by seeds, they may still be effectively part of the natural tree population. If domestication and selection proceed, genetic diversity can be expected to decrease.

In this chapter we explore how current understanding on the structure and function of African savannah might be used to design or improve agroforestry systems for sub-Saharan Africa (Tothill and Mott, 1985; Walker, 1987; Belsky and Amundson, 1998). Ong and Leakey (this volume) compare the resource capture and influence of trees on crops in agroforestry systems using evidence from the savannah. In this chapter we address:

- the abiotic and biotic environment of savannah systems;
- constraints on total resource usage within such systems;
- the possibility of risk reduction through increased diversity;

and conclude by discussing the value of the mimicry hypotheses as a basis on which new or existing savannah systems might be improved.

2. Structure and function of African savannahs

2.1. *Vegetation and climate*

There is nothing particularly new in the recognition of ecological zones in Africa and the need to take into account information of soils, vegetation, climate and locally developed farming systems when planning agriculture and forestry (De Vos, 1975; Van Noordwijk, 1984). Separate terminologies persist for west, east and southern Africa for similar vegetation types and ecological zones. Within the main savannah zone spanning from Africa's west coast to the Ethiopian highlands, one commonly distinguishes a Sahel, Sudano- and Guinean savannah zone, with additional man-made derived savannah in the wetter forest zone. In eastern and southern Africa the term miombo woodland is used for a vegetation similar to the Guinea savannah (Campbell et al., 1998). Rainfall as fraction of potential evapotranspiration for the various savannah types in western Africa is < 0.21 for the Sahel, 0.21–0.40 for Sudan savannah, 0.40–0.66 for Northern Guinea savannah, 0.65–0.88 for Southern Guinea savannah, and 0.75–1.0 for derived savannah (Sanford and Isichei, 1986). The major rivers flowing through the savannah zones (Niger, Nile, Zambezi) derive their water from rainfall on mountains in the forest zone and, thereby provide additional resources of water and nutrients on that small fraction of the land, on which a considerable proportion of the total human population subsists.

The standard deviation of annual rainfall is remarkably constant across climatic zones of the savannah, so that the coefficient of variation (standard deviation divided by the mean) increases almost inversely proportional to a decreasing mean. Thus, El-Tom (1975) calculated that zones of the Sudan, with mean annual rainfall values of 1300, 900, 600, 400, 150 and 50 mm had coefficients of variation of 11, 15, 25, 30, 60 and 100%, respectively. Although water resources in the savannah zone are considerable in magnitude, they fluctuate widely in time and can be equally variable in space. This selects for a balance between two opposed strategies within indigenous floristic components, namely an opportunistic strategy of rapid expansion when resources are present and surviving in seed or other dormant forms in less favourable periods, and a complementary strategy of slow, steady growth, with ample internal storage and survival mechanisms.

Walker (1987) defined savannahs as those tropical (or sub-tropical) ecosystems characterised by strong seasonality (wet summer/dry winter) related to water stress, dominated by a continuous herbaceous vegetation consisting principally of grasses (usually C_4 type species) and sedges, and a few scattered trees and shrubs. For the purpose of this chapter we will describe the structure and function of savannah ecosystems in terms of the flow of water and energy through the system rather than dwell in detail on the botanical composition and behaviour of plant species. Figure 1 is the model proposed by Solbrig et al. (1996) to predict the function of tropical savannahs in terms of water, nutrients, fire and herbivory. According to this model, mesic woody species would be expected to dominate when water is relatively abundant, but give way to moist forest at the extreme wet and to dry desert regions at the ends of the biome. Nutrient supply can modify the vegetation expected for a given rainfall zone, so that, for example, a combination of low rainfall and high nutrients (e.g. Serengeti in Kenya) can support a savannah grassland rather than desert, whereas a decrease in nutrient supply would be expected to induce desertification without change in rainfall. Local effects of soil types and topography compounded with effects of fire and herbivory modify the density of the tree layer, and productivity and rates of water and nutrient flow through each system.

Judging from the relatively sparse data available, belowground productivity of savannah, ignored in many earlier studies, can exceed the aboveground part. For example, total productivity of grassland of the Nairobi National Park

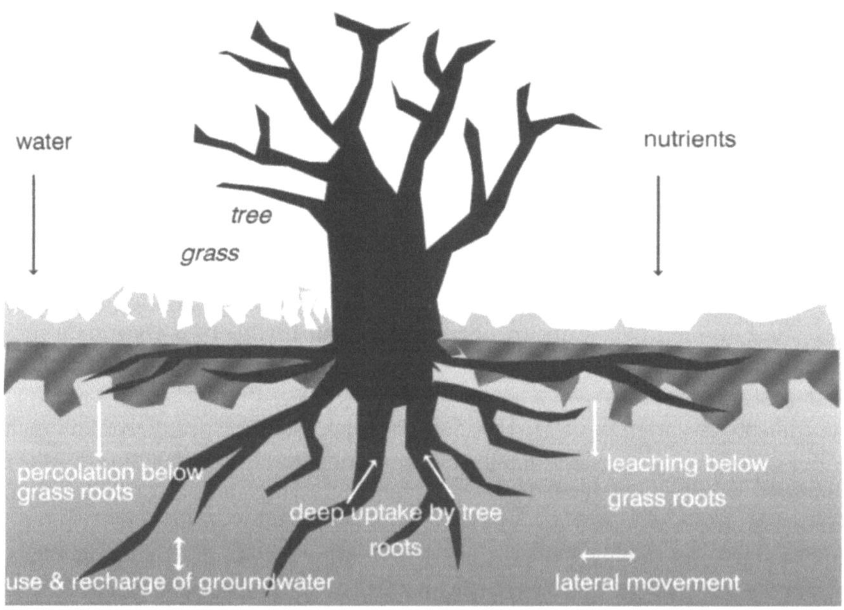

Figure 1. Water and nutrient cycling in savannah. Deep tree roots obtain water and nutrients from lower layers of the soil. Tree leaf and litter fall decompose on the upper layers where they are utilised by the grass layer. Source: Adapted from Solbrig et al. (1996).

(annual rainfall of 800 mm) was estimated to be 18.8 t ha^{-1} y^{-1} of which only 8 t ha^{-1} y^{-1} was above ground (Long et al., 1992). In Rwenzori National Park (rainfall of 900 mm), Uganda, total productivity of savannah grasses proved to be greater at 23 t ha^{-1} y^{-1}, with two-thirds as below ground biomass (Strugnell and Piggot, 1978). The values indicate that the efficiency of production of the perennial C_4 type grass vegetation can be much higher than previously estimated. Recent studies indicate that a large number of grasses and tree species allocate a high proportion of their photosynthate and nutrient reserves to roots and underground organs, largely as an adaptation to herbivory and fire. In general, the biomass contribution from shrubs and trees tends to be very low comprising only 1 to 5% of the total biomass of grasses.

2.2. *Water use*

The herbaceous layer comprises the dominant component of all savannah and is accordingly likely to have greatest impact on the flow of water through the soil-plant-atmosphere continuum and on the net primary productivity and mineral cycling of the system in relation to fire and herbivory. Grass production is tightly linked to seasonal rainfall since grasses are generally shallow-rooted. For instance, as shown in Figure 2a, leaf area index matches rainfall distribution at the Nairobi National Park. During the short rains (October–January) both above and below ground biomass was quite similar but below ground biomass was highest during November. Then, in the long rains (February-June) shoot growth was greatest. Virtually all the grass root biomass (Figure 2b) was present in the top 15 cm of the soil. Trees are deciduous at this site, as both trees and crops depend on the same soil water supply from seasonal rainfall for their growth. Only in riverine habitats where trees access deeper water can such species afford to be evergreen. This situation is similar to the sandy savannahs of southern Africa where trees and grasses compete strongly for soil water, and do not have distinct rooting zones as previously assumed (Knoop and Walker, 1985; Walker, 1987). Thus, at sites like Nairobi National Park, where a high proportion of rainfall never penetrates beyond the root zone of grasses, the community is grass dominated. In South Africa, Adcock (1983) demonstrated a negative correlation between the proportion of trees and water holding capacity of the topsoil. In contrast, (some) neotropical savannahs of South America show two distinct patterns of soil drying and root distribution, where trees maintaining high transpiration rates throughout the year but grass roots are confined to the topsoil and desiccate in the dry season (Goldstein and Sarmiento, 1987).

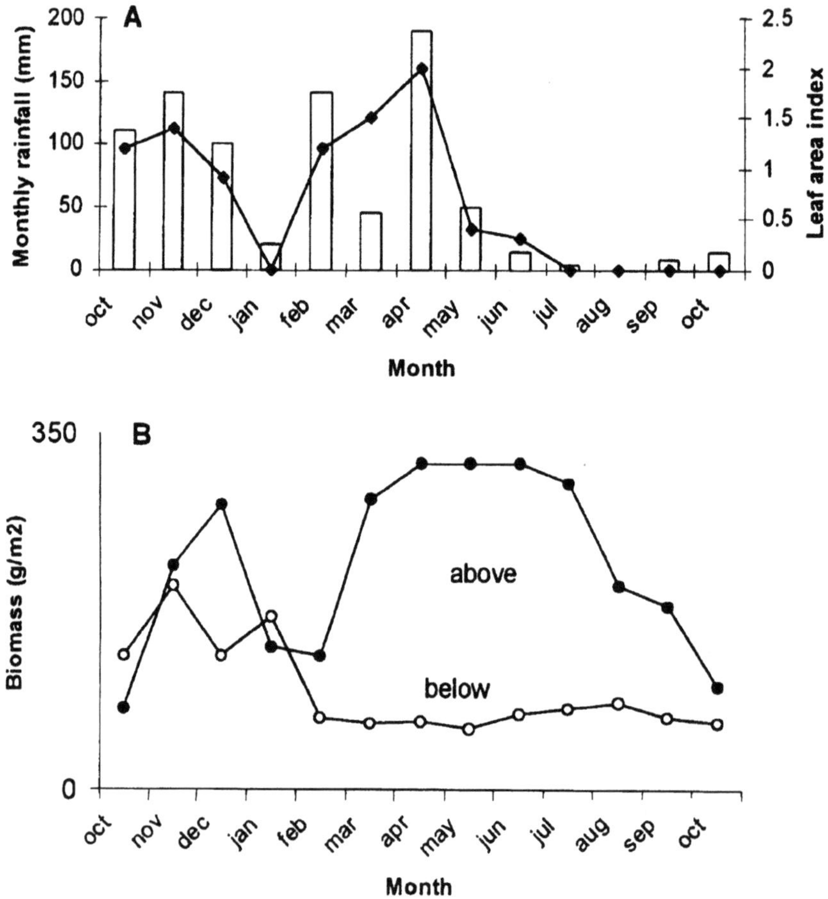

Figure 2. Monthly rainfall and leaf area index (a) and above- and below-ground biomass production (b) of grassland at Nairobi National Park, Kenya. *Source:* Adapted from Kinyamario and Imbamba (1992).

3. Structure and function of agroecosystems in the savannah zone

3.1. *Human livelihood strategies in the savannah*

Human hunter/gatherer strategies in savannah have traditionally depended on roots and tubers as main staple, enriched by fruits and grass seeds in season and supplemented opportunistically by the highly variable outcomes of hunting fauna. Thus, the hunter/gatherer is both a generalist and an opportunist. By contrast, keeping livestock in migratory herds (or in a semi-sedentary transhumance system) can develop progressively into a recognisable specialisation, as it exploits a resource which itself is of generalist nature, exploiting grazing resources when and where they exist and not depending, therefore, on any single location throughout the year. Traditional crop production in the savannah zone relies heavily on risk-prone grain crops rather than tubers and has typically utilised a much lesser diversity of species than is utilised in the forest zone, where it became the dominant livelihood strategy rather than a complement to livestock keeping. Introduction in colonial times of cassava has provided an attractive root crop for the wetter parts of the savannah zone and facilitated highly specialised production of this species. Part of the tree resources of the savannah have been maintained in what developed into parkland systems, and may have actually benefited from increased water availability wherever grass was replaced by food crops.

People have modified the biotic environment of the savannah zone by:

- removing predators and competitors to allow *Homo sapiens* to emerge as prime carnivore;
- reducing the population density of grazers competing with domesticated livestock;
- removing trees from the landscape which supply nesting sites for granivorous (weaver) bird pests;
- removing game animals acting as a host to *Trypanosoma* and trees ('bush') which provide cover for the tse-tse fly which act as vector for this disease. These two strategies at managing trypanomosiasis partly counter this major constraint to keeping livestock in wetter parts of Africa (Delaney and Happold, 1979).

When conservation of wildlife became an issue of global interest, a segregated pathway of development emerged as the obvious choice in which the richness of animal populations was maintained in national parks (generally on agriculturally marginal lands) while agricultural development was focused elsewhere. Well-defined and articulated boundaries were thus established between the two. There has, however, also been some interest in pursuing an integrated pathway of development and conservation where game ranching is managed as a semi-domesticated ecosystem. But this closest of approaches to mimicry has failed to provide for more than a small niche market. The current approach to conservation accepts segregation of activities as an implicit element, encouraging at the same time social integration through wider participation by local communities in wildlife protection and thereby sharing benefits from tourism.

3.2. *Agroforestry as derived agroecosystems or mimics?*

Parkland savannah systems are agroecosystems which have selectively maintained a considerable share of the mature trees of the natural savannah, but have replaced most of the original understorey with food crops, often alternating with fallow periods when the regenerating grass and shrub vegetation are grazed. These parkland systems extend into the dry Sudano-savannah zone. Model predictions of agroforestry systems with fast growing trees in such water limited environment indicate that trees have a strong negative impact on crop production (Mobbs et al., 1998; Cannell et al., 1998). Cannell et al. (1998) concluded from their model that although total resource capture will nearly always be higher if trees are added, simultaneous agroforestry systems are inappropriate in regions with less than 800 mm rainfall due to:

a) the low water use efficiency (dry matter production per unit transpiration) of C_3 trees at dry sites, and
b) the sensitivity of C_4 crop yield to shading.

Food security in this zone may be seriously jeopardised by (re)introducing trees unless tree roots have access to the water table, and at the same time, improve soil fertility and/or produce biomass of high value (fuel, gum, fodder, fruits).

The main agroforestry system of the parklands has elements of sequential (fallow-crop rotations, Torquebiau and Kwesiga, 1996) as well as simultaneous systems, as low tree density is maintained throughout the cropping period, while shrub and tree biomass increases during the fallow phase. Traditional systems of this type, such as the gum arabic (*Acacia senegal*) fallow rotation of the Sudan, are generally in decline but embrace an interesting balance by deriving value across and within fallow and crop stages. Tree species in the parklands vary with rainfall, from the desert date (*Balanites aegyptica*) in the drier parts to shea-butter nut (*Vitellaria paradoxa*, *Butyrospermum* sp.) and néré (*Parkia biglobosa*) in wetter zones. Aboveground tree density in parklands may be low, but all land area between trees is normally exploited by tree roots, which may extend up to 40 m from the tree trunk (Van Noordwijk et al., 1996).

Results from five-year-old *Grevillea robusta*/maize agroforestry trials at Machakos, Kenya, a site with similar rainfall to Nairobi National Park, showed that 90% of the biomass from tree and maize was produced during rainy seasons (Lott, 1998). Utilisation of deep water by the trees was small but gradually depleted the shallow soil profile (1.5 m). Indeed, production of biomass during the long dry season, July–October, accounted for only 10% of the total annual biomass produced by the tree. This low productivity was due to the small amount of residual water after the harvest of the crop, low tree root density below 1 m depth and a daily tree growth rate of about one-third of the value attained during the rainy season. A major difference between the savannah and the agroforestry systems was the lower biomass of the maize (usually about half that reported for savannah grassland) and strong negative competition effect of trees in the relatively young agroforestry systems on crops, reducing yields by 30–70% compared to a no-tree control.

3.3. *Leaf area fraction as an indicator of water use per unit shade*

Ong and Leakey (this volume) conclude that the positive influence of trees on grass productivity in African savannah requires considerable time to become apparent because of its association with slow growing trees. In the absence of convincing indirect benefits for crop production, trees must be proven to provide direct benefits to farmers if adopted in widespread fashion. However, agroforestry literature on tree-crop interactions is largely based on studies with *young* trees, with the exception of some studies of parkland systems, while literature on tree-grass interactions in savannah mainly considers *mature* trees. There is considerable evidence that mature scattered trees in such environments are associated with improved soil nutrient supply in both savannah and parkland systems, but it is uncertain whether trees actually increase the total pool of nutrients in the system or simply re-distribute existing soil nutrients more effectively across the landscape (Belsky and Amundson, 1998). If higher nutrients under trees are due to redistribution, no net gain in production at the landscape scale should result. It might still be worthwhile to the farmer, however, if it would allow the same family food requirements to be met by either farming a smaller area of land or using less labour. Unfortunately, few studies have sampled soil nutrients randomly over sufficiently large areas to test these suggestions. For example, Chivaura-Mususa and Campbell (1998) reported increased soil fertility under mature scattered trees in both the savannah and arable lands in Zimbabwe but their study only compared understorey soil to open soil 15 m away from the tree trunk. In such environments, tree roots are known to spread laterally well beyond their canopies (e.g. up to seven times the canopy dimensions of *Burkea africana* as reported by Rutherford, 1982) so it would remain uncertain as to whether the contrast in soil fertility resulted from redistribution or truly from addition of nutrients.

Van Noordwijk (1996) presented an algebraically-solvable model of tree-soil-crop interactions for situations where shade (light capture) and mulch (nutrient supply) are the dominant interactions and water availability is not limiting. This 'shade and mulch' model pointed to the ratio of light captured per unit nutrient recycled as the main criterion for testing effectiveness of fast growing trees for early succession in agroecosystems such as hedgerow intercropping. Apart from reducing light availability, tree canopies may also reduce temperature and increase humidity at crop level, a vital property in view of the fact that savannah temperatures during the growing season can become elevated well above the optimum for crops. This potentially additional advantage of trees will, of course, be (partly) offset by water use by the tree. In similar fashion to the 'mulch per unit shade ratio' concept, we may use the 'shade per unit water use ratio' as an indicator of suitability of trees for situations where microclimatic effects and water use dominate a tree-soil-crop interaction.

Mobbs et al. (1998) in their agroforestry model did not include a direct effect of tree canopies on the temperature in the crop zone, but their crop model did suggest that a decrease in crop temperature by 1 °C and a parallel increase of air humidity for the understorey vegetation increased crop grain yield by 15% at all three sites tested. Whereas crop production in parkland systems may benefit from lower temperatures under tree canopies, water use by trees providing this shade may often prevent crops from benefiting. In older trees, however, a considerable part of the total shade can be provided by branches rather than leaves, and thus not be directly linked counter-productively to transpiration.

To further explore this relationship, we extended earlier descriptions of above- and belowground tree architecture based on self-repeating (fractal) branching (Van Noordwijk and Purnomosidhi, 1995), with leaves (or fine roots) conceived as end structures of the system. This functional branch analysis scheme is based on the following entities:

- a parameter for the change (or conservation) of cross-sectional area in each branching point;
- a parameter for the allocation of cross sectional area over the branches (links) produced as offspring at each branching point;
- the average number of offspring and the minimum diameter of branch links;
- two parameters to describe the relation between link diameter and link length; the average distance between leaves and the area per leaf; and
- two parameters indicating leaf turn-over (by describing the link diameter at which leaf density starts to decline by leaf fall and the diameter at which it reaches zero).

All parameters can be measured non-destructively. This functional branch analysis scheme was used to evaluate the fraction of total surface area which is comprised of leaves. Results (Figure 3) show that the leaf area fraction decreases with tree stem diameter (age), especially in the 0–10 cm stem diameter range. As might be expected, leaf area fraction also decreases with wider spacings between leaves, smaller area per leaf and more rapid leaf turnover. For older trees, the leaf area fraction may fall below 50% for some of the parameter combinations tested. Preliminary measurements in Machakos (Kenya) (Table 1) suggest that for *Balanites aegyptiaca*, the leaf area fraction can be less than 60%, compared with more than 90% for *Melia volkensii* and *Gliricidia sepium*. Fast growing trees introduced for agroforestry trials differ in leaf area fraction for similar tree diameters to those of trees which farmers have traditionally spared when establishing field crops. Maintaining well-established trees may thus convey advantages which are not easily replicated by planting new trees.

3.4. *A model for predicting agroforestry water use across a rainfall gradient*

The WaNuLCAS (Water, Nutrient and Light Capture in Agroforestry Systems) model, developed by Van Noordwijk et al. (1998) and Van Noordwijk and Lusiana (1998, 1999) incorporates predicted interactions and effects of a four-layer soil profile, with four spatial zones, a water and nitrogen balance, and uptake by a crop and a tree. The model can be applied to simultaneous as well

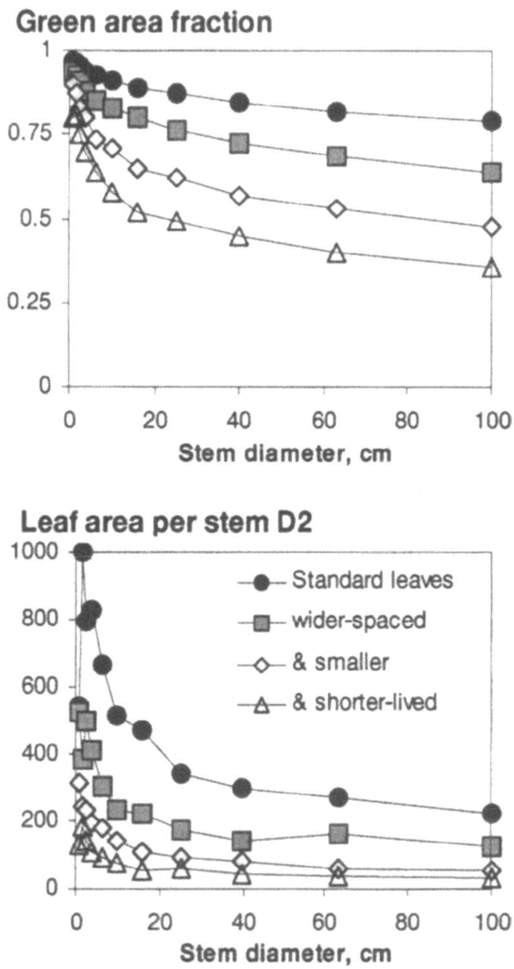

Figure 3. Calculations with a fractal branching model of the relationship between tree diameter and (a) the part of the total leaf + branch projection area formed by leaves; and (b) the relationship between leaf area and stem diameter^2. The various simulation runs were based on a stepwise change from a 'standard' parameter set, by doubling the distance between leaves, halving the area per leaf and halving the life span of leaves.

as sequential agroforestry systems and may accordingly help towards evaluating the continuum of options ranging from 'improved fallow' to relay planting of tree fallows to rotational and simultaneous forms of hedgerow intercropping through to isolated trees in parklands. It also explicitly incorporates management options such as tree spacing, pruning regime and choice of species or provenance and reflects various functional tree characteristics such as root distribution, canopy shape, litter quality, maximum growth rate and speed of recovery after pruning.

Table 1. Tree stem diameter (D_o), minimum branch diameter (D_{min}) and leaf area per unit branch length for four isolated trees at ICRAF's research station in Machakos (Kenya), and calculated values for the green area fraction, using the functional branch analysis scheme.

Species	Age (yrs)	D_o (cm)	D_{min} (cm)	Leaf size (cm^2)	Inter-leaf distance (cm)	Leaf area/ branch length (cm^2 cm^{-1})	Green area fraction
Balanites aegyptiaca	16	15.6	0.17	2	1.9	1.04	0.46
Acacia tortilis	15	13.2	0.13	1	0.72	1.38	0.60
Melia volkensii	4	19.9	0.80	305	8.2	37.3	0.92
Gliricidia sepium	11	15.4	0.32	160	1.6	102	0.98

Source: Van Noordwijk (in prep.).

The first step, in the case of parklands, would be to analyse the spatial pattern of trees in the landscape and derive from this a set of 'equivalent circle' representations, which represent the cumulative distribution of distances of random points to the nearest tree. Whether tree distribution is clustered, random or regular, it is always possible to find a combination of circles which effectively embraces this specification. The system as a whole can then be represented by a weighted average of the results from different radii of a model of circular geometry. As a first order approach, we can then explore the interactions of such circular zones, using an amount of detail in the interactions within each cell of the model which would not be possible for discretization of the field as a whole. Limitations to this approach are that we essentially have to assume that every part of the soil is influenced by the nearest neighbour tree and not by others, and that tree size and age are independent of spatial position within the field. If the latter assumption does not hold, lateral interactions between neighbouring trees can not be treated on the basis of a symmetry rule. While accepting these shortcomings, we can still use the model to explore effectively the density of trees compatible with zones of different rainfall pattern (Figure 4). An advantage in total biomass production for agroforestry is associated with a larger water uptake and less drainage or groundwater recharge, especially at low-to-intermediate rainfall.

The WaNuLCAS model can also explore the consequences of differences in root distribution (both horizontally and vertically), rainfall pattern and soil physical parameters on the water balance of a system at patch scale. The effects of water redistribution on the soil surface by run-off/run-on can be simulated with the model, and by appropriate combinations of model runs, run-off/run-on relations between patches can be investigated. We have included options to consider hydraulic lift and the reverse flow from wet topsoils to dry subsoils, when trees are 'harvesting water' after the first rains (Ong and Leakey, Pate and Dawson, Pate and Bell, this volume). The total quantity of water involved in this reverse flow appears to be relatively small for reasonable parameter values (results not shown here) suggesting that the main function as far as the tree component is concerned is that this process

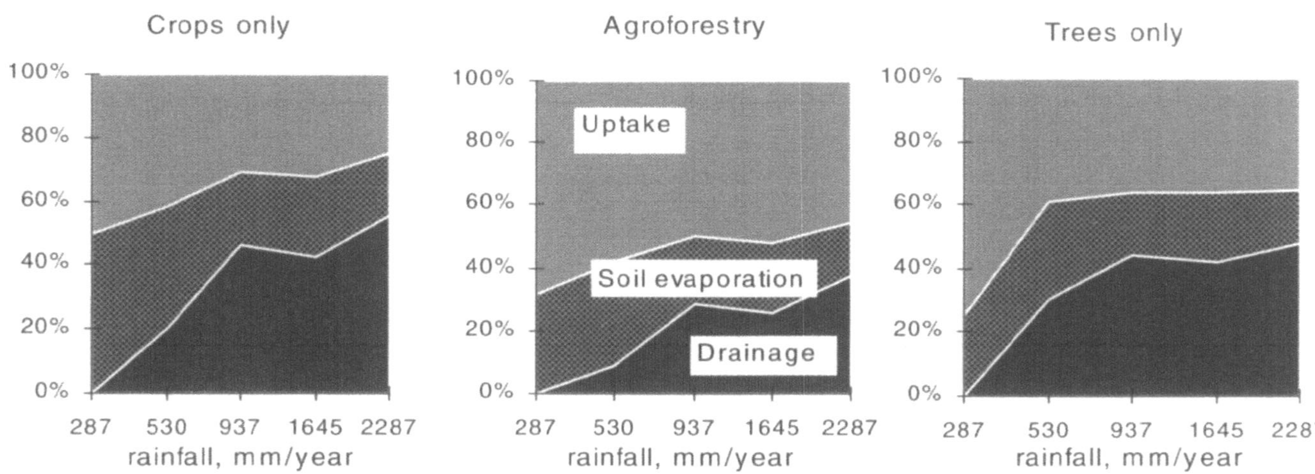

Figure 4. Water balance for WaNuLCAS calculations for a parkland system, or a separate crop or tree system, across a rainfall gradient. *Source:* Van Noordwijk and Lusiana (1998).

allows its roots to expand into dry soil layers while on their way to the groundwater table.

The main strength of the model appears to be its ability to explore longer time scales than available in any experimental data set, and thereby predict responses to stochastic variations in rainfall. Experimental evidence in Machakos indicates that in years with less-than-average rainfall, trees survive by gradually depleting belowground stores of water, whereas in years with more-than-average rainfall these water stocks would be replenished. Without trees, extra rainfall would, of course, translate to extra flow of rivers. A major challenge ahead of us is to explore the landscape scale consequences of what we now know at tree and plot level. Landscape mosaics at intermediate scale may radically differ from the weighted sum of their components, because lateral resource flows (above or below the soil surface) and lateral resource capture (e.g. scavenging tree roots) cause neighbourhood effects (Van Noordwijk and Ong, 1996). Subsurface water flows allow specific strips in a catena to use additional resources. Such effects probably underly the success of line plantings of *Gliricidia* in Mali (ICRAF Annual Report, 1994), and this success must not be mistaken as indication that the technologies concerned can be increased successfully to operations at larger scale. Trees can buffer some of the inter-annual fluctuations in agroecosystems, just as they do in natural savannah systems, but the value of this to the farmer depends on the products which are obtained from the tree.

4. Reducing production risk through diversity

4.1. *Diversity and stability*

Knoop and Walker (1985) attributed the remarkable functional stability and resilience (ability to recover from change) of savannahs to unique properties of species and their interaction at system level. Special attributes include seed dormancy through seed bank recovery, vegetative reproduction, the presence of underground reserves and related coppicing ability, phenology of different species in relation to rainfall distribution and adaptations to drought and grazing. Parallel attributes of systems include the slow release of inorganic nitrogen, conservation of nutrients by vegetation, reduction of anaerobic conditions by trees and the trophic complementarity between species in system concerned. However, recent reviews suggest that some savannah systems have become degraded below a biotic threshold which prevents them from recovering to their initial 'climax' once the disturbance is eliminated. This then leads towards a different 'stable' vegetation state after disturbance (Belsky and Amundson, 1998). They cited several processes such as loss of topsoil, losses of dominant or keystone species (including mycorrhizal partners), invasion by weedy species, loss of soil microorganisms and severe soil compaction as possible causes for such thresholds.

A principal reason why the debate on stability and complexity of natural and agroecosystems is so confused is that insufficient recognition has been given to the hierarchy of systems considered. Stability at higher system levels does not imply stability of its constituent populations, at lower levels or *vice versa*. The implicit assumption is often made that stability at lower hierarchical levels is necessary before stability will accrue at higher level. In fact, the reverse may be true when the higher level is a 'generalist' for whom different resources are interchangeable. A number of examples may illustrate this type of stabilisation for generalists at different system levels:

a) A plant which roots in superficial as well as deep layers of soil may maintain near constant rates of transpiration despite wild fluctuations in the soil water content of the topsoil and usually more damped fluctuations in the subsoil.
b) The genetic diversity in a plant or animal population can be maintained despite substantial selective pressures and resulting shifts in gene frequencies, so long as the selection pressures involved fluctuate and favour different genes under different circumstances.
c) A mixed plant community may maintain resource flows at a stable level, while its component species may be fluctuating widely, for example in an internally generated cyclical succession or in response to external disturbances.
d) From the perspective of a grazer interacting with a mixed plant community, a stable feed supply can be achieved with highly variable components, e.g. in a grass – legume (e.g. clover) mixture maintained in a competitive balance in terms of soil N level; the 'meadow' as a whole may be stable while patches within it are fluctuating in a cyclical manner, yet the grazing cow consumes all.
e) At field level, farmers can achieve stable yields by intercropping different cultivars or crops, especially when the relative performance of the crops depends on external factors which can not be predicted with any degree of certainty at the time of sowing. This unpredictability may be due to subtle unrecognised spatial heterogeneity within the system, as shown for barley-oat mixtures on soils with patches of low pH (De Wit, 1960), or to events happening after a crop is established, as in the case of maize-sorghum mixtures of the floodplain of the Nile in S. Sudan (Struif Bontkes, 1986). In these examples the various food crops may be of comparable value to the farmer for maintaining subsistence diets and/or for sale on the market.
f) At farm level, a farmer can try to balance different field-level enterprises which are attractive but risky, by maintaining a portfolio of complementary activities, e.g. a combination of market- and subsistence-oriented activities.
g) A successful 'human livelihood' farm-urban system can achieve stable food supplies simply by linking very diverse and potentially risky specialized

farming enterprises, leading to the paradox that the food choice in the 'developed' part of the world is more diverse than at any time in human history while all of this is produced on farms which are less diverse than at any time before. The more specialized farms are exposed to substantial risks, but pay for protection by insurance schemes and social networks.

The above examples spanning a wide range of systems have a common theme: a 'generalist system' can switch between resources and achieve stability on the basis of a diverse set of relatively unstable components. In many earlier versions of the 'diversity-stability' hypothesis no distinction was made between the various ways of evaluating a system – from its component parts or from an overall 'generalist' perspective, by integrating its various components.

A common feature of the above examples is that specialisation on less resources may increase the efficiency of using these resources and acquiring them in competition with other factors, and thus increase average yield or success. Nevertheless, it increases risks which can then be partly avoided by focusing (specialising) on stable resources, or subverted by using 'several baskets to carry eggs to the market'. In human societies, the scale at which risks can be managed has been generally shown with increased sophistication in economic development from subsistence to market integration. Market forces drive farmers in the direction of more specialised and risky farms, but also provide opportunity to externalise the risk from a strict farm-based system to broader social-financial systems (Vandermeer et al., 1998). Economic portfolio theory is well developed in financial markets to reduce risks while benefiting from the specialist advantage in taking risks.

4.2. *How much diversity is needed?*

Risk reduction in diverse systems is based on effects of mixtures on the expected yield (mean) and its variance. If the sum of all components reflects the 'generalist' perspective of the success of the system, the variance of this sum follows a simple rule for normally distributed variates: it equals the sum of variances of all components (weighted for their share in the total system), plus a term reflecting the covariance, or the degree to which success of components tends to coincide with successes or failures of other components. This rule for the variance of multi-component systems is essentially more favourable than the rules for deriving averages. Average total yield for a mixture can be higher, equal to, or less than the proportionate sum yields of the components (Figure 5a). However, the variance can only be equal to or less than this proportionate variance. There are thus many opportunities for variance reduction contributing to risk reduction (Figure 5c). For example, for a two-component system with a correlation coefficient of –1, one can construct a system which has zero variance (Schweigman, 1985). Under such conditions risk reduction is hardly a valid argument for maintaining diversity, as

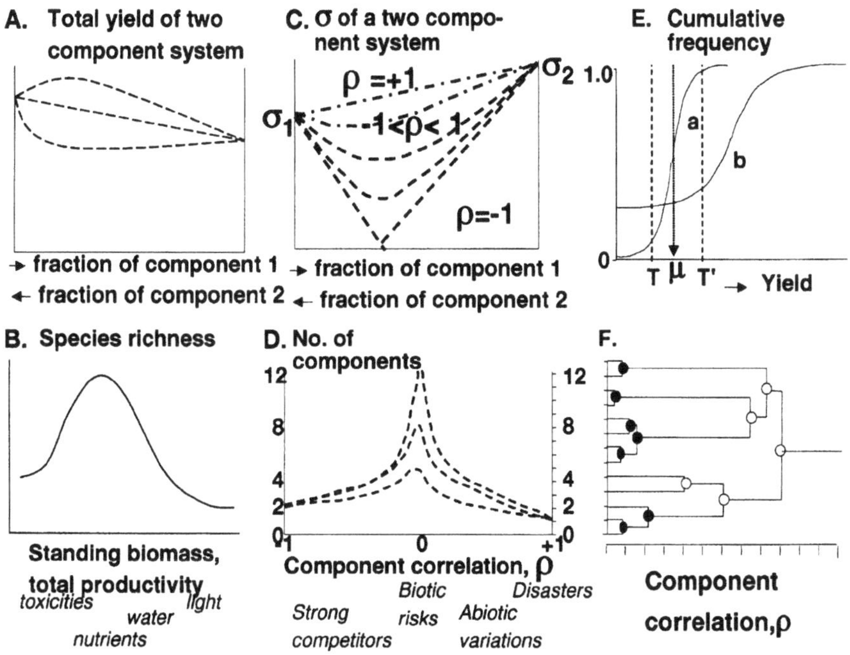

Figure 5. Replacement series (Yin-Yang figures) for average and variance of total production (a and c) and the possible species richness in relation to total production and dominant growth limiting factors (b) and to the covariance between components and the nature of the external source of variation (d); risk can be defined as the probability of not meeting a target T and the choice between component a and b can depend on the target as well as on the probability of yields (e); by classifying production options by their similarity, a rigorous procedure for selecting components of a risk-minimising portfolio of activities can be constructed (e).

a well-designed two-component system can reduce the variance to zero. However, if the covariance between the various components is zero, and all components fluctuate independently of each other, the variance of the sum will decrease with 1/N, where N is the number of components (compare with Trenbath, this volume). Once N has reached a value of four or eight, variance will have been reduced to 25 or 12.5% of the variance for a single component system, respectively, and further increases in number of components beyond such a value might not be worthwhile.

The diversity of plants in temperate grasslands has been linked to the total standing biomass or productivity, as envisaged by the 'Grime corridor' depicted in Figure 5b. Tentatively, this can be understood as indicating that there are more opportunities for niche differentiation and complementarity in the lower middle part of the range. Here nutrient shortage, rather than extreme conditions (lowest part), water shortage (upper middle), or light (higher part of the range), are the main limits to plant production (Figure 5b). Speculating in the same mode, we can interpret the X-axis of the graph linking optimum system diversity to correlation between components (Figure 5d) as referring to a range of sources of external variation, *viz*:

- disasters, affecting all components to a similar degree leading to a high correlation of components and giving little opportunities for risk reduction by diversity;
- weather related abiotic risks, allowing differences in response and adaptations and leading to partial, but generally positive correlations in performance;
- biotic risks based on outbreaks of specific pests and diseases, leading to virtual independence in response and maximum scope for reducing risk by increasing diversity;
- risks specifically affecting strong competitors in the system, leading to negative correlations in performance and compensatory production of the component least affected by the external agent. The situation carrying a correlation of –1 may arise as an extreme case of this type of relation.

In this view, risk reduction by diversity is least effective against 'disasters' and most effective (leading to the lowest system variance) when a small number of strong competitors are present, which fluctuate primarily in relation to specific biotic variables.

4.3. *The combination of options most reductive of risk*

The theory discussed so far gives a first indication of what advantages can be expected from maintaining component diversity ('how many baskets') – but it does not tell which ones to select (nor 'how to allocate the eggs'). Van Noordwijk et al. (1994) used a model to apply these concepts for water-limited sorghum production in the Sudan savannah zone. For a large number of production options (combinations of soil position in the catena, crop duration and N fertilisation) yield predictions were made on the basis of 30-year rainfall records. Through a cluster analysis technique, a dendrogram was constructed (Figure 5f) indicating functional groupings of production options at each level of similarity (covariance). By testing whether the reduction in variance possible at each level in this dendrogram is worthwhile and choosing the representative of each group showing highest average production, an objective routine can be derived for choosing both the number and type of options to be used. However, this is possible only if the source of external variation is known and the response of each possible component to this variation is predictable.

This procedure is in marked contrast, however, with most current practices aimed at yield stabilisation by testing crop genotypes across a large number of sites and years, and selecting the one with the best average and lowest variance as the most stable producer. This latter procedure presupposes that the best system will have a single component – as soon as one accepts that the optimum number may be one rather than two, the procedure misses out on opportunities provided by two components of high variance but performing well in contrasting years. For example, late and early maturing varieties are likely to fail in a stability test at component level, but may form a very stable pair, performing reasonably well on average over a number of seasons regardless of variations in rainfall. It may be better in such a case to keep the differently performing cultivars in separate fields, rather than as intercrops, as they might interact negatively at plot level, even though their use separately in an ecosystem might reduce overall risks at farm level.

Spatial heterogeneity of soils is normally seen as a negative attribute, especially if one aims at a homogenous crop producing close to the theoretical maximum without nutrient and water limitations. Heterogeneity of the soil, however, may contribute to yield stability if the different patches engender different response of crops to variation in rainfall. Brouwer et al. (1993) has made an interesting analysis of effects of soil variability from this perspective in Niger.

4.4. *Risk transmission in specialist – generalist chains*

The examples given above, where a generalist can be relatively stable while depending on fluctuating components, span a wide range of systems. We can now turn to consider what happens when we arrange these systems in a hierarchy.

If we simplify the choice to a G (generalist) and S (specialist) strategy, and assign a higher efficiency (production under suitable conditions) to S and a smaller variance to G, we can consider four basic sequences: S-S, G-G, S-G and G-S, with the G component always based on at least two resources. The S-S sequence may be too risky for the upper level, while the G-G sequence may be sub-optimal in efficiency, so the S-G and especially G-S sequence may achieve the best outcome from the perspective of the upper level. If we extend the analysis to chains of three or more levels, we find that S-G-S or G-S-G sequences provide attractive opportunities for combining efficiency and stability.

If we apply this concept to farmers in a subsistence economy, we find analogy with the G-S link in specialised animal husbandry, where grazers have choice of a wide range of food resources, and a S-G link in crop fields where farmers grow a large number of crops in a mixture, embodying a large degree of exchangeability between food sources in the human diet.

Finally, if the G-S-G sequence proves superior, it is interesting to note what happens if a new level with sufficient influence on the underlying levels is added on top of the hierarchy. One possible result is that it may induce all of the components to reorganise (flip over) from a G to an S strategy (or *vice versa*), thus bringing the whole system in a confused state of transition for a while as it undergoes structural readjustment. Economic development and market integration essentially add hierarchical levels of this type to a given chain, with farmers no longer comprising the main driving force of the system, but becoming dependent on policies and market forces generated in towns. The influence of the town market place is to have a generalist tendency and implicitly or explicitly induce farms to specialise (Vandermeer et al., 1998). This specialisation reduces farm-level diversity and, therefore, increases risk to farmers, unless an institutional framework develops to provide a buffer and safety-net to protect specialist farmers. Here we come right back at the existing policy environment, which does not sufficiently favour smallholder rural development and does not provide sufficient financial incentives for farmers to intensify their land use (Sanchez and Leakey, 1997).

4.5. *Diversity and priority setting*

The value of agro-diversity for risk reduction is most obvious where the constraints involved act independently across a number of possible systems. Clearly, the value of diversity as insurance against risk increases with the lifespan of the components, and with the time needed to replace them were they to be wiped out by pests or diseases. The arguments for tree diversity on farms are clearly stronger than those for diversity in annual crops since the latter can be replaced relatively easily and quickly should adverse conditions prevail. However, the cost of maintaining diversity remains, wherever components are being maintained which do not live up to their expected utility value.

If all components had equal value per unit resource use, competition would not be harmful and might contribute to stability of overall yields. The whole issue of weeds and competition thus starts with differences in value per unit resource use, between desirable and less desirable components of the system. Efforts to increase the value of 'desirables' inevitably increases the contrast in the above values, and therefore increases the real and perceived problems involved. Efforts at domesticating trees can help to maintain agroecosystem diversity, but may equally induce a specialisation on a few tree species. If domestication were to aim at the agroforest rather than at the tree, priority might have to be given to adding value to trees which grow well but have little value, instead of concentrating on tree species which are already rated as most valuable components. This perspective may stimulate us to re-think priority-setting exercises within a given system. Conventionally, for example, priority rating starts by listing options and ranking them in terms of utility value. Entities at the top of this list of values would then be normally considered to be priority choices for further research or action. However, when we wish to stimulate conditions favouring diversity, we may well have to focus on the middle part of our value listing to find priorities for action. If the overall aim is to be able to grow plants that have value, we may as well adopt a policy of adding value to plants which grow well, rather than ascertaining how and where to grow plants which attain a high value. There is no consensus as to how returns from effort in domestication relate to current value. Possibly an S-shaped curve is involved, with a difficult start for low-value products, followed by a phase of near-constant returns to investment before diminishing returns set in. Initial gains in performance by selection are often supposed to be larger than can be obtained in subsequent steps in selection. Product values and markets are conservative, however, and it is much easier to benefit incrementally from existing success of products than to create new types of use and markets for those products. The scope for focusing efforts on middle-value options remains to be tested.

5. Concluding remarks

We have focused in this chapter on two mimicry hypotheses, one relating to resource use and the other to diversity. Clear indications exist for the first hypothesis to be relevant in that there are clear advantages in productivity and less downstream effects if man-made land use systems do not deviate far from the resource use patterns exhibited by natural ecosystems within the same climatic zone. However, with complete technical control of resources, additional input supply and removal of excess outputs a measure of flexibility becomes possible. At a field or plot scale there may be considerable scope for utilising more resources, but only for a limited period of time, and so long as the environmental buffer stocks last. Alternatively, improvements may accrue from capture of resources from neighbouring plots and fields, and this may indeed be part of the reason why border plantings of trees are so common in the savannah zone (Van Noordwijk and Ong, 1996).

The value of diversity (second hypothesis) strongly depends on our perspective and in particular on the ability of farmers to derive value from a large number of components. The currently prevailing negative perception of competition in mixed systems remains valid if one component dominates the total value the farmer expects to derive from a system.

The question remains as to whether these mimicry hypotheses are of value as such. Agroecosystems can either gradually evolve locally by increasing the intensity of managing natural systems (as happened with the parklands), or be designed and synthesised *de novo*. In the first scenario the boundaries of what is possible are likely to be heavily constrained but can be explored on a trial and error basis. However, old, slow growing trees cannot easily be reconstituted in a system once they have been removed, so prospective analysis of likely constraints may help in avoiding calamities. In the case of *de novo* synthesis, the mimicry hypotheses can not only provide inspiration, but also indicate the overall constraints for a particular mimic solution to remain sustainable. However, prescriptions for agroecosystems derived from comparisons with natural systems and their diversity should not be accepted as anything more than useful hypotheses for action. Higher level stability at landscape or regional scale may well be achieved by allowing a degree of specialisation, limiting farm-level diversity. The best precondition for farmers to maintain diversity in their agroecosystem is probably the availability of a broad basket of choices, with deliberate avoidance of prescribing clear winners or 'best bets'.

6. Acknowledgements

The authors are grateful for the opportunity to jointly develop the ideas presented here during the very enjoyable workshop in September 1997 and discuss them with other participants; Dennis P. Garrity, Roger Leakey and the volume editors provided valuable comments on earlier versions of the text.

7. Note added in proof

The discussion of variance in Figure 5 and Section 4.2 assumes the absence of interaction effects on the means.

8. References

Adcock K (1983) The relation between water holding capacity of the top-soil, and the ratio of woody vegetation to grass in five savanna communities at Nylsvley. Botany 3 project report, University of the Witwatersrand, South Africa

Badiane O and Delgado CL (1995) A 2020 vision for food, agriculture and the environment in sub-Saharan Africa. Food, Agriculture and the Environment Discussion Paper 4, IFPRI, Washington, 56 pp

Belsky AJ and Amundson RG (1998) Influence of savannah trees and shrubs on understorey grasses and soils: new directions in research. In: Bergstrom L and Kirchman H (eds) Carbon and Nutrient Dynamics in Natural and Agricultural Tropical Ecosystems, pp 153–171. CAB-International, Wallingford, UK

Breman H and Kessler JJ (1995) Woody Plants in Agroecosystems of Semi-arid Regions (with an Emphasis on the Sahelian Countries). Springer-Verlag, Berlin, 340 pp

Breman H and Kessler JJ (1997) The potential benefits of agroforestry in the Sahel and other semi-arid regions. European Journal of Agronomy 7: 25–33

Brouwer J, Fussell LK and Herrmann L (1993) Soil and crop growth micro-variability in the West African semi-arid tropics: a possible risk-reducing factor for subsistence farmers. Agriculture, Ecosystems, and Environment 45: 229–238

Cannell MGR, Van Noordwijk M and Ong CK (1996) The central agroforestry hypothesis: the trees must acquire resources that the crop would not otherwise acquire. Agroforestry Systems 33: 1–5

Cannell MGR, Mobbs DC and Lawson GJ (1998) Complementarity of light and water use in tropical agroforests. II. Modelled theoretical tree production and potential crop yield in arid to humid climates. Forest Ecology and Management 102: 275–282

Campbell BM, Swift MJ, Frost PGH and Kirchman H (1998) Comparative ecosystem characteristics of a Miombo woodland and an adjacent agricultural field (Zimbabwe). In: Bergstrom L and Kirchman H (eds) Carbon and Nutrient Dynamics in Natural and Agricultural Tropical Ecosystems, pp 201–226. CAB-International, Wallingford, UK

Chivaura-Mususa CC and Campbell BM (1998) The influence of scattered *Parinari curatellifolia* and *Acacia sieberana* trees on soil nutrients in grassland pasture and in arable fields. In: Bergstrom L and Kirchmann H (eds) Carbon and Nutrient Dynamics in Natural and Agricultural Tropical Ecosystems, pp 191–200. CAB International

Delaney MJ and Happold DCD (1979) Ecology of African Mammals. Longman, London and New York
De Vos A (1975) Africa, the Devastated Continent? Dr W. Junk Publ., The Hague
De Wit CT (1960) On Competition. Verslagen Landbouwkundige Onderzoekingen 66.8, 82 pp
El-Tom MA (1975) The rains of the Sudan: mechanisms and distribution. Khartoum University Press, Khartoum
Goldstein G and Sarmiento G (1987) Water relations of trees and grasses and their consequences for the structure of savannah vegetation. In: Walker BH (ed) Determinants of Tropical Savannahs, pp 13–38. International Union of Biological Sciences Monograph No. 3
Hadley M (1985) Comparative aspects of land use and resource management in savannah environment. In: Tothill JC and Mott JJ (eds) Ecology and Management of the World's Savannah, pp 142–158. CAB, UK
Jones M, Sinclair FL and Grime VL (1998) Effect of tree species crown pruning on root length and soil water content in semi-arid agroforestry. Plant and Soil 201: 197–207
Kinyamario JI and Imbamba SK (1992) Savannah at Nairobi National Park. In: Long SP, Jones MB and Roberts MJ (eds) Primary Productivity of Grass Ecosystems of the Tropics and Subtropics, pp 25–69. Chapman and Hall, London
Knoop WT and Walker BH (1985) Interactions of woody and herbaceous vegetation in a southern African savannah. Journal of Ecology 73: 235–254
Long SP, Jones MB and Roberts MJ (1992) Primary Productivity of Grass Ecosystems of the Tropics and Subtropics. Chapman and Hall, London
Lott JE (1998) Resource utilisation by trees and crops in agroforestry systems. Unpublished PhD thesis, University of Nottingham
Michon G and De Foresta H (1997a) Agroforestry: pre-domestication of forest trees or true domestication of forest ecosystems? Netherlands Journal of Agricultural Science 45: 451–462
Michon G and De Foresta H (1997b) Agroforests as an alternative to pure plantations for the domestication and commercialisation of NTFP's. In: Leakey RRB and Vantomme P (eds) Domestication and Commercialization of Non-Timber Forest Products in Agroforestry systems, pp 160–175. FAO, Rome
Mobbs DC, Cannell MGR, Crout NMJ, Lawson GJ, Friend AD and Arah J (1998) Complementarity of light and water use in tropical agroforests. I. Theoretical model outline, performance and sensitivity. Forest Ecology and Management 102: 259–274
Palm CA, Myers RJK and Nandwa SM (1997) Combined use of organic and inorganic nutrient sources for soil fertility maintenance and replenishment. In: Buresh RJ, Sanchez PA and Calhoun F (eds) Replenishing Soil fertility in Africa, SSSA Spec. Publ. Nr. 51, pp 193–217. Soil Science Society of America, Madison, Wisconsin, USA
Rutherford MC (1982) Woody plant biomass distribution in Burkia africana savannahs. In: Huntley BJ and Walker BH (eds) Ecology of Tropical Savannahs, pp 120–142. Springer-Verlag, Berlin
Sae-Lee S, Vityakon P and Prachaiyo B (1992) Effects of trees on paddy bund on soil fertility and rice growth in Northeast Thailand. Agroforestry Systems 18: 213–223
Sanchez PA (1995) Science in agroforestry. Agroforestry Systems 30: 5–55
Sanchez PA and Leakey RRB (1997) Land use transformations in Africa: three determinants for balancing food security with natural resource utilization. European Journal of Agronomy 7: 15–23
Sanchez PA, Shepherd KD, Soule MJ, Place FM, Buresh RJ, Izac AMN, Mokwunye AU, Kwesiga FR, Ndiritu CG and Woomer PL (1997) Soil fertility replenishment: an investment in natural resource capital. In: Buresh RJ, Sanchez PA and Calhoun F (eds) Replenishing Soil Fertility in Africa, SSSA Spec. Publ. Nr. 51, pp 1–46. Soil Science Society of America, Madison, Wisconsin, USA
Sanford WW and Isichei AO (1986) Savannah. In: Lawson GW (ed) Plant Ecology in West Africa, pp 95–149. John Wiley, Chichester
Schweigman C (1985) Operations Research Problems in Agriculture in Developing Countries. Khartoum University Press, Khartoum, Sudan, 361 pp

Solbrig OT, Medina E and Silva JF (1996) Biodiversity and tropical savannah properties: a global view. In: Mooney HA, Cushman JH, Medina E, Sala OE and Schulze ED (eds) Functional Roles of Biodiversity, pp 185–212. Wiley

Strugell RG and Pigott CD (1978). Biomass, shoot-production and grazing of two grasslands in Rwenzori National Park, Uganda. Journal of Ecology 66: 73–96

Struif Bontkes TE (1986) Sorghum-rice intercropping, a security system for the poorly drained soils of Southern Sudan. Netherlands Journal of Agricultural Science 34: 193–198

Torquebiau EF and Kwesiga F (1996) Root development in as *Sesbania sesban* fallow – maize system in Eastern Zambia. Agroforestry Systems 34: 193–211

Tothill JC and Mott JJ (1985) Ecology and Management of the World's Savannah. CAB, UK

Van Keulen H and Breman H (1990) Agricultural development in West African Sahelian regions: a cure against land hunger? Agriculture, Ecosystems and Environment 32: 177–197

Vandermeer J, Van Noordwijk M, Ong C, Anderson J and Perfecto Y (1998) Global change and multi-species agroecosystems: concepts and issues. Agriculture, Ecosystems and Environment 67: 1–22

Van Noordwijk M (1984) Ecology Textbook for the Sudan. Khartoum University Press, Khartoum, 280 pp

Van Noordwijk M (1996) A simple model to quantify mulch and shade effects. In: Ong CK and Huxley PA (eds). Tree-Crop Interactions – a Physiological Approach, pp 51–72. CAB International, Wallingford

Van Noordwijk M and Purnomosidhi P (1995) Root architecture in relation to tree-soil-crop interactions and shoot pruning in agroforestry. Agroforestry Systems 30: 161–173.

Van Noordwijk M and Ong CK (1996) Lateral resource flow and capture – the key to scaling up agroforestry results. Agroforestry Forum 7: 29–31

Van Noordwijk M, Dijksterhuis G and Van Keulen H (1994) Risk management in crop production and fertilizer use with uncertain rainfall: how many eggs in which baskets. Netherlands Journal of Agricultural Science 42: 249–269

Van Noordwijk M, Lawson G, Groot JJR and Hairiah K (1996) Root distribution in relation to nutrients and competition. In: Ong CK and Huxley PA (eds) Tree-Crop Interactions – a Physiological Approach, pp 319–364. CAB International, Wallingford

Van Noordwijk M, Tomich TP, de Foresta H and Michon G (1997) To segregate – or to integrate: the question of balance between production and biodiversity conservation in complex agroforestry systems. Agroforestry Today 9: 6–9

Van Noordwijk M, Hairiah K, Lusiana B and Cadisch G (1998) Tree-soil-crop interactions in sequential and simultaneous agroforestry systems. In: Bergstrom L and Kirchman H (eds) Carbon and Nutrient Dynamics in Natural and Agricultural Tropical Ecosystems, pp 173–190. CAB-International, Wallingford, UK

Van Noordwijk M and Lusiana B (1998) WaNuLCAS 1.1, Backgrounds of a model of water, nutrient and light capture in agroforestry systems. Working Document, International Centre for Research in Agroforestry (ICRAF), Bogor, Indonesia, 98 pp

Van Noordwijk M and Lusiana B (1999) WaNuLCAS, a model of water, light and nutrient capture in agroforestry systems. Agroforestry Systems 43: 217–242

Walker BH (1982) Structure and function of savannahs: an overview. In: Huntley BJ and Walker BJ (eds) Ecology of Tropical Savannahs, pp 83–91. Springer Verlag, Berlin

Walker BH (1987) A general model of savannah structure and function. In: Walker BH (ed) Determinants of Tropical Savannahs, pp 1–12. International Union of Biological Sciences Monograph Series No. 3, Paris

Wiersum KF (1997) From natural forest to tree crops, co-domestication of forests and tree species, an overview. Netherlands Journal of Agricultural Science 45: 425–438

Yaker L (1993) Constraints and opportunities for sustainable food production in sub-Saharan Africa. In: Mohan Raj DR (ed) Sustainable food production in sub-Saharan Africa. 2 Constraints and opportunities. IITA, Ibadan, Nigeria, 116 pp

Agroforestry Systems **45**: 159–185, 1999.

Soil community composition and ecosystem processes
Comparing agricultural ecosystems with natural ecosystems

D. A. NEHER
Department of Biology, University of Toledo, Toledo, OH 43606, USA; E-mail: dneher@uoft02.utoledo.edu

Key words: decomposition, environmental monitoring, fertiliser, mineralisation, nitrogen, pesticides

Abstract. Soil organisms play principal roles in several ecosystem functions, i.e. promoting plant productivity, enhancing water relations, regulating nutrient mineralisation, permitting decomposition, and acting as an environmental buffer. Agricultural soils would more closely resemble soils of natural ecosystems if management practices would reduce or eliminate cultivation, heavy machinery, and general biocides; incorporate perennial crops and organic material; and synchronise nutrient release and water availability with plant demand. In order to achieve these goals, research must be completed to develop methods for successful application of organic materials and associated micro-organisms, synchronisation of management practices with crop and soil biota phenology, and improve our knowledge of the mechanisms linking species to ecosystem processes.

1. Introduction

Jackson (1995) suggests that modern agriculture operates in a 'paradigm of ignorance.' This concept is appropriate for the discipline of soil ecology, which has been recognised as a scientific discipline for only 20 years. Some have labelled soil ecology as a 'last frontier' (Andre et al., 1994). About 10% of soil species have been identified (Hawksworth and Mound, 1991). Of the world species, insects, fungi and nematodes are three groups that have not been identified fully (Table 1). Our knowledge of soil organisms has been limited by our ability to extract organisms from soil efficiently and by an ability to appropriately identify juvenile stages. Furthermore, microbiology and soil biology are often ignored by ecologists. Consequently, modern studies of soil and sediment ecology are several steps behind those of other subdisciplines of ecology. Many aspects of decomposer ecology are not well characterised for terrestrial soil or sediments in lakes, streams or oceans. Soil and sediment ecologists are still completing the taxonomy and systematics of soil organisms, revealing life history strategies, and just beginning to understand relationships between organisms and their contribution to ecosystem function (Crossley et al., 1992). One exception is that of earthworms and nitrogen-fixing bacteria whose relationship to ecosystem function has been known for decades. This apparent lack of knowledge does not, however, diminish the importance of soil organisms. Evolutionary and geological

Table 1. Numbers of species in the world.

Organism group	Described species ($\times 10^3$)	Estimated species ($\times 10^3$)	% of estimated species described
Dicot plants	170		
Monocot plants	50		
Ferns	10		
Mosses and liverworts	17		
Fish	19	21	95
Birds	9	9.1	99
Mammals	4	4	100
Reptiles and amphibians	9	9.5	95
Algae	40	60	67
Protozoa	30	100	30
Nematodes	15	500	3
Bacteria	3	30	10
Viruses	5	130	4
Fungi	69	1,500	5
Insects	800	2,000–10,000	0.08–0.4

Source: Hawksworth and Mound (1991).

evidence suggests that soil organisms have considerably longer histories on earth than organisms that have received more attention such as mammals and vascular plants (Main, this volume; Van Noordwijk and Ong, this volume). Their longevity alone suggests they play an essential role(s) in ecosystem function.

This paper will introduce the discipline of soil ecology, with an emphasis on characterising members of the soil community (Table 2) and their respective roles in ecosystem function. Second, approaches to design and management of soil communities to optimise ecosystem function will be discussed.

Table 2. Hierarchy of size and abundance of organisms inhabiting soil.

Class	Example(s)	Biomass ($g\ m^{-2}$)	Length (mm)	Populations (m^{-2})
Microflora	Bacteria, fungi, algae, actinomycetes	1–100	n.a.[a]	10^6–10^{12}
Microfauna	Protozoa	1.5–6.0	0.005–0.2	10^6–10^{12}
Mesofauna	Nematodes, arthropods, enchytraeids, mites, springtails	0.01–10	0.2–10	10^2–10^7
Macrofauna	Insects	0.1–2.5	10–20	10^2–10^7
Megafauna	Earthworms	10–40	≥ 20	0–10^3

[a] Not applicable.
Source: Dindal (1990) and Lal (1991).

Finally, research priorities will be summarised. Most discussion will focus on nematodes, springtails (Collembola), and mites because they predominate in total numbers, biomass and species of fauna in soil (Harding and Studdart, 1974; Samways, 1992).

2. Soil food web

Mesofauna occupy all trophic levels within the soil food web (Figure 1) and affect primary production directly by root-feeding and indirectly through their contribution to decomposition and nutrient mineralisation (Crossley et al., 1992). In soil, the basic food web structure is similar to other food webs by containing primary producers, consumers and detritivores. Numbers and biomass per volume of soil organisms decrease by orders of magnitude from bottom to top positions in the food chain. In contrast to other food webs, soil food webs tend to have longer food chains, greater incidences of omnivory, and possibly greater complexity than other food webs. Further, all fauna depend on primary producers (e.g. for litter).

Plants and organic debris provide habitats for soil organisms. Plants affect soil biota directly by generating inputs of organic matter above- and below-ground and indirectly by the physical effects of shading, soil protection and uptake of water and nutrients by roots. Energy and nutrients obtained by plants eventually become incorporated in detritus that provide the resource base of a complex soil food web. Plant roots, for example, exude amino acids and sugars that serve as a food source for micro-organisms (Curl and Truelove, 1986). Soil micro- and mesofauna are often aggregated spatially which is probably indicative of the distribution of favoured resources, such as plant roots and organic debris (Swift et al., 1979; Goodell and Ferris, 1980; Barker and Campbell, 1981; Noe and Campbell, 1985; Griffiths, 1994).

Three basic energy pathways exist in soil: those of roots, bacteria and fungi (Moore et al., 1988). The root pathway includes primary herbivores such as pathogenic fungi, bacteria, nematodes, protozoa and their consumers. These organisms decrease primary productivity by altering uptake of water and nutrients, and may create abnormalities in root morphology and/or physiology. The bacterial pathway includes saprophytic and pathogenic bacteria and their respective consumers (e.g. protozoa, bacterial-feeding nematodes). The fungal pathway includes saprophytic, mycorrhizal and pathogenic fungi and their respective consumers (e.g. fungal-feeding nematodes, oribatid mites and springtails). The root, bacterial, and fungal pathways unite at levels higher in the food chain, i.e. omnivores and predators. Many microarthropods and nematodes are omnivores and feed on a variety of food sources, such as algae, fungi, bacteria, small rotifers, enchytraeids and small nematodes. Soil mesofauna are often categorised by specific feeding behaviours and are often depicted as microbial-feeders. However, many organisms are at least capable of feeding of other trophic groups. As a result, omnivory in soil communi-

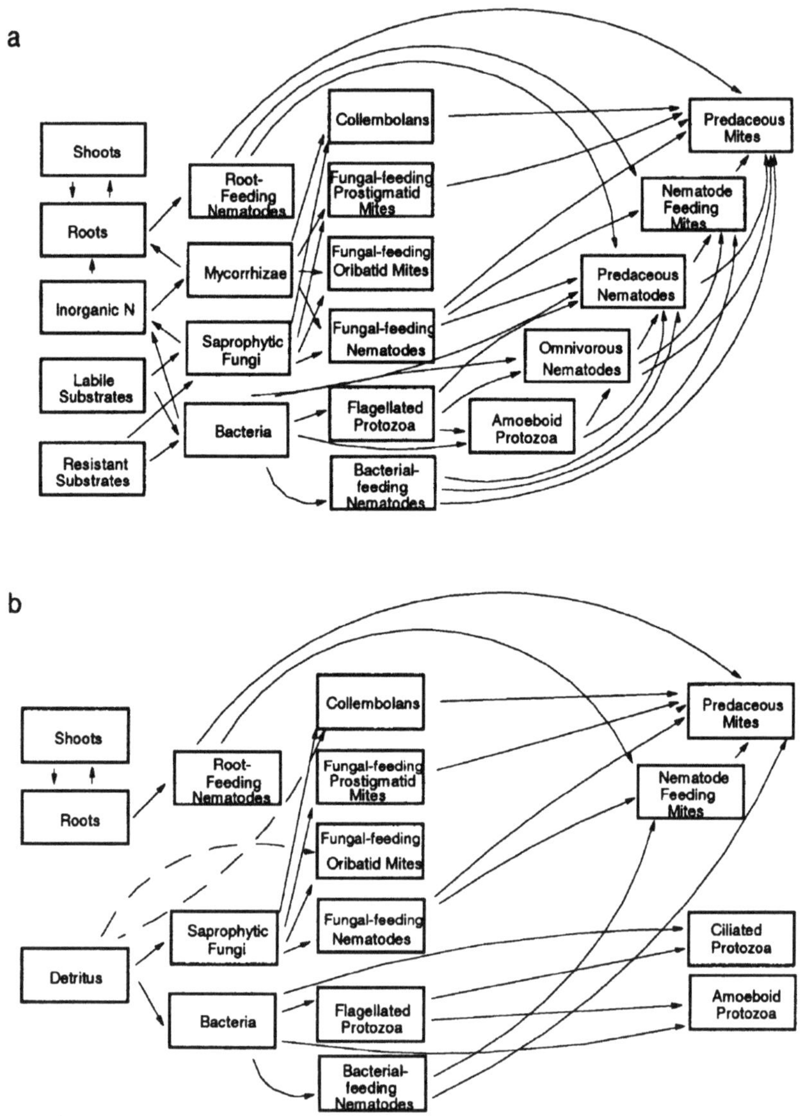

Figure 1. Soil food webs for a) grassland and b) cultivated wheat. Solid and dashed lines represent potential feeding relationships that were and were not quantified, respectively. *Source:* After Moore and de Ruiter (1991).

ties may be more prevalent than assumed previously (Walter, 1987; Walter et al., 1986, 1988; Walter and Ikonen, 1989; Mueller et al., 1990; Bengtsson et al., 1995). Predators include secondary, tertiary and quarternary consumers, including certain nematodes, beetles, fly larvae, centipedes, spiders and mites. Some mesofauna, such as nematodes and protozoa, may serve as predators or prey depending on the other species in the community (Griffiths, 1994; Yeates and Wardle, 1996). Soil microarthropods can be important predators on small arthropods and their eggs (e.g. proturans, pauropods, enchytraeids), nematodes, and on each other (Dindal, 1990).

Soil food web structure varies with geography and climate. In North America, shortgrass prairie (*Bouteloua gracilis*), lodgepole pine (*Pinus contorta* ssp. *latifolia*), and mountain meadow (*Agropyron smithii*) have similar food web structure (Hunt et al., 1987; Ingham et al., 1989). Although structure is conserved, relative abundance of organisms within trophic or functional groups may vary by ecosystem type. In Poland, bacterial-feeders and root-feeding nematodes are most abundant in agricultural soils, omnivory is more common in grasslands than agriculture, and fungal-feeders are relatively more abundant in forest than agricultural soils (Table 3). In contrast, relative numbers of organisms in each functional group differ in Swedish soils. For example, ratios of organisms in fungal to bacterial pathways are greatest in fertilised barley (*Hordeum vulgare* L.) followed in descending order by meadow fescue (*Festuca pratensis* L.) and lucerne ley (*Medicago sativa* L.) (Beare, 1997). In the Netherlands, disturbances such as cultivation and addition of mineral fertilisers eliminate certain functional groups such as predacious nematodes, omnivorous nematodes, and mycorrhizae that would otherwise be present in undisturbed grasslands (Figure 1). Furthermore, testate amoebae are reduced in abundance by at least 50% in agricultural compared to natural ecosystems (Foissner, 1997). Other organisms, such as the Enchytraeidae are less sensitive to cultivation than seasonal changes in climate (van Vliet et al., 1995).

Soil organisms vary in size by several orders of magnitude (Table 2). Microflora and microfauna are the smallest in size and most abundant. Mesofauna are moderate in size and abundance. Mesofauna generally do not have the ability to reshape the soil and, therefore, are forced to use existing

Table 3. Number of nematode genera per trophic group (percent of mean abundance), species diversity, species richness, and numbers of nematodes in soils of Poland.

Trophic Group	Annual crop (rye, potato)	Perennial crop (alfalfa)	Grassland	Forest
Bacterial-feeders	9–15 (41)	15–16[a]	8–18 (29)	11–18 (39)
Fungal-feeders	2–4 (16)	4[a]	1–4 (6)	2–5 (21)
Root-feeders	6–11 (37)	11[a]	9–14 (38)	7–11 (23)
Omnivores/predators	2–7 (6)	11–14[a]	7–17 (27)	2–8 (18)
Species diversity[b]	3.1–4.2	–[a]	3.9–4.9	3.2–4.3
Species richness	33–34	87–100	74	34–68
Mean no. nematodes[c]	3.5–5.0	–[a]	2.3–3.3	2.3–3.7

[a] Not available
[b] Shannon's index of diversity
[c] 10^6 per m^2
Source: After Wasilewska (1979).

pore spaces, water cavities or channels for locomotion within soil. Most microfauna and mesofauna inhabit soil pores of 25–100 μm diameter. Protozoa (flagellates and small amoebae) occupy pores as small as 8 μm diameter (Griffiths, 1994). Macrofauna and megafauna are the largest and least abundant per unit area. Their size exceeds that of many soil pores and pore necks. Therefore, their movement and activity re-form soil and create burrows or channels.

Habitable pore space (voids of sufficient size and connectivity to support mesofauna) accounts for a small portion of total pore space (Hassink et al., 1993). Microfaunal community composition becomes increasingly dominated by smaller animals as average pore volume decreases (as in compacted soil or soils dominated by fine clays). Within the habitable pore space, microbial and mesofaunal activity are influenced by the balance between water and air. Maximum aerobic microbial activity occurs when 60% of the pore volume is filled with water (Linn and Doran, 1984). Saturation (water-logging) and drought are detrimental to soil faunal communities, because these conditions result in anaerobiosis and dehydration, respectively. Microbes and small fauna (e.g. nematodes, protozoa) depend on water films to live and move through the soil system (Griffiths, 1994; Lavelle et al., 1995). In aerobic environments, nematodes are more abundant when amoebae are present as food. This suggests that amoebae feed on bacteria in pores inaccessible to nematodes and then emerge to act as food for nematodes (Foster and Dormaar, 1991; Griffiths, 1994).

3. Soil function

The 'first-link' hypothesis can partly explain the origin of biodiversity in soil but the relationship between biodiversity on soil function remains untested (Lavelle et al., 1995). The hypothesis originated from observed changes in structure of earthworm communities along thermo-latitudinal gradients and extrapolation of observed patterns to plants based on the similarities observed in the general function of both drilosphere and rhizosphere systems (Lavelle et al., 1995). Janzen (1985) asserts that 'plants wear their guts on the outside'; they produce exudates that trigger microbial activity and subsequent mineralisation of nutrients. In guts of earthworms, intestinal mucus and movement of soil through the gut are functional equivalents of root exudates and elongation through soil, respectively. It has been demonstrated that the mutualistic digestion system of earthworms becomes increasingly more efficient with increasing temperatures (Lavelle et al., 1995). It is assumed that increased temperatures in soils give roots access to a greater volume of nutrient resources because of more efficient mutualisms between soil microflora and plant roots. This would be the first link of a cascade process in which the species richness in the food web of consumers and decomposers would become larger in the tropics than in colder temperate or arctic areas.

In some geographic areas, richness of species composition in grassland and forest soils exceeds that of arable soils with annual crops (Table 3). In contrast, Wardle (1995) reports several cases in which conventional agricultural practices stimulate diversity. For example, the diversity of nematode genera in soils within asparagus (*Asparagus officinale* L.) systems was not affected by management practices such as addition of sawdust mulch, cultivation or herbicide applications. Soil-associated beetle diversity, however, was increased significantly by mulching and (sometimes) high weed densities.

Disturbance certainly plays a role in altering diversity. Perhaps intermediate disturbance promotes macrofaunal diversity, and extreme or lack of disturbance reduces diversity relative to undisturbed systems (Petraitis et al., 1989; Hobbs and Huenneke, 1992). The 'intermediate disturbance hypothesis' (Connell, 1978) could explain why some groups of organisms are more abundant in no-till (i.e. intermediate disturbance) than either conventionally-tilled (i.e. extreme disturbance) or old-field (i.e. no disturbance) systems (Wardle, 1995). If disturbance is common or harsh, only a few taxa that are resistant to disruption will persist (Petraitis et al., 1989). If disturbance is mild or rare, soil communities will approach equilibrium and be dominated by a few taxa that can out-compete all other taxa. However, attainment of steady-state equilibrium in agricultural or natural ecosystems is rare (Richards, 1987). There is little data to support this hypothesis but temporal patterns in diversity appear consistent with patterns detected during natural succession in plant communities (Whittaker, 1975).

Perhaps, it is more important to mimic the diversity of ecosystem function observed in natural systems than to mimic diversity of community composition (Main, this volume; Van Noordwijk and Ong, this volume). For example, an index of trophic diversity may serve as a measure of functional diversity in soil communities (Figure 2). Reducing the frequency of cultivation (Hendrix et al., 1986) and including perennial crops in agricultural systems (Wasilewska, 1979; Freckman and Ettema, 1993; Neher and Campbell, 1994) are two ways to increase trophic diversity in arable soils. Use of a trophic diversity index assumes that greater diversity (an integration of taxa richness and evenness) of trophic groups in soil food webs and longer food chains correspond with improved ecosystem function. In order to test the validity of such assumptions, it is important to identify ecological functions of soil and how soil organisms are involved in those functions. To date, five ecological functions of soil have been identified (*sensu* Larson and Pierce, 1991):

- promote plant growth;
- receive, hold and release water;
- recycle carbohydrates and nutrients through mineralisation;
- transfer energy in the detritus food chain; and
- act as an environmental buffer.

Individual taxa may have multiple functions and several taxa appear to have similar functions. However, taxa may not necessarily be redundant because

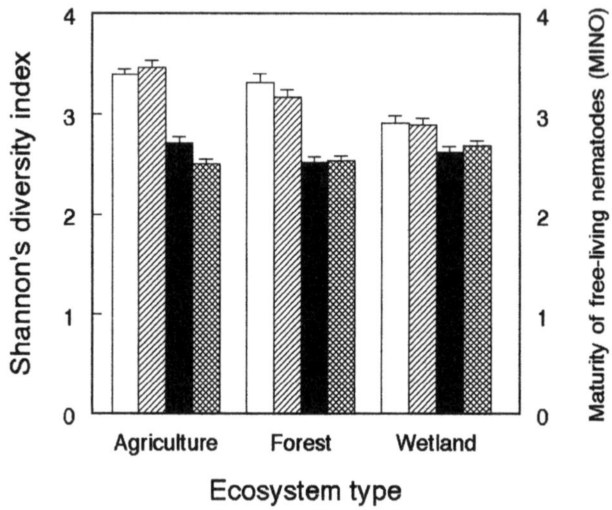

Figure 2. Means and standard errors of two measures of nematode community composition: 1) trophic diversity in disturbed (open bars) and undisturbed (hatched bars) and 2) successional maturity of free-living nematodes in disturbed (solid bars) and undisturbed (stippled bars) soils of North Carolina, USA. Undisturbed is defined as 10+ year-old pastures, 30+ year-old forests, and functioning wetlands. Disturbed is defined as annually cultivated arable soils, one to three year-old forests and wetlands converted to agricultural production. *Source:* Neher et al. (unpubl. data).

taxa performing the same function are often isolated spatially, temporally, or by microhabitat preference (Beare et al., 1995). Ettema (1998) suggests that the extent of nematode functional redundancy in soil has been greatly overestimated. Although redundancy of single functions is common, distinct physiological and environmental requirements drive species of the same functional group to play widely different roles in soil ecosystem processes.

3.1. *Promoting plant growth*

Growth may be enhanced for plants in soils containing multiple functional groups. For example, in North American grasslands containing blue gamma grass (*Bouteloua gracilis*), soil food webs containing primary decomposers and microbial grazers had greater primary productivity than systems limited to only primary decomposers (Ingham et al., 1985). Increases in plant growth have been observed for plants grown in soil containing protozoa and/or nematodes (Verhoef and Brussaard, 1990; Griffiths, 1994; Alphei et al., 1996). In glasshouse experiments, blue gamma grass withdrew more nitrogen from fertilised soil in the presence of amoebae than in their absence (Zwart et al., 1994). Protozoan grazing is necessary to release nitrogen from bacterial biomass for plant uptake (Clarholm, 1985). Finally, shoot production was enhanced 1.5 times in birch (*Betula pendula*) and 1.7 times in Scots pine

(*Pinus sylvestris*) when seedlings were grown in soils containing a more complex food web (bacterial-feeding nematodes, fungal-feeding nematodes, omnivores, springtails and oribatid mites) than when grown in simple systems containing only bacterial- and fungal-feeding nematodes (Setälä et al., 1995).

Based on a simulation model of shortgrass prairie (Hunt et al., 1987), 14% of nitrogen extracted by plants is accounted for by predation of bacteria by amoeba (Zwart et al., 1994). Having similar carbon, nitrogen and phosphorus content as their prey, protozoa incorporate only 10–40% of prey carbon, respire the remaining carbon, and excrete excess (20–60%) nitrogen and phosphorus into soil mostly as inorganic forms that can be assimilated by plants (Griffiths, 1994; Zwart et al., 1994).

Soil fauna not only alter the availability of nutrients for plants, but also alter relative distributions of carbon and nitrogen within plants. For example, plants grown in soils with only protozoa have less carbon in shoots and more carbon in roots than plants grown without protozoa. The opposite pattern was observed for plants grown in soils containing only nematodes (Alphei et al., 1996). Soil fauna generally affect amounts of nitrogen in roots more than in shoots (Alphei et al., 1996). One method of protozoa affecting nitrogen supply to roots is by consuming *Rhizobium* spp. and, consequently, reducing nodulation in the rhizospheres of common garden bean (*Phaseolus vulgaris*) (Zwart et al., 1994).

The effect of soil fauna on plant growth cannot be attributed entirely to an increased supply of nutrients to plants because nutrient leaching may also increase in the presence of soil fauna (Alphei et al., 1996). Protozoa may further stimulate plant growth by altering concentrations of plant hormones (e.g. auxin, tryptophan) in the rhizosphere and/or suppressing pathogenic bacteria (Jentschke et al., 1995; Alphei et al., 1996). Hormonal substances may derive directly from protozoa or indirectly from lysis of bacterial cells grazed by protozoa.

3.2. *Receiving, holding and releasing water*

Dawson and Pate, Hatton and Nulsen, and Dunin et al. (this volume) stress the importance of water in plant physiology and hydrogeology in designing agricultural systems that mimic nature. Soil serves as an interface between plants and geology. Soil water may have positive or negative impacts on soil organisms. Soil microbial and megafaunal populations improve water infiltration by altering soil physical structure. For example, bacteria produce polysaccharide adhesives and fungi produce thread-like hyphae that bind soil particles into stable aggregates and reduce potential soil losses by erosion (Gupta and Germida, 1988; Eash et al., 1994; Beare, 1997). Enchytraeidae (van Vliet et al., 1995) and earthworms (Edward and Bohlen, 1996) create burrows to improve infiltration and improve aeration. Van Vliet et al. (1995) hypothesise that enchytraeids have more influence on soil structure in agricultural fields than in forested areas.

As matric potential in soils declines to about –3 or –6 bars, bacterial respiration declines rapidly and is negligible at –20 bars (Griffin, 1981). Fungi often tolerate matric potential in excess of –30 bars, conditions unsuitable for growth of most bacteria except actinomycetes such as *Strepomyces* spp. Under these dry conditions, diseases caused by fungal pathogens, such as *Fusarium culmorum* on wheat (*Triticum aestivum*), become more severe. *F. culmorum* thrives at matric potentials that reach –100 bars at the surface and –30 bars in the rhizosphere (Griffin, 1981). Other fungi such as the take-all pathogen on wheat, *Gaeumannomyces graminis*, may predominate in irrigated soils (Griffin, 1981).

Ecosystem processes relate directly to the water content in soil. For example, a negative linear relationship occurs between relative nitrogen mineralisation and the logarithm of water potential. Decomposition of organic matter is also influenced by soil matric potential. An initial rapid decrease in decomposition within the –0.3 to –10 bar range is followed by another region where decomposition decreases linearly with decreasing water availability. The role of most bacteria is probably minimal once soils reach –15 bar water potential resulting in actinomycetes (filamentous, gram- positive bacteria) and fungi being the major decomposers in soils. Dry soils reduce the ability of substrate molecules to diffuse to the bacterial cell and the ability of bacteria to move to new substrates (Sommers et al., 1981). At the other extreme, water potential influences decomposition where saturated conditions result in the depletion of oxygen (O_2) and the development of anaerobic conditions. Under these conditions, anaerobic bacteria are the predominant organisms responsible for decomposition. Comparison of decomposition rate to soil water potential must incorporate metabolic shifts that occur in the transition from aerobic to anaerobic conditions. In the former, carbon dioxide (CO_2) is the major end product and in the latter, CO_2 may underestimate microbial activity because methane (CH_4) and other reduced carbon compounds may be significant end-products under saturation (Sommers et al., 1981).

3.3. *Recycling carbohydrates and nutrients through mineralisation*

Fauna may contribute directly to mineralised forms of nitrogen by excretion of ammonium (e.g. nematodes and protozoans) or nitrate (e.g. springtails) (Anderson et al., 1983; Ingham et al., 1985; Teuben and Verhoef, 1992; Darbyshire et al., 1994; Griffiths, 1994). As reservoirs of nutrients, microflora and microfauna also contribute indirectly to mineralisation. Net increases in nitrogen concentration in soil caused by mesofauna grazing on microbes have been demonstrated in petri dish experiments (Trofymow and Coleman, 1982). Nutrients immobilised in microbes are mineralised and subsequently become available to plants (Seastedt et al., 1988; Söhlenius et al., 1988).

Model simulations of soil food webs estimate that 97 and 99% of total nitrogen flux can be attributed to the bacterial pathway in integrated farming and conventional farming systems, respectively (Beare, 1997). Griffiths (1994) estimates from several independent food web studies that protozoa, especially amoebae, are responsible for 20–40% of net nitrogen mineralisation under field conditions. Estimates of protozoan contribution to net nitrogen mineralisation vary by geographic location and farming system. For example, protozoa are estimated to mineralise about 54 and 90 kg of nitrogen per year in no-till and conventionally tilled soils (Beare, 1997). In The Netherlands, protozoa mineralise approximately 30 and 43 kg of nitrogen per year in conventional and integrated farming trials (Beare, 1997). These estimates likely underestimate total contributions to net mineralisation by protozoa because bacteria simultaneously mobilise nutrients rapidly (Griffiths, 1994).

In addition to protozoa, bacterial-feeding and predatory nematodes are estimated to contribute (directly and indirectly) about 8 to 19% of nitrogen mineralisation in conventional and integrated farming systems (Beare, 1997). Although counter-intuitive, predatory nematodes sometimes demonstrate stronger relationships with the amount of microbial food source than the abundance of microbial-feeding fauna, which may serve merely as a conduit by which resources pass from the bottom to top trophic levels (Wardle and Yeates, 1993).

Protozoa may out-compete nematodes for consumption of bacteria because they have much shorter generation times (one to two days compared to four to seven days) (Griffiths, 1994). However, it is assumed that nematodes and protozoa do not compete directly. Although the abundance of protozoa may correlate inversely with numbers of nematodes, nematodes are thought to have a stimulating effect on protozoa (Alphei et al., 1996) by increasing bacterial production (Griffiths, 1994). This hypothesis is supported by faster accumulation of ammonium from cultures containing bacteria, ciliates and nematodes than only bacteria and ciliates (Griffiths, 1994). When protozoa are active in absence of nematodes, nitrate is the major ion leached and is accentuated in the presence of earthworms (Alphei et al., 1996). Mites may stimulate amoebal numbers by increasing the availability of bacteria and decreasing numbers of nematodes in the presence of a large abundance of amoebae (Griffiths, 1994). The latter hypothesis is supported by evidence that nitrogen mineralisation is increased by amoebae feeding on flagellates in the presence of mites (Griffths, 1994), or increased nitrate leaching with protozoa and earthworms (Alphei et al., 1996). The presence and abundance of earthworm activity have also been linked to increased leaching of inorganic nitrogen in field and laboratory conditions (Filser et al., 1995). Soil oxidation state was correlated positively with nitrate leaching in conventionally tilled soils (Doran, 1980). In no-till soils, both nitrification and denitrification were greater in surface soil (0–5 cm) compared to conventionally tilled soils, with the reverse pattern at greater depths (5–21 cm) (Beare, 1997).

The bacterial pathway is not always the dominant force in net nitrogen mineralisation. In Sweden, fungal contributions to nitrogen immobilisation are about 3.5 times more than bacterial contributions (Beare, 1997). It is likely that fungal-feeding microarthropods are more important in mobilising nitrogen from surface residues through grazing in no-till soils than bacteria and their consumers (Beare, 1997).

Not only may soil fauna stimulate net mineralisation of nutrients but they may also promote nutrient immobilisation. In Sweden, more nitrogen was immobilised in surface applied than buried straw in soils cropped to meadow fescue or lucerne. Perhaps competition for available nitrogen occurred between plant roots and residue-borne decomposers (Beare, 1997). Furthermore, phosphorus is immobilised in the presence of nematodes and more so when both protozoa and nematodes are present (Alphei et al., 1996). Earthworms may temporarily immobilise nutrients in microbes while they pass through the earthworm gut (Filser et al., 1995). The relative balance of nutrient mineralisation and immobilisation is not solely a function of soil faunal activities but also depends on the presence of nitrifying bacteria and the mobility of nutrients ions and loading of exchange surfaces (Alphei et al., 1996).

3.4. *Transferring energy in the food chain*

Bacteria (including actinomycetes), fungi, algae and protozoa are primary decomposers of organic matter. Nematodes, large protozoa, springtails and mites feed on microbes in contact with decaying organic matter, but not on the organic matter itself. Microbial-grazing mesofauna affect growth and metabolic activities of microbes and alter community composition, thus regulating decomposition rate of organic matter (Wasilewska et al., 1975; Trofymow and Coleman, 1982; Whitford et al., 1982; Yeates and Coleman, 1982; Seastedt, 1984). Protozoa alter the composition and activity of the microbial community by selectively grazing on small, slow-growing cells which act to maintain taxonomic and metabolic diversity of microflora. It is assumed that this type of grazing preference maintains a bacterial population in a 'youthful' state and maintains decomposition activity (Griffiths, 1994). Amoebae apparently have a stimulatory effect when added to bacterial cultures based on evidence of increased respiration, ammonification and siderophore production (Griffiths, 1994). Other fauna have indirect effects on decomposition. For example, microarthropods and Enchytraeidae fragment detritus and increase surface area for further microbial attack (Berg and Pawluk, 1984; Van Vliet et al., 1995). Subsequently, protozoa re-inoculate new substrates by phoretic transport or excretion of viable bacteria (Griffiths, 1994). Examples of invertebrate-microbe mutualisms are prevalent in soil. For instance, most soil invertebrates (e.g. termites and earthworms) do not seem to possess suitable enzymes to directly digest most resources in soil, such as cellulose, lignin, tannin and humic complexes. Instead, the enzymes seem to be produced by ingested microbes rather than by the invertebrate itself (Lavelle et al., 1995).

Carbon and nitrogen dynamics are tightly linked in terrestrial ecosystems suggesting that classic food web theory is insufficient to analyse relations between food web structure and ecosystem processes. Food web structure may influence decomposition rate even if trophic interactions are controlled by the rate of detritus input (Bengtsson et al., 1995). An increased number of trophic levels would increase decomposition rate given that consumers do not influence each others consumption rate or mortality (Bengtsson et al., 1995). This hypothesis is supported by several laboratory studies (Setälä and Huhta, 1991; Bengtsson et al., 1995).

Perhaps ecosystem function depends more on the presence and abundance of particular functional groups rather than complexity of food webs or biodiversity alone (Andrén et al., 1995). The 'snow chain' hypothesis predicts a succession of decomposer organisms that respond to changes in substrate quality. This hypothesis is analogous to the addition of snow chains to tires when making the transition from clear to snow-covered roads. Evidence of succession occurs among soil animals. Initially, a peak of enchytraeids and bacterial-feeding nematodes are present. Once the substrate is decomposed somewhat and changed qualitatively, numbers of microarthropod and nematode omnivores and predators increase with a subsequent decline in numbers of bacterial-feeding nematodes and enchytraeids (Andrén et al., 1995).

In desert ecosystems containing creosote-bush (*Larrea tridentata*), food webs with additional trophic groups correspond with faster rates of decomposition than food webs with fewer trophic groups (Whitford et al., 1982). Decomposition rates were slower for systems without mites (treated with the insecticide, chlordane) than with mites (no insecticide). Without mites, nematode grazing may have decreased the decomposition rate. The presence of mites reduced nematode populations and released microbial populations from predation, which resulted in faster decomposition rates. Consequently, predatory mites maintain nematode grazer populations at levels that presumably increase microbial activity and decomposition rates.

3.5. *Acting as an environmental buffer*

Soils act as an environmental buffer physically, chemically and biologically. Physically, soils may filter materials such as pesticides and fertiliser through its porous matrix and reduce the transfer of leachates into ground and surface water. Fulvic and humic acids of organic matter play a major role in the binding of pesticides in soil. Pesticides, or their degradation intermediates, can also be polymerised or incorporated into humus by the action of soil microbial enzymes (Bollag et al., 1992). Additional chemical buffering functions of soil are responsible for regulating the availability of nutrients by cation exchange capacity (principally influenced by organic matter and clay content), serve as a sink for carbon dioxide, and are a source of alleopathic chemicals. Carbon compounds are stored in plant, microbe and animal bodies, in debris and in abiotic carbonate compounds. During decomposition processes, respiration releases carbon dioxide as a by-product. Cultivation releases carbon dioxide to the atmosphere, both through burning of fossil fuels to run the tractor and disturbing the soil, which increases decomposition rates.

Soil microbes, especially fungi, may secrete allelopathic substances that suppress germination of seeds or make seeds unpalatable to predators (Janzen, 1977). This has been demonstrated in the California chaparral with *Adenostoma fasiculatum* (Wicklow, 1981). Fire eliminates the alleopathic effect temporarily. However, it returns after re-establishment of the microbial community. Agronomists have utilised fire to manage soil fungal pathogens of wheat, barley and rice.

The presence of particular groups of organisms has been associated with the suppression of plant disease. First, the springtails *Proisotoma minuta* and *Onychiurus encarpatus* consume the soilborne fungal plant pathogen *Rhizoctonia solani* which causes damping-off disease on cotton seedlings. Springtails prefer feeding on the fungal pathogen in soil compared to the biocontrol fungi *Laetisaria arvalis*, *Trichoderma harzianum*, and *Gliocladium virens* (Curl et al., 1988). Springtail species can also distinguish and graze selectively on different species of vesicular-arbuscular mycorrhizae (Thimm and Larink, 1995). Second, amoebae of the *Vampyrellidae* perforate conidia of *Cochliobolus sativus* on barley (Old, 1967). Protozoa have been reported to play an active role in disease suppression by consuming pathogenic nematodes, bacteria and fungi (Zwart et al., 1994). Generally, plant pathogens have faster growth rates than protozoan predators, so it is unlikely that protozoa will suppress plant pathogens completely in soil. However, for crops grown in liquid media, this balance may tip the other way (Zwart et al., 1994). Finally, abundances of the bacterium *Pseudomonas fluorescens* increase and may act as a biocontrol agent of the take-all pathogen *Gaeumannomyces gramminis* under conditions of no-till, continuous wheat production (Ryder et al., 1990).

4. Designing and maintaining soil communities for optimum function

Modern agricultural methods have replaced or substituted natural and seasonal patterns of ecological processes with inorganic materials. These substitutions of nature are likely to have environmental and ecological costs due to the loss of natural cycles among organisms and their environment through time. The paucity of studies quantifying these costs compels some scientists to declare that generalisations such as 'conventional farming destroys life in the soil' or 'ecofarming stimulates soil life' are not supported by adequate evidence (Foissner, 1992; Wardle, 1995; Giller et al., 1997).

Regardless, we have evidence that increased detrital food webs containing more groups of organisms are associated with faster decomposition and nutrient turnover rates (Setälä et al., 1991) and increased primary production (Setälä and Huhta, 1991). Balanced with organisms that immobilise nutrients, the net effect is regulation of nutrient availability to plants and other organisms. We cannot conclude that soil biodiversity *causes* ecosystem function because there is insufficient evidence to support this claim.

To better mimic nature, soil ecosystems must be designed and managed to foster food webs containing multiple functional groups or taxa (i.e. complex community) and be relatively mature successionally. This can be achieved by incorporating four types of activities:

- reduction or elimination cultivation, heavy machinery and general biocides;
- incorporating perennial crops and organic material;
- synchronising nutrient release and water availability with plant demand; and
- monitoring biological activity.

4.1. *Reduction or elimination of cultivation, use of heavy machinery and general biocides*

Decomposition rates are sensitive to physical disturbances such as cultivation. For example, decomposition rates are 1.4 to 1.9 times faster in conventionally tilled than in no-till soils with surface residues (Beare, 1997). Fungi are more likely to regulate decomposition of surface-applied residues in no-till soils and bacteria regulate decay rates in incorporated residues in conventionally tilled soils (Beare, 1997).

Soils managed by conventional, reduced, or no tillage practices have distinct biological and functional properties (Doran, 1980; Hendrix et al., 1986). Plant residue is distributed throughout the plough layer in fields managed with conventional tillage. Under these conditions, organisms with short generation times, small body size, rapid dispersal and generalist feeding habits thrive (Steen, 1983). These soils are dominated by bacteria and their predators such as bacterial-feeding nematodes and astigmatid mites (Andrén and Lagerlöf, 1983; Yeates, 1984; Hendrix et al., 1986; Beare et al., 1992; Beare, 1997) and are considered to be in an early stage of succession. Many microarthropods have omnivorous feeding habits in systems which are cultivated frequently (Beare et al., 1992). Alternatively, reduced tillage leaves most of the previous crop residue on the soil surface and results in changes in physical and chemical properties of the soil (Blevins et al., 1983). Surface residues retain moisture, dampen temperature fluctuations, and provide a continuous substrate that promotes fungal growth. Furthermore, the dominance of fungi in early stages of decay of surface residues can be explained by the initially high lignin to nitrogen ratio. Bacteria move in as secondary colonisers after decay is initiated (Beare, 1997). Relative abundance of fungi and their consumers in no-till soils represent a more mature successional state, more closely resembling undisturbed natural ecosystems (Yeates, 1984; Boström and Söhlenius, 1986; Hendrix et al., 1986; Holland and Coleman, 1987; Walter, 1987; Neher and Campbell, 1994; Beare, 1997).

Sometimes agricultural disturbances, such as cultivation, do not reduce soil biodiversity. Instead, cultivation appears to affect species composition but not diversity and more so for macrofauna than microfauna (Wardle, 1995).

Diversity of microfauna is mostly unresponsive to cultivation and there is no general trend for macrofauna. Some species are effected negatively, and others positively, by cultivation (Wardle, 1995). The lack of response of the diversity of groups of small organisms to cultivation may reflect their slight overall response to cultivation or possibly their poorer taxonomic resolution in most studies.

Soil fumigation with general biocides has short and/or long-term impacts on soil community composition and abundance. Biocides may either impact the microbes or fauna directly with toxic effects or indirectly through reduced vegetation or other non-target effects (Andrén and Lagerlöf, 1983). Methyl bromide and benomyl have direct toxic effects on nematodes and earthworms, respectively (Yeates et al., 1991; Ettema and Bongers, 1993; Edwards and Bohlen, 1996). Phenoxy acetic acid herbicides (e.g. 2,4-D, 2,4,5-T, MCPA) do not depress soil fauna directly with toxic effects, but indirectly through reduced vegetation and smaller additions of organic matter to soil (Andrén and Lagerlöf, 1983). Conversion of continuous hop (*Humulus lupulus*) production to a cereal rotation of winter wheat (*Triticum aestivum*) and spring barley (*Hordeum vulgare*) demonstrate the long-lasting effects of intensive biocide input, especially fungicides with copper, which accumulate and remain in soil for more than three years (Filser et al., 1995). Heavy metals, such as copper, eliminate sensitive species such as earthworms and the springtails *Folsomia quadrioculata* and *Isotomiella minor*. Other springtails tolerate heavy metals and increase in dominance, for example *Mesaphorura krausbaueri* and *Onychiurus armatus* (Filser et al., 1995). Broad-spectrum insecticides such as chlorpyrifos, isofenphos, and aldicarb are toxic to non-target insects, such as predaceous and parasitic arthropods (Koehler, 1992; Potter, 1993). Insecticides are usually more toxic than herbicides or fungicides and disturb soil protozoa critically, i.e. populations often do not fully recover within 60 days (Foissner, 1997). Edwards and Bohlen (1996) and Neher and Barbercheck (1998) provide more thorough reviews of effects of biocides on soil community composition.

Benefits of eliminating general biocides are illustrated by two case studies (Ingham, 1998) involving strawberry (*Fragaria* × *Ananassa* Duch.). The first study was in California, USA, where strawberries are grown as an annual crop in monoculture. Annual applications of methyl bromide were used to decrease the population density of certain root pathogens. However, there were major environmental costs. For example, nitrate leached into ground water reaching concentrations of 150 ppm (3 ppm is toxic to humans). Natural disease suppression was lost which may partly be explained by a 50-fold reduction in numbers of bacterial species. A remediation program was implemented to reverse the apparent environmental problems. Spent mushroom compost (high fungal content) was applied to soil at a rate of 60 tons per hectare. Abundances of microflora and microfauna increased and a balance of fungi and bacteria (and their respective consumers) returned to the soil after four to five years, accompanied by nitrate leaching and increased natural disease suppression

(Ingham, 1998). The second study was in Costa Rica where strawberries were grown as perennials. Disease management was maintained and pesticides replaced by transplanting strawberry seedlings in plugs containing compost colonised by a mixture of bacteria, fungi and their respective consumers. This compost promoted the establishment of a complex food web defined as a balance of bacteria and fungi, and their respective grazers. This practice transfers a relatively complex soil food web to the field and obviates the need to manipulate the entire field (Ingham, 1998). In the long-term, managing soil to achieve relatively balanced fungal and bacterial components will reduce fertiliser requirements because soil organisms will maintain processes of nutrient mineralisation and decomposition at sustainable levels.

4.2. *Incorporating perennial crops and organic matter*

Introduction of perennial crops into agricultural systems has several advantages. Root growth is more extensive and less ephemeral than with annual crops and supports a soil community with many omnivores and predators. Soil community composition with perennial crops resembles that of natural ecosystem soils more closely than is the case with annual crops (Ferris and Ferris, 1974; Wasilewska, 1979; Freckman and Ettema, 1993; Neher and Campbell, 1994). Differences between soils with perennial (e.g. meadow fescue) and annual (e.g. barley) crops may be less pronounced for perennial crops younger than three years old than with more mature crops (Boström and Söhlenius, 1986). Furthermore, perennial plants restore function in water infiltration and reduce compaction and, thus, increase rooting depth.

Organic matter can be added to soil as crop residues, urban sludge, manure, green manure, cover crops or compost. Addition of organic matter to soil has multiple benefits. First, organic matter adds nutrients and cation exchange sites to retain nutrients. Second, organic matter increases porosity to soil (Andrén and Lagerlöf, 1983). Third, organic matter includes microbes and microbial grazers (Andrén and Lagerlöf, 1983; Weiss and Larink, 1991) which restore ecosystem processes such as decomposition, nutrient mineralisation and disease suppression. For example, populations of fungal-feeding nematodes (Weiss and Larink, 1991), potworms (enchytraeids), springtails and sometimes mesostigmatid mites increase with applications of manure which provide a better substrate for growth (Andrén and Lagerlöf, 1983). However, there are exceptions, as in Dutch polder soil (van de Bund, 1970).

Although it is difficult to separate their individual effects, concurrent additions of organic matter and elimination of general biocides have the benefit of allowing omnivores and predaceous species to increase in prevalence, which promotes natural disease suppression. In these cases, natural community interactions such as competition, antagonism, competitive exclusion and niche partitioning must be understood for effective results. There is a fine balance between disease suppressive agents and pathogens. Some pathogenic fungi may be stimulated initially by additions of organic residues.

Caution is advised when considering simple solutions to disease and nutrient management. For example, application of mycorrhizal fungi is often considered a favourable practice. A parallel good practice is the promotion of fungal-feeding nematodes, which are usually less abundant than root-feeding or bacterial-feeding species in agricultural soils. However, fungal-feeding organisms such as springtails, nematodes and mites may actually reduce the beneficial effect of promoting mycorrhizae by consuming mycorrhizae. Despite the reduced benefit of mycorrhizae, the plant-fungus symbiosis remains given the soil food web contains interactions among saprophytic microbes and their faunal grazers to ensure efficient mobilisation of nutrients in the rhizosphere (Setälä, 1995).

As with any practice, there are risks associated with applications of organic matter. For example, high doses of manure or sludge fertilisers can harm mesofauna because of toxicity (e.g. anhydrous ammonia), high osmotic pressure due to salt accumulation (Andrén and Lagerlöf, 1983), or heavy metal accumulation (Weiss and Larink, 1991). The repellent nature of ammonium can affect soil invertebrates adversely (Potter, 1993). However, potential for toxic effects can be decreased by applying composted manure or sludge (Ott et al., 1983). One must consider the period of time material is composted prior to incorporating it into soil. For example, the plant-pathogenic fungus *Rhizoctonia solani* may cause damping-off disease in soil when fresh or immature compost material high in cellulose content is added. However, in aged compost, cellulose is degraded and the biocontrol fungus *Trichoderma* spp. can grow and parasitise the pathogen effectively, thus suppressing disease (Chung et al., 1988).

4.3. *Synchronising nutrient and water availability with plant demand*

Fertiliser addition affects soil microflora and indirectly impacts soil mesofauna by changing their food resources (Weil and Kroontje, 1979). Additions of nitrogen may acidify soil and, consequently, inhibit microbial growth and activity. Nitrogen may also affect the quality of microbes as a food source for mesofauna (Darbyshire et al., 1994). Booth and Anderson (1979) grew two species of fungi in liquid media with 2, 20, 200, or 2000 ppm nitrogen and determined the fecundity of the springtail *Folsomia candida* while feeding on the fungi. Fecundity increased with increasing nitrogen content up to 200 ppm, although *F. candida* did not show a preference for feeding on fungi with a greater or lesser nitrogen content.

The effect of fertilisation on microarthropod species diversity and abundance within taxa, and the subsequent impact on decomposition and nutrient mineralisation processes, are not well understood. For example, synthetic fertilisers increase nematode diversity but applications of manure decrease nematode diversity (Wasilewska, 1989). Applications of synthetic nitrogen fertiliser on Swedish arable soils growing spring barley (*Hordeum distichum*

L.) changed community composition but not numbers and biomass of nematodes, springtails and mites (Andrén et al., 1988).

Appropriate timing of water is important for disease management. For example, frequent irrigation episodes may increase potential of root rot diseases caused by *Phytophthora* spp. The reproductive biology of *Phytophthora* spp. is stimulated by changes in the matric potential of soil (Duniway, 1983). Water drainage stimulates production of asexual spores such as sporangia, whereas subsequent saturation (i.e. irrigation event) stimulates the sporangia to germinate indirectly by producing multiple zoospores that are flagellated and may move great distances in surface water (Figure 3). Phytophthora root rots have been well documented world-wide, including those on jarrah (*Eucalyptus marginata*) and *Banksia* in Australia and on many vegetable and tree crops in the US (Erwin et al., 1983).

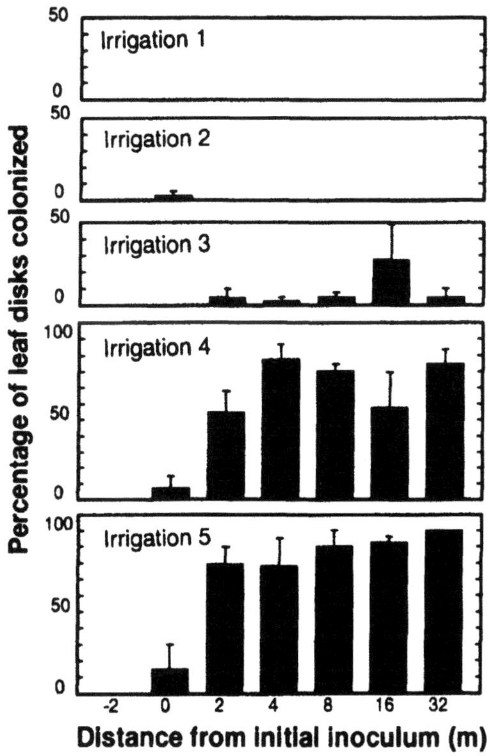

Figure 3. Percentages of tomato leaf disks colonised during five successive irrigations plotted as a function of the distance from the point source where *Phytophthora parasitica* was buried in irrigation furrows. Means for four replicate furrows and standard errors are shown. *Source:* Neher and Duniway (1992).

4.4. *Monitoring biological activity*

The successional status of a soil community may reflect the history of disturbance. Succession in cropped agricultural fields begins with depauperate soil after cultivation and clearing of native vegetation which acts like an island to which organisms migrate. First, opportunistic species, such as bacteria and their predators, are colonists of soil. Subsequently, fungi and their predators migrate into the area (Boström and Söhlenius, 1986). Microarthropods, such as springtails, mites and fly maggots can colonise nearly bare ground and rise quickly in population density. Top predator microarthropods, such as predaceous mites and nematodes, become established later and may have a function similar to keystone predators in other community food webs (Elliott et al., 1988). Finally, macro- and megafauna such as earthworms, millipedes, slugs, centipedes, wood lice, sow bugs and pill bugs join the soil community (Strueve-Kusenberg, 1982).

Succession can be interrupted at various stages by agricultural practices, such as cultivation and applications of fertiliser and pesticides (Ferris and Ferris, 1974; Waskilewska, 1979). The quantification of successional stage using a 'maturity index' (Bongers, 1990) proves promising as a monitoring tool of community composition and function (Freckman and Ettema, 1993; Neher and Campbell, 1994; Neher et al., 1995; Neher and Barbercheck, 1998). Maturity indices are based on the principles of succession and relative sensitivity of various taxa to stress or disruption of the successional pattern. Maturity indices, based on life strategy characteristics, were developed originally for nematode communities, but have potential for adaptation to other groups of organisms. Interpretation of maturity indices depends on the type of ecosystem (Figure 2). Successional maturity was greater in older forests (> 30 years) and functional wetlands than young forests (< 3 years) and wetlands converted to conventionally tilled agriculture, respectively (Figure 2). The opposite pattern was observed for agricultural soils, where successional maturity was greater in conventionally tilled soil with annual crops (i.e. disturbed) than permanent pastures (i.e. undisturbed). Indices that describe associations within biological communities, such as a maturity index, are less variable than measures of abundance of a single taxonomic or functional group, and are more statistically reliable as measures of ecosystem condition (Neher et al., 1995; Neher and Campbell, 1996). Because index values for New Zealand (Yeates, 1994) and the US (Neher and Campbell, 1996) were greater than those published in European studies, it is suggested that biogeography may be a confounding factor in interpreting index values. Maturity and trophic diversity indices measure different aspects of soil communities and, together, are complementary.

5. Essential research

Our challenge is to understand concepts and mechanisms that mimic nature, qualitatively and quantitatively, at appropriate spatial (centimetres to hectares), ecological (population, community, ecosystem and landscape) and temporal (seconds to centuries) scales. Most studies have focused on single factors in an effort to reveal underlying mechanisms, resulting in a lack of understanding of how multiple and interacting environmental and biotic factors affect soil biodiversity, nutrient cycling, pest populations and plant productivity. Future research should include studies on productivity of soil animals under various management systems, the analysis of single factors to elucidate causative mechanisms, and studies on the relationship between soil animals, crop production and sustained yield (Foissner, 1992). Holistic systems and their dynamics must be understood to effectively design agricultural systems in concordance with nature. With this information, we should be able to tailor agricultural practices to positively affect beneficial soil organisms and the functions they perform to optimise crop productivity. To achieve the ultimate goal of designing and managing agricultural systems as mimics of nature, the following research goals must be achieved.

5.1. *Methods for successful application of compost materials and/or biocontrol agents*

Simple techniques are usually favoured but may also disregard ecological interactions among organisms, added and between those added and those already existing in the soil. Basic natural history and fundamental niche requirements must be understood at the individual and population level. At the community level, potential competitive exclusion, predation and/or antagonism by organisms already occupying that niche or utilising the resources must be considered. At the ecosystem level, methods to optimise the role that soil organisms play in nutrient cycling, energy flow, and disease management must be evaluated.

5.2. *Timing of management implementation*

While many studies have examined the impact of additions or removals of materials, they have not considered the seasonal impacts or time lags that occur between implementation and response, whether the treatment is biotic or abiotic. By understanding appropriate temporal relationships, nutrient and water supplements can be scheduled according to plant and soil community use. Interactions of intra- and interspecific crop phenology and root architecture relative to soil community composition and function must be integrated to avoid intense competition for nutrients between microbes and plants.

5.3. *Explicit relationships between soil organisms and ecosystem function*

Current understanding is limited to trophic or functional group resolution. However, resolution at a species-level is desirable. Additionally, a more thorough understanding of the sequence of community succession relative to soil function dynamics would be useful in making long-term predictions of community composition associated with ecologically sound agricultural systems.

6. Conclusions

Clearly, soil microbes and fauna play important roles in ecosystem function. Unfortunately, many modern agricultural practices correspond with a decline in abundance and alter the composition of soil communities, which subsequently impacts ecological processes. Interruption to the cycling of carbon, nitrogen, phosphorus and/or water prevents crops from obtaining all requirements for primary productivity. Production deficiencies are replaced by fossil-fuel based inputs that eventually replace natural cycles and processes. To restore ecosystem functions of soil organisms, agricultural systems must be designed to reduce or eliminate cultivation, heavy machinery, and general biocides. In addition, systems should incorporate perennial crops and increase soil organic material. In order to achieve these goals, more research is needed to determine the impact of multiple and interacting management practices on biodiversity, nutrient cycling, decomposition, pest populations, and their concurrent impact on agricultural productivity. With this information, we can maximise our ability to tailor agricultural practices to optimise crop productivity while positively affecting beneficial soil organisms and the functions they perform.

7. References

Alphei J, Bonkowski M and Scheu S (1996) Protozoa, Nematoda and Lumbricidae in the rhizosphere of *Hordelymus europaeus* (Poaceae): faunal interactions, response of microorganisms and effects on plant growth. Oecologia 106: 111–126

Anderson RV, Gould WD, Woods LE, Cambardella C, Ingham RE and Coleman DC (1983) Organic and inorganic nitrogenous losses by microbivorous nematodes in soil. Oikos 40: 75–80

Andre HM, Noti MI and Lebrun P (1994) The soil fauna: the other last biotic frontier. Biodiversity and Conservation 3: 45–56

Andrén O, Bengtsson J and Clarhom M (1995) Biodiversity and species redundancy among litter decomposers. In: Collins HP, Robertson GP and Klug MJ (eds) The Significance and Regulation of Soil Biodiversity, pp 141–151. Kluwer Academic Publishers, Dordrecht, The Netherlands

Andrén O and Lagerlöf J (1983) Soil fauna (microarthropods, enchytraeids, nematodes) in Swedish agricultural cropping systems. Acta Agriculturae Scandinavica 33: 33–52

Andrén O, Paustian K and Rosswall T (1988) Soil biotic interactions in the functioning of agroecosystems. Agriculture, Ecosystems and Environment 24: 57–67

Barker KR and Campbell CL (1981) Sampling nematode populations. In: Zuckerman BM Rohde RA (eds) Plant Parasitic Nematodes, Vol III, pp 451–474. Academic Press, New York, NY

Beare MH (1997) Fungal and bacterial pathways of organic matter decomposition and nitrogen mineralization in arable soil. In: Brussaard L and Ferrera-Cerrato R (eds) Soil Ecology in Sustainable Agricultural Systems, pp 37–70. Lewis Publishers, Boca Raton, LA

Beare MH, Coleman DC, Crossley DA Jr, Hendrix PF and Odum EP (1995) A hierarchical approach to evaluating the significance of soil biodiversity to biogeochemical cycling. In: Collins HP, Robertson GP and Klug MJ (eds) The Significance and Regulation of Soil Biodiversity, pp 5–22. Kluwer Academic Publishers, Dordrecht, The Netherlands

Beare MH, Parmelee RW, Hendrix PF, Cheng W, Coleman DC and Crossley DA Jr (1992) Microbial and faunal interactions and effects on litter nitrogen and decomposition in agroecosystems. Ecological Monographs 62: 569–591

Bengtsson J, Zheng DW, Agren GI and Persson T (1995) Food webs in soil: an interface between population and ecosystem ecology. In: Jones C and Lawton J (eds) Linking Species and Ecosystems, pp 159–165. Chapman and Hall, New York

Berg NW and Pawluk S (1984) Soil mesofaunal studies under different vegetative regimes in North Central Alberta. Canadian Journal of Soil Science 64: 209–223

Blevins RL, Smith MS, Thomas GW and Fry WW (1983) Influence of conservation tillage on soil properties. Journal of Soil and Water Conservation 38: 301–304

Bollag JM, Myers CJ and Minard RD (1992) Biological and chemical interactions of pesticides with soil organic matter. Science Total Environment 123/124: 205–217

Bongers T (1990) The maturity index: an ecological measure of environmental disturbance based on nematode species composition. Oecologia 83: 14–19

Booth RG and Anderson JM (1979) The influence of fungal food quality on the growth and fecundity of *Folsomia candida* (Collembola: Isotomidae). Oecologia 38: 317–323

Boström S and Söhlenius B (1986) Short-term dynamics of nematode communities in arable soil: Influence of a perennial and an annual cropping system. Pedobiologia 29: 345–357

Chung YR, Hoitink HAH and Lipps PE (1988) Interactions between organic-matter decomposition level and soilborne disease severity. Agriculture, Ecosystems and Environment 24: 183–193

Connell JH (1978) Diversity in tropical rain forests and coral reefs. Science 199: 1302–1310

Clarholm M (1985) Interactions of bacteria, protozoa and plants leading to mineralization of soil nitrogen. Soil Biology and Biochemistry 17: 181–187

Crossley DA Jr, Mueller BR and Perdue JC (1992) Biodiversity of microarthropods in agricultural soils: relations to processes. Agriculture, Ecosystems and Environment 40: 37–46

Curl EA and Truelove B (1986) The Rhizosphere. Springer-Verlag, New York, NY, 288 pp

Curl EA, Lartey R and Peterson CC (1988) Interactions between root pathogens and soil microarthropods. Agriculture, Ecosystems and Environment 24: 249–261

Darbyshire JF, Davidson MS, Chapman SJ and Ritchie S (1994) Excretion of nitrogen and phosphorus by the soil ciliate *Colpoda steinii* when fed the soil bacterium *Arthrobacter* sp. Soil Biology and Biochemistry 26: 1193–1199

Dindal DL (1990) Soil Biology Guide. John Wiley, New York, NY, 1349 pp

Doran JW (1980) Soil microbial and biochemical changes associated with reduced tillage. Soil Science Society of America Journal 44: 765–771

Duniway JM (1983) Role of physical factors in the development of *Phytophthora* diseases. In: Erwin DC, Bartnicki-Garcia S and Tsao PH (eds) *Phytophthora*: Its Biology, Taxonomy, Ecology, and Pathology, pp 175–187. American Phytopathological Society, St Paul, MN

Eash NS, Karlen DL and Parkin TB (1994) Fungal contributions to soil aggregation and soil quality. In: Doran JW, Coleman DC, Bezdicek DF and Stewart BA (eds) Defining Soil Quality for a Sustainable Environment, pp 221–228. SSSA Special Publication No. 35, Madison, Wisconsin

Edwards CA and Bohlen PJ (1996) Biology and Ecology of Earthworms, 3rd ed. Chapman and Hall, London

Elliott ET, Hunt HW and Walter CE (1988) Detrital food web interactions in North American grassland ecosystems. Agriculture, Ecosystems and Environment 24: 41–56

Erwin DC, Bartnicki-Garcia S and Tsao PH (1983) *Phytophthora*: Its Biology, Taxonomy, Ecology, and Pathology. American Phytopathological Society, St Paul, MN, 392 pp

Ettema CH (1998) Soil nematode diversity: species coexistance and ecosystem function. Journal of Nematology 30: 159–169

Ettema CH and Bongers T (1993) Characterisation of nematode colonisation and succession in disturbed soil using the Maturity Index. Biology and Fertility of Soils 16: 79–85

Ferris VR and Ferris JC (1974) Inter-relationships between nematode and plant communities in agricultural ecosystems. Agro-Ecosystems 1: 275–299

Filser J, Fromm H, Nagel RF and Winter K (1995) Effects of previous intensive agricultural management on microorganisms and the biodiversity of soil fauna. Plant and Soil 170: 123–129

Foissner W (1992) Comparative studies on the soil life in ecofarmed and conventionally farmed fields and grasslands of Austria. Agriculture, Ecosystems and Environment 40: 207–218

Foissner W (1997) Protozoa as bioindicators in agroecosystems, with emphasis on farming practices, biocides, and biodiversity. Agriculture, Ecosystems and Environment 62: 93–103

Foster RC and Dormaar JF (1991) Bacteria-grazing amoebae in situ in the rhizosphere. Biology and Fertility of Soils 11: 83–87

Freckman DW and Ettema CH (1993) Assessing nematode communities in agroecosystems of varying human intervention. Agriculture, Ecosystems and Environment 45: 239–261

Giller KE, Beare MH, Lavelle P, Izac A-MN and Swift MJ (1997) Agricultural intensification, soil biodiversity and agroecosystem function. Applied Soil Ecology 6: 3–16

Goodell P and Ferris H (1980) Plant-parasitic nematode distributions in an alfalfa field. Journal of Nematology 12: 136–141

Griffin DM (1981) Water potential as a selective factor in the microbial ecology of soils. In: Parr JF, Gardner WR and Elliott LF (eds) Water Potential Relations in Soil Microbiology, pp 141–151. Soil Science Society of America, Special Publication No. 9, Madison, Wisconsin

Griffiths BS (1994) Soil nutrient flow. In: Darbyshire J (ed) Soil Protozoa, pp 65–91. CAB International, Wallingford, Oxon, UK

Gupta VVSR and Germida JJ (1988) Distribution of microbial biomass and its activity in different soil aggregate size classes as affected by cultivation. Soil Biology and Biochemistry 20: 777–786

Harding DJL and Studdart RA (1974) Microarthropods. In: Dickinson DH and Pugh GJF (eds) Biology of Plant Litter Decomposition, pp 489–532. Academic Press, New York, NY

Hassink J, Bouwman LA, Zwart KB and Brussaard L (1993) Relationships between habitable pore space soil biota and mineralization rates in grassland soils. Soil Biology and Biochemistry 25: 47–55

Hawksworth DL and Mound LA (1991) Biodiversity databases: the crucial significance of collections. In: Hawksworth DL (ed) The Biodiversity of Microorganisms and Invertebrates: Its Role in Sustainable Agriculture, pp 17–29. CAB International, Wallingford, UK

Hendrix PF, Parmelee RW, Crossley DA Jr., Coleman DC, Odum EP and Groffman PM (1986) Detritus food webs in conventional and no-tillage agroecosystems. BioScience 36: 374–380

Hobbs RJ and Huenneke LF (1992) Disturbance, diversity, and invasion: implications for conservation. Conservation Biology 6: 324–337

Holland EA and Coleman DC (1987) Litter placement effects on microbial and organic matter dynamics in an agroecosystem. Ecology 68: 425–433

Hunt HW, Coleman DC, Ingham ER, Ingham RE, Elliott ET, Moore JC, Rose SL, Reid CPP and Morley CR (1987) The detrital food web in a shortgrass prairie. Biology and Fertility of Soils 3: 57–68

Ingham ER (1998) Managing the soil foodweb to benefit plant growth: care and feeding of soil microherds. Journal of Nematology 29: 585

Ingham ER, Coleman DC and Moore JC (1989) An analysis of food-web structure and function in a shortgrass prairie, a mountain meadow, and a lodgepole pine forest. Biology and Fertility of Soils 8: 29–37

Ingham RE, Trofymow JA, Ingham ER and Coleman DC (1985) Interactions of bacteria, fungi, and their nematode grazers: effects on nutrient cycling and plant growth. Ecological Monographs 55: 119–140

Jackson W (1985) New Roots for Agriculture. University of Nebraska Press, Nebraska, 150 pp

Janzen D (1977) Why fruits rot, seeds mold, and meats spoil. American Naturalist 111: 691–713

Janzen D (1985) Natural history of mutualism. In: Boucher DH (ed) Biology of Mutualism. Ecology and Evolution, pp 40–99. Oxford University Press, New York, NY

Jentschke G, Bonkowski M, Godbold DL and Scheu S (1995) Soil protozoa and forest tree growth: non-nutritional effects and interactions with mycorrhizae. Biology and Fertility of Soils 20: 263–269

King KL and Hutchinson KJ (1976) The effects of sheep stocking intensity on the abundance and distribution of mesofauna in pastures. Journal of Applied Ecology 13: 41–55

Kitazawa Y (1971) Biological regionality of the soil fauna and its function in forest ecosystem types. In: Proceedings of the Brussels Symposium 1969, Ecology and Conservation No. 4. Productivity of Forest Ecosystems, pp 485–498. UNESCO

Koehler HH (1992) The use of soil mesofauna for the judgement of chemical impact on ecosystems. Agriculture, Ecosystems and Environment 40: 193–205

Lal R (1991) Soil conservation and biodiversity In: Hawksworth DL (ed) The Biodiversity of Microorganisms and Invertebrates: Its Role in Sustainable Agriculture, pp 89–104. CAB International, London, UK

Larson WE and Pierce FJ (1991) Conservation and enhancement of soil quality. In: IBSRAM Proceedings 12 (2). Evaluation for Sustainable Land Management in the Development World. Volume 2. Bangkok, Thailand. International Board for Soil Research and Management

Lavelle P, Lattaud C, Trigo D and Barois I (1995) Mutualism and biodiversity in soils. Plant and Soil 170: 23–33

Linn DC and Doran JW (1984) Effect of water-filled pore space on carbon dioxide and nitrous oxide production in tilled and non-tilled soils. Soil Science Society of America Journal 48: 1267–1272

Moore JC and de Ruiter PC (1991) Temporal and spatial heterogeneity of trophic interactions within below-ground food webs. Agriculture, Ecosystems and Environment 34: 371–397

Moore JC, Walter DE and Hunt HW (1988) Arthropod regulation of micro- and mesobiota in below-ground detrital food webs. Annual Review of Entomology 33: 419–439

Mueller BR, Beare MH and Crossley DA Jr (1990) Soil mites in detrital food webs of conventional and no-tillage agroecosystems. Pedobiologia 34: 389–401

Neher D and Duniway JM (1992) Dispersal of *Phytophthora parasitica* in tomato fields by furrow irrigation. Plant Disease 76: 582–586

Neher DA and Campbell CL (1994) Nematode communities and microbial biomass in soils with annual and perennial crops. Applied Soil Ecology 1: 17–28

Neher DA and Campbell CL (1996) Sampling for regional monitoring of nematode communities in agricultural soils. Journal of Nematology 28: 196–208

Neher DA and Barbercheck ME (1998) Diversity and role of soil mesofauna. In: Collins W (ed) Importance of Biodiversity in Agroecosystems, Lewis Publishers, Chelsea, Michigan (in press)

Neher DA, Peck SL, Rawlings JO and Campbell CL (1995) Measures of nematode community structure for an agroecosystem monitoring program and sources of variability among and within agricultural fields. Plant and Soil 170: 167–181

Noe JP and Campbell CL (1985). Spatial pattern analysis of plant-parasitic nematodes. Journal of Nematology 17: 86–93

Old KM (1967) Effects of natural soil on survival of *Cochliobolus sativus*. Transactions of the British Mycological Society 50: 615–624

Ott P, Hansen S and Vogtmann H (1983) Nitrates in relation to composting and use of farmyard

manures. In: Lockeretz W (ed) Environmentally Sound Agriculture, pp 145–154. Praeger, New York, NY
Petraitis PS, Latham RE and Niesenbaum RA (1989) The maintenance of species diversity by disturbance. Quarterly Review of Biology 64: 393–418
Potter DA (1993) Pesticide and fertiliser effects on beneficial invertebrates and consequences for thatch degradation and pest outbreaks in turfgrass. In: Racke KD and Leslie AR (eds) Pesticides in Urban Environments: Fate and Significance, pp 331–343. ACS Symposium Series No. 522, American Chemical Society, Washington, DC
Richards BN (1987) The Microbiology of Terrestrial Ecosystems. Longman Scientific and Technical, New York, New York, 399 pp
Ryder MH, Brisbane PG and Rovira AD (1990) Mechanisms in the biological control of take-all of wheat by rhizosphere bacteria. In: Hornby D (ed) Biological Control of Soil-Borne Plant Pathogens, pp 123–130. CAB International, Wallingford, UK
Samways MH (1992) Some comparative insect conservation issues of north temperate, tropical and south temperate landscapes. Agriculture, Ecosystems and Environment 40: 137–154
Seastedt TR (1984) The role of microarthropods in decomposition and mineralization processes. Annual Review of Entomology 29: 25–46
Seastedt TR, James SW and Todd TC (1988) Interactions among soil invertebrates, microbes and plant growth in the tallgrass prairie. Agriculture, Ecosystems & Environment 24: 219–228
Setälä H (1995) Growth of birch and pine seedlings in relation to grazing by soil fauna on ectomycorrhizal fungi. Ecology 76: 1844–1851
Setälä H and Huhta V (1991) Soil fauna increase *Betula pendula* growth: laboratory experiments with coniferous forest floor. Ecology 72: 665–671
Setälä H, Tyynismaa M, Martikainen E and Huhta V (1991) Mineralisation of C, N and P in relation to decomposer community structure in coniferous forest soil. Pedobiologia 35: 285–296
Söhlenius B, Boström S and Sandor A (1988) Carbon and nitrogen budgets of nematodes in arable soil. Biology and Fertility of Soils 6: 1–8
Sommers LE, Gilmour CM, Wildung RE and Beck SM (1981) In: Parr JF, Gardner WR and Elliott LF (eds) Water Potential Relations in Soil Microbiology, pp 97–117. Soil Science Society of America, Special Publication No. 9, Madison, Wisconsin
Steen E (1983) Soil animals in relation to agricultural practices and soil productivity. Swedish Journal of Agricultural Research 13: 157–165
Strueve-Kusenberg R (1982) Succession and trophic structure of soil animal communities in different suburban fallow areas. In: Bornkamm R, Lee JA and Seaward CRD (eds) Urban Ecology, pp 89–98. Blackwell Scientific, Oxford, UK
Swift CJ, Heal OW and Anderson JC (1979). Decomposition in Terrestrial Ecosystems. University of California, Berkeley, California, 372 pp
Teuben A and Verhoef HA (1992) Direct contribution by soil arthropods to nutrient availability through body and faecal nutrient content. Biology and Fertility of Soils 14: 71–75
Thimm T and Larink O (1995) Grazing preferences of some Collembola for endomycorrhizal fungi. Biology and Fertility of Soils 19: 266–268
Trofymow JA and Coleman DC (1982) The role of bacterivorous and fungivorous nematodes in cellulose and chitin decomposition in the context of a root/rhizosphere/soil conceptual model. In: Freckman DW (ed) Nematodes in Soil Ecosystems, pp 117–138. University of Texas, Austin
Van de Bund CF (1970) Influence of crop and tillage on mites and springtails in arable soil. Netherlands Journal of Agricultural Science 18: 308–314
Van Vliet PCJ, Beare MH and Coleman DC (1995) Population dynamics and functional roles of Enchytraeidae (Oligochaeta) in hardwood forest and agricultural ecosystems. Plant and Soil 170: 199–207
Verhoef HA and Brussard L (1990) Decomposition and nitrogen mineralization in natural and agro-ecosystems: The contribution of soil animals. Biogeochemistry 11: 175–211
Walter DE (1987) Trophic behaviour of (mycophagous) microarthropods. Ecology 68: 226–229

Walter DE and Ikonen EK (1989) Species, guilds, and functional groups: taxonomy and behaviour in nematophagous arthropods. Journal of Nematology 21: 315–327

Walter DE, Hudgens RA and Freckman DW (1986) Consumption of nematodes by fungivorous mites, *Tyrophagus* spp. (Acarina: Astigmata: Acaridae). Oecologia 70: 357–361

Walter DE, Hunt HW and Elliott ET (1988) Guilds or functional groups? An analysis of predatory arthropods from a shortgrass steppe soil. Pedobiologia 31: 247–260

Wardle DA (1995) Impacts of disturbance on detritus food webs in agro-ecosystems of contrasting tillage and weed management practices. Advances in Ecological Research 26: 105–185

Wardle DA and Yeates GW (1993) The dual importance of competition and predation as regulatory forces in terrestrial ecosystems: evidence from decomposer food-webs. Oecologia 93: 303–306

Wardle DA, Yeates GW, Watson RN and Nicholson KS (1995) The detritus food-web and the diversity of soil fauna as indicators of disturbance regimes in agro-ecosystems. Plant and Soil 170: 35–43

Wasilewska L (1979) The structure and function of soil nematode communities in natural ecosystems and agrocenoses. Polish Ecological Studies 5: 97–145

Wasilewska L (1989) Impact of human activities on nematodes. In: Charholm C and Bergstrom L (eds) Ecology of Arable Land, pp 123–132. Kluwer Academic, Dordrecht, The Netherlands

Wasilewska L, Jakubczyk H and Paplinska E (1975) Production of *Aphelenchus avenae* Bastian (Nematoda) and reduction of mycelium of saprophytic fungi by them. Polish Ecological Studies 1: 61–73

Weil RR and Kroontje W (1979) Effects of manuring on the arthropod community in an arable soil. Soil Biology and Biochemistry 11: 669–679

Weiss B and Larink O (1991) Influence of sewage sludge and heavy metals on nematodes in an arable soil. Biology and Fertility of Soils 12: 5–9

Whitford WG, Freckman DW, Santos PF, Elkins NZ and Parker LW (1982) The role of nematodes in decomposition in desert ecosystems. In: Freckman DW (ed) Nematodes in Soil Ecosystems, pp 98–115. University of Texas, Austin

Wicklow DT (1981) Interference competition and the organization of fungal communities. In: Wicklow DT and Carroll GC (eds) The Fungal Community: Its Organization and Role in the Ecosystem, pp 351–375. Marcel Dekker, New York, NY

Whitaker RH (1975) Communities and Ecosystems. 2nd ed. Macmillian, New York, 385 pp

Yeates GW (1984) Variation in soil nematode diversity under pasture with soil and year. Soil Biology and Biochemistry 16: 95–102

Yeates GW (1994) Modification and qualification of the nematode maturity index. Pedobiologia 38: 97–101

Yeates GW and Wardle DA (1996) Nematodes as predators and prey: relationships to biological control and soil processes. Pedobiologia 40: 43–50

Yeates GW and Coleman DC (1982) Nematodes in decomposition. In: Freckman DW (ed) Nematodes in Soil Ecosystems, pp 55–80. University of Texas, Austin

Yeates GW, Bamforth SS, Ross DJ, Tate KR and Sparling GP (1991) Recolonization of methyl bromide sterilized soils under four different field conditions. Biology and Fertility of Soils 11: 181–189

Zwart KB, Kuikman PJ and VanVeen JA (1994) Rhizosphere protozoa: their significance in nutrient dynamics. In: Darbyshire J (ed) Soil Protozoa, pp 93–121. CAB International, Wallingford, Oxon, UK

Agroforestry Systems **45**: 187–202, 1999.

The problem of irrigated horticulture: matching the biophysical efficiency with the economic efficiency

R. J. STIRZAKER
CSIRO Land and Water, P.O. Box 1666, Canberra 2601, Australia;
E-mail: richard.stirzaker@cbr.clw.csiro.au

Key words: drainage, irrigation, leaching, soil degradation

Abstract. Modern irrigated horticulture provides society with a wide variety of quality produce at a cheap price. Set against this economic efficiency is the leakage of water and nutrients to the environment coupled with degradation of the resource base upon which future production depends. This paper illustrates why we might expect irrigated horticulture to spawn environmental problems related to excess water, excess nutrients and land degradation. The solutions to these problems will entail much more than just better management by farmers. If we cannot manage or engineer away the problems of irrigated horticulture, we must then address the very way the agro-ecosystem functions, in the areas of water and nutrient capture and soil management. This will require the design of new agricultural systems where the sources and sinks of water and nutrients are more evenly matched in time and space.

1. Introduction

Fifty years ago, the majority of horticultural crops in Australia were grown on small farms in peri-urban areas. A variety of different crops were grown on the same farm, production was highly labour intensive, fertility was largely managed through animal manures and crop rotation, and few chemicals were available for weed, disease and insect control. Developments in the areas of irrigation, agrochemicals, mechanisation, transport, refrigeration and volume marketing systems have underpinned the rapid transition to large scale specialised farms. Increasingly, these are not situated near population centres, but where climate, soils or water supplies are most favourable for production of specific crops. This trend has seen the industry become more efficient in terms of yields per hectare and output per farmer, and the consumers have benefited from a greater variety of quality produce at lower prices.

As the immediate production constraints of water, nutrients and pests have been lifted, a suite of new problems has emerged on irrigated lands. Whilst only ~15% of the world's total cultivated land is irrigated, it produces up to one third of the world's food. Of this irrigated land, FAO and UNESCO estimate that as much as half is under the influence of salinity, sodicity and waterlogging (Szabolcs, 1994). Other problems highlighted in the 'Australia: State of the Environment' report include the degradation of soil structure, pollution of aquifers with salt, nutrients or pesticides, and the destruction of riverine ecosystems (Department of Environment Sport and Territories, 1996).

These problems have long incubation periods, and in their initial manifestation have little effect on farm productivity. Worst of all, the input based solutions we have grown so used to seem to be impotent in the face of such issues.

The new problems have their roots in the spectacular mismatch between the economic efficiency of irrigated horticulture and its biophysical inefficiency. Biophysical efficiency is defined here as the amount of inputs that are used by the crop (water and nutrients) or degraded on site (pesticides) compared to the amount that leave the farm and become potential pollutants on-farm or in some other part of the ecosystem. Under the general heading of biophysical efficiency is also the degradation of the resource base. Intensively cultivated soils often become structurally degraded; degradation becomes part of a spiralling feedback loop as further cultivation, irrigation and fertiliser additions are needed to overcome the structural limitations to production. In the short term yields can be maintained as management masks degradation, but ultimately practices which impair the productive capacity of the land will undermine its economic viability.

The history of irrigated agriculture is the history of an enormous struggle to master the hydrological cycle. Despite many failures throughout the centuries, and especially this one, the question is still asked 'Can irrigated agriculture be made sustainable?' The answer seems to be 'Yes, but at what cost?' (Gardner, 1993).

2. The problem of irrigated horticulture

Horticultural crops flourish in a moist and fertile environment. Irrigation removes the primary constraint to productivity, but nutrition is quickly revealed as the next constraint of fast growing, often shallow rooted crops grown in soils leached by irrigation. Once the two major abiotic factors which limit production have been taken care of, the way is opened for a new set of biotic constraints; the now lush but defenceless horticultural crops must face the onslaught from competition (weeds), herbivory (insects) and parasitism (diseases) (Ewel, this volume). Thus a whole array of other inputs are needed to back up the investment in irrigation.

Even though water is the primary constraint to production, it is one of the cheapest to remedy. In the state of New South Wales, for example, irrigation costs account for only 2–4% of the total variable costs of production (Table 1; NSW Agriculture, 1996). The cheap price of irrigation relative to the value of the crop gives rise to a steep input response curve. A good farmer would want to operate somewhere around point B in Figure 1, with perhaps a little extra input to cover the uncertainty of where the flat part of the curve really begins. Herein lies the farmer's dilemma. Available water and nutrients, particularly nitrogen, fluctuate widely over short periods. Though the technology is available to monitor these changes, few farmers have the time,

Table 1. Percentage of variable costs attributed to major operations involved in growing a selection of vegetable crops in New South Wales, Australia.

	Broccoli	Capsicum	Cucumber	Lettuce	Tomato	Potato
Planting	15	19	12	14	2	12
Machinery	6	5	5	5	4	8
Irrigation	1	2	2	2	1	4
Fertiliser	4	14	10	10	1	11
Pesticides	4	2	8	8	1	10
Casual labour	4	3	1	1	1	17
Harvesting	36	30	34	34	58	7
Marketing	30	25	26	26	33	29

Source: New South Wales Agriculture (1996).

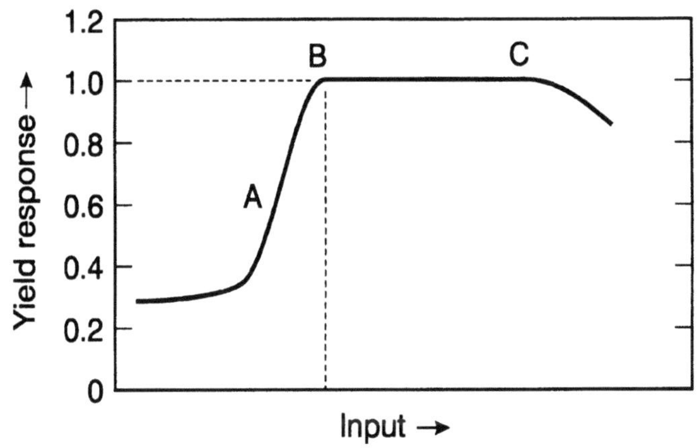

Figure 1. An input response curve for water or fertiliser. The uncertainty of position B, and the cheap cost of inputs, makes it economically attractive to operate on the wide plateau between points B and C.

money or skill to carry out such monitoring. Because of the uncertainty of position B, and the fear of sliding down the steep response curve towards point A, one strategy is to operate midway between points B and C. This is why well drained soils, with a wide plateau between B and C, are so desirable for horticulture. An extra 100 mm of water plus 50 kg of nitrogen during the season is cheap insurance for a crop with a gross value of between \$6000 and \$30,000 ha^{-1}.

3. To manage or to mimic?

The problem of horticulture can be summarised thus: it is too difficult to know exactly how much water or fertiliser or pesticide to apply, and too risky to apply too little. Since inputs are cheap, they can be used in excess to provide insurance. If the excesses of horticulture degrade other resources, the cost of this strategy will be paid for by the community and the environment. The question now arises: if farmers operated with perfect knowledge (knew the exact water and nutrient requirements and pest and disease thresholds) – could horticulture pass the test of biophysical efficiency? In other words, can this agro-ecosystem be expertly managed from the outside via physical and chemical inputs and remain both productive and non polluting? If the above statement is true, then research should focus primarily on the technology and training that would make management decisions more precise. If not, we will have to address the very biological functioning of the agro-ecosystem as well.

The thesis of this volume is that, while the natural ecosystem may not be productive in economic terms, it has passed the test of enduring sustainability. Thus, we may be able to bring a level of sustainability to the agriculturally productive ecosystem by trying to mimic the natural ecosystem it replaced, in function if not in structure or form. In the case of Australian horticulture, the original ecosystems were open woodland, clay plains (mostly grassland) and heathlands. These ecosystems have been so altered by inputs of water and nutrients, that they cease to be suitable standards. The ecosystem which may be the most useful biological standard for horticulture is the rainforest, since rainforests sustain high primary productivity in wet and leaching environments through efficient cycling of water and nutrients. They also minimise biotic stresses through diversity. The most obvious lessons from the rainforest are:

1. a high year round leaf area which maximises transpiration and minimises deep drainage;
2. major nutrient store in standing biomass and litter;
3. dimorphic root systems with high density of feeder roots near the surface and sinker roots to use sub soil water;
4. a high input of litter protects and enhances soil structure;
5. a high diversity of species, including microorganisms.

At first sight the rainforest analogy may seem most applicable to perennial horticulture (fruit/nuts), because the rainforest does not undergo the regular disturbance suffered by annual horticulture. However the rainforest remains a suitable standard for annual horticulture because it provides the encouragement to design an annual system which behaves in a more perennial manner. For example, to achieve a modicum of functional mimicry with the rainforest, it will be necessary to include species in the vegetable system which are not subject to external inputs. The role of the additional species is to be the specialist conservers and recyclers – to use water and nutrients left behind by horticultural crops, so that deep drainage is reduced and nutrients become trapped in the organic fraction. Further aims of these species are to protect the soil surface, enhance soil structure and provide an input of organic matter. These conserver and recycler species can be grown in rotation with the vegetable crop as cover crops, or simultaneously as intercrops. In the case of perennial horticulture they can be grown together with the economic crop in the inter-row region.

This paper evaluates the biological efficiency of irrigated horticulture under perfect management to test the hypothesis that sustainable horticulture can be attained by focusing on inputs alone. The 'perfect management' option is compared to alternative farming systems, inspired by principles of the rainforest described above, that change the way that the agro-ecosystem functions in the areas of:

1. Managing excess water;
2. Managing excess nutrients;
3. Managing soil degradation.

3.1. *Managing excess water*

Horticultural crops are sensitive to short periods of water stress, particularly around flowering and early fruit growth when yields can be devastated by a slip up in irrigation management (e.g. Rudich et al., 1977). For over fifty years, soil physicists and plant physiologists have tried to come up with strategies that will help the farmer apply water in accordance with plant needs. Progress has been made in relating plant water use to atmospheric demand, various indicators of plant water stress, and soil based measurements such as tensiometers, neutron scattering, time domain reflectometry and capacitance sensors. However, few of these methods are employed by irrigators, due to some combination of their cost, complexity of operation, and problems associated with soil variability, sensor calibration and maintenance, and interpretation of output. Clearly the difficulties of irrigation scheduling, from the farmer perspective, outweigh the benefits.

3.1.1. *Excess water under current practice*

If an irrigation scheduling device was invented with sufficient accuracy to keep the farmer around point B in Figure 1 and was cheap and straight forward enough to be embraced by farmers, would the problem of excess water be solved? The answer seems to be no. This point can be demonstrated by running a simple water balance model, which irrigates a crop with precisely the correct amount of water at a prescribed deficit against the background of historical weather data. The model is a simple daily water balance calculator which includes plant water use and soil evaporation, a two layer soil, and a routine to irrigate the soil back to the upper drained limit when a deficit of 10, 20 or 40 mm has been reached.

The model was calibrated to a 550 day data set collected from an alluvial soil near Camden, New South Wales (Stirzaker, 1996) and applied using the same plant and soil characteristics over a 17 year period using historical weather data (Table 2). Camden has an average annual rainfall of 799 mm and pan evaporation of 1408 mm and is characterised by high monthly and annual

Table 2. Predicted average annual excess water produced at two sites when irrigation is perfectly scheduled at soil water deficits of 10, 20 or 40 mm using seventeen years weather data in a model based on data collected at Camden, New South Wales, Australia.

Location	Deficit[1] (mm)	Irrigation[2] (mm y^{-1})	E_{top}[3] (mm y^{-1})	E_{sub}[4] (mm y^{-1})	Rain eff[5] (%)	Excess water[6] (mm y^{-1})
Camden	10	853	1055	0	26	491
	20	809	1055	0	32	482
	40	659	988	67	52	376
Tatura	10	906	1040	0	29	243
	20	835	1040	0	44	212
	40	727	965	75	68	157

[1] Irrigation was scheduled when the soil had deficits of 10, 20 and 40 mm.
[2] The amount of water which must be applied to ensure the prescribed deficit is not exceeded during the period crops are in the ground.
[3] E_{top} is the evapotranspiration from the top 0.6 m of soil.
[4] E_{sub} is the amount of water transpired from the 0.6–1.8 m depth.
[5] Rain eff is the efficiency with which rainfall is used and is calculated as the plant water use less irrigation divided by rainfall.
[6] Excess water is defined as that which cannot be stored in the soil profile, and either runs off the surface or drains below the root zone.
Source: Stirzaker (1996).

rainfall variability. The model was then applied to Tatura, Victoria, a drier site with winter dominant rainfall (average annual rainfall 472 mm, pan evaporation of 1391 mm). For the simulations shown in Table 2, two vegetable crops were grown per year. They were planted on 1 April and 1 October, each of 100 day duration, with a bare fallow between them. Actual daily rainfall and pan evaporation values were used over the period 1980 to 1996 from data collected at the two sites.

When irrigation was perfectly scheduled to a deficit of 20 mm at Camden, the average excess water was a remarkable 482 mm y^{-1} (Table 2). Since the soil store was kept near full by irrigation, rain that cannot be stored in the soil must either drain below the root zone or run off the surface. Given the high permeability of most horticultural soils, much of the excess water would be deep drainage (it is important to note here that drip irrigation, which does not wet the entire field, results in dry areas to buffer against rainfall). Even allowing for deficits of 40 mm, which forces the plants to use water from the subsoils and would induce plant stress, the average excess water at Camden was still 376 mm y^{-1}. At Tatura, with over 300 mm less rainfall than Camden, only about 60 mm less irrigation was required. The excess water was around half that of Camden, but still alarmingly high. The efficiency of rainfall use is the proportion of rainfall which is used by the crop. It ranged from a low of 26% in the variable Camden climate, with a maximum in-crop deficit of 10 mm, to 68% in Tatura when the soil water deficit was allowed to fall to 40 mm during the cropping period.

Similar enormous drainage values have been found experimentally. Using water balance calculations, Dougherty and Wells (1998) found up to 882 mm deep drainage under irrigated vegetables at a site north of Camden over an 18 month period. The examples in Table 2 are also applicable to irrigated perennial horticulture. McNab et al. (1994) measured surface runoff and collected water from tile drains beneath an orchard in the Goulburn Murray valley. Their experiment differed from the model scenario above in that they studied one summer season's irrigation in a wetter than average year, and it is most unlikely that tile drains intercepted all the drainage water. Nevertheless they measured 170 mm of excess water over a 12 month period, comprising 130 mm from tile drains and 40 mm of surface run-off.

Table 2 illustrates that the problem of deep drainage cannot be resolved by good irrigation management alone; none of the excess water was generated by over-irrigation. Nor can deep drainage be sufficiently reduced by increasing the irrigation interval, because sensitive shallow rooted crops would be devastated by the degree of deficit required. The practice of regulated deficit irrigation for pome and stone fruit does offer exciting possibilities for maintaining an important buffer for storing water during part of summer (Chalmers et al., 1981). It is clear, however, that annual and perennial systems could both benefit from a specialist plant to dry out the soils at strategic times.

3.1.2. *Excess water under an alternative system*

Stirzaker (1996) grew vegetable crops and lucerne, *Medicago sativa* L., as monocultures and intercropped in strips 1.3 m wide. The vegetable crops were drip irrigated and the lucerne unirrigated, so that wet and dry strips developed through the intercropped field (Figure 2). The soil water content was monitored to a depth of 1.8 m for a period of 550 days. Over this period the water stored in the subsoil (0.6–1.8 m depth) of the deep rooted lucerne monocrop never exceeded 60% of the total available water storage. The subsoil in the vegetable monocrop remained at or near the upper drained limit for the entire period, and deep drainage was a frequent event. In contrast, the subsoil of the lucerne/vegetable intercropped treatments fluctuated between 40 and 80% of total available water. Two reasons contributed to the drier subsoil beneath intercropped vegetables. First, a few lucerne roots grew about 1 m laterally into the vegetable beds, and used water both in and below the vegetable root zone. This was most obvious in the period between vegetable crops when there was no irrigation, so that the subsoil of intercropped vegetable rows commenced the season in a drier state. The second reason for dry subsoils in the intercropped rows was that drainage did not proceed vertically through the soil profile. This occurred because the lucerne had dried the soil profile to several metres on either side of the intercropped vegetables. Water draining through the intercropped rows was pulled sideways towards the dry lucerne profile under the influence of capillary forces, at the same time as it moved downwards under the influence of gravity. Theoretical calculations showed that this effect could operate for a distance of 1 to 2 metres laterally from

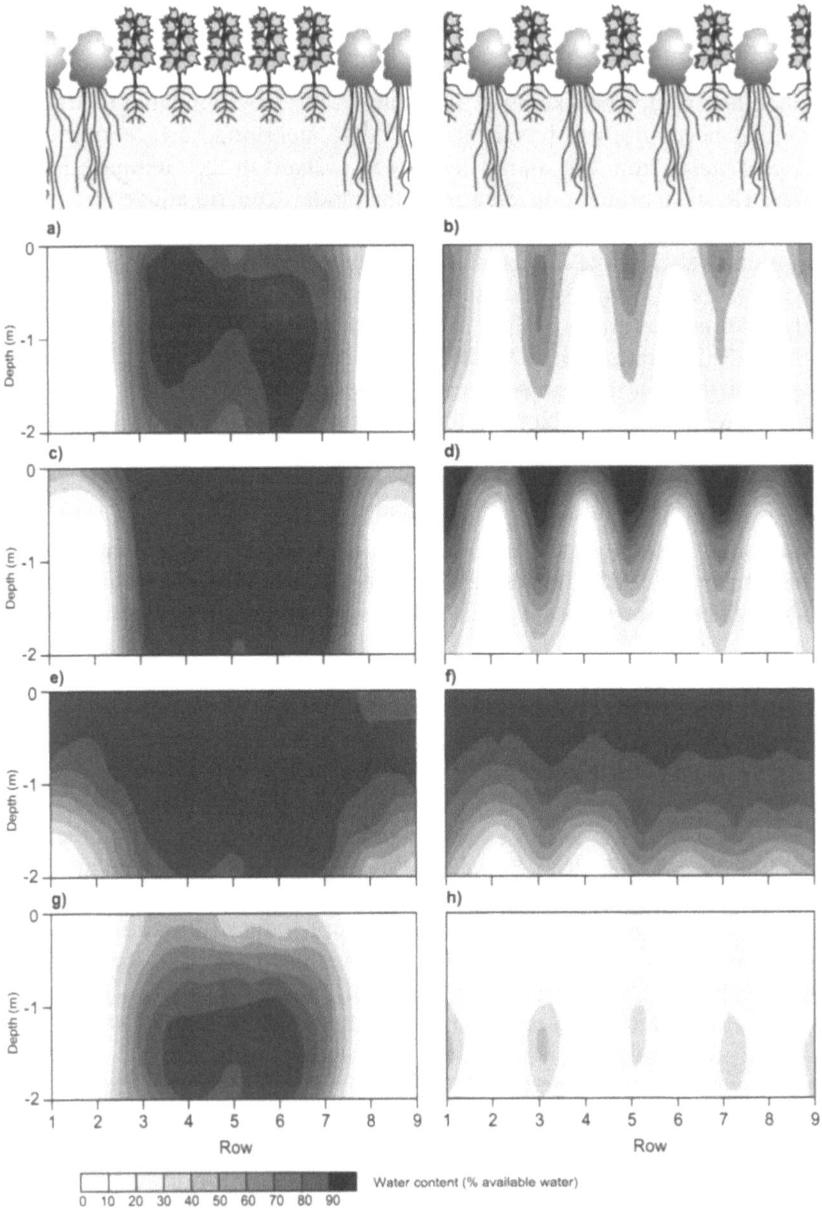

Figure 2. Cross sections through nine rows of monocrop and intercrop treatments to a soil depth of 2 m at Camden, News South Wales. Left hand column shows monocrop treatment with tomatoes planted in rows 3, 4, 5, 6, and 7 bounded by lucerne in rows 1, 2, 8 and 9. Right hand column shows intercrop treatment with lucerne in rows 2, 4, 6 and 8, and tomatoes in rows 1, 3, 5, 7 and 9. Measurements were made a) and b) before planting on 11/11/93; c) and d) after irrigation commenced on 2/12/93; e) and f) after harvest on 25/3/94; g) and h) after seven months with no crop or irrigation on 26/10/94.

the lucerne rows in this soil type (Stirzaker et al., 1999 provide a detailed explanation).

Figure 2 shows cross sections of soil water content through a vegetable monocrop treatment and an intercrop treatment over a period of 12 months. Soil water content is shown as a percentage of total available water, i.e. the water content at the upper drained limit minus that at the lower limit expressed as a percentage. Tomatoes were grown in adjacent rows 3, 4, 5, 6 and 7 in the monocrop treatment (left hand column), bounded by lucerne on either side (rows 1, 2 and 8, 9). Lucerne and tomatoes were grown in alternate rows in the intercrop treatment, with the tomatoes in rows 1, 3, 5, 7, 9. Rows were 1.3 m apart, with one neutron probe access tube in the centre of each row and water content measured at depths of 0.2, 0.3, 0.4, 0.5, 0.6, 0.8, 1.0, 1.2, 1.4, 1.6, 1.8 m. Each diagram is a composite of 3 replicates.

Before the crop was planted on 11 November 1993, the soil was wet in the vegetable monocrop treatment, relatively dry in the intercrop vegetable rows and dry in the lucerne rows (Figure 2a,b). The tomato crop was planted in late November and the irrigation necessary to establish seedlings, together with the water already in the soil, brought the entire monocrop profile to full point within two weeks (Figure 2c). Wet soil over dry soil was maintained in the intercrop treatment – the ideal scenario for irrigated agriculture (Figure 2d). The tomato crop was harvested in early March 1994. Prolonged rain following harvest in mid March replenished the water in the lucerne rows, but the soil below 1.5 m in the intercropped vegetable rows remained below 60% of the storage capacity (Figure 2e,f). The vegetable rows were then left fallow for seven months between March and October (autumn to spring). Soil evaporation in the monocrop treatment resulted in a profile with moderately wet topsoil above wet subsoil (Figure 2g). The lucerne in the intercrop treatment not only removed all the water from its own rows during this period, but substantially dried the vegetable rows as well (Figure 2h). The next irrigation season would, therefore, commence on a dry profile.

Whether intercropping is viable as a productive agricultural system depends on the degree of competition and complementarity between the lucerne and vegetable components for finite resources. When lucerne removes water from the vegetable crop root zone during the vegetable season, the two species are clearly in competition with each other. When the perennial lucerne uses water between vegetable crops, the two species are complementary (Ong et al., 1996). Similarly, when excess water moves sideways under capillary forces from the vegetable row to the lucerne row the two species are complementary. Stirzaker (1996) found that the lucerne extracted water very slowly from adjacent vegetable rows, even when the lucerne rows were dry, because the lucerne root length density was very low in vegetable beds. This meant that competition could be easily managed via irrigation. When the lucerne was cut and placed on the vegetable beds as a mulch, there were substantial yield benefits.

3.2. *Managing excess nutrients*

Rising watertables in irrigation areas provide evidence that water is lost below the crop root zone. The case for nutrients being provided in excess of requirement is similarly straight forward. Consider the advice given to farmers by local departments of agriculture. These agencies set expected yields and advise farmers how much fertiliser to apply to each crop, based on regional field trials. The efficiency of nutrient use is simply the amount of nutrient harvested in the product divided by that applied in the fertiliser. Advice must be general, and farmers are given low and high application rates based on their perceptions of soil fertility and soil type.

Table 3 gives an example of the average values for recommended fertiliser additions divided by the average removal in harvested product for a selection of common leaf, fruit and root crops. Phosphorus is the nutrient applied to the greatest excess with potassium generally under supplied. Nitrogen efficiency depends largely on crop type; the deeper rooted crop (tomato) or crops taken to physiological maturity (potato) having the highest efficiencies.

The fertiliser recoveries in Table 3 are most likely over-estimates. Trials reproducing district farmer practice conducted by the NSW Department of Agriculture show an estimated 633 kg N ha^{-1} was leached in an 18 months period during which capsicum and cabbage crops were grown, giving an overall crop nitrogen use efficiency of 13%. Recoveries were improved to 51% under conditions of best management practice (Dougherty and Wells, 1998). Low recoveries have been found in other countries; Greenwood et al. (1974) present N:P:K recoveries of 7%, 2% and 8% for lettuce and 65%, 6% and 55% for potatoes under UK conditions. Expected nitrogen recoveries for perennial horticultural crops based on agency recommendations are in the range of 30–50 % (A. Boland pers. comm., 1998) and experimental evidence reveals enormous losses. McNab et al. (1994) recovered 44–70 kg N ha^{-1} y^{-1} from tile drains beneath an orchard and Gerritse et al. (1994) found that 20–50% of nitrate applied to orchards was lost in stream flow.

Table 3. The percentage of nitrogen, phosphorus and potassium exported in the product as a function of recommended application (100% recovery means fertiliser additions equal harvest removal, > 100% means that the crop is mining the soil store).

Crop	Nitrogen	Phosphorus	Potassium
Broccoli	30	8	30
Capsicum	22	7	156
Lettuce	34	6	120
Tomato	74	26	139
Potato	191	10	388

Source: NSW Department of Agriculture Agfact series; Reuter and Robinson (1986); Pierce (1987).

There are several reasons why we may expect low nitrogen use efficiencies in horticulture. Crops for which fruit is the economic product are generally harvested when the plant still has a complete leaf canopy. Thus, there is little opportunity for remobilisation from the vegetative structures to the fruit. Nutrients from the unharvested portions of the crop are returned to the soil and are often not taken into account when fertilising the next crop. A second reason relates to the high concentrations of nutrients needed in the soil to sustain the rapid growth rates required for a quality product. For example a lettuce crop would have a fresh weight of at least 40 t ha^{-1}, half of which would be put on during the last 20 days. If the lettuce were 4% dry matter and contained 4% nitrogen, then 32 kg N ha^{-1} would be taken up in 20 days. If the average transpiration rate was 5 mm d^{-1}, and the nitrate taken up passively in the transpiration stream, then the concentration of nitrate in the soil solution would have to be over 30 mg L^{-1}. But the main reason for low nitrogen use efficiency must lie in the irrigation problem outlined in the previous section. High soil nitrate concentrations coupled with the excess water due to rainfall (Table 2) and the inevitable periods of over irrigation, combine to make leaching inevitable.

Although fertigation is a management tool that can improve fertiliser recovery, again there is a strong case for altering the way the agroecosystem functions in addition to how it is managed. In a literature survey covering a range of cover crops in different environments, Meisinger et al. (1991) found that cover crops reduced the amount of nitrogen leached by 20–80% compared to a crop-fallow rotation. The result was a combination of the cover crops keeping the soil drier, thus reducing deep drainage, and the conversion of soil nitrate to less mobile organic forms of nitrogen. The release of this organic nitrogen to subsequent crops has been shown in a range of studies to reduce the requirement of the inorganic fertiliser nitrogen by, on average, 40 to 120 kg h^{-1} (Smith et al., 1987). The replacement of inorganic fertiliser does not, however, equate to the amount of nitrogen supplied to the crop from an organic source; there are important non-nitrogen benefits of rotations which are hard to disentangle, and inorganic nitrogen is obviously not 100% efficient. Studies with 15N have shown that less than a quarter of the nitrogen in residues is likely to end up in the subsequent crop (Ladd et al., 1983; Harris and Hesterman, 1990).

3.3. *Managing soil degradation*

The soil is the bank to which occasional deposits of water and nutrients are made and from which the crop makes its daily withdrawal. What we call a good soil is simply a soil that is able to buffer the daily plant demand against the variable inputs. A good soil is permeable so that water enters with minimal runoff, has a high water holding capacity so that water does not move too deeply, and can be quickly and extensively colonised by roots. It can hold nutrients and exchange them with the soil solution and contains low concentrations of chemicals which are detrimental to plant growth. Good crops can be grown on poor soils, but only if management is tighter, so that the role of the soil as a buffer is reduced. In fact improved water and fertiliser management can and do mask soil degradation, as more inputs are needed to get acceptable yields from soils that are becoming hard and impermeable.

It is self evident that the soil should be managed so that it remains the best storage for water and nutrients it can be, within the limits set by its particle size distribution, organic matter content and clay mineralogy. Most vegetable crops require frequent, intense cultivation because the rotations are short, and fine seedbeds are required for accurate plant densities of small seeded crops. Under these conditions, the weight of experimental evidence predicts that soils will be structurally degraded. Reduced tillage, protection of the surface soil, inputs of organic matter and crop diversity are the practices which can slow degradation or even improve some soils. Even though these good husbandry practices are well accepted, they are slow, unwieldy and sometimes unreliable for the modern vegetable farmer who has already made an investment in machinery and agrochemicals. Yet there is a growing realisation among many farmers that biological methods of land management can and must be integrated with conventional methods. In this section, case studies of reduced tillage and rotation crops are presented as a means to slow down the degradation of intensively managed soils.

Stirzaker et al. (1993b) argued that reduced tillage should be well suited to horticulture, because the length of roots required to pick up water and nutrients decreases as the supply of these resources increases; even if root growth is reduced by higher soil strengths, this should be more than offset by the regular inputs of water and fertiliser. This hypothesis was tested in a series of 15 experiments on three soil types where the yields of vegetable crops under no-tillage were compared with yields obtained when the soil was thoroughly cultivated to a depth of at least 20 cm (Table 4). In addition to a no-tillage treatment with a bare fallow between crops, the experiments also included mulch crops such as subterranean clover *Trifolium subterraneum* L. and ryegrass *Lolium rigidum* Gand. These crops were grown through the winter months, and either died naturally or were killed with herbicide, and vegetable crops were grown through the residues without tillage. A further treatment involved the use of lucerne, grown as an intercrop, where the residue was cut and placed around no-till vegetables on adjacent beds. The bulk densities in Table 4 were measured within the top 10 cm of soil, and the yields were expressed relative to that obtained under cultivated conditions.

Several general guidelines about no-tillage can be drawn from Table 4. First, no-tillage with a fallow between vegetable crops was not successful for leaf crops (lettuce and cabbage) in the lighter textured soils. These soils were largely apedal and tended to compact to bulk densities > 1.5 t m^{-3}. Second, fruit crops (tomato, bean) and storage crops (beetroot) performed satisfactorily under no-tillage. Third, the efficacy of no-tillage changed with time. In a sequence of crops in the clay loam soil (crops 8–11), yields under

Table 4. Normalised vegetable crop yield under no-tillage following no cover crop, a clover cover crop, a ryegrass cover crop or a lucerne intercrop for 15 experiments on three soil types (yield under no-tillage has been divided by that obtained under cultivated conditions)

No.	Soil type	Density ($t\ m^{-3}$)[a]	Crop	none	Cover crop		Intercrop
					Clover	Ryegrass	Lucerne
1	Sandy loam	1.58	Lettuce	0.88[b]			
2	Sandy loam	1.55	Lettuce	0.81[b]			
3	Sandy loam	1.55	Lettuce	0.67[b]	0.91		
4	Sandy loam	1.55	Tomato	0.98	1.06		
5	Sandy loam	1.55	Tomato	0.97	0.96		
6	Sandy loam	1.44	Lettuce	0.93	1.04		
7	Sandy loam	1.66	Lettuce	0.39[b]	0.84[b]		
8	Clay loam	1.49	Lettuce	0.93	0.87	0.57[b]	
9	Clay loam	1.44	Lettuce	0.98	1.04	0.83[b]	
10	Clay loam	1.43	Cabbage	0.85[b]	0.78[b]	0.65[b]	
11	Clay loam	1.44	Lettuce	0.76[b]	0.80[b]	0.72[b]	
12	Loamy sand	1.57	Cabbage	0.82[b]			1.18
13	Loamy sand	1.57	Cucumber	0.75[b]			1.07
14	Loamy sand	1.57	Beetroot	0.93			1.34
15	Loamy sand	1.57	Beans	0.97			1.21

[a] In all cases the cultivated soil had a significantly lower bulk density than the no-till soil.
[b] Denotes yield under no-tillage is significantly lower than yield under cultivation ($P = 0.05$).
Source: Experiment No. 1, 2, 3 – Stirzaker et al. (1993b); No. 4, 5 – Stirzaker et al. (1993a); No. 6, 7 – Stirzaker and White (1995); No. 8, 9 – Stirzaker et al. (1995); No. 9, 10 – Sutton (1997); No. 12, 13, 14, 15 – Stirzaker et al. (1996).

no-tillage were initially similar to the cultivated treatment (crops 8 and 9) but fell below the cultivated treatment thereafter. Over the same time period the soil physical properties improved, as measured by infiltration under ponded and 10 mm suction using a disk permeameter. This suggests that non-physical factors, such as root diseases and phytotoxicity, dominated the outcome in an experiment designed to test the importance of physical factors.

In most of the experiments in Table 4, an attempt was made to ameliorate a no-tilled soil with cover crops. While cover crops are growing they protect surface soil structure, and improve soil structure by stimulating soil fauna and producing biopores with their roots (Stirzaker and White, 1995). After death, the cover crops modify soil temperature, water content and mechanical impedance of the surface soil. The potential nutritional effects of these cover crops were eliminated in the above experiments by supplying high levels of fertiliser to all treatments. On the sandy loam soil, the yield of lettuce was improved by growing a clover cover crop prior the vegetable crop compared to having a fallow (crops 3 and 7). In the case of crop 7 which was grown on very compacted soil, the clover cover crop caused a twofold yield increase compared to no-tillage without a cover crop, but yields were still better when the soil was cultivated. The clover mulch did not improve yields on the clay loam soil (crops 10 and 11) where the problem of no-tillage appeared to relate to phytotoxicity or root disease rather than hard soil.

A ryegrass mulch proved to be a better soil improver of soil physical properties than clover, but resulted in lower crop yields, probably because it released phytotoxic substances (Stirzaker et al., 1995; Stirzaker and Bunn, 1996). Work on wheat by Roget (1995) showed that certain soil borne diseases were problematic in undisturbed compared to cultivated soils. However, after a number of years the no-till soils became suppressive of the disease as the soil biology adjusted to the new conditions. These examples reveal some of the complexity unleashed when making relatively minor alterations to a production system. The success of reducing tillage and using cover crops depended on the crop type, the soil type, cover crop species, the vegetable crop sequence, the season in which the crops were grown and the years the new system had been operating. Added to this is the skill required to manage such systems. Little wonder that a farmer struggling to survive in the short term feels more secure with the plough, the herbicide and the bag of fertiliser.

4. Conclusion

The real problem of horticulture is, in essence, structural, as much as it is a problem of poor management. This is not to say that management must not be improved. Huge gains in the efficiency of water and fertiliser use would be attained if farmers had simple and robust indicators upon which to base routine decisions. However, the best managed lands are still likely to suffer soil degradation and leak water, nutrients and pesticides into the surrounding environment. To address these problems we must strive to get our managed farming systems to function, in part, like the natural systems which have proven resilience. No doubt engineering solutions will play their part, as we need to collect, transport and dispose of the waste products of irrigated agriculture. But the plans of engineers will often be undone in the face of the extraordinary complexity of the hydrological cycle: witness the problems of toxic quantities of selenium from the San Joaquin Valley contaminating wildlife. Sometimes the scale of the problem may overwhelm the engineering solution, as in the Murray-Darling Basin of Australia.

To rely solely on the traditional agronomic solution (more accurate management of inputs) or the engineering solution (out of sight out of mind) would be to repeat the mistakes of the past. In his critique of irrigated agriculture, Gardner (1993) warned that 'We must be more critical of ourselves, of the quality of our science and the narrowness of our scholarship'. It remains to be seen if we can broaden our scholarship to create agroecosystems which, in addition to being productive, can also incorporate the functions of trapping nutrients, de-watering the subsoil, and biological ploughing.

As is the case for all complex farming systems, the complementarity between the components must outweigh the competition. This will dictate how the components are integrated – as intercrops (simultaneous systems), long or short rotations (sequential systems). As a start, it is essential to include dryland crops into current rotations of irrigated crops, and to minimise the time the land is fallow. Since many environmental problems manifest at catchment level, the correct integration will also be at this scale. Though a significant polluter at a point scale, the area of irrigated land in Australia is small relative to other land use. The most sensible option may be to confine irrigated cropping to the most favourable soils and to buffer these areas with forestry or dryland agriculture in an attempt to soak up the excesses of irrigated land.

5. References

Chalmers DJ, Mitchell PD and Van Heek L (1981) Control of peach tree growth and productivity by regulated water supply, tree density and summer pruning. Journal American Society Horticultural Science 106: 307–312

Department of Environment Sport and Territories (1996) Australia: State of the Environment. CSIRO Publishing, Collingwood, Australia

Dougherty WJ and Wells AT (1998) Nitrate leaching under various vegetable production systems at Somersby. In: Environmental benefits of soil management, pp 186–191. National Soils Conference, Brisbane 27–29 April, Australian Soil Science Society Incorporated

Gardner WR (1993) The future of irrigated Agriculture. In: Buxton DR, Shibles R, Forsberg RA, Blad BL, Asay KH, Paulsen GM and Wilson RF (eds) International Crop Science I., pp 97–99. Crop Science Society of America, Wisconsin

Gerritse RG, Adeney JA and Engel R (1994) Impact of horticulture on water quality in the Darling Range of Western Australia. In: Nutrient and fertiliser management in perennial horticulture, pp 80–86. Land and water Resources Research and Development Corporation Occasional paper 07/94

Greenwood DJ, Cleaver TJ and Turner MK (1974) Fertiliser requirements of vegetable crops. Proceedings of the Fertiliser Society 145: 16–29

Harris GH and Hesterman OB (1990) Quantifying the nitrogen contribution from Alfalfa to soil and two succeeding crops using Nitrogen-15. Agronomy Journal 82: 129–134

Ladd JN, Amato M, Jackson RB and Butler JHA (1983) Utilization by wheat crops of nitrogen from legume residues decomposing in the field. Soil Biology and Biochemistry 18: 417–425

Meisinger JJ, Hargrove WL, Mikkelsen RL, Williams JR and Benson VW (1991) Effects of cover crops on groundwater quality. In: Cover-crops for clean water. The proceedings of an international conference, West Tennessee Experiment Station, 9–11 April, Jackson, Tennessee, pp 57–68. Soil and Water Conservation Society, Ankeny, IA

McNab S, Jerie P, O'Connor R and MacDonald P (1994) Efficient fertiliser application techniques and nutrient losses in irrigated horticulture. In: Nutrient and fertiliser management in perennial horticulture, pp 30–41. Land and water Resources Research and Development Corporation Occasional paper 07/94

New South Wales Department of Agriculture (1996) Farm Budget Handbook. NSW Vegetables. NSW Agriculture, 56 pp

Ong CK, Black CR, Marshall FM and Corlett JE (1996) Principles of resource capture and utilization of light and water. In: Ong CK and Huxley P (eds) Tree-crop interactions: a physiological approach, pp 73–158. CAB International, Wallingford, UK

Pierce LC (1987) Vegetables. Characteristics, production and marketing. John Wiley and Sons, New York, 433 pp

Reuter DJ and Robinson JB (1986) Plant analysis. An interpretation manual. Inkata Press, Melbourne, 218 pp
Roget DK (1995) Decline in root rot (Rhizoctonia solani AG-8) in wheat in a tillage and rotation experiment at Avon, South Australia. Australian Journal of Experimental Agriculture 35: 1009–1013
Rudich J, Kalmar D, Geizenberg C and Harel S (1977) Low water tensions in defined growth stages of processing tomato plants and their effects on yield and quality. Journal of Horticultural Science 52: 391–399
Smit MS, Frye WW and Varco JJ (1987) Legume winter cover crops. Advances in Soil Science 7: 96–139
Stirzaker RJ, Sutton BG and Collis-George N (1993a) Soil management for vegetable production. 1. The growth of processing tomatoes following soil preparation by cultivation, zero-tillage and an in-situ grown mulch. Australian Journal of Agricultural Research 44: 817–829
Stirzaker RJ, Passioura JB, Sutton BG and Collis-George N (1993b) Soil management for vegetable production 2. Possible causes for slow vegetative growth of lettuce associated with zero-tillage. Australian Journal of Agricultural Research 44: 830–844
Stirzaker RJ and White I (1995) Cover crop amelioration for no-tillage vegetable production. Australian Journal of Agricultural Research 46: 553–568
Stirzaker RJ, Bunn DG, Cook FJ, Sutton BG and White I (1995) No-tillage vegetable production using cover crops and alley cropping. In: Cook H and Lee H (eds) Soil management for sustainable agriculture, pp 466–474. Wye College Press
Stirzaker RJ, Mosena ML and Etherington RJ (1996) Soil and irrigation management for small-scale commercial vegetable farmers in the Northern Province of South Africa. Centre for Environmental Mechanics Technical report No. 112
Stirzaker RJ and Bunn DG (1996) Phytotoxicity of ryegrass and clover cover crops, and a lucerne alley crop for no-till vegetable production. Biological Agriculture and Horticulture 13: 85–103
Stirzaker RJ (1996) Lucerne/Vegetable Alley Cropping to reduce leaching. In: Proceedings of the 8th Australian Agronomy Conference, pp 522–525
Stirzaker RJ, Cook FJ and Knight JH (1999) Approximate solutions for where to plant trees on cropping land for control of dryland salinity. Agricultural Water Management 39: 115–133
Sutton BG (1997) Use of in-situ mulches in horticulture. Final report for the Rural Industries Research and development Corporation, Canberra
Szabolcs I (1994) Soils and salinisation. In: Pessarakali M (ed) Handbook of plant and crop stress, pp 3–11. Marcel Dekker, New York

Agroforestry Systems **45**: 203–214, 1999.

Towards achieving functional ecosystem mimicry with respect to water cycling in southern Australian agriculture

T. J. HATTON[1,*] and R. A. NULSEN[2]
[1] *CSIRO Land and Water, Private Bag, P.O. Wembley 6014, Australia;* [2] *Agriculture Western Australia, 3 Baron Hay Court, South Perth 6151, Australia (*Author for correspondence: E-mail: tom.hatton@per.clw.csiro.au)*

Key words: dryland salinity, ecohydrology, groundwater, hydrology

Abstract. The southern Australian landscape is characterised by large climatic variability and an old, weathered surface with generally low hydraulic gradients. Many of these landscapes are prone to salinisation. The original ecosystems have strategies to cope with this environment in ways which minimise runoff and groundwater recharge. Replacement of these ecosystems with conventional agriculture has resulted in large-scale disruption of the natural water cycle with severe environmental, economic and social consequences. Sustainable agricultural systems aimed at mimicking natural hydrology must either minimise groundwater recharge over most of this landscape or intercept groundwater discharge in lower parts of the landscape prior to salinisation to restore the original hydrological balance. Empirical and theoretical analyses suggest that however vegetation is managed in the landscape, effective water balance control will be achieved only at a leaf area index approaching that of the natural state, indicating revegetation of most or all parts of the catchment with either trees or plants with similar ecohydrological characteristics as trees. There are at present few economically-attractive agricultural systems able to achieve hydrologic mimicry likely to be adopted at the necessary scale.

1. Introduction

The landscape of southern Australia has been changed by European settlers to a remarkable degree in a relatively short time. While this process began 200 years ago, vast portions of the continent have only been converted from native bush to agriculture in the past 50 years. Nevertheless, the consequences of this conversion are already having dramatic impacts on the health and productivity of the Australian landscape. Governments and community groups recognise these impacts and the imperative is now to arrest degradation and, implicitly, to restore the original ecosystem functions and the environmental benefits they provide while maintaining rural communities and the economically important role they fill for the nation.

This challenge is on a temporal and spatial scale perhaps no other nation has faced, and the impediments are not only biophysical but economic, cultural and political. In this article we emphasise biophysical considerations and examine the limits to what can be accomplished in restoring ecosystem function and the times over which we may see returns on an investment in restorative practice. More specifically, we focus on the key aspects of the southern Australian natural hydrological cycle which must be recaptured by agriculture. This hydrological emphasis is wholly appropriate in that most of the continent's environmental degradation results from changes in the water balance. The geographical emphasis is equally appropriate in that the vast majority of Australia's development, agriculture and people are in the southern half of the continent.

We first examine how water originally moved through the ecosystems of southern Australia, and then look at how these temporal and spatial patterns of cycling have changed with the development of agriculture and the nature and extent of the associated impacts. Finally, we look at how the original hydrological functions might be mimicked in an agricultural system and the feasibility of doing so.

2. Water cycling in the natural landscape

Two key aspects of southern Australia must be understood to fully appreciate the hydrological cycle as it existed at the time of European settlement. The first, and most important of these, is that Australia's geology has been relatively quiet over the past 60 million years; in some cases, what uplifting has occurred has actually restricted drainage. Little renewal of surface material means that soils are old and nutritionally poor, profiles are deeply weathered, and geomorphological units result from differential erosion of an ancient surface (McArthur, 1993). This geologic history has resulted in the flattest continent on earth with generally low hydraulic gradients and transmissivities. Water (and solutes) cannot move quickly through the Australian landscape as surface water or groundwater.

The deeply weathered regolith typically encompasses several, often interacting, aquifers including a shallow, seasonal perched system, a local semiconfined aquifer, and often a deeper, confined regional system (McFarlane et al., 1993; George et al., 1994). In the native state, these latter systems may be only rudimentary and normally water tables (if present) are quite deep (Salama et al., 1993). The overlying unsaturated zone typically contains a large amount of accumulated salt of largely atmospheric origin stored in the soil water, mostly below the major rooting zone, which is largely immobile prior to land clearance.

The second key aspect of southern Australian hydrology is the high climatic and hydrologic variability. McMahon et al. (1992) showed that in terms of mean annual runoff, Australia is not particularly distinctive (420 mm vs 620 mm for the world), but that the variability of streamflow is far higher than the rest of the world with the possible exception of South Africa. This variability is in part the result of the El Niño Southern Oscillation phenomenon. Australia is the only continent where the overwhelming influence on climate is non-annual climatic variation (Flannery, 1994). That is, the high degree of predictability associated with seasonal cycles elsewhere on earth is small in Australia relative to variations on a longer cycle. This lack of predictability has major implications for the evolutionary adaptation of Australian biota and the productivity of Australian ecosystems. It also puts southern Australia into the Ewel (this volume) category of land with low agricultural utility, in this case due to aridity.

The seasonality in climate in the southwestern part of the continent tends toward a mediterranean distribution of rainfall, ranging between 250 and 1200 mm annually. In the southeast, the distribution is more uniform with a similar range. In both regions, potential evaporation exceeds rainfall in most months and in all months in the lower rainfall zones.

In response to these geological and climatic conditions, Australian vegetation has adapted in a number of ways which distinguish its ecohydrology. By far the most distinctive characteristic is sclerophylly. There is debate as to whether this is in response to low water or nutrient availability (Adams, 1995). The other aspect which characterises the vegetation is that it is generally dominated by a perennial, evergreen, woody component, ranging from shrublands and heaths to woodland and forest. Understorey vegetation normally consists of grasses and shrubs, but in the drier areas can be largely bare ground for most of the year. Specht (1972) showed that the leaf area index of a wide range of Australian sclerophyll communities have a similar relationship to an index of evaporation and rainfall. Pook (1985) and Hatton and Wu (1995) showed that these communities vary their leaf area index with the changing availability of water over long wet-dry cycles. There is, in general, an ecohydrological equilibrium between leaf area index and climate (*sensu* Eagleson, 1982) in southern Australian ecosystems (Pierce et al., 1993). The theory and background underlying this principle are reviewed and discussed by Hatton et al. (1998).

The evergreen woody perennials typically have a dimorphic root system, with a set of shallow roots spreading as far as 30 m from the trunk (Jacobs, 1955), and deeper roots which may extend downwards to 20 m or more (Kimber, 1974; Carbon et al., 1980; Dell et al., 1983). These rooting systems allow the fullest possible exploitation of water in the regolith and thereby sustain vegetation through long, periodic droughts. The aerial architecture of the upper storey is effective at capturing rain falling on the canopy and directing it down the stem to infiltrate around the bole. Nulsen et al. (1986) showed that the infiltrating water percolated to considerable depth down the annular pathway around the roots. Thus the water is stored at depth and is thus available during the dry period. Removal of the stems deactivates this mechanism and dramatically changes the hydrology on a micro scale.

In the natural state, the hydrological cycle of the agricultural areas of southern Australia has the following functional generalities which result from the combination of the above attributes:

- Virtually all of annual rainfall is evaporated or transpired;
- The vast majority of rainfall reaching the soil infiltrates locally and thus surface runoff to streams is small, fresh and generally episodic in nature (Nulsen et al., 1986; McFarlane et al., 1993);
- Water use by native vegetation is somewhat conservative in winter, when

water is quite plentiful, with much more physiological activity toward summer (Farrington et al., 1992);

- Net groundwater recharge to the semi-confined and confined systems is a very small proportion of rainfall, usually less than 1 mm yr^{-1} except in the higher rainfall areas (Nulsen et al., 1986; Allison et al., 1990; George, 1992);
- What little groundwater discharge and runoff that is generated supports limited areas of riparian vegetation, which are able to switch their source of plant water from local rainfall to groundwater over the dry season (Thorburn et al., 1993; Walker et al., 1996);
- In some cases there is sufficient groundwater discharge to form natural saline surface features, but alluvial systems were often fresh at the surface (Farrington and Salama, 1996);
- There is generally little or no natural base flow of rivers in streams originating in the agricultural zone.

In short, the conservative water use of the native vegetation of southern Australia resulted in systems in which very little rainfall was discharged in liquid form. As one consequence, millennia of atmospherically-deposited salts built up in the unsaturated zone at some depth below the root zone. The exchange of water between the soil and the atmosphere can be characterised as mostly *vertical*, with the rain that falls on a given piece of ground evaporated from that same surface.

3. Changes to the water cycle following clearance for agriculture

The hydrological impacts of land clearance for agriculture in southern Australia are documented and reviewed in detail by a number of authors (Peck and Hurle, 1973; Williamson and Bettenay, 1979; Schofield, 1990; McFarlane et al., 1993; Nulsen, 1993; Farrington and Salama, 1996). These impacts result from the replacement of the native sclerophyll vegetation with annual crops and pastures over vast areas, largely during this century. For instance, Walker et al. (1993) estimate that 12 to 20 billion trees were removed, without replacement, from the Murray-Darling Basin alone. It should be noted that some of these impacts were recognised early on (e.g. Bleazby, 1917; Teakle and Burvill, 1938) but with little influence on land policy.

The pattern of water use of these agricultural systems varies dramatically from that of the original vegetation. The cool-season annual crops and pastures are active during the wet winter and early spring period, and senesce thereafter. Some perennial pasture systems such as lucerne (alfalfa) persist longer into the summer. In the case of pasture systems, the leaf area index is reduced by grazing. Nevertheless, during the growing season, leaf area index of these systems can exceed that of the native vegetation (Figure 1). Physiological activity during this season is not accompanied by particularly high rates of evaporation, for atmospheric demand is quite low, but again can exceed that of the native systems (Figure 2). McFarlane et al. (1993) asserted that annual crops evaporate more water than annual pastures, at least in southwestern Australia; this was demonstrated by Nulsen and Baxter (1982) and Farrington et al. (1992). In either case, however, annual evaporation falls significantly short of that under native vegetation (Greenwood and Beresford, 1982; Greenwood et al., 1985; Farrington et al., 1992). Nulsen and Baxter (1986) and Carbon et al. (1982) demonstrated large increases in annual evaporation from lucerne over wheat and annual pasture, respectively.

Other aspects of the hydrological cycle can change in addition to the seasonality and amount of evaporation. Tillage and grazing can have a direct impact on the hydraulic properties of soils, while soil erosion has obvious and

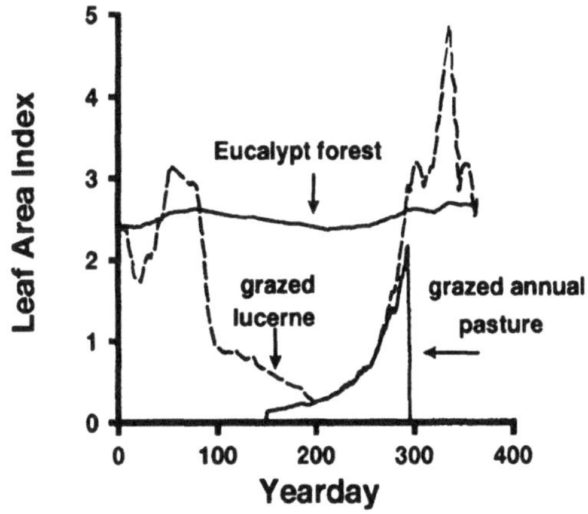

Figure 1. Leaf Area Index of eucalypt forest, annual pasture and lucerne, simulated for a site near Wagga Wagga, New South Wales, Australia. *Source:* Hatton and Francis (1993).

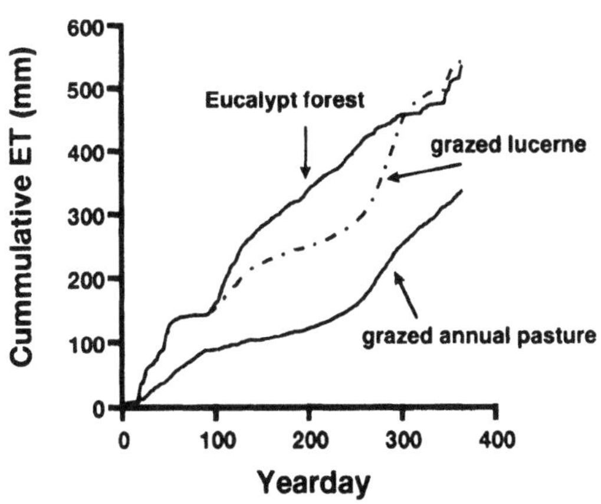

Figure 2. Cumulative evaporation from eucalypt forest, grazed lucerne and grazed annual pasture simulated for a site near Wagga Wagga, New South Wales, Australia. Note that pasture rates exceed woodland rates at times. *Source:* Hatton and Francis (1993).

dramatic impacts on fertility, structure and water-holding capacity. Land degradation issues such as soil acidification, soil structural decline and waterlogging have strong impacts on the hydrological cycle, and impact more regional scale phenomena such as salinisation and streamflow (Williams and Bettenay, 1979; Nulsen et al., 1986; Flavell et al., 1987; Nulsen, 1993).

A major consequence of these changes is a dramatic increase in groundwater recharge, particularly under annual crops and pastures; two orders of magnitude increases are typical (Peck and Hurle, 1973; George, 1992; Kennett-Smith et al., 1992a,b). Lucerne tends to have much less recharge (Figure 3); Kennett-Smith et al. (1990) showed that the rates of recharge under lucerne grown in South Australia approached that of the original mallee vegetation (1 mm y^{-1}). In duplex soils in more mediterranean environments, recharge control with lucerne may be less effective as sustaining summer rains become less frequent.

In summary, the hydrological characteristics of conventional agricultural systems in southern Australia can be summarised as follows:

- A dominance of physiological activity and evaporation during the wet (winter-spring) season;
- Soil moisture surplus during the growing season;
- Enhanced groundwater recharge and rising groundwater tables;
- The development of perched and/or active near-surface aquifer systems, and the activation of deeper aquifers and the mobilisation of stored salts;
- The increased discharge of water and salt to streams.

The social and environmental consequences of these changes in hydrology are immense. For instance, Ferdowsian et al. (1996) suggested that by the middle of the next century we will lose one-third of the Western Australian wheatbelt to salinisation; this region currently produces about one-half of

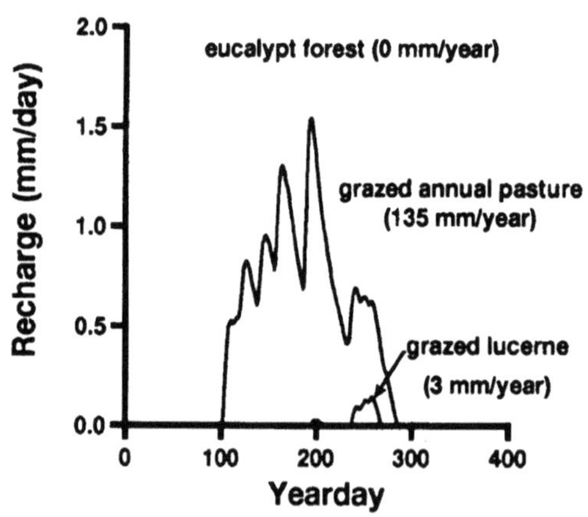

Figure 3. Recharge under eucalypt woodland, lucerne and annual pasture simulated for a site near Wagga Wagga, New South Wales, Australia. *Source:* Hatton and Francis (1993).

Australia's grain. Australia's largest river system, the Murray-Darling, is salinising at an accelerating rate (Williamson et al., 1997); due to other hydrological changes resulting from land clearing and settlement it is also eutrophying and subject to increased toxic algal blooms (Davis, 1997). This river not only supports one of southern Australia's most diverse and important set of ecosystems, but also supplies most of the domestic and industrial water for South Australia.

4. Achieving functional mimicry

The achievement of functional mimicry in southern Australian hydrology must be examined at two scales:

a) the local scale, where one considers the details of plant-soil-atmosphere interactions; and
b) the regional scale.

At the local scale, the re-introduction of perennials with summer activity is clearly essential in reducing runoff and groundwater recharge to pre-clearing levels. There is ample evidence that putting trees or shrubs (native and non-native) back into the landscape can achieve this result (Nulsen et al., 1986; Bell et al., 1990; Greenwood et al., 1992). It is less clear as to whether landscape position matters over the actual proportion of the catchment planted for recharge control (Bell et al., 1990; McFarlane et al., 1993; Zhang et al., 1997). Arguments based on the results of Specht (1972), Pierce et al. (1993) and Hatton and Wu (1995) suggest that however trees are put back into the landscape, effective control of recharge will be achieved only at a leaf area index approaching that of the natural state, involving revegetation of most or all parts of the catchment. Questions remain, however, regarding the potential for management to break the apparent long-term constraints on the leaf area index – water balance relationship at the landscape scale. For instance, it appears that the management of the dehesa systems of the Iberian Peninsula at times involve the lowering of tree leaf area index from natural woodland levels to a savannah state which may provide more stable agricultural production with respect to long-term fluctuations in climate (Joffre, this volume).

A major challenge is to reconstruct ecosystems in such a way as to provide farm income. One option is alley farming with an edible shrub, such as tagasaste (*Chamaecytisus palmensis*); cropping or pasture can be maintained between the rows with sufficiently narrow alleys to minimise recharge. Alley farming is under investigation across Australia at present (Lefroy and Stirzaker, this volume), but overseas agroforestry research in water-limited environments indicate significant negative trade-offs between tree performance and crop yield (Ong, 1996; Ong and Leakey, this volume).

Given that presently there appears to be limited (economic) agroforestry potential for much of southern Australia, non-tree alternatives to restoring the original hydrological balance are attractive if they can work. Chief among these are perennial pastures, especially lucerne. The key to its success in this regard is the ability to achieve the rooting depths and activity achieved by the original vegetation at a given site. Many of Australia's subsoils have properties hostile to lucerne root growth (e.g. sodicity, acidity, density), but where sites are favourable lucerne presents a viable option, at least as far as hydrological mimicry is concerned. Economic realities of the animal based industries will dictate the adoption of this option. Additional functional mimicry in terms of biodiversity is, of course, not achieved with lucerne. It is also becoming clear that enhancement of winter pastures (e.g. through fertilisation or reduced grazing) can only have modest impacts toward restoring the water balance (Figure 4).

If we take a catchment-scale perspective on functional mimicry, the focus can be shifted toward the minimisation of saline discharge. While recharge control remains an option in this regard, there is the additional option of intercepting shallow (fresh) groundwater before it interacts with deeper, saline aquifers. At face value, this is a more economically and socially attractive idea, for potentially less land may have to be aforested to achieve the same result. The idea is to place trees or similarly deep-rooted vegetation in landscape positions at which they can access not only local rainfall but shallow, fresh groundwater from upslope as well. Thus, recharge can be intercepted before it reaches saline discharge areas. Where aquifers are sufficiently fresh and transmissive, this option is feasible. For instance, George (1990) demonstrated the potential for drying up sandplain seeps supplied by a perched hillslope aquifer by planting trees in an interception belt. Schofield et al. (1989) reported a similar result. The long-term effectiveness of such treatments is less clear.

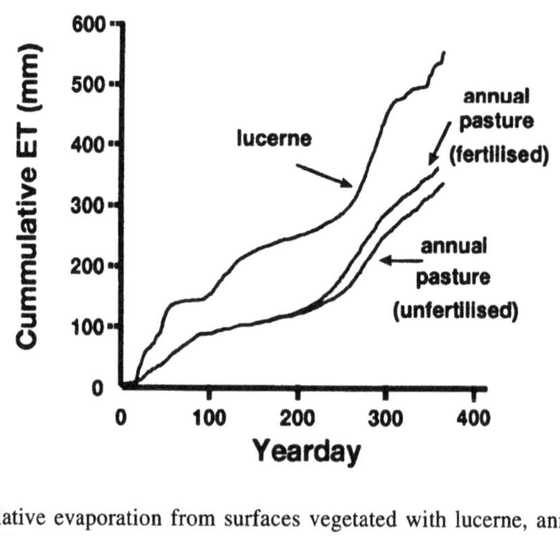

Figure 4. Cumulative evaporation from surfaces vegetated with lucerne, annual pasture with 0.25 optimum fertility, and annual pasture with optimum fertility, simulated for a site near Wagga Wagga, New South Wales, Australia. *Source*: Hatton and Francis (1993).

The key feature of the groundwater interception strategy is the placement of trees in landscape positions in which:

- The groundwater is reasonably fresh;
- The groundwater is reasonably close to the surface;
- The aquifer has reasonable transmissivity and gradient; and
- The saturated thickness of the unconfined aquifer is limited (< 10 m).

This generally precludes the planting of trees on saline discharge areas *per se* and restricts application to a limited fraction of affected landscapes in southern Australia (generally sand country). The lack of sustainability of trees at saline sites without sufficient periodic leaching of salts which otherwise accumulate in the root zone is well documented and modelled (Thorburn et al., 1995).

Finally, achieving functional mimicry with respect to minimising waterlogging and land salinisation at the catchment scale can involve engineering. The installation of shallow interceptor drains has been promoted to accelerate the removal of surface waters and to some extent drain shallow perched aquifers (McFarlane and Cox, 1990; McFarlane et al., 1990). When successful, these drains reduce waterlogging and thus lead indirectly to recharge control. As a most extreme engineering measure, groundwater pumping and deep drainage (~ 2 m) to reduce land salinisation has been widely practised at scales from managing local discharges (George and Frantom, 1990) to regional systems (Otto and Salama, 1994). An important limitation to engineering groundwater discharge is the disposal of drainage waters, particularly when they are saline. We also recognise that such drainage works do not restore the original streamflow or stream water quality.

5. Discussion

Water cycling in the natural southern Australian landscape is dominated by the *in situ* evaporation of local rainfall. This results from deeply weathered regolith, low hydraulic gradients, high potential evaporation relative to rainfall, and a conservative, deeply-rooted flora which remains physiologically active during summer. Current agricultural systems generally have vegetation which is not active in summer and is shallow-rooted. The hydrological consequences of the resulting soil degradation and excess water are immense.

The structural and functional characteristics of the original system are well-enough understood to specify the essential hydrologic attributes of an agricultural mimic, at least at the local scale. While there is hope that alternative production systems possessing these functional attributes can be proven and implemented, it is difficult to envisage sustainable systems not involving trees or other perennial vegetation. While profitable agroforestry options exist in some environments, the scale of the tree planting required to significantly restore the original water balance across southern Australia is daunting. Few hydrologically effective and economically attractive options exist for the vast proportion of southern Australia's agricultural region.

Equally daunting is the response time of many hydrological systems to such changes. While remediation of local saline seeps using trees can show results in a few years, the control of salt loads to Australia's major southern river systems may take hundreds of years to achieve following revegetation. This is due to the low gradients and long length scales of these regional systems.

It is unlikely that the full complement of hydrological functions can ever be restored with revegetation, even using the original genetic material. Some changes in the hydraulic and hydrochemical characteristics of the system may be irreversible. The most pessimistic assessment suggests that Australia's southern landscape will not be renewed until the next geologic orogeny or a large change in climate. Nevertheless, there is an ethical compulsion to bring to our agriculture as much of the original hydrologic function as possible.

6. References

Adams MA (1995) Distribution of eucalypts in Australian landscapes: landforms, soils, fire and nutrition. In: Attiwill PM and Adams MA (eds) Nutrition of the Eucalypts, pp 61–76. CSIRO, Melbourne

Allison GB, Cook P, Barnett SB, Walker GE, Jolly ID and Hughes MW (1990) Land clearing and river salinisation in the western Murray Basin. Journal of Hydrology 119: 1–20

Anon. (1996) Salinity: A situation statement for Western Australia. Report to the Minister for Primary Industries and Minister for the Environment, Western Australia, Perth, 37 pp

Bell RW, Schofield NJ, Loh IC and Bari MA (1990) Groundwater response to reforestation in the Darling Range of Western Australia. Journal of Hydrology 115: 297–317

Bleazby R (1917) Railway water supplies in Western Australia – difficulties caused by salt in soil. Institute of Civil Engineers London, Proceedings 203: 394–400

Carbon BA, Bartle GA, Murray AM and MacPherson DK (1980) The distribution of root length, and limits to flow of soil water to roots in a dry sclerophyll forest. Forest Science 26: 656–664

Carbon BA, Bartle GA, Farrington P and Beresford JD (1982) Deep drainage and water use of forests and pastures grown on deep sands in a Mediterranean environment. Journal of Hydrology 55: 53–64

Davis JR (ed) (1997) Managing Algal Blooms. CSIRO Land and Water Report, Canberra, 113 pp

Dell B, Bartle JR and Tacey, WH (1983) Root occupation and root channels of jarrah forest subsoils. Australian Journal of Botany 31: 615–627

Eagleson PS (1982) Ecological optimality in water-limited natural soil-vegetation systems. 1. Theory and hypothesis. Water Resources Research 18: 325–340

Farrington P, Salama RB, Watson GD and Bartle GA (1992) Water use of agricultural and native plants in a Western Australian wheatbelt catchment. Agricultural Water Management 22: 357–367

Farrington P and Salama RB (1996) Controlling dryland salinity by planting trees in the best hydrogeological setting. Land Degradation and Development 7: 183–204

Ferdowsian R, George R, Lewis F, McFarlane D, Short R and Speed R (1996) The extent of dryland salinity in Western Australia. Proceedings 4th National Conference and Workshop on the Productive Use and Rehabilitation of Saline Lands, pp 89–97. Promaco Conventions

Flannery TF (1994) The Future Eaters. Reed Books, Chatswood, 423 pp

Flavell DJ, Martin DK and Belstead BS (1987) Flood estimation procedures for Western Australia. Western Australian Main Roads Department Technical report 50T, Perth

George RJ (1990) Reclaiming sandplain seeps by intercepting perched ground water with Eucalypts. Land Degradation and Rehabilitation 2: 13–25

George RJ (1992) Estimating and modifying the effects of agricultural development on the groundwater balance of large wheatbelt catchments, Western Australia. Journal of Applied Hydrogeology 1: 41–54

George RJ and Frantom PWC (1990) Using pumps and siphons to control salinity at a saline seep in the Wallatin Creek Catchment. Western Australian Department of Agriculture Division of Resource Management Technical Report 91

George RJ, Cochrane DL and Bennett DL (1994) Groundwater systems responsible for dryland salinity in the Lake Towerrining catchment, Western Australia. Proceedings, Water Down Under '94, Congress of the International Association of Hydrogeologists and International Hydrology and Water resources Symposium, Adelaide, 2: 355–360

Greenwood EAN and Beresford JD (1982) Evaporation from vegetation in landscapes developing secondary salinity using the ventilated chamber technique. 4. Evaporation from a regenerating forest of *Eucalyptus wandoo* on land formerly cleared for agriculture. Journal of Hydrology 58: 357–366

Greenwood EAN, Klein L, Beresford JD and Watson GD (1985) Differences in annual evaporation between grazed pasture and *Eucalyptus* species in plantations on a saline farm catchment. Journal of hydrology 78: 261–278

Greenwood EAN, Milligan A, Biddiscombe EF, Rogers AL, Beresford JD, Watson GD and Wright KD (1992) Hydrologic and salinity changes associated with tree plantations in a saline agricultural catchment in southwestern Australia. Agricultural Water Management 22: 307–323

Hatton TJ and Francis J (1993) Dryland salinity. 6. The role of vegetation management. Salt Action and the National Soil Conservation Program, Department of Conservation and Land Management, Sydney, 12 pp

Hatton TJ and Wu H (1995) Scaling theory to extrapolate individual tree water use to stand water use. Hydrological Processes 9: 527–540

Hatton TJ, Salvucci GD and Wu H (1998) Eagleson's optimality theory of an ecohydrological equilibrium: Quo vadis? Functional Ecology (in press)

Jacobs MR (1955) Growth habits of the eucalypts. Commonwealth Government Printer, Canberra

Kennett-Smith A, Budd GR, Cook PG and Walker GR (1990) The effect of lucerne on the recharge of cleared mallee lands. Centre for Groundwater Studies Report No. 27

Kennett-Smith A, O'Leary GJ and Thorne R (1992a) Estimation of recharge from soil water profiles under dryland agriculture, eucalypt plantation, and natural woodland in the Wimmera District of Victoria. CSIRO Division of Water Resources Report 92/2, 14 pp

Kennett-Smith A, Budd GR and Walker GR (1992b) Groundwater recharge beneath woodlands cleared for grazing, south western New South Wales. CSIRO Division of Water Resources Report 92/1, 17 pp

Kimber PC (1974) The root system of jarrah (Eucalyptus marginata). Research paper No. 10, Forests Department of Western Australia, Perth, 5 pp

McArthur WM (1993) History of landscape development. In: Hobbs RJ and Saunders DA (eds) Reintegrating Fragmented Landscapes, pp 10–22. Springer-Verlag, New York

McFarlane DJ and Cox JW (1990) Seepage interceptor drains for reducing waterlogging and salinity. Journal of Agriculture – Western Australia 31: 66–69

McFarlane DJ, Negus TR and Ryder AT (1990) Shallow drains for reducing waterlogging and salinity on clay flats. Journal of Agriculture – Western Australia 31: 70–73

McFarlane DJ, George RJ and Farrington P (1993) Changes in the hydrologic cycle. In: Hobbs RJ and Saunders DA (eds) Reintegrating Fragmented Landscapes, pp 146–186. Springer-Verlag, New York

McMahon TA, Finlayson BL, Haines AT and Srikanthan R (1992) Global Runoff – Continental Comparisons of Annual Flows and Peak Discharges. Catena Verlag, Cremlingen-Destedt, West Germany, 166 pp

Nulsen RA and Baxter IN (1986) Water use by some crops and pastures in the southern agricultural areas of Western Australia. Division of Resource management Western Australia Department of Agriculture Technical Report No. 32

Nulsen RA, Bligh KJ, Baxter IN, Solin EJ and Imrie DH (1986) The fate of rainfall in a mallee and heath vegetated catchment in southern Western Australia. Agricultural Water Management 4: 173–186

Nulsen RA (1993) Changes in soil properties. In: Hobbs RJ and Saunders DA (eds) Reintegrating Fragmented Landscapes, pp 1107–1145. Springer-Verlag, New York

Ong CK (1996) A framework for quantifying various effects of trees crop interactions. In: Ong CK and Huxley P (eds) Tree Crop Interactions – A Physiological Approach. CAB International, Oxford, 386 pp

Otto C and Salama RB (1994) Linked enhanced discharge – Evaporative disposal systems. In: Reeve R and Watts J (eds) Groundwater – Drought, Pollution and Management, pp 35–44. AA Balkema, Rotterdam

Peck AJ and Hurle DH (1973) Chloride balance of some farmed and forested catchments in southwestern Australia. Water Resources Research 9: 648–657

Pierce LL, Walker J, Dowling TI, McVicar T, Hatton TJ, Running SW and Coughlan JC (1993) Hydroecological changes in the Murray-Darling Basin: Part 3 – A simulation of regional hydrological changes. Journal of Applied Ecology 30: 283–294

Pook EW (1985) Canopy dynamics of *Eucalyptus maculata* Hook. III. Effects of drought. Australian Journal of Botany 33: 65–79

Salama RA, Farrington P, Bartle GA and Watson GD (1993) Salinity trends in the wheatbelt of Western Australia: results of water and salt balance studies from Cuballing catchment. Journal of Hydrology 145: 41–63

Schofield NJ (1990) Water interactions with land use and climate in South Western Australia. Report No. WS 60, Water Authority of Western Australia, Perth, 121 pp

Schofield NJ, Loh IC, Scott PR, Bartle JR, Ritson P, Bell RW, Borg H, Anson B and Moore R (1989) Vegetation strategies to reduce stream salinity of water resource catchments in southwest Western Australia. Water Authority of Western Australia Report WS 33

Specht RL (1972) Water use by perennial evergreen plant communities in Australia and Papua New Guinea. Australian Journal of Botany 20: 273–299

Teakle LJH and Burvill GH (1938) The movement of soluble salts in soils under light rainfall conditions, Journal of Agriculture Western Australia (2nd series) 15: 218–245

Thorburn PJ, Hatton TJ and Walker GR (1993) Combining measurements of tree water use and stable isotopes of water to determine groundwater discharge from forests. Journal of Hydrology 150: 563–587

Thorburn PJ, Walker GR and Jolly ID (1995) Uptake of saline groundwater by plants: an analytical model for semi-arid and arid areas. Plant and Soil 175: 1–11

Walker G, Jolly ID and Jarwal SD (eds) (1996) Salt and water movement on the Chowilla flood-plain. CSIRO Division of Water Resources Series No. 15, Canberra

Walker J, Bullen F and Williams BG (1993) Ecohydrological changes in the Murray-Darling Basin. I. The number of trees cleared over two centuries. Journal of Applied Ecology 30: 265–273

Williamson DR and Bettenay E (1979) Agricultural land use and its effect on catchment output of salt and water – Evidence from southern Australia. Progress in Water Technology 11: 463–480

Williamson DR, Gates GWB, Robinson G, Linke GK, Seker MP, amd Evans WR (1997) Salt trends. Historic trend in salt concentration and saltload of stream flow in the Murray-Darling Drainage Division. Murray-Darling Basin Commission Dryland Technical report No. 1, Canberra, 64 pp

Zhang L, Dawes WR, Hatton HJ, Reece PH, Beale GTH and Packer I (1997) Estimation of soil moisture and groundwater recharge using the TOPOG_IRM model. Water Resources Research (in press)

Agroforestry Systems **45**: 215–244, 1999.

Nutrient cycling and growth in forest ecosystems of south western Australia

Relevance to agricultural landscapes

P. F. GRIERSON* and M. A. ADAMS
*Botany Department, The University of Western Australia, Nedlands, WA 6907, Australia (*Author for correspondence: E-mail: pfgblue@cyllene.uwa.edu.au)*

Key words: biodiversity, disturbance, nitrogen, phosphorus, resilience, resistance

Abstract. Nutrient and hydrologic cycles in harvested native forests in southern Australia are largely balanced. For example, we have little or no evidence of any decline in nutrient capital or availability in harvested forests. Short-term and small-scale reductions in evapotranspiration due to loss of leaf area after harvesting are adequately balanced at the landscape scale by large areas of regenerating or older-age forest. In contrast, agricultural systems on similar soils are a) dependent on large inputs of fertilisers to maintain growth and b) frequently subject to increasing salinity and waterlogging or other forms of degradation. The large-scale replacement of long-lived communities of perennial and often deep-rooting native species with annual crops or other communities of shallow-rooting species might be better managed within the framework of knowledge developed from studies of native plant communities. However, application of such a mimic concept to systems of low natural productivity is limited when agricultural systems require continued high productivity. Nonetheless, the mimic concept may help in developing sustainable management of agriculture on marginal lands, and contribute to the nutritional resilience of agroecosystems. Relevant characteristics for mimic agroecosystems in south western Australia include: high species diversity, diversity of rooting attributes, utilisation of different forms of nutrients (especially of N and P) in space and time, and the promotion of practices which increase soil organic matter content.

> 'Man cannot influence nature, or convert its forces to his own use, unless he understands the natural laws according to their relations of measure and number.'
> Alexander von Humboldt (1769–1859)

1. Introduction

The replacement of perennial vegetation with annual crops and pastures throughout much of southern Australia has significantly altered the hydrologic cycle. Increased fluxes of water and nutrients and erosion have resulted in problems of salinisation, loss of fertility and eutrophication of drainage systems (Hamblin and Williams, 1995; Hatton and Nulsen, Lefroy and Stirzaker, Hobbs and O'Connor, this volume). Remnant natural systems within agricultural mosaics are also under constant pressure from surrounding areas as a consequence of these changes (Yates and Hobbs, 1997).

In south western Australia, the challenge is to develop agricultural ecosystems that are sustainable where climate and water supply are variable and soil fertility is naturally low (Leeper, 1964). In almost all definitions of sustainable ecosystems, based on time scales of hundreds to thousands of years, maintenance of nutrient cycling (along with carbon cycling, the hydrologic cycle and soil formation) ranks amongst the most important of ecosystem processes (Solbrig, 1991; Hobbs, 1992; Woodward, 1993). In sustainable ecosystems, nutrient inputs should largely balance losses due to harvest (of either agricultural or forest products) to the atmosphere and to the deeper soil profile, drainage systems and groundwater. Furthermore, the current state of knowledge suggests that processes which regulate, limit or replace those non-harvest losses (e.g. volatilisation, denitrification, mineralisation-immobilisation, N-fixation), particularly in the surface centimetres of the soil, are highly susceptible to anthropogenic influence and can be 'overloaded' (Vitousek et al., 1997).

For forests, the biological processes of nutrient cycling (including plant uptake, internal cycling of nutrients, litterfall, litter decomposition and microbial transformations of C, N and P) help to retain added nutrients and minimise losses due to leaching and erosion. In agricultural systems, intensive management practices (including application of fertilisers and herbicides, cultivation, seeding and harvest) cause almost constant disruption (or disturbance) of these processes, with significant consequences for the capacity of the system to retain nutrients and water. However, even in forests, these processes can be seriously disrupted. Forests in central Europe and parts of North America are examples of the deleterious consequences of an imbalance between inputs and outputs and overloading of nutrient retention processes (Aber et al., 1989; Aber, 1992; Schulze and Mooney, 1993). Excessive inputs of N from the polluted atmosphere in these areas are largely responsible for tree decline and for increases in nutrient outputs to streams and lakes (Ägren and Boscatta, 1988; Dise and Wright, 1995), often with disastrous consequences (Vitousek et al., 1997).

One way to develop sustainable agroecosystems may be to imitate or 'mimic' the structure and functioning of natural plant communities, an approach which has been adopted with varying success in the tropics (Ewel, 1986; Ewel and Bigelow, 1996; Ewel, this volume) and in cool temperate grasslands (Jackson and Bender, 1984; Jackson, 1985). The mimic approach seeks to maximise the use of available resources (nutrients, water, light), and confer resistance to invasion by unwanted plant species or pests. In order to do so, emphasis is placed on recreating both the *effects* on ecosystem processes a group of species may have, as well as the *responses* of those species to environmental factors, i.e. mimicking *functional groups* (Catovsky, 1998). Mimic agroecosystems may prove to be sustainable when primary production (and its distribution above- and belowground) and the efficiency of resource capture are similar to the natural vegetation (Lefroy et al., 1993). At present, few agricultural systems in southern Australia even approach the functional mimicry of natural systems which Ewel (1986) demonstrated was possible in the tropics. Further, without restoration of some, if not all, of the attributes which affect resource capture and transfer in natural ecosystems, there is serious doubt that other elements of the landscape (e.g. biodiversity) can be conserved or restored.

We discuss here some of the main attributes of growth and nutrient cycling of forest and woodland ecosystems of southern Australia, and how an understanding of these may contribute to improved management of the agricultural ecosystems which have replaced them in the landscape. Particular emphasis is given to mediterranean-climate regions of south western Australia as these are areas of great concern to land managers and epitomise many of the problems facing agriculture globally (see Saunders et al., 1987 and chapters therein for detailed discussion). The south western region has considerable endemism, containing almost one-third of all plant species in Australia, with about 75% of species occurring nowhere else (Hopper, 1992). Nonetheless, only about 10% of the original valley woodlands remain in the central wheatbelt of Western Australia (Beard and Sprenger, 1984; Yates et al., 1994). This over-clearing has resulted in a substantial loss of biodiversity, and in major hydrological changes which have increased salinity and waterlogging (Ferdowsian et al., 1996; Hatton, this volume).

2. Resilience and resistance – a basis for sustainability

Elsewhere in this volume readers will find discussion of the wide variety of elements which may form the basis of functional mimicry, such as lifeform, biodiversity, above- and belowground productivity, and hydrologic fluxes, to add to those which we have chosen here. All these elements influence each other to some extent and can rarely be considered independently – for example, the cycling of nutrients is closely-linked to the hydrologic cycle – and can be considered within the general categories of resource capture and resource transfer (including inputs to balance losses due to harvest). A further aspect of functional mimicry is the capacity of an ecosystem to tolerate disturbance and recover at a relatively rapid rate. In contrast to disturbances of agricultural ecosystems, natural ecosystems are more likely to be burnt or disturbed by other sudden, catastrophic phenomena. Studies of the response of forests and woodlands to such disturbances have added greatly to our understanding of their functional ecology. Equally, those studies have yielded a set of hypotheses about the functional attributes of natural ecosystems (measured as resistance and/or resilience) which are testable. Included in these attributes are measures of the rate of recovery following disturbance and the degree to which ecosystem processes may be displaced by disturbance (Main, this volume).

There are many inconsistencies in the use of terminology in the literature. Vitousek et al. (1982) defined nutritional *resistance* as the extent to which nutrient cycles of an ecosystem are displaced by disturbance, and *resilience* to be the rate of recovery of the ecosystem. These definitions have been used in some subsequent analyses of Australian forests (Polglase et al., 1986; Attiwill, 1991). An alternative is the use of the concept of 'nutritional resilience' which was considered by O'Connell and Grove (1991) to be the degree of stability following disturbance which results in changes to the nutrient status of the ecosystem (where nutritional status is characterised by the size of, and fluxes between, soil and plant components and the balance between inputs and outputs). In this discussion, we shall adhere to the simpler definition of nutrient resilience or resistance provided by Vitousek and others.

In the more productive (wetter) forests in southern Australia, rapid establishment and regeneration, nutrient uptake by the regenerating vegetation, nutrient immobilisation by microorganisms, and the potential for significant biological fixation are all important processes contributing to nutrient resilience after fire (Polglase et al., 1986; Attiwill, 1991; O'Connell and Grove, 1991). Nutrient fluxes (inputs and outputs) during and after fire are comparatively large and take place via external processes of plant death, organic matter decomposition and mineralisation in the soil. Quick establishment of regeneration and rapid growth following disturbance prevents loss of mineralised nutrients, including those from killed trees. In the lower productivity (and generally drier) forests growing on poor soils (e.g. jarrah forests of the south western Australia, and the box-ironbark forests and woodlands of south eastern Australia), nutrient fluxes are smaller and these forests tend to show more attributes of resistance. For example, trees are generally not killed by fire and nutrients are redistributed internally to new epicormic growth. After disturbance, forest ecosystems with attributes of resistance exhibit smaller changes in nutrient cycling (including rates of mineralisation and litter decomposition) than resilient ecosystems, and often have a high proportion of resprouters but a low occurrence of N-fixers in a diverse understorey. Either pattern, resilience or resistance, can provide protection of key ecosystem attributes with respect to growth and nutrient cycling.

There are no records of a loss of productivity for any native eucalypt forests which have been managed for timber production (Stewart et al., 1985). However, of the total forest area which is logged using clear-felling and slash-burning, the largest proportion has been logged only once, with much smaller proportions logged twice or more. Hence, any potential losses of productivity which might be due to harvesting-associated losses of nutrients (Judd, 1996) are unlikely to be realised for tens or maybe hundreds of years. Nonetheless, the present situation in forests contrasts with the obvious and large-scale loss of productive land in agricultural landscapes owing to a variety of changes in attributes of physical, chemical and biological activity (White, 1993). A sustainable agroecosystem is one which has attributes of nutrient resilience or resistance as exemplified by native forests, and where inputs and outputs are in balance, i.e. through integrated nutrient management (Smaling and Fresco, 1993; Smaling and Braun, 1996). Given the requirements for high productivity from agricultural land, the model of resilience is more applicable (or acceptable) to the design of mimic agroecosystems. To some extent, this idea has recently been applied to improving the management and function of Australia's rangelands (Ludwig et al., 1997).

3. Growth and productivity of forest and agro-ecosystems

In order to assess whether agriculture could mimic natural ecosystems in regions of relatively low rainfall, we first compare growth and productivity of agriculture with that of natural ecosystems.

Rates and patterns of growth in forests and rainfed agriculture of south western Australia are reasonably well-known (Table 1). Net primary productivity (NPP) in native forests seldom exceeds 10–15 t ha^{-1} yr^{-1} in south western Australia, even in the best karri (*Eucalyptus diversicolor*) forests (Rayner, 1991; O'Connell and Grove, 1996). In jarrah (*Eucalyptus marginata*) forests, rates of growth are often an order of magnitude less (Hingston et al., 1981; Abbott and Loneragan, 1983) and are comparable with other eucalypt forests in eastern Australia growing in regions of similarly low rainfall and soils of low fertility (Grierson et al., 1992). In contrast, agricultural production may be as great as 8–10 t ha^{-1} yr^{-1} (but generally less) in areas where productivity of the undisturbed native vegetation would be less than 1–2 t ha^{-1} yr^{-1} and certainly in areas where rainfall is less than that in either jarrah or karri forests (Table 1).

It is likely that the NPP of agroecosystems will only be sustainable at a lower rate than present if our model for functional mimicry is based on the native vegetation. Such a reduction in productivity is almost certain to be politically and socially unacceptable and many scientists will argue that we can not justify granting rights to nature but should instead find ways to maintain and even increase rates of production. There are many difficulties combining productivity in the short-term with sustainability in the long-term. Growth rates in tall-open forests in south-east Australia may exceed 25 t ha^{-1} yr^{-1} (Attiwill, 1991) but attempts to establish agriculture by clearing these forests often failed, owing largely to low temperatures and low availability of nutrients such as phosphorus (Skinner and Attiwill, 1981). Much of the land cleared has now been returned to forest or tree plantations and rates of growth are again comparable with those in native forests (Bennett et al., 1997). Presumably, if farmers had found a crop which coped with the low temperatures and low nutrients, then productivity may have been similar. However, there is no suggestion it could have been better than that of the native forest.

In comparison to introduced crop species, the native vegetation of much of southern Australia displays various adaptations to drought and poor soils. All of these adaptations have the common cost of reducing productivity – resources which might be used to grow leaves with high photosynthetic

Table 1. Aboveground biomass accumulation for some forest species and wheat in southern Australia.

Vegetation type	Age (yrs)	Biomass		MABI ($t\ ha^{-1}\ yr^{-1}$)	Reference
		Stemwood ($t\ ha^{-1}$)	Total ($t\ ha^{-1}$)		
Western Australia					
C. calophylla	4	13.2	22.7	5.7	Ward and Pickersgill (1985)
C. calophylla	60	37.9	54.9	0.9	Hingston et al. (1981)
E. diversicolor	4–11	23.5–33.3	31.2–43.5	4.4–7.8	Grove and Malajczuk (1985)
E. diversicolor	36	232.2	455.5	12.7	Grove and Malajczuk (1985)
E. diversicolor	40	183.8	225	5.6	Hingston et al. (1979)
E. diversicolor/C. calophylla	Mixed	207.9	284.1		Hingston et al. (1979)
E. maculata	4	15.1	26.1	6.52	Ward and Pickersgill (1985)
E. marginata	60	148.3	206.1	3.43	Hingston et al. (1981)
Wheat	Annual	5–7	5–7		Tennant et al. (1991)
Eastern Australia					
Mallee	0–60			0.5–0.7	Grierson et al. (1992)
Box-ironbark	0–60			1.8–< 0.1	Grierson et al. (1992)
Wheat	Annual		8.2–13.7	8.2–13.7	Tennant et al. (1991)

capacities or to develop a large aboveground biomass are instead used to maintain metabolism during periods of limited availability of water, including a large root:shoot ratio (Pate and Beard, 1984; Pate and Bell, this volume). Native plants of south western Australia are nearly all 'drought tolerators' to some extent and the co-evolution of the flora with the climate and soils is likely to have produced plant communities in which growth is optimised for available resources (Chapin, 1980; Cowling et al., 1996).

Several possible advantages of an inherently low growth rate in an infertile habitat were described by Chapin (1980). Such species are more likely to be functioning close to optimal growth and metabolic rates and thus more fit in a low-nutrient environment (i.e. more resistant) than a rapidly-growing species that experiences a large reduction in yield under the same infertile conditions. For example, in south western Australia, wheat will reach a maximum yield of less than 20% of its potential without addition of phosphate fertilisers (Bolland et al., 1991). A traditional crop/pasture rotation in south western Australia maintains a net nutrient level of less than 60% that of the natural ecosystem (Russell, 1980). In contrast, slow-growing species may absorb nutrients in excess of immediate requirements, and use these reserves to support growth after soil reserves become limited through low water availability, nutrient depletion or other reasons (Specht and Groves, 1966; Jeschke and Pate, 1995). Bieleski (1973) argued that luxury consumption is most developed in slow-growing species because rates of absorption differ less among species than rates of growth.

On a global scale, soils with low concentrations of labile N and P are dominated by slow-growing perennial rather than annual species (Grime and Hunt, 1975). Seasonal storage of minerals and carbohydrates in non-synthetic tissues is an essential feature of these perennials in which the same nutrient capital can serve several functions throughout the growing season (e.g. transfer of nutrients from stems or leaves to reproductive parts). The disjunct phenology of above- and belowground components, in particular, becomes more pronounced with increasing water and nutrient stress (Specht et al., 1983) and is an important attribute when considering life-form mimics for inclusion in agroecosystems.

The growth of perennial tree crops, such as plantations of Tasmanian Bluegum (*Eucalyptus globulus*), makes an interesting comparison with other tree species of natural ecosystems in south western Australia. Bluegum plantations in Western Australia were expected to grow at rates as great as 40 m^3 ha^{-1} yr^{-1}, or about 25 t ha^{-1} yr^{-1} (e.g. Shea and Bartle, 1988). Mostly, plantations have been established on land previously cleared for agriculture, and often on land which has a long history of fertiliser application and/or rotation with legumes and consequently high levels of residual nutrients in the soil. Unfortunately, the expected rates of growth appear optimistic, considering available evidence from elsewhere in Australia was that growth of *E. globulus* on much better soils and under more favourable climate conditions, seldom approaches such a figure (Bennett et al., 1997; Table 2).

Measurements of growth in a variety of plantations across a range of annual rainfall of between 700 and 1000 mm and a range of soil types and land-use histories suggest that rates of growth will be, on average, less than half of the figures quoted above (Figure 1; Table 3). Even on sites where growth is enhanced by abundant water and nutrients (e.g. via the application of sewage effluent), rates of growth are unlikely to reach more than 15–20 t ha^{-1} yr^{-1} or 30 m^3 ha^{-1} yr^{-1}. Some plantations established on the best soils in the higher rainfall regions of south western Australia (e.g. around Manjimup/Pemberton and near Albany) may yield 40 m^3 ha^{-1} yr^{-1} in small areas, but the greater the area considered, the less likely that figure will be maintained owing to variations in soil properties across the landscape, especially those that influence the availability of water (Beard, 1990; Wardell-Johnson and Horwitz, 1996). The likely average rates of growth will be similar to rates of growth of the native forest species (karri), especially if rates are averaged over several rotations since it is expected that rates will decline due to nutrient exports and to a reduction in water storage in soil profiles.

Belowground, root growth and root sloughing are poorly described for Australian forest ecosystems. However, even in moderately productive forests, belowground cycling may account for up to half of the total nutrient turnover and nutrient requirements of the vegetation. On a global basis, the proportion of carbon and nutrient turnover due to root growth and decomposition increases as total productivity declines (Bazilevich and Rodin, 1968; Vogt et al., 1996). Recent estimates of fine roots in the northern jarrah forest of south western Australia growing on lateritic soils suggest a fine root biomass of around 15–20 t ha^{-1} in the top 20 cm of the soil (Grierson, unpubl. data) and a large proportion of this will be turned over annually. In contrast, aboveground biomass in the same ecosystems is about 150 tonnes ha^{-1} (Abbott and Loneragan, 1983) but annual turnover of aboveground litter will be less than

Table 2. Growth rates of Eucalyptus globulus in plantations of southern Australia.

Locality	Age (yrs)	MABI[a] (t ha^{-1} yr^{-1})	Reference
Victoria	4	3.2	Cromer et al. (1975)
Victoria	9.5	2	Cromer et al. (1975)
Tasmania	3	5	Judd et al. (unpubl. data)
Tasmania	5	14	Judd et al. (unpubl. data)
Tasmania	7	13	Judd et al. (unpubl. data)
Tasmania	7.5	11	Judd et al. (unpubl. data)
Tasmania	7.5	2	Judd et al. (unpubl. data)
Victoria	6	7.6 (max 8.3)	Bennett et al. (1997)
Victoria	6	10.6 (max 12)	Bennett et al. (1997)
Victoria	6	10.0 (max 11)	Bennett et al. (1997)
Victoria	6	13.7 (max 18.2)	Bennett et al. (1997)
Victoria	6	14.3 (max 17.5)	Bennett et al. (1997)

[a] Mean Annual Biomass Increment.

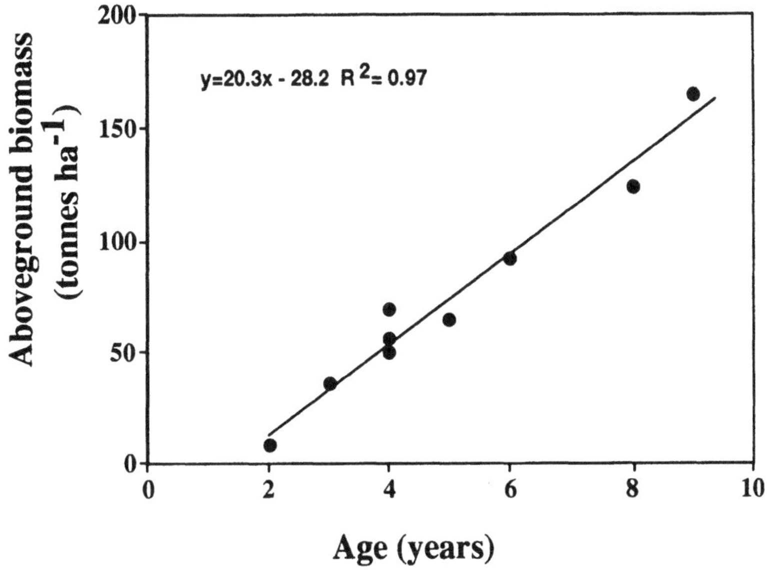

Figure 1. Relationship between total aboveground biomass and age for a range of *E. globulus* plantations in south-west Western Australia. Details of sites are given in Table 3.

Table 3. Age and climate of a range of *E. globulus* plantations in south-west Western Australia. P/E is the ratio of precipitation to pan evaporation.

Site	Age (yrs)	Longitude (°E)	Latitude (°S)	Rainfall (mm)	P/E
Oldham	2	116.37	34.08	752	0.56
Albany	3	117.80	34.95	1016	0.75
Wandalup	4	115.82	32.49	942	0.54
Meares	4	116.38	34.00	693	0.51
Wheatley	6	116.27	33.96	777	0.58
Lindsay	7	115.75	33.98	1008	0.77
Wren	8	116.13	34.27	1090	0.88

3 t ha^{-1} (Todd et al., unpublished data). These results support the contention that plants growing in, and adapted to, nutrient-poor conditions invest a higher proportion of growth into the development of root systems than species growing on better soils (Fitter, 1997). It is likely that in forests of south western Australia, more than half of the C and nutrient turnover and a similar proportion of the total nutrient requirements, are met from belowground turnover of fine roots.

The plasticity of response to both spatial and temporal variation in resources is an important attribute of native plant communities (Grime, 1997). When trees and other native species are used for land rehabilitation (e.g. after mining, in reforestation projects), the spatial distribution of plants and the physical properties of the land surface have marked effects on nutrient availability, and consequent growth and turnover of plant organic matter. For example, when mounding is used to prepare the soil surface, fine root biomass is likely to be greater in the furrows than mounds due to litter accumulation and preferential flow of water (Jackson and Caldwell, 1989; Caldwell et al., 1992). Several studies have also shown that plants can alter rates of nutrient uptake when a localised supply of nutrients is increased experimentally (e.g. Jackson et al., 1990), although few of these studies have been on tree species, and even fewer on species with a distinct seasonality in fine root production.

Both jarrah (Shea and Dell, 1981) and *Banksia* spp. (common throughout south western Australia, often as a forest understorey species) have extensive proliferations of lateral roots at times of increased inputs of nutrients and water (Jeffrey, 1967; Lamont, 1981; Grierson and Attiwill, 1989). While the relative importance of altered root morphology (root hairs, length, degree of branching, clustering of roots) *versus* uptake kinetics is as yet unresolved, it is clear that many plant species are capable of rapidly adjusting both their morphology and physiology in order to acquire limiting resources that become available in a localised patch of soil (Caldwell and Pearcy, 1994). For example, legumes which grow as understorey species in forests of south western Australia and which produce cluster roots, may alter the distribution of root growth in response to the availability and form of P in soil (Figure 2). In glasshouse experiments with a range of species, root growth (excluding the growth of the tap root) was greatest where P had been added (Figure 2). However root growth of some species was greatest when P was supplied in ash and in others when P was supplied as iron phytate. Cluster roots developed most prolifically when P was supplied in ash. Root nodules developed best without added P, suggesting root clusters and root nodules to be mutually exclusive. This plasticity in response to sudden and localised environmental change is a desirable attribute for species selected for agroecosystems, particularly where water and nutrients are limiting and where supply may be erratic. Deep, long-lived root systems are important in nutrient-poor substrates because they increase the volume of resources exploited by vegetation. However, there is only limited understanding of the role of belowground productivity in water and nutrient acquisition, and the cycling of nutrients, in forests generally, and in the forests and woodlands of Australia specifically.

4. Nutrient cycling and bioavailability

In natural ecosystems, only a small fraction of the total nutrient pool is in the soil solution (in either inorganic or organic forms) at any one time. By far, the majority (> 95%) is stored in the biomass and bound to organic and inorganic complexes in the soil (Binkley and Vitousek, 1989). All native forest systems depend on cycling of nutrients to meet the nutritional demands of

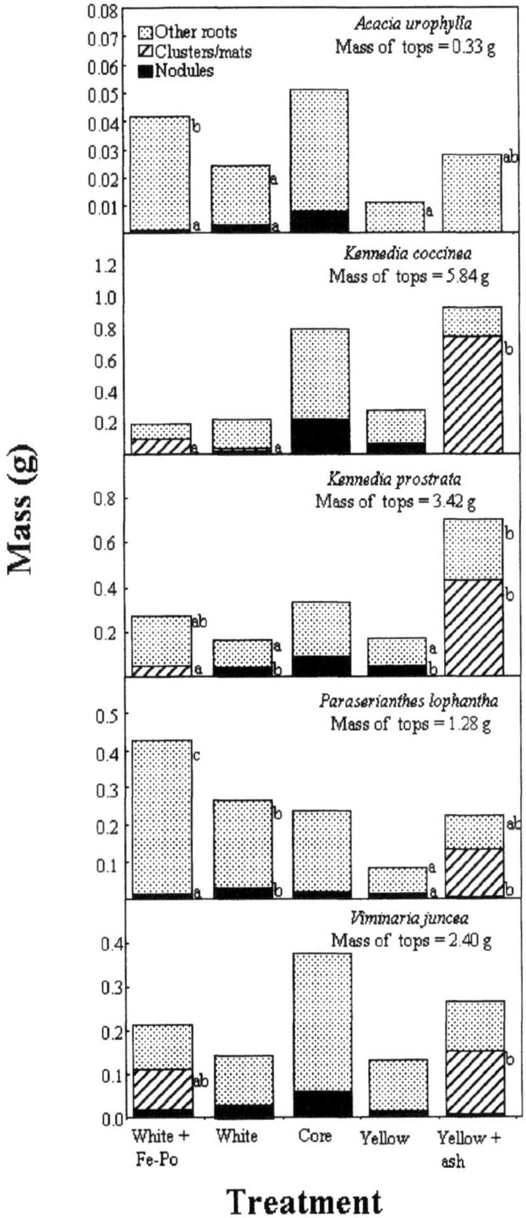

Figure 2. Plasticity of root growth and growth of specialised root structures (nodules and cluster roots) for a range of legumes from south western Australia. Plants were grown in 4 L containers which were divided into quadrants around a central (~400 mL) core of clean, white-quartz sand. Each quadrant contained either white quartz sand (no P-fixation capacity) or a yellow, clayey sand (with limited P-fixation capacity) or yellow sand plus 100 μg P g^{-1} in the form of ash (from *Eucalyptus marginata* and *Corymbia calophylla*) or white sand plus 100 μg P g^{-1} in the form of iron-phytate. Source: Adams and Pate (unpubl. data).

the growing plants, and this is particularly so of native ecosystems in Australia which have developed on highly weathered, ancient soils (Attiwill and Adams, 1993; Adams, 1996).

Nutrient bioavailability can be viewed as a function of the rate of nutrient supply and nutrient limitation (Binkley and Vitousek, 1989). Natural ecosystems vary widely in supply rate because of differences in rates of decomposition, mineralisation (Swift et al., 1979) and other processes. Productivity of individuals or whole ecosystems may be reduced by an inadequate supply of mineral nutrient (Chapin et al., 1986) or altered because of adaptive strategies of individual species. In agricultural systems, nutrient supply rate and nutrient limitation are closely linked via management, in contrast to natural ecosystems where differences in species, plant life stage and moisture supply can uncouple low nutrient supply rates from nutrient limitation (Specht et al., 1983; Binkley and Vitousek, 1989).

4.1. *Litterfall and decomposition*

The major cycle of N and P in native forests, woodlands, heathlands and grasslands is in litterfall, both above- and belowground. The significance of nutrient cycling increases as both rainfall and growth decline, so that in arid and semi-arid environments, litterfall accounts for 90–100% of the external cycles of these major nutrients (Attiwill and Adams, 1993). Decomposition of organic matter and mineralisation processes are responsible for virtually all of the nutrients taken up by vegetation. The effects of plant species richness and diversity in determining the nature of relationships between biodiversity and ecosystem properties has not been widely studied. The most likely effect will be expressed via the C:N ratio of litter as a fundamental control of litter decomposition and N mobilisation patterns (Swift et al., 1979; Wardle et al., 1997). Generally, when the C:N ratio of soil organic matter is less than 32, some N will be mineralised and the *rate* will increase with a decreasing ratio; at ratios greater than 32, there will be net immobilisation of N in the soil (Troeh and Thompson, 1993).

Like growth, aboveground litterfall may be predicted on the basis of climate. Relationships between climate and litterfall have been demonstrated for tropical forests (Vitousek, 1984), temperate forests (Bray and Gorham, 1964) and eucalypt forests (Attiwill and Adams, 1991; Figure 3). As for growth, rates of nutrient cycling in Australian ecosystems also tend to increase at higher rainfall and temperature (Attiwill and Adams, 1993).

A distinctive feature of the natural landscape in south western Australia is the high degree of variation in species composition and rates of growth brought about by subtle changes in the landform and more distinct changes in soil types. It is common to find two- or three-fold differences in rates of growth across distances as short as 1–200 metres. These spatial patterns are mirrored by variation in rates of nutrient turnover and availability. Despite the magnitude of such variation, an understanding of temporal variation may be of

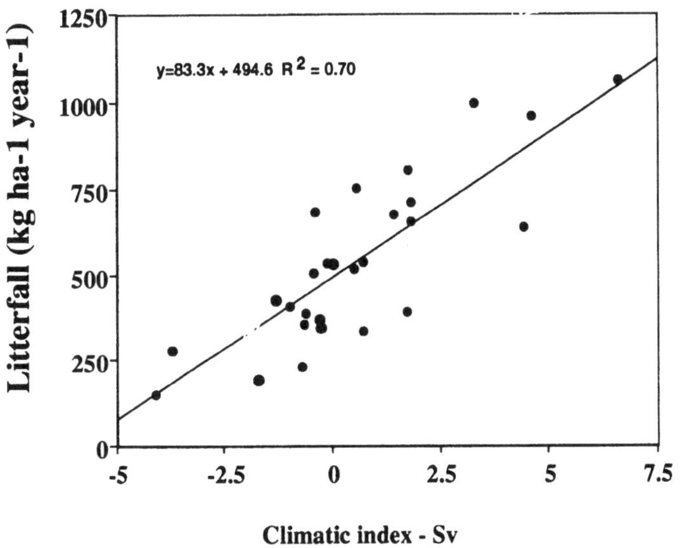

Figure 3. Relationship between litterfall and an index of climate (Sv) for a range of Australian eucalypt forests. Sv (Vitousek, 1984) uses a scalar (TEMP) for temperature °C (which is based on elevation and latitude) and rainfall and is of the form: Sv = TEMP × Log (rainfall, metres). *Source:* Redrawn from Adams and Attiwill (1986).

greater significance when applied to agroecosystems where the spatial variation in litter quality and quantity will be defined directly by management, i.e. by choice of species and arrangement in the landscape (Boerner et al., 1998).

Nutrient concentrations in the soil solution and, therefore, nutrient absorption by plants fluctuate considerably during the year. In non-agricultural soils, there is a predictable spring increase in nutrient availability due to increases in microbial activity and flushes of nutrients associated with wetting and drying cycles (Grierson et al., 1998). In some areas, there are also autumn and winter flushes and other less predictable flushes associated with leaching and breakdown of leaf litter. In infertile habitats, it is likely that much of the nutrient requirements of plants are met by uptake during such relatively short periods, particularly late winter and early spring, rather than by steady-state absorption under constant conditions. Until estimates are made of the size and frequency of the non steady-state absorption of nutrients, it will remain difficult to predict their overall contribution to the nutrient balance of an ecosystem at a variety of scales.

In natural forest ecosystems, nutrients are retained or restored after disturbance across a scale of intensity. For example, the importance of litterfall, the development of a litter layer and their contribution to nutrient availability can be illustrated by a recent study of rehabilitation of jarrah forest after bauxite mining (Todd et al., unpubl. data). The soil surface of rehabilitated areas is not uniform but instead consists of mounds and broad furrows. Litter collects in the furrows and around perennial species, producing a heterogenous pattern and the distribution of plant-available forms of N (and P) is directly and closely related to the distribution of litter (Figure 4). Similarly, semi-arid woodland and grassland ecosystems in Australia show considerable patchiness in resource distribution, where fertile patches favour the establishment, growth and survival of vegetation (Ludwig and Tongway, 1996).

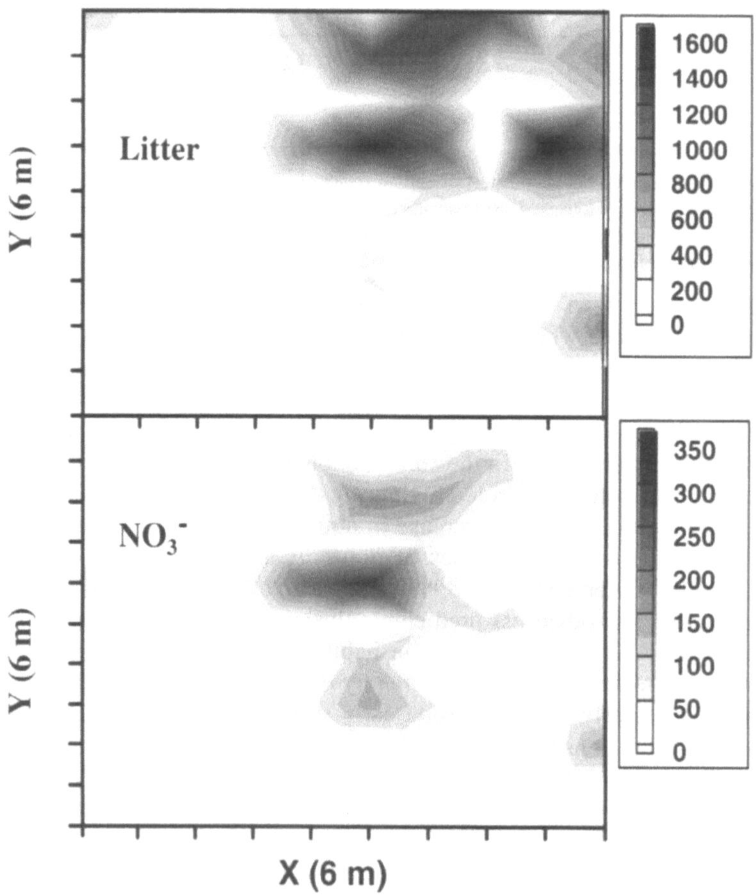

Figure 4. Distribution of leaf litter (g) and NO_3^- (0-5 cm, μg g^{-1} soil) accumulated in furrows (parallel with X-axis) of a rehabilitated mine site, illustrating the effect of microtopography on nutrient distribution. *Source:* Todd et al. (unpubl. data).

4.2. *Phosphorus*

Phosphorus deficiency is a major constraint to agricultural production in many regions of the world, including Australia, tropical Africa, Latin America and Asia and is often due to soils low in native P and/or high fixation capacities (Mokwunye et al., 1986; Warren, 1992). Many of the soils of southern Australia are characterised by low concentrations of extractable P and many native plant species respond negatively to extra P (Ozanne and Specht, 1977) or remain unaffected (Attiwill, 1991). Some species respond positively but at the expense of longevity (Specht, 1963) or tolerance to drought (Heddle and Specht, 1975).

The cycling of P in natural ecosystems is dominated by internal redistribution within the plant and by the external transfer of P from plant to soil via litterfall. In general, the processes responsible for mineralisation of the organically-bound P in litter have received scant attention, mainly due to technical difficulties associated with low concentrations and slow rates. By comparison, there has been a considerable amount of research into specific root attributes which increase the availability of soil (and litter) P to plants. Root adaptations which affect P availability and uptake on impoverished soils have been well-described for Australian ecosystems (e.g. Bowen, 1981).

Australian native plants exhibit mechanisms that may provide direct access to the many forms of soil P. Most common are adaptations such as cluster or proteoid roots, and mycorrhizal associations, which effectively increase the ratio of root surface area to soil volume, thereby reducing the diffusion path for phosphate ions to the root surface and exploring a greater soil volume. Also common among species adapted to low P supply, is the capacity to obtain P from sparingly-soluble organic and inorganic forms in soils dominated by Fe and Al. One adaptation is the production of large amounts of root exudates, including organic acids and extracellular enzymes, which have strong P-solubilisation or P-mineralisation properties (Grierson and Attiwill, 1989; Grierson, 1992; Dinkelaker et al., 1995).

Both mycorrhizal (Dodd et al., 1987; MacFall et al., 1991) and non-mycorrhizal roots (Dinkelaker and Marschner, 1992) produce phosphatase enzymes which catalyse the mineralisation of organic P, releasing inorganic P (Tarafdar and Claasen, 1988). Root-mediated changes in pH, production of phytosiderophores and phytochelates, and root-mediated microbiological changes may increase nutrient availability in the rhizosphere. For example, *Banksia* spp. modify the soil-root environment by lowering pH, and through the production of organic acids and phosphatase enzymes (Grierson and Attiwill, 1989; Grierson, 1992) and there is strong evidence that enhanced P uptake is a key function of proteoid roots (Jeschke and Pate, 1995). Oxalate production by ectomycorrhizal roots of jarrah may also enhance P uptake in poor soils (Malajczuk and Cromack, 1982). In wetter periods, ectomycorrhizae may store P as polyphosphates in hyphae for use for growth when soils are drier (Coleman and Specht, 1981). Leake and Miles (1996) also demonstrated direct uptake of organic P by some plant species (i.e. in a manner analogous to that of uptake of organic N, see below). Overall there is a tremendous diversity of means of acquisition of soil P.

The formation of proteoid (or cluster) roots and infection by mycorrhizal fungi appear to be alternative strategies for enhanced uptake of water and nutrients (Lamont, 1986). Other species possess hairy roots or have rapid establishment of a large root system, which maximise uptake of nutrients (and water) from the soil. For example, *Eucalyptus gummifera*, a species indigenous to low P soils, rapidly establishes an extensive root system as a seedling (Mulligan, 1988). Similarly, *Acacia cyclops* and *A. saligna* have early vigorous production of roots (Witowski, 1994) while species such as *E. pilularis* may store P in their roots; an attribute which appears to be most strongly expressed in provenances from low P soils (Mulligan and Sands, 1988). This is an important point, because it suggests that provenance selection of certain species can impact on the efficacy of root systems to acquire P.

In addition to root modifications, many Australian native species are efficient at retaining P within the plant, partly through internal recycling and partly due to retention of leaves for up to several seasons. Redistribution of P from senescing leaves is greatest in species that have evolved where soils have the least P content. If species are to be considered for inclusion in agricultural ecosystems, the accumulation of P (and other elements) in leaves increases the potential of these species as sources for improvement of soil fertility, as has been shown for mulching species used in agroforestry in Kenya (Nziguheba et al., 1998). Some crop species also have root modifications, and produce root exudates high in organic acids (e.g. *Lupinus albus*, *Brassica napus*). When used as an intercropping or fallow species, the growth of other crops may be enhanced (Gardner et al., 1982).

Owino (1996) recently noted that there is a pressing need for research into the identification and utilisation of multipurpose trees and shrubs for production and function in agroecosystems and that, when considering adaptability in natural populations and exotic species, consideration should be given to constraining factors such as low P and low pH, and the plasticity of response to these factors (Figure 2). Such plasticity may explain why simple models of P uptake by roots which have been applied successfully to high-input cropping systems in temperate regions (Van Noordwijk et al., 1990) clearly underestimate P acquisition by species growing on poor soils. For example, models of nutrient acquisition in heterogenous environments (Jackson and Caldwell, 1996) suggest that plasticity may be relatively unimportant for the uptake of nitrate whereas root proliferation always results in additional net uptake of P.

4.3. *Nitrogen*

Most nitrogen in natural ecosystems of southern Australia is held in the soil organic matter, although the litter and understorey may contribute a significant proportion of the total, depending on fire frequency (O'Connell and Grove, 1996). Understorey species, such as *Acacia pulchella*, can contribute significant amounts of N to the system post-fire through symbiotic fixation (Hansen and Pate, 1987). However, rates of fixation greater than 1–2 kg ha^{-1} y^{-1} are unlikely to persist for more than a few years after disturbance, after which the bioavailability of P returns to pre-fire levels (Attiwill and Adams, 1993; Adams, 1996) and limits N-fixation. In many jarrah forests, the distribution and density of N-fixing flora is variable and, even after fire, inputs are unlikely to ever be greater than a few kg ha^{-1} y^{-1}. The understorey, in contrast to the higher productivity forests of the south-west (karri forests), tends to be dominated by genera belonging to the Proteaceae (particularly *Banksia grandis*), which have cluster roots rather than mycorrhizal associations, and which do not fix N. However, some species in the Proteaceae have cluster (proteoid) roots which are enriched in ^{15}N (Schmidt and Stewart 1997), suggesting an undescribed role in nitrogen acquisition, possibly the utilisation of organic N.

Recently, a number of authors have confirmed the earlier suggestion of Abuzinaddah and Read (1986) that organic N may be taken up by native plants in a range of nutrient poor ecosystems (Chapin et al., 1993; Kielland, 1994; Nadelhoffer et al., 1996; Raab et al., 1996), including wallum (wet heathland) ecosystems in the sub-tropical region of southern Queensland (Schmidt and Stewart, 1997). In such systems, some species use nitrate, others use ammonium and a few may utilise organic N. The diversity of species provides a comprehensive range of use of all available N-sources and is a classic example of the benefits of diversity to nutrient conservation.

A considerable number of woody species and perennials used in agricultural systems throughout the world are N-fixing (Nair, 1993), including *Lupinus* spp., *Medicago* spp., *Sesbania sesban*, *Calliandra calothyrsus* and *Leucaena leucocephala*. Rates of N-fixation are generally greater than in Australian forests, presumably because of greater availability of P, although Sprent (1995) has noted the large variability in rates of fixation by *Acacia* spp. Nitrogen-fixation in agricultural systems in south western Australia is also highly variable (Unkovich et al., 1995). This variation may be due to differences in P-availability among species and sites, but it is more likely the result of end-product inhibition. For example, nitrate is a potent inhibitor of N-fixation and strong nitrification is characteristic of most agricultural soils in south western Australia. Nitrate may accumulate through the summer when growth and nutrient demand are limited either by water or by the absence of plants (Sanford et al., 1995; Unkovich et al., 1995). The contribution of N fixed by legume fallows to cropped-N is also highly variable, dependent in part on the availability of inorganic N in the soil.

The quantity of N fertiliser used in the wheat belt of south western Australia has increased steadily with demand for higher cereal crop yields and in response to declining soil fertility (Dalal and Mayer, 1986). Concurrent with increased fertiliser applications are substantial inputs of N from legume pasture leys. Incorporation of a pasture (Holford, 1980) or legume rotation (Marcellos, 1984) in south western Australia has reduced the need for the application of N-fertilisers and the loss of N from farming systems through repeated cropping. Nonetheless, maintenance of yields and concentrations of protein in grain has required the increased use of nitrogenous fertilisers (McDonald, 1992), despite a variable response to added N which has been attributed to the effects of residual soil N as well as the influence of variation in moisture availability and temperature (Russell, 1980). In sandy soils, nitrate leaching may (i) reduce the efficiency of fertiliser N, particularly where nitrification patterns and crop N uptake are asynchronous (McInnes and Fillery, 1989) and (ii) contribute to serious environmental problems as a consequence of excessive leaching of nitrate into groundwater and waterways (Anderson et al., 1998).

The difference in dominance of form of N between natural and agro-ecosystems has a major influence on N-uptake and leaching, i.e. the balance of inputs and outputs to the system. The dominance of nitrification in agricultural soils is a major factor in their acidification and resultant need for applications of lime (Leeper, 1964). Forest soils may also be moderately acidic (pH 5.5–6.5) but, with the exception of the most highly productive systems where nitrification is common (Attiwill et al., 1996), this lower pH is due to weakly acidic organic anions (e.g. HCO_3^-) derived from the accumulation of organic matter, instead of strong acid anions such as nitrate.

Such pH differences between forest and agricultural soils have further implications for the bioavailability and forms of N, P, and S in agriculture compared to forested ecosystems. Acidic forest soils tend to show a predominance of available NH_4^+, with NO_3^- increasing as pH rises. The solubility of P, if not the absolute amount of P available, tends to follow the NH_4^+ curve. At one extreme, tolerance of NH_4^+ ensures an adequate supply of N for plants on acidic soils, although NH_4^+ may be one of the major deterrents to the establishment of plants from other soil types (Rorison, 1987), i.e. for many crop species. Under less extreme conditions, plants may survive on a mixture of N sources with the absolute supply varying with season.

The history of agriculture in south-west western Australia is such that broadacre productivity is totally dependent on fertiliser application and on winter-annual species which avoid the summer drought. Balancing nutrient inputs and outputs has seldom been a priority of agriculture in the past. Instead, the major aim of most agriculture has been to increase productivity, generally through intensification of management and often with far-reaching consequences for much of the primary resource (water and soil). This is evidenced by increased concentrations of nitrate in groundwater, elevated concentrations of nitrate and phosphorus in waterways and estuaries, increased acidification of soil, waterlogging, and increased salinisation of land and waterways (Lovering and Crabb, 1998). Importantly, annual production has been increased by a reduction in the natural variation of soil properties due to repeated tillage and repeated application of fertilisers and herbicides

(Boerner et al., 1998). In terms of nutrients, for example, nitrate production and uptake dominates the soil-plant N cycle in many agricultural systems, with few species using ammonium or organic N, to the extent that successful weeds of agricultural crops and pastures are generally nitrophilous species and may even store nitrate at high concentrations in leaf tissue (Unkovich et al., 1998).

4.4. *Other nutrients*

While the cycling of C, P and N have dominated the studies of nutrient cycling in forests, the limited supply of other nutrients has had a major impact on the development of agriculture on Australian soils. Sulphur, for example, is applied to agricultural systems primarily as a component of superphosphate fertiliser. Depending on crop and soil, some requirements for sulphur are at least as great as they are for phosphorus (Leeper, 1964).

Micronutrients have also been applied extensively in agriculture, in some cases not just to improve plant growth but also to improve production of grazing animals. Most of the nutrient deficiencies observed in introduced crops are associated with sandy soils, and often on alkaline soils high in free carbonates. For example, combined deficiency of copper and zinc are fairly common on the Esperance sandplain of Western Australia. Molybdenum (Mo) is essential for all plants and needed in greatest amounts for nitrogen fixation and in non-legumes, and is required for the function of nitrate reductase. It appears that Mo is only essential for many native non-legumes where inorganic N is taken up as nitrate (Dell, 1996). Consequently, if native species absorb N in other forms (see section 4.3) then symptoms of Mo-deficiency are unlikely to appear. In contrast, pasture systems where subterranean clover and other legumes are used on poor acid soils, require substantial applications of sodium molybdate in order to achieve sufficient growth (Leeper, 1964).

Deficiency symptoms are rarely reported for native species, which appear well-adapted to the soils on which they naturally grow. Eucalypts in their native landscape, for example, rarely show symptoms of nutrient disorders (Dell, 1996). However, in eucalypt plantations (of species exotic to other regions), a range of nutrient deficiencies are seen (Dell et al., 1983; Wallace et al., 1986; Dell and Robinson, 1993).

4.5. *Biodiversity effects on nutrient cycling*

The mediterranean-climate regions of southern Australia are moderately species-rich (Cowling et al., 1996) and parts of south western Australia are highly species-rich (Hopper, 1992). It is worthwhile to explore the relationships between species richness, growth and nutrient cycling.

There is good evidence that species diversity is inversely related to productivity or other measures of nutrient supply. This is particularly true for south western Australia where edaphic complexity and associated specialisation has been invoked as the principle determinant of the high regional diversity of this area (Hopper, 1992). The areas of greatest diversity are generally associated with the poorest soils (deep sands, podsolic soils). Given the frequently severe summer drought and reasonably frequent disturbance by fire, it is no surprise that rates of growth in these areas are slow and that there is a wide variety of life forms as well as morphological and physiological adaptations. Many of these adaptations improve the capacity of plants to obtain and conserve nutrients, and survive drought (Hooper and Vitousek, 1998).

The diversity of species and of their form and function is a powerful argument that mono-specific communities lack the necessary diversity of function to cope with the exacting environment. Theoretically, it has been argued that the relative diversity of mediterranean plant communities results from low rates of competitive displacement as a result of low nutrients and summer drought, coupled with a moderately high frequency of disturbance from fire and grazing (Cowling et al., 1996).

Clearly, it is possible to classify species within functional groups. The basis for classification might be, for example, mycorrhizal status or rate of growth (Sala et al., 1996). We might then rank functional groups according to their abundance and their contribution to ecosystem processes such as nutrient cycling, and construct a rank-abundance diagram for each. The decrease in ecosystem processes should be greatest when deleting first the most abundant group. For example, if we were to remove the species which makes the greatest contribution to the lifting of water from deep in the profile (see Dawson and Pate, this volume) then the availability of water to the whole plant community may well be significantly reduced. Similarly, if species which have the greatest rates of growth (and, by definition, the greatest rates of litterfall) were removed, then the turnover and availability of nutrients for the whole community would be altered. The effects of adding or deleting individuals, species, communities or landscape units or processes such as transpiration and nutrient mineralisation, depends on the sequence in which species, communities or landscapes are added or deleted and as productivity rises, biodiversity ultimately decreases.

Different species also modify the rates of element and energy transfer within ecosystems in species-specific ways. For example, there are substantial differences in litter quality, effects on soil temperature and moisture, exudation from roots, all of which will effect rates of nutrient mineralisation (see earlier discussion). Microbial functional groups that mediate nutrient cycling processes through their effect on decomposition of organic matter, mineralisation processes and nutrient transfer also differ in their effect on nutrient availability and loss from ecosystems (Wardle et al., 1997). The traits exhibited by species found in the mediterranean-climate regions in Australia which are most likely to affect nutrient cycling are, therefore, those that (i) modify the availability, capture and use of soil resources such as water and nutrients (for example, specialised root adaptations), (ii) affect the feeding relationships (trophic structure) within a community, and (iii) influence the frequency, severity and extent of disturbances such as fire. The introduction or loss of a species that alters resource availability therefore has the potential to markedly affect nutrient cycling. For example, a species could otherwise alter resource supply by tapping an otherwise unused source as is the case where mycorrhizal fungi are introduced into rehabilitated minesites where they enhance P availability to their host plant, or when N-fixers are included into a system where they did not previously grow. These species shifts can have major implications for runoff, salinisation or recovery of degraded ecosystems, but are rarely considered.

The concepts of resilience and resistance in the design of functional analogues for agricultural systems can be extended if we consider that higher diversity in natural ecosystems might also increase the stability of ecosystem processes. Greater species diversity provides a greater diversity of trophic interactions, including alternative pathways for transfer of nutrients and reduced susceptibility to invasion after disturbance, i.e. increase resilience and resistance (Chapin et al., 1997; see also Main, Neher, this volume). The inclusion of species that have similar effects on nutrient cycling or other processes but which differ in their environmental responses to, say water stress or fire, can improve agroecosystem resilience by increasing the capacity to adjust or recover from perturbation (such as drought). For example, in south western Australia, functionally similar *Acacia* species have different temperature thresholds for germination and different species colonise after fires of differing intensities (Atkins and Hobbs, 1995). The replacement of nitrogen after fire across a broad range of burn conditions is thereby assured. In order to confer the desirable traits of resilience and resistance (sustainability), such conditions must also be included in the design of functional analogues for agricultural systems.

5. Designing a mimic for agriculture in south western Australia

In order to assess whether nutrient cycling processes in native forests can be used as a model for agricultural systems, it is essential that we first characterise the natural ecosystems which we would like to mimic in terms of:

- Community structure (life-forms);
- Species composition;
- Net primary productivity;
- Rates of organic matter accumulation, decomposition and mineralisation;
- Rates of nutrient acquisition and sequestration.

Despite the still substantial gaps in our knowledge, we can summarise some of the major characteristics important to maintenance of native ecosystems of south western Australia. These include:

- Extensive and deep penetration of root systems;

- Specialised root structures, mycorrhizal root associations, and cluster root formation, and the seasonal formation of surface root mats;
- Slow growth rates and conservative nutrient cycling;
- Disjunct phenology of growth, nutrient uptake and cycling above- and belowground;
- Adaptations to recover or re-establish after fire;
- Spatial and temporal heterogeneity in resource availability;
- High species diversity.

A mimic agroecosystem should include some or all of these attributes in order to develop analogous processes (Pate and Bell, this volume). Most attempts at mimicry emphasise that differences in life-forms or functional types (e.g. woody perennials, annual grasses) may have the greatest impact on ecosystem processes. However, classification by life-form alone (Ewel and Bigelow, 1996), or by life-form plus effects on forest dynamics (Denslow, 1996), should also include some understanding of the dependence of this classification on the type and scale of that response. For example, relative response of species to creation of a gap in the canopy of a closed tropical forest (and therefore increased availability of nutrients, water and light) may be more significant to ecosystem processes than life-form in the short-term. That is, relative adjustments in physiology and plant form to an increase in light and nutrient availability are likely to have greater impacts on the recovery of ecosystem processes, such as productivity and biogeochemical cycling, than whether a seedling will become an understorey shrub or overstorey tree. Such arguements reiterate the importance of selecting species for mimic agroecosystems on the basis of their plasticity of response to change, as well as on lifeform. The importance of differences in life-form is likely to increase with time where, for example, the short-term effects of replacement of perennial native species with a monoculture of annual species may be negligible, but the long-term effects very great.

The success of the mimic concept in tropical ecosystems (see Ewel, this volume) is largely attributable to the natural system being of similarly high productivity to the analogue. In order to create an acceptable model for agriculture in south western Australia and elsewhere where the aim has been to maintain high rates of production in a system of naturally *low* productivity, a resilient ecosystem (a natural ecosystem of greater fertility and productivity) may be a more acceptable analogue. For example, careful selection of N-fixers (some *Acacia* species are better than others, for example) that are fast-growing, in combination with longer-lived tree species (perhaps slower-growing), and planted to features of the landscape (rather than in rows or along fence-lines), could be incorporated with species that survive on low levels of available P and which may enhance availability of P to crops, e.g. *Lupinus albus* (Gardner et al., 1983). However, any significant increase in soil fertility will probably impact negatively on native plant growth, particularly when there is an imbalance between N and P (Handreck, 1996). Similarly, higher productivity will almost always be at the cost of biodiversity although, somewhat paradoxically, in low nutrient environments increased biodiversity seems to enhance productivity (Odum, 1998).

Numerous studies have shown that agricultural management often decreases soil biodiversity, alters the structure of soil biological communities, decreases soil organic matter and reduces the soil microbial biomass as compared with native forest or grassland ecosystems (e.g. Lavelle and Pashanasi, 1989; Sparling et al., 1985; Crossley et al., 1992; Neher, this volume). These reductions are attributable mostly to an increase in the frequency and intensity of soil perturbation that result from land-use change and a reduction in the quantity and diversity of organic resources returned to the soil. In addition, the use of xenobiotic compounds such as industrial fertilisers and pesticides, have frequently adverse impacts on soil communities. Improving soil organic matter content is crucial, for example, to the maintenance of soil fertility of many deeply weathered and leached soils and sands of south western Australia. The inclusion of shade trees, mixed cropping of species that differ in phenology (both above- and belowground), controlling weeds and pests through crop rotation, dense spacing, mulching and application of manure rather than through cultivation, are all likely to result in an increase in soil organic matter.

Finally, there is a need to establish long-term trials on representative sites in order to evaluate the effects of agricultural practices on ecosystem processes at various spatial and temporal scales. These sites must adequately represent the dominant soil types, the major climatic regions and the range of agricultural practices used (or suggested) in the regions studied. In order to do so, we need to develop a better understanding of the biological interactions which promote key processes of production and nutrient cycling. In addition to specific processes (such as residue decomposition, N, P and C mineralisation), experiments should include aspects of ecosystem properties, e.g. resilience and stability, that are likely to be important in achieving true sustainability of our agricultural systems.

6. Acknowledgements

We would like to thank Ted Lefroy for organising the workshop which has formed the basis for this article, plus the many contributions of the postgraduate students and researchers of the Ecosystems Research Group in the Botany Department at The University of Western Australia. Much of the work described in this paper was made possible through funding from the Australian Research Council. Our manuscript was improved by comments from J. Noble, R. J. Hobbs, E. C. Lefroy and D. Neher.

7. References

Abbott I and Loneragan O (1983) Growth rate of jarrah (*Eucalyptus marginata*) in relation to site quality in cut over forest, Western Australia. Australian Forester 46: 91–102

Aber JD (1992) Nitrogen cycling and nitrogen saturation in temperate forest ecosystems. Trends in Ecology and Evolution 7: 220–224

Aber JD, Nadelhoffer KJ, Steudler P and Melillo J (1989) Nitrogen saturation in northern forest ecosystems. BioScience 39: 378–386

Abudzinadah SM and Read DJ (1986) The role of proteins in the nitrogen nutrition of mycorrhizal plants. III. Protein utilization in mycorrhizal association with *Hebeloma crustuliniforme*. New Phytologist 103: 507–514

Adams MA (1996) Distribution of eucalypts in Australian landscapes: landforms, soils, fire and nutrition. In: Attiwill PM and Adams MA (eds) Nutrition of Eucalypts, pp 61–76. CSIRO Publishing, Collingwood, Australia

Adams MA and Attiwill PM (1991) Nutrient balance in forests of northern Tasmania. I. Atmospheric inputs and within stand cycles. Forest Ecology and Management 44: 93–113

Agren GI and Boscatta E (1988) Nitrogen saturation of terrestrial ecosystems. Environmental Pollution 54: 185–198

Atkins L and Hobbs RJ (1995) Measurement and effects of fire heterogeneity in south-west Australian wheatbelt vegetation. CALM Science Supplement 4: 67–76

Anderson GC, Fillery IRP, Dunin FX, Dolling PJ and Asseng S (1998) Nitrogen and water flows under pasture-wheat and lupin-wheat rotations in deep sands in Western Australia 2. Drainage and nitrate leaching. Australian Journal of Agricultural Research 49: 345–361

Attiwill PM (1991) Nutrient resilience of wet sclerophyll forest recovery after disturbance. In: Ryan PJ (ed) Third Australian Forest Soils and Nutrition Conference, pp 172–179. Forestry Commission of NSW, Sydney

Attiwill PM and Adams MA (1993) Nutrient cycling in forests. New Phytologist 124: 561–582

Attiwill PM, Polglase PJ, Weston CJ and Adams MA (1996) Nutrient cycling in forests of south-eastern Australia. In: Attiwill PM and Adams MA (eds) Nutrition of Eucalypts, pp 191–227. CSIRO Publishing, Collingwood, Australia

Bazilivech NI and Rodin LE (1968) Reserves of organic matter in the underground sphere of terrestrial phytocoenoses. In: Methods of Productivity Studies in Root Systems and Rhizosphere Organisms. Proceedings International Symposium, USSR, Biddles, UK

Beard JS (1990) Plant Life of Western Australia. Kangaroo Press, Kenthurst, NSW, 319 pp

Beard JS and Sprenger BS (1984) Geographical data from the vegetation survey of Western Australia. Vegetation survey of Western Australia, occasional paper 2. Vegmap Publications, Perth, Western Australia

Bennett LT, Weston CJ and Attiwill PM. (1997) Biomass, nutrient content and growth response to fertilisers of six-year-old *Eucalyptus globulus* plantations at three contrasting sites in Gippsland, Victoria. Australian Journal of Botany 45: 103–121

Bieleski RL (1973) Phosphate pools, phosphate transport and phosphate availability. Annual Revue of Plant Physiology 24: 225–252

Binkley D and Vitousek P (1989) Soil nutrient availability. In: Pearcy RW, Ehrleringer J, Mooney HA, and Rundel PW (eds) Plant Physiological Ecology, pp 75–96. Chapman and Hall, New York

Boerner REJ, Scherzer AJ and Brinkman JA (1998) Spatial patterns of inorganic N, P availability, and organic C in relation to soil disturbance: a chronosequence analysis. Applied Soil Ecology 7: 159–177

Bolland MDA, Bowden JW, Brennan RF, Gartrell JW, Mason MG and Porter WM (1991) Nutrition. In: Perry M and Hillman B (eds) The Wheat Book, pp 55–77. Department of Agriculture Bulletin 4196, Western Australia

Bowen GD (1981) Coping with low nutrients. In: Pate JS and McComb AJ (eds) The Biology of Australian Plants, pp 33–164. University of Western Australia Press, Nedlands, Australia

Bray JR and Gorham E (1964) Litter production in forests of the world. Advances in Ecological Research 2: 101–157

Caldwell MM, Dudley LM and Lilieholm B (1992) Soil solution phosphate, root uptake kinetics and nutrient acquisition: implications from a patchy soil environment. Oecologia 89: 305–309

Caldwell MM and Pearcy RW (eds) (1994) Exploitation of Environmental Heterogeneity By Plants. Academic Press, San Diego, 429 pp

Catovsky S (1998) Functional groups: clarifying our use of the term. Bulletin of the Ecological Society of America 79: 126–127

Chapin FS (1980) The mineral nutrition of wild plants. Annual Review of Ecology and Systematics 11: 233–260

Chapin FS, Moilanen L and Kielland K (1993) Preferential use of organic nitrogen for growth by a non-mycorrhizal Arctic sedge. Nature 361: 150–153

Chapin FS, Van Cleve K and Vitousek P (1986) The nature of nutrient limitation in plant communities. American Naturalist 127: 148–158

Chapin FS, Walker BH, Hobbs RJ, Hooper DU, Lawton JH, Sala OE and Tilman D (1997) Biotic control over the functioning of ecosystems. Science 277: 500–504

Coleman and Specht RL (1981) Mineral nutrition of heathlands: the possible role of polyphosphate in the phosphorus economy of heathland species. In: Specht RL (ed) Ecosystems of the World. 9B. Heathlands and Related Shrublands, pp 197–207. Elsevier, Amsterdam

Cowling RM, Rundel PW, Lamont BB, Arroyo MK and Arianoutsou M (1996) Plant diversity in mediterranean-climate regions. Trends in Ecology and Evolution 11: 362–366

Cromer RN, Raupach M, Clarke ARP and Cameron JN (1975) Eucalypt plantations in Australia: the potential for intensive production and utilisation. Appita 29: 165–173

Crossley DA, Mueller BR and Rerdue JC (1992) Biodiversity of microarthropods in agricultural soils: relations to processes. Agricultural Ecosystems and Environment 40: 37–46

Dalal RC and Mayer RJ (1986) Long term trends in fertility of soils under continuous cultivation and cereal cropping in southern Queensland. V. Rate of loss of total nitrogen from the soil profile and changes in carbon: nitrogen ratios. Australian Journal of Soil Research 24: 493–504

Dell B (1996) Diagnosis of nutrient deficiencies in eucalypts. In: Attiwill PM and Adams MA (eds) Nutrition of Eucalypts, pp 417–440. CSIRO Publishing, Collingwood, Australia

Dell B and Robinson JM (1993) Symptoms of mineral nutrient deficiencies and the nutrient concentration ranges in seedlings of *Eucalyptus maculata* Hook. Plant and Soil 113: 287–290

Dell B, Bartle JR and Tacey WH (1983) Root occupation and root channels of jarrah forest subsoils. Australian Journal of Botany 31: 615–627

Denslow JS (1996) Functional group diversity and responses to disturbance. In: Orians GH, Dirzo R and Cushman JH (eds) Biodiversity and Ecosystem Processes in Tropical Forests, Ecological Studies 122, pp 127–151. Springer-Verlag, Heidelberg

Dinkelaker B and Marschner H (1992) In vivo demonstration of acid phosphatase activity in the rhizosphere of soil-grown plants. Plant and Soil 144: 199–205

Dinkelaker B, Römheld V and Marschner H (1989) Citric acid excretion and precipitation of calcium citrate in the rhizosphere of white lupin (*Lupinus albus* L). Plant Cell and Environment 12: 285–292

Dinkelaker B, Hengeler C and Marschner H (1995) Distribution and function of proteoid roots and other root clusters. Botanica Acta 108: 183–200

Dise NB and Wright RF (1995) Nitrogen leaching from European forest in relation to nitrogen deposition. Forest Ecology and Management 71: 153–161

Dodd JC, Burton CC, Burns RG and Jeffries P (1987) Phosphatase activity associated with the roots and the rhizosphere of plants infected with vesicular-arbuscular mycorrhizal fungi. New Phytologist 107: 163–172

Ewel JJ (1986) Designing agricultural ecosystems for the humid tropics. Annual Review of Ecology and Systematics 17: 245–271

Ewel JJ and Bigelow SW (1996) Plant life-forms and tropical ecosystem functioning. In: Orians

GH, Dirzo R and Cushman JH (eds) Biodiversity and Ecosystem Processes in Tropical Forests, Ecological Studies 122, pp 101–126. Springer-Verlag, Heidelberg

Ferdowsian R, George RJ, Lewis MF, McFarlane DJ, Short R and Speed R (1996) The extent of dryland salinity in Western Australia. In: Proceedings of the Productive Use of Saline Lands Conference, pp 89–97. Promaco Conventions, Perth

Fitter A (1997) Nutrient acquisition. In: Crawley MJ (ed) Plant Ecology, 2nd ed, pp 51–72. Blackwell Scientific, Oxford

Gardner WK, Parbery DG and Barber DA (1982) The acquisition of phosphorus by *Lupinus albus* L. I. Some characteristics of the soil-root interface. Plant and Soil 68: 19–32

Grierson PF (1992) Organic acids in the rhizosphere of *Banksia integrifolia* Lf. Plant and Soil 144: 259–265

Grierson PF and Attiwill PM (1989) Chemical characteristics of the proteoid root mat of *Banksia integrifolia* Lf. Australian Journal of Botany 37: 137–143

Grierson PF, Adams MA and Attiwill PM (1992) Biomass of Victoria's forests. Australian Journal of Botany 40: 631–640

Grierson PF, Comerford NB and Jokela EJ (1998) Phosphorus mineralization kinetics and response of microbial phosphorus to drying and rewetting in a Florida Spodosol. Soil Biology and Biochemistry 332: 1323–1331

Grime JP and Hunt R (1975) Relative growth rate: its range and adaptive significance in a local flora. Journal of Ecology 63: 393–422

Grime JP et al (1997) Functional types: testing the concept in northern England. In: Smith TM, Shugart HH and Woodward FI (eds) Plant Functional Types: Their Relevance to Ecosystem Properties and Global Change, pp 122–152. Cambridge University Press

Grove TS and Malajczuk N (1985) Biomass production by trees and understorey shrubs in an age-series of *Eucalyptus diversicolor* F. Muell stands. Forest Ecology and Management 11: 59–74

Hamblin A and Williams J (1995) Soils: key issues in sustaining the land resource base. In: Sustaining the Agricultural Resource Base, pp 31–39. Office of the Chief Scientist, Department of Prime Minister and Cabinet, Canberra

Handreck KA (1996) Phosphorus requirements of Australian native plants. Australian Journal of Soil Research 35: 241–289

Hansen AP and Pate J S (1987) Comparative growth and symbiotic performance of seedlings of *Acacia* spp. in defined pot culture or as understorey components of a eucalypt forest in SW Australia. Journal of Experimental Botany 38: 13–25

Heddle EM and Specht RL (1975) Dark Island Heath (Ninety-Mile Plain South Australia). VIII. The effect of fertilisers on composition and growth, 1950–1972. Australian Journal of Botany 23: 151–164

Hingston FJ, Turton AG and Dimmock GM (1979) Nutrient distribution in karri (*Eucalyptus diversicolor* F. Muell.) ecosystem in south-west Western Australia. Forest Ecology and Management 1: 133–158

Hingston FJ, Dimmock GM and Turton AG (1981) Nutrient distribution in a jarrah (*Eucalyptus marginata* Donn ex Sm.) ecosystem in south-west Australia. Forest Ecology and Management 3: 183–207

Hobbs RJ (1992) Function of biodiversity in mediterranean ecosystems in Australia: definitions and background. In: Hobbs RJ (ed) Biodiversity of Mediterranean Ecosystems in Australia, pp 1–25. Surrey Beatty & Sons, Chipping Norton, NSW, Australia

Holford ICR (1980) Effects of duration of grazed lucerne on long-term yields and nitrogen uptake of subsequent wheat. Australian Journal of Agricultural Research 31: 239–250

Hopper SD (1992) Patterns of plant diversity at the population and species level in south-west Australian mediterranean ecosystems. In: Hobbs RJ (ed) Biodiversity of Mediterranean Ecosystems in Australia, pp 27–46. Surrey Beatty & Sons, Chipping Norton, NSW, Australia

Hooper DU and Vitousek PM (1998) Effects of plant composition and diversity on nutrient cycling. Ecological Monographs 68: 121–149

Hubbell SP and Foster RB (1986) Canopy gaps and the dynamics of a neotropical forest. In: Crawley MJ (ed) Plant Ecology, pp 77–96. Blackwell Scientific Publications, Oxford

Jackson W (1985) New Roots for Agriculture. University of Nebraska Press, Nebraska

Jackson W and Bender M (1984) Investigations into perennial polyculture. In: Jackson W, Berry W, and Colman B (eds) Meeting the Expectations of the Land. Essays in Sustainable Agriculture and Stewardship, pp 183–194. North Point Press, San Francisco

Jackson RB and Caldwell MM (1989) The timing and degree of root proliferation in fertile microsites for three cold-desert perennials. Oecologia 81: 149–154

Jackson RB, Manwaring JH and Caldwell MM (1990) Rapid physiological adjustment of roots to localized soil enrichment. Nature 344: 58–60

Jackson RB and Caldwell MM (1996) Integrating resource heterogeneity and plant plasticity: modelling nitrate and phosphate uptake is a patchy environment. Journal of Ecology 84: 891–903

Jeffrey DW (1967) Phosphate nutrition of Australian heath plants. I. The importance of proteoid roots in *Banksia* (Proteaceae). Australian Journal of Botany 15: 403–411

Jeschke WD and Pate JS (1995) Mineral nutrition and transport in xylem and phloem of *Banksia prinotes* (Proteaceae) a tree with dimorphic root morphology. Journal of Experimental Botany 46: 895–905

Judd TS (1996) Simulated nutrient losses due to timber harvesting in highly productive eucalypt forests and plantations. In: Attiwill PM and Adams MA (eds) Nutrition of Eucalypts, pp 249–258. CSIRO Publishing, Collingwood, Australia

Kielland K (1994) Amino acid absorption by Arctic plants: implications for plant nutrition and nitrogen cycling. Ecology 75: 2373–2383

Lamont BB (1981) Specialized roots of non-symbiotic origin in heathlands. In: Specht RL (ed) Ecosystems of the World. 9B. Heathlands and Related Shrublands, pp 183–195. Elsevier Sci Pub, Amsterdam

Lamont BB (1986) The significance of proteoid roots in proteas. Acta Horticulturae 185: 163–170

Lavalle P and Pashanasi B (1989) Soil macrofauna and land management in Peruvian Amazonia (Yurimaguas, Loreto). Pedobiologia 33: 282–291

Leake JR and Miles W (1996) Phosphodiesters as mycorrhizal P sources I. Phosphodiesterase production and the utilisation of DNA as a phosphorus source by the ericoid mycorrhizal fungus *Hymenoscyphus ericae*. New Phytologist 132: 435–443

Leeper GW (1964) Introduction to Soil Science. Melbourne University Press, Melbourne.

Lefroy EC, Hobbs RJ and Scheltema M (1993) Reconciling agriculture and nature conservation: toward a restoration strategy for the Western Australian wheatbelt. In: Saunders DA, Hobbs RJ and Ehrlich PR (eds) Nature Conservation 3: Reconstruction of Fragmented Ecosystems, pp 243–257. Surrey Beatty & Sons, Chipping Norton, NSW, Australia

Lovering JF and Crabb P (1998) Australia's 200-year experiment in agricultural sustainability. Agricultural Science 11: 17–25

Ludwig JA and Tongway DJ (1996) Rehabilitation of semiarid landscapes in Australia. II. Restoring vegetation patches. Restoration Ecology 4: 398–406

Ludwig JA, Tongway DJ, Freudenberger DO, Noble JC and Hodgkinson KC (eds) (1997) Landscape Ecology, Function and Management: Principles From Australia's Rangelands. CSIRO Publishing, Collingwood, 158 pp

MacFall J, Slack SA and Iyer J (1991) Effects of *Hebeloma arenosa* and phosphorus fertility on root acid phosphatase activity of red pine (*Pinus resinosa*) seedlings. Canadian Journal of Botany 69: 380–383

Malajczuk N and Cromack K Jr (1982) Accumulation of calcium oxalate in the mantle of ectomycorrhizal roots of *Pinus radiata* and *Eucalyptus marginata*. New Phytologist 92: 527–531

Marcellos H (1984) Influence of prior crops of chickpeas, faba beans and lupins on wheat. Journal of the Australian Institute of Agricultural Science 50: 111–113

McDonald GK (1992) Effects of nitrogenous fertilizer on the growth, grain yield and grain protein concentration of wheat. Australian Journal of Agricultural Research 43: 949–967

McInnes KJ and Fillery IRP (1989) Modelling and field measurements of the effect of nitrogen source on nitrification. Soil Science Society of America Journal 53: 1264–1269

Mokwunye AU, Chien SH and Rhodes E (1986) Phosphorus reaction with tropical African soils In: Mokwunye AU and Vlek PLG (eds) Management of Nitrogen and Phosphorus Fertilisers in Sub-Saharan Africa, pp 253–281. Martinus Nijhoff Publishers, Dordrecht, The Netherlands

Mulligan DR (1988) Phosphorus concentrations and chemical fractions in *Eucalyptus* seedlings grown for a long period under nutrient-deficient conditions. New Phytologist 110: 479–486

Mulligan DR and Sands R (1988) Dry matter phosphorus and nitrogen partitioning in three *Eucalyptus* species grown under a nutrient deficit. New Phytologist 109: 21–28

Nadelhoffer KJ, Shaver G, Fry B, Giblin A, Johnson L and McKane R (1996) ^{15}N natural abundance and N use by tundra plants. Oecologia 107: 386–394

Nair PK (1993) An Introduction to Agroforestry. Kluwer Academic Publishers, Dordrecht, The Netherlands, 499 pp

Nziguheba G, Palm CA and Buresh RJ (1998) Soil phosphorus fractions and adsorption as affected by organic and inorganic sources. Plant and Soil 198: 159–168

O'Connell AM and Grove TS (1991) Processes contributing to the nutritional resilience or vulnerability of jarrah and karri forests in Western Australia. In: Ryan PJ (ed) Third Australian Forest Soils and Nutrition Conference, pp 180–197. Forestry Commission of NSW, Sydney

O'Connell AM and Grove TS (1996) Biomass production, nutrient uptake and nutrient cycling in the jarrah (*Eucalyptus marginata*) and karri (*Eucalyptus diversicolor*) forests of south western Australia. In: Attiwill PM and Adams MA (eds) Nutrition of Eucalypts, pp 155–189. CSIRO Publishing, Collingwood, Australia

Odum EP (1998) Productivity and biodiversity: a two-way relationship. Bulletin of the Ecological Society of America 79: 125

Owino F (1996) Selection for adaptation in multipurpose trees and shrubs for production and function in agroforestry systems. Euphytica 92: 225–234

Ozanne PG and Specht RL (1981) Mineral nutrition of heathlands: Phosphorus toxicity. In: Specht RL (ed) Ecosystems of the World Vol 9B. Heathlands and Related Shrublands. Descriptive Studies, pp 209–213. Elsevier Scientific, Amsterdam

Pate JS and Beard JS (eds) (1984) Kwongan, Plant Life of the Sandplain: Biology of a South-West Australian Shrubland Ecosystem. University of Western Australia Press, Nedlands, 284 pp

Polglase PJ, Attiwill PM and Adams MA (1986) Immobilization of nitrogen following wildfire in two eucalypt forests of south-eastern Australia. Oecologia Plantarum 7: 261–272

Raab TK, Lipson DA and Monson RK (1996) Non-mycorrhizal uptake of amino acids by roots of the alpine sedge *Kobresia myosuroides*: implications for the alpine nitrogen cycle. Oecologia 108: 488–494

Rayner M (1991) KARSIM – Modelling the growth and yield of Karri. Agricultural Systems and Information Technology Newsletter, Bureau of Rural Research 3: 59–60

Rorison IH (1987) Mineral nutrition in time and space. New Phytologist 106: 79–92

Russell JS (1980) Crop sequences, crop pasture rotations and soil fertility. In: Proceedings Australian Agronomy Conferences

Sala OE, Lauenroth WK, McNaughton SJ, Rusch G and Zhang X (1996) Biodiversity and ecosystem functioning in grasslands. In: Mooney, HA, Cushman JH, Medina E, Sala OE and Schulze E-D (eds) Functional Roles of Biodiversity: A Global Perspective, pp 129–149. John Wiley and Sons Ltd, Chichester

Sanford P, Pate JS, Unkovich MJ and Thompson AN (1995) Nitrogen fixation in grazed and ungrazed subterranean clover pasture in south-west Australia assessed by the ^{15}N natural abundance technique. Australian Journal of Agricultural Research 46: 1427–1443

Saunders DA, Arnold GW, Burbridge AA and Hopkins AJM (eds) (1987) Nature Conservation:

The Role of Remnants of Native Vegetation, Surrey Beatty & Sons, Chipping Norton, NSW, Australia, 410 pp

Schmidt S and Stewart SR (1997) Waterlogging and fire impacts on nitrogen availability and utilization in a subtropical wet heathland (wallum). Plant Cell and Environment 20: 1231–1241

Schulze E-D and Mooney HA (1993) Biological Diversity and Ecosystem Functioning. Springer, New York

Shea SR and Bartle JR (1988) Restoring nature's balance. The potential for major reforestation of south Western Australia. Landscope 3: 3–14

Shea SR and Dell B (1981.) Structure of the surface root system of *Eucalyptus marginata* Donn ex Sm. and its infection by *Phytophthora cinnamomi* Rands. Australian Journal of Botany 29: 49–58

Skinner MF and Attiwill PM (1981) The productivity of pine plantations in relation to previous land use. I. Growth responses in agricultural and forest soils. Plant and Soil 60: 161–176

Smaling EMA and Braun AR (1996) Soil fertility research in sub-Saharan Africa – new dimensions, new challenges. Communications in Soil Science and Plant Analalysis 27: 365–386

Smaling EMA and Fresco LO (1993) A decision-support model for monitoring nutrient balances under agricultural land use (NUTMON). Geoderma 60: 235–256

Solbrig OT (1991) From genes to ecosystems: a research agenda for biodiversity. IUBS, Paris

Sparling GP, White KN and Ramsay AJ (1985) Quantifying the contribution from the soil microbial biomass to the extractable P levels of fresh and air-dried soils. Australian Journal of Agricultural Research 23: 613–621

Specht RL (1963) Dark Island Heath (Ninety-Mile Plain South Australia) VII The effect of fertilisers on composition and growth, 1950–1960. Australian Journal of Botany 11: 67–94

Specht RL (1972) The Vegetation of South Australia. Government Printer, Adelaide, 328 pp

Specht RL and Groves RH (1966) A comparison of the phosphorus nutrition of Australian heath plants and introduced economic plants. Australian Journal of Botany 14: 201–221

Specht RL, Moll EJ, Pressinger F and Sommerville J (1983) Moisture regime and nutrient control of seasonal growth in Mediterranean ecosystems. In: Kruger FJ, Mitchell DT and Jarvis JUM (eds) Mediterranean-Type Ecosystems, pp 120–132. Springer-Verlag, Berlin

Sprent JI (1995) Legume trees and shrubs in the tropics: N2 fixation in perspective. Soil Biology and Biochemistry 27: 401–407

Stewart HTL, Flin DW and Hopmans P (1985) On harvesting and site productivity in eucalypt forests. Search 16: 206–208

Swift MJ, Heal OW and Anderson JM (1979) Decomposition. In: Terrestrial Ecosystems. University of California Press, Berkeley, California, 372 pp

Tarafdar JC and Claassen N (1988) Organic phosphorus compounds as a phosphorus source for higher plants through the activity of phosphatases produced by plant roots and microorganisms. Biology and Fertility of Soils 5: 308–312

Tennant D, Siddique KHM and Perry MW (1991) Crop growth and water use. In: Perry M and Hillman B (eds) The Wheat Book, pp 39–51. Department of Agriculture Bulletin 4196, Western Australia

Troeh FR and Thompson LM (1993) Soils and Soil Fertility, 5th ed. Oxford University Press, New York, 462 pp

Unkovich MJ, Pate JS, Armstrong EL and Sanford P (1995) Nitrogen economy of annual crop and pasture legumes in southwest Australia. Soil Biology and Biochemistry 27: 585–588

Unkovich MJ, Sanford P, Pate JS and Hyder M (1998) Effects of grazing on plant and soil nitrogen relations of pasture-crop rotations. Australian Journal of Agricultural Research 49: 475–485

Van Noordwijk M, De Willigen P, Ehlert PAI and Chardon WJ (1990) A simple model of P uptake by crops as a possible basis for fertilizer recommendations. Netherlands Journal of Agricricultural Science 38: 317–332

Vitousek PM (1984) Litterfall, nutrient cycling, and nutrient limitation in tropical forests. Ecology 65: 285–298

Vitousek PM, Aber JD, Howarth RW, Likens GE, Matson PA, Schindler DW, Schlesinger WH and Tilman GD (1997) Human alteration of the global nitrogen cycle: causes and consequences. Issues in Ecology 1: 1–15

Vitousek PM, Gosz JR, Grier CC, Melillo JM and Reiners WA (1982) A comparative analysis of potential nitrification and nitrate mobility in forest ecosystems: a field experiment. Ecological Monographs 52: 155–177

Vogt K A, Vogt DJ, Palmiotto PA, Boon P, O'Hara J and Asbjornsen H (1996) Review of root dynamics in forest ecosystems grouped by climate, climatic forest type and species. Plant and Soil 187: 159–219

Wallace IM, Dell B and Loneragan JF (1986) Zinc nutrition of jarrah (*Eucalyptus marginata* Donn ex Sm.) seedlings. Australian Journal of Botany 34: 41–51

Ward SC and Pickersgill GE (1985) Biomass and nutrient distribution in eucalypt plantations growing on rehabilitated bauxite mines. Australian Journal of Ecology 10: 111–124

Wardell-Johnson G and Horwitz P (1996) Conserving biodiversity and the recognition of heterogeneity in ancient landscapes: a case study from south western Australia. Forest Ecology and Management 85: 219–238

Wardle DA, Bonner KI and Nicholson KS (1997) Biodiversity and plant litter: experimental evidence which does not support the view that enhanced species richness improves ecosystem function. Oikos 79: 247–258

Warren G (1992) Fertiliser phosphorus sorption and residual value in tropical African soils. NRI Bulletin 37, Natural Resources Institute, Chatham, England

White R (1993) Is a decline in soil fertility contributing to land degradation in Australia? Agricultural Science 6: 45–48

Witowski ETF (1994) Growth of seedlings of the invasives *Acacia saligna* and *Acacia cyclops* in relation to soil phosphorus. Australian Journal of Ecology 19: 290–296

Woodward FI (1993) How many species are required for a functional ecosystem? In: Schulze E-D and Mooney HA (eds) Biodiversity and Ecosystem Function. Springer-Verlag, New York

Yates CJ and Hobbs RJ (1997) Woodland restoration in the Western Australian wheatbelt: a conceptual framework using a state and transition model. Restoration Ecology 5: 28–35

Yates CJ, Hobbs RJ and Bell RW (1994) Landscape-scale disturbances and regeneration in semi-arid woodlands of south western Australia. Pacific Conservation Biology 1: 214–221

Agroforestry Systems **45**: 245–275, 1999.

Assessing the performance of woody plants in uptake and utilisation of carbon, water and nutrients

Implications for designing agricultural mimic systems

J. S. PATE[1,*] and T. E. DAWSON[2]
[1] *Department of Botany and Centre for Legumes in Mediterranean Agriculture, The University of Western Australia, Nedlands 6907, Australia;* [2] *Section of Ecology and Systematics and the Laboratory for Isotope Research and Analysis at Cornell University, It, USA (*Author for correspondence: darth@cyllene.uwa.edu.au)*

Key words: carbon assimilation, nutrient transport, resource utilisation, stable isotope techniques, water-use efficiency

Abstract. This article introduces a series of innovative experimental approaches for monitoring current and past performances of woody plant species in utilisation of key resources within ecosystems. We demonstrate how such information might be useful for monitoring the progress of any novel agricultural system which attempts to mimic such natural systems. The experimental data presented principally involve the use of stable isotope techniques coupled to collection and analysis of substances transported in the sap of xylem and phloem. Study species examined are sugar maple (*Acer saccharum*), tagasaste (*Chamaecytisus proliferus*), blue gum (*Eucalyptus globulus*), and acorn banksia (*Banksia prionotes*). Topics covered include investigations of (a) seasonal changes in water source utilisation using hydrogen stable isotope analyses, (b) water loss from tree roots to the surrounding soil (termed hydraulic lift) using hydrogen stable isotope analyses, (c) relationships between carbon isotope discrimination of plant dry matter and whole-plant water-use efficiency, (d) phloem sap analysis as a means for assessing aspects of a plant's water and nutrient status, (e) carbon isotope ratios of foliage and annual growth rings as they relate to seasonal changes in carbon isotope ratios of phloem translocate, and (f) the composition of simultaneously collected samples of xylem and phloem sap as a means of monitoring current plant performance in uptake and mobilisation of key mineral resources. The chapter concludes with discussion of prospective uses of the above-mentioned techniques in study of managed ecosystems in which woody species play pivotal roles in maintenance of functional stability.

1. Introduction

Whether studying relatively simple herbaceous tree/plant alley-cropping systems or the complex assemblages of species typical of natural vegetation, it is imperative that ecophysiologists develop suitably definitive experimental means for evaluating performances of key species in utilisation of critical limiting resources within the systems under study. As far as woody species are concerned, the sheer size, longevity and morphological heterogeneity of above- and below-ground biomass present daunting challenges when assessing how various species contribute in relative and absolute terms to resource capture and annual net productivity, and how each responds to various stresses on daily, seasonal and year-to-year time scales. Traditionally, tree growth rates are not estimated by direct harvesting and weighing, but assessed indirectly using allometrically-calibrated measurements of bole diameter and trunk height, or, over longer time frames, by analyses of growth rings. Similarly, nutrient status is usually measured by targeted foliar analyses rather than on a whole tree basis, while parallel measurements of water and CO_2 exchanges and responses to soil water deficits are mostly confined to leaf cuvette-based assays, heat pulse-based measurements of water flow in trunks and pressure chamber-based measurements of daytime and pre-dawn water potentials.

Relatively recently the application of stable isotope techniques has provided novel classes of information complementing the above types of studies and, in so doing, may be regarded as revolutionising ecophysiological studies of many woody taxa (Ehleringer et al., 1993; Ehleringer and Rundel, 1988; Tiezen and Boutton, 1988). Foremost among such studies are those using carbon isotope (^{13}C:^{12}C) ratios of tissues as indices of water stress and water-use efficiency (Farquhar et al., 1989), whether by analysis of the carbon of dry matter of whole plants, foliage or the dry matter laid down in growth rings of trunk wood during the lifetime of a tree (e.g. see Levitt and Long, 1988; Guy et al., 1988; Read and Farquhar, 1991; Waring and Silvester, 1994). Also of considerable value has been the use of D:H and ^{18}O:^{16}O stable isotope analyses of rainfall, ground water, soil water and xylem (sap) to assess reliance on different water sources by trees during specific periods of growth (Ehleringer and Dawson, 1992; Dawson, 1993b). Hydrogen stable isotope analyses have also been used specifically to test whether water obtained from depth is being excreted from superficial lateral roots into the surrounding soil by a phenomenon known as 'hydraulic lift' (see Dawson, 1993a; Dawson and Pate, 1996; Horton and Hart, 1998).

Another novel and equally incisive set of experimental techniques relevant to plant resource usage revolves around analyses of transport fluids of xylem and phloem collected from target regions of a tree or shrub (Pate and Jeschke, 1993, 1995). Such analyses provide instructive commentaries on the amounts of different organic and inorganic solutes mobilised through specific transport pathways. Commonly, the xylem sap of woody species is obtained by mild vacuum extraction of root systems, twigs and main shoot branches, and the resulting samples usually referred to as 'tracheal sap' (Bollard, 1960; Pate, 1980). This technique allows monitoring of current performance of roots in mineral uptake from the soil and in the subsequent synthesis and recycling of certain organic compounds within the above ground system. Unfortunately matching information on contents of translocatory streams of phloem of woody taxa is sparse, being restricted to those few species whose phloem 'bleeds' when shallow incisions are made in bark, or where it has proved possible to collect stylet exudates from aphids feeding on twigs (Pate, 1975, 1986; Fisher, 1983). Not surprisingly, only a few studies have been able to compare the composition of xylem sap with that of phloem sap collected simultaneously from the same woody species. Examples of such data will be provided later.

The objective of this paper is to highlight the above-mentioned techniques for assessing physiological attributes of trees, and to do so within the context of their potential role in functional studies of the various agroforestry mimic systems considered elsewhere in this volume. Continuity and authenticity will be provided by concentrating on species which we have studied in detail in our respective laboratories. These are:

1. sugar maple (*Acer saccharum*), a North American temperate deciduous tree which engages in hydraulic lift of water and also comprises a valuable source of commercially exploited xylem-derived sugar (maple syrup);
2. tagasaste (*Chamaecytisus proliferus*), a leguminous tree from the Canary Islands and now planted in farmlands on deep sands in southern Australia as summer fodder for sheep and cattle and with the long term objective of rehabilitating land experiencing dangerously rising water tables;
3. the Tasmanian blue gum (*Eucalyptus globulus*), a plantation tree crop widely planted for wood chip and timber production in southern Australia;
4. the acorn banksia (*Banksia prionotes*), a deep-rooted proteaceous tree native to sandplain regions of Western Australia.

The six case studies presented impinge upon some of the major issues which should be addressed and the types of information which one needs to obtain when studying integrated resource use by woody plants.

2. Stable isotopes of carbon, hydrogen and oxygen – terminology and biological application

Before venturing further, readers unfamiliar with the general background to and terminology used in stable isotope technologies in respect of carbon, oxygen and deuterium:hydrogen should study this section.

The natural abundances of the heavy stable isotopes of carbon hydrogen and oxygen (^{13}C, D(^{2}H) and ^{18}O respectively) relative to the respective lighter isotopes ^{12}C and H and ^{16}O in a given sample are traditionally expressed by the numerical ratio of atoms of heavy to light isotope.

2.1. *Carbon isotope values in plants*

For carbon, the ^{13}C:^{12}C ratio is about 1.1 percent. Because natural variations in this ratio span only a few percent of this 1.1 percent, natural carbon isotope composition is usually expressed as deviation in per mil (‰ = parts per thousand) from the isotope composition of a particular standard. These deviations, or relative differences, are symbolized by $\delta^{13}C$, or δ, and defined as:

$$\delta^{13}C = [R_{sample}/R_{standard} - 1] \times 1000, ‰ \qquad (1)$$

where R = ^{13}C:^{12}C. The usual standard against which all measurements are referred is PDB (Pee Dee Belemnite), namely CO_2 obtained from the carbonate shell of a Cretaceous mollusc, *Belemnitella americana* from the Pee Dee Formation in South Carolina. Since organic matter and atmospheric CO_2 are depleted in ^{13}C when compared with carbonates, $\delta^{13}C$ values turn out as negative numbers. The $\delta^{13}C$ values of standard and sample are obtained by high precision mass spectrometric techniques.

Carbon dioxide fixed initially by RUBP carboxylase [Rubisco] (C_3 plants) and by PEP carboxylase (C_4 and CAM plants) show different $\delta^{13}C$ values, since the former enzyme discriminates 27 to 30‰, the latter only 0 to 2‰. $\delta^{13}C$ values for C_3 plants range from –20 to –34‰ and for C_4 plants from –9 to –16.8‰. CAM plants which always operate in this mode have $\delta^{13}C$ values similar to C_4 plants, but many CAM plants are able to shift faculatively back and forth between C_3 and CAM photosynthesis depending on aridity conditions. In these 'Facultative CAM' species, $\delta^{13}C$ values are intermediate between those typical of C_3 and C_4 plants.

Historically, carbon isotope values for plants have been commonly expressed using the 'small delta' notation (δ, see equations 1 and 5). However, in 1984 Farquhar and Richards formulated a more direct expression for the biological fractionation which occurs in plant leaves during the carbon fixation process. They suggested using 'big delta' (Δ), to assess actual discrimination by the leaf. This expression measures the leaf isotopic composition relative to the incident source of carbon dioxide (atmospheric CO_2) rather than that relative to the arbitrary PDB standard. The isotope effect (α) associated with carbon accumulation by a leaf is then defined as,

$$\alpha = R_{source}/R_{product} \; (= R_a/R_p) \qquad (2)$$

where R_a and R_p are the $^{13}C{:}^{12}C$ composition of the air and the plant.

Discrimination by the plant leaf is accordingly given in terms of deviation from unity.

$$\Delta = \alpha - 1 \qquad (3)$$

When the isotopic composition of the leaf is expressed as Δ, plants consistently show positive discrimination against ^{13}C in the case of C_3 species, for which a typical discrimination value would be 22×10^{-3} or 22‰. In terms of how Δ is related to δ (relative to the PDB standard), we use the following equation,

$$\Delta = (\delta_a - \delta_p) \,/\, (1 + \delta_p) \qquad (4)$$

where *a* and *p* denote air and plant, respectively. It is important to note, that when actually using the equation above, δ should be expressed in 'absolute' rather than 'relative units' units (0.022 or 22×10^{-3}, but not –22). The advantage of reporting carbon isotope values as Δ is that it directly expresses the consequences of biological processes, that is where the isotope ratio reflects both the source air isotope composition and biological activity.

For plants with C_3 photosynthesis, we commonly express the carbon isotope ratio as,

$$\delta^{13}C_{plant} = \delta^{13}C_{air} - a - (b - a) \times (ci/ca) \quad (5)$$

where a = 4.4‰ and represents the isotope fractionation associated with differential diffusivities of $^{13}CO_2$ to $^{12}CO_2$ in air through plant stomata, *ci* and *ca* refer respectively to ambient CO_2 concentrations in the internal gas space of the leaf and the surrounding atmosphere, b = 27‰ and denotes the net fractionation by Rubisco during the photosynthetic carbon fixation reaction, and $\delta^{13}C_{air}$ = –8‰ (see Farquhar et al., 1988). Referring alternatively to Δ notation, the above expression becomes,

$$\Delta = a + (b - a) \times (ci/ca) \quad (6)$$

The ratio *ci/ca* is determined by the interaction of CO_2 assimilation rate and the extent of stomatal opening (i.e. by the supply of CO_2 to and utilization of this CO_2 at sites of carboxylation). A plant with relatively closed stomata will tend to have a low *ci/ca* and smaller Δ values (towards 4.4‰) than one with relatively open stomata and therefore higher *ci/ca* and larger Δ values (towards 27‰). To date, *ci/ca* values have been shown to range from 0.4 to 0.8 or a Δ range of 13 to 22‰ (Ehleringer, 1988; Ehleringer and Rundel, 1988; Ehleringer et al., 1991) or in $\delta^{13}C$ terminology a range from –21 to –30‰.

Dawson and Ehleringer (1993) noted that there is a relationship between Δ and water-use efficiency (the molar ration of photosynthesis (A) to transpiration (E), or A/E). They wrote, 'since the carbon incorporated into leaf tissues is assimilated over a considerable period of time and under more than one set of environmental conditions, measurements of Δ provide an average *estimate* of *ci/ca* and, therefore, an index to time-integrated (and flux-weighted) plant metabolism. Because both Δ and A/E are a function of *ci/ca* (Equation 6), Δ can be used to estimate time-integrated or flux-weighted – long-term – A/E for the same foliage from which gas exchange observation are being made'. This is only true provided that:

1. the leaf-to-air vapour pressure gradients (v) are equivalent and known;
2. each plant leaf being compared is experiencing the same range of v during tissue synthesis, and
3. plant species do not differ in the fraction of the carbon fixed during photosynthesis which they subsequently lose during respiratory processes (Farquhar and Richards, 1984).

Because it is unlikely that assumptions 1–3 stated above will be strictly met, Δ-values are unlikely to provide a direct, quantitative measure of long-term A/E. As such, Δ values should be considered absolute indices of integrated *ci/ca* but only relative indicies of A/E. Secondly, the reader should note that Δ is inversely proportional to A/E (Farquhar and Richards, 1984) so that high water-use efficiency in leaves inevitably leads to low Δ. This is of course the opposite of $\delta^{13}C$ based measurements, where good water use efficiency would be associated with high (less negative) values.

The two carbon isotope expressions (Equations 5 and 6) are related to one another as,

$$\Delta = (-8 - \delta^{13}C_{plant}) \,/\, [1 + (\delta^{13}C_{plant}/1000)] \qquad (7)$$

Leaves collected for the determination of Δ or $\delta^{13}C$ are dried to constant mass and homogenized to a very fine powder. The carbon isotope composition is obtained from the cryogenically purified CO_2 produced by combustion of the dried sample in a sealed tube containing cupric oxide and a silver foil catalyst (see Ehleringer and Osmond, 1989; Smedley et al., 1991 for details) and then measured on an isotope ratio mass spectrometer (IRMS). Overall precision of the preparation and analysis is generally about ±0.10 to 0.30‰, though some of the newer IRMS models can give even higher precision (e.g. ±0.01 to 0.08‰).

Throughout the chapter we will express the carbon isotope discrimination values for our study species in the more widely used $\delta^{13}C$ rather than according to the above mentioned alternative Δ notation.

2.2. *Hydrogen isotope and oxygen values in water*

For hydrogen or oxygen stable isotopes in water the isotope ratios are similar to that shown above for carbon in Equation 1, but where $\delta^{18}O$ and δD are expressions which contain R as either the D.H. The international standard for water is generally SMOW or V-SMOW (Standard Mean Ocean Water or Vienna-SMOW, respectively) and the isotope ratios of these standards are rated as 0.00‰ for both H and O (see Dawson, 1993a).

Freshwater is usually depleted in the heavier isotope D and ^{18}O relative to SMOW so that most values one sees in the literature are negative (Gat, 1996). There are many factors which contribute to the over 400‰ variation observed in the δD and corresponding 70‰ variation in ^{18}O of meteoric waters. Perhaps the most important fractionation effects are associated with phase changes of water (e.g. evaporation and condensation) and thermal effects, both influencing equilibrium and kinetic isotope fractionation processes. A discussion of these factors can be found in Dawson (1993a, 1993b), Dawson and Ehleringer (1998), and Gat (1996). Briefly, evaporation and condensation processes approximate Rayleigh distillation (Gat, 1996).

Condensation depletes the vapour (cloud) phase while the liquid phase (rain or snow) becomes enriched in ^{18}O and D because these isotopic species have lower vapour pressures (see below). As the process continues, the $\delta^{18}O$ and

δD values of water vapour in an air mass (the cloud) become progressively more negative (isotopically lighter). This means that δ values of the liquid or solid precipitation will also become progressively more negative as the airmass looses more and more water. Other factors, such as atmospheric humidity and, as mentioned above, temperature, can have a marked influence on this process, leading to substantial local variation in the isotopic values of water that might be used by plants. In arid and semi-arid regions of the world, strong deviations from what is predicted for the δD and $\delta^{18}O$ of waters are observed because of excessive evaporation. These deviations are caused primarily by kinetic (unidirectional) isotope effects taking place during evaporation when (a) water moves into an unsaturated atmosphere (much less than 100% relative humidity), and (b) extensive re-evaporation of water from the surface of raindrops beneath clouds interacting with surface water in the dry ground. One can accordingly find apparently unaccountably wide differences in δD and $\delta^{18}O$ values of rain water between sites and seasons in a region such as Western Australia.

Vapour pressure, freezing-point, and temperature also influence the δD and $\delta^{18}O$ of waters. A great deal of the variation observed in δD and $\delta^{18}O$ in water is caused by vapour pressure effects, and to a lesser degree by freezing-point differences of water that contains different 'species' of water (e.g. $H_2{}^{18}O$ vs. $H_2{}^{16}O$) as a function of temperature. For example, in a mixture of water, both HDO and H_2O exist, but the vapour pressure of HDO is slightly lower than that of H_2O and the concentration of D accordingly higher in the liquid than vapour phase. For this reason, water vapour formed by evaporation of liquid water is enriched in the lighter isotopes ^{16}O and H while the remaining liquid water is enriched in the heavy isotopes ^{18}O and D. Because temperature influences the fractionation factor (a) and is the ratio of vapour pressures between liquid and vapour phases, temperature is a major factor in determining the isotopic composition of water.

Because plants do not fractionate water during the uptake process, if one characterizes the isotopic variation in waters that might be used by plants and then also measures the isotope variation in water extracted from the xylem sap, one can determine the water sources used by the vegetation (see Ehleringer and Dawson, 1992; Dawson, 1993a, 1993b; Dawson and Pate, 1996; Dawson and Ehleringer, 1998; White, 1988 for reviews). However, it is imperative that xylem sap is obtained far removed from transpiring surfaces of a plant since evaporation from leaves causes radical changes in δD. For example in *Eucalyptus globulus* during the day the δD of xylem sap of trunks is typically in the range –20 to –30‰, whereas that of actively transpiring foliage may be +10 to +20‰ (T. E. Dawson and J. S. Pate, unpubl. data).

3. Case studies

3.1. *Combining hydrogen isotope, sap flow and dye injection techniques to test for hydraulic lift of water by roots*

Hydraulic lift by sugar maple in natural woodland near Cornell University, USA has been studied by collecting and analysing soil and ground water, water extracted from the wood of trunks and roots of trees, and water obtained from roots of understorey herbaceous species and soil within and outside the rooting catchment of a study tree (see Dawson, 1993a, 1996; Emerman and Dawson, 1996). Figure 1 summarises a typical set of results for high summer, with all values shown expressed in delta notation (δD) as described in section 2. The δD values obtained (Figure 1) indicate that ground water recently absorbed by the deeply-penetrating tap roots of the tree moves distally outwards into the superficially extending lateral roots and is then released into surrounding soil. Shallow-rooted understorey herbaceous species then obtain this 'hydraulically-lifted' water and appear to benefit measurably so in terms of improved water relations and growth during summer (Dawson, 1993a). Potential advantages to the tree engaging in such hydraulic lift would include:

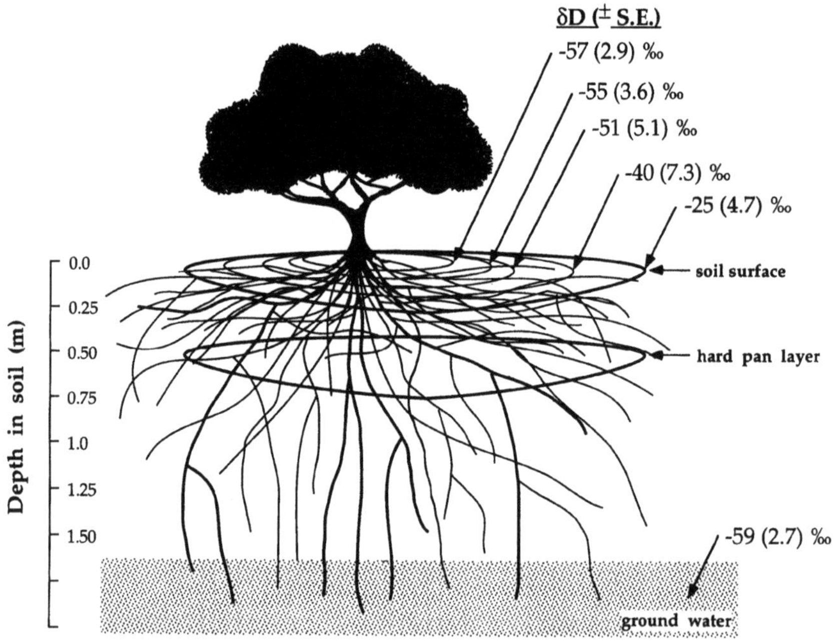

Figure 1. The average (+/– s.d.) hydrogen stable isotope ratios (ΔD) of soil water at 20 cm below the soil surface collected at regular spacings outwards from the bole of a mature sugar maple tree. The tree in question was concluded to be engaging in hydraulic lift, i.e. night time loss from its surface roots of water transported upwards via tap roots from deeper, moist, layers. *Source:* Redrawn from Dawson (1993a).

a) effective exploitation of valuable ground water resources;
b) amelioration of water stress in summer (Dawson, 1995, 1996);
c) maintenance of soil moisture sufficient to promote solubilisation of soil nutrients (Dawson, 1998);
d) sponsoring of microbial processes releasing nutrients from organic and mineral fractions of the soil; and
e) maintenance of integrity of symbiotic mycorrhizal associations through periods when the soil is dry (see Caldwell et al. (1998) and the review by Horton and Hart (1998) on hydraulic lift in plants).

In a recent test for hydraulic lift in acorn banksia in native bush in Western Australia we have incorporated information from heat pulse probes inserted into tap roots, lateral roots and trunks of trees to follow the magnitude and direction of flow within these complementary xylem streams of a tree (S. S. O. Burgess, J. S. Pate and T. E. Dawson, unpubl. data). Data for early summer of 1997 demonstrated appreciable net outward flow of tap root-derived water into certain lateral roots during the night, and for one root during both night and day. The simple in vivo technique of syringing basic fuchsin or fast green dyes into the xylem of lateral roots was also used to assess directions of flow in a number of trees. This provided instances of lateral roots of certain trees moving water from lateral root to trunk during the day while reversing flow away from the trunk at night and also, during high summer, of certain laterals continuously drawing water outwards from the rest of the root system. It was not known of course from such studies whether definite excretion of the distally moving ground water (i.e. true hydraulic lift) was occurring from such laterals into the soil or whether the instances of outward flow observed merely represented recharge of root tissues following loss of water through the periderm of mature parts of the root.

Acorn banksia possesses dense superficial groupings of nutrient-absorbing roots termed proteoid or cluster roots (see Pate and Bell, this volume; Pate et al., 1995, 1998). In a recent study conducted on the species in early summer (November, 1997) we bagged discrete clusters of these roots at dusk while taking great care to ensure that each enclosed cluster remained connected at its base to its parent lateral root. As shown in Table 1, water condensing in the bag overnight carried a δD signature very similar to that present in the xylem of the lateral root to which the cluster was attached. Moreover, the δD of bag water of the larger of the two trees studied turned out to be very different from that present in adjacent soil not exploited by plant roots (Table 1). The same applied to lesser extent to the smaller tree whose growth had been stunted by shading of surrounding trees.

The experimental approaches outlined above clearly provide strong evidence that hydraulic lift does indeed occur in acorn banksia during summer but the extent to which this might be of adaptive value in extending the functional life of nutrient-absorbing proteoid roots has yet to be evaluated.

Table 1. Tree size (diameter at breast height), plant community type, and percent change over a 12-hour night in the water content of silica-gel in a bag enclosing still attached groups of proteoid root clusters of 35-year old Banksia prionotes trees at Moora, Western Australia. Mean (± s.d.) stable hydrogen isotope ratios (δD), are given for xylem water of the main lateral root, water absorbed by the silica-gel and water extracted from root-free bulk soil adjacent to the bagged roots. Different superscripts indicate that δD values are significantly different from each other [Student's *t*-test, $P < 0.05$]. All δD values expressed in 'per mil' (‰) notation with negative values denoting depletion and positive values enrichment in the heavier (D) isotope.

Tree size (dbh–cm)	Plant community type	% change in water content of bag overnight	Root δD (‰)	Bag δD (‰)	Soil δD (‰)
4.9–5.7	Small stunted tree in dense, closed woodland	5.5 ± 2.0	–7.6 ± 3.0 a	–10.4 ± 0.3 a	-3.1 ± 1.7 b
14.6–17.1	Large rapidly growing tree in mature, open woodland	6.9 ± 1.2	–47.5 ± 0.2 c	–47.1 ± 2.0 c	+13.1 ± 12.0 d

Source: T. E. Dawson, J. S. Pate and S. S. O. Burgess (unpubl. data).

3.2. *Hydrogen isotope analysis of rainfall, xylem sap, soil and ground water as a means of tracking seasonal changes in plant utilisation of different water sources*

A recent study (Dawson and Pate, 1996) defining seasonalities in utilisation of different water sources by *Banksia prionotes* in its native habitat is summarised in Figure 2A. The data obtained over an 18 month period show changes in the δD of xylem (tracheal) sap obtained from:

a) lateral roots accessing soil water from the top 0.5 m of the soil profile,
b) sinker (tap) roots tapping ground water at 2 to 3 m depth, and
c) trunk xylem presumed to be deriving varying amounts of water from one or both water sources.

Seasonal variations in δD of rainwater and ground water are included in the figure to illustrate the consistently different ranges of variations in δD values of these two sources during the study period.

Seasonal changes in δD of water within the xylem sap of the main stem indicated that superficial, lateral roots supplied most water used by the canopy through the wet winter and early spring but that, with the onset of the dry summer months, the tree switched towards using ground water through its deep tap (sinker) root. Then in mid-summer through to early autumn, when the upper soil was almost completely dehydrated, δD values of xylem water of lateral roots and trunk matched closely those of ground water moving up through the tap root. We therefore concluded that ground water must have been vital for maintaining transpiration and photosynthesis during summer

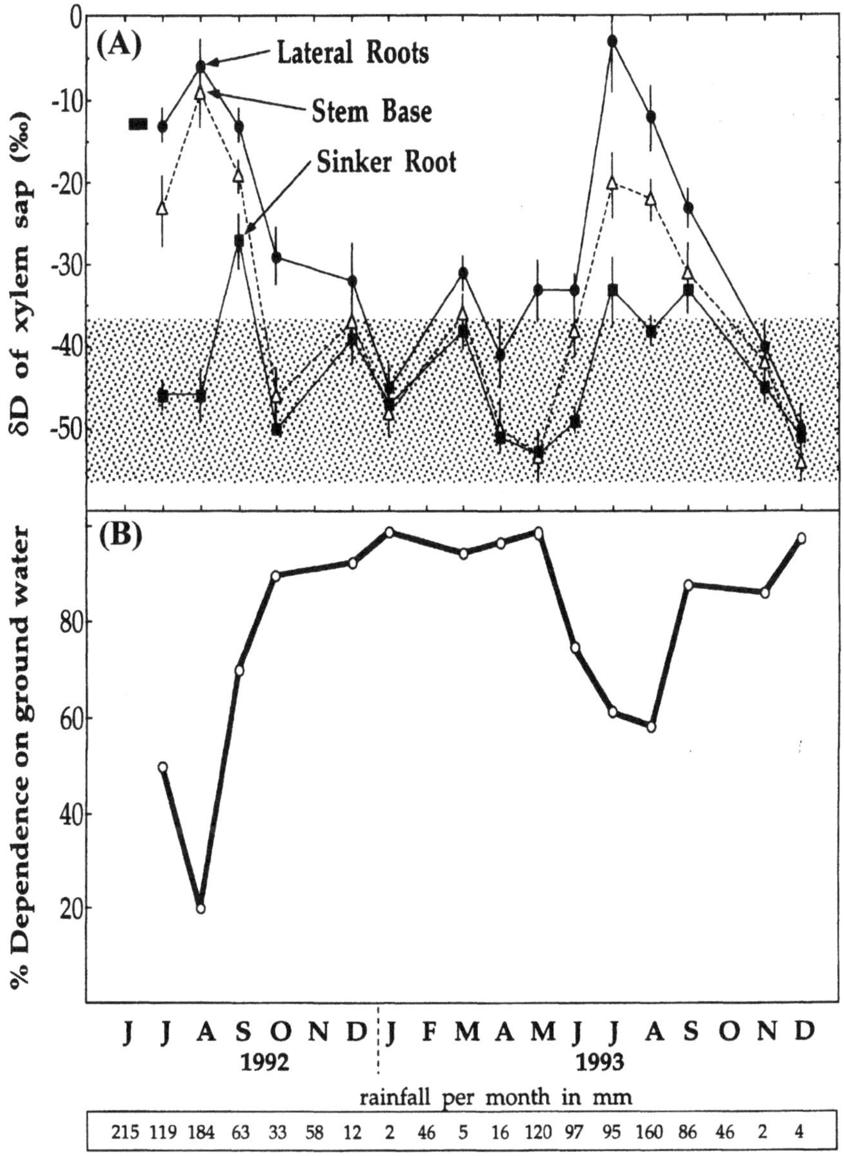

Figure 2. A) Seasonal changes in the δD of tracheal xylem sap extracted from lateral roots, main tap (sinker) root and main stem (trunk) base of trees of *Banksia prionotes* in native bushland in south western Australia. Rainfall collected at the site carried δD values between 0 and –20 ‰ over the study period. The comparable range of δD values of ground water is shown by the stippled bar at the bottom of the figure (δD notation defined as in section 2), B) Estimates of seasonal changes in proportional dependence of the same *Banksia* trees on ground water, deduced from the data shown in A and the 'mixing model' approach outlined in the text. Monthly rainfall (mm) data for the study period were as shown at the base of the figure. *Source:* Redrawn from Dawson and Pate (1996).

extension growth of the shoot, while also keeping lateral roots well hydrated throughout the long dry summer (see also discussion by Pate et al., 1998). The data are consistent with the earlier mentioned conclusion that acorn banksia has the capacity to engage in hydraulic lift at certain times of the year.

Use of a simple two-water-source mixing model has enabled us to use the δD data shown in Figure 2A to follow dependence of the banksia trees on surface versus ground water during the study period (Figure 2B). Unfortunately such good discrimination between isotope signals of ground and soil water is not always manifest in an ecosystem, especially where complex soil profiles are involved (see White et al., 1985; White, 1988; Walker and Richardson, 1991; Dawson, 1993b, 1995; Thorburn and Walker, 1993; Brunel et al., 1995; Dawson and Ehleringer, 1998 for discussions on the use and limitations of isotopic mixing models).

Working also on deep sand ecosystems, where we were again lucky to find predictable patterns of change in δD soil and ground water, we undertook a similar δD based study on tagasaste (*Chamaecytisus proliferus*) grown in the alley crop and plantation systems described by Lefroy and Stirzaker (this volume). Unfortunately the complexity of the root system of this species did not permit us to distinguish which parts of a root system were superficially extending roots and which terminated in tap roots. δD measurements were accordingly restricted to rain and ground water, trunk tracheal xylem sap from the trees themselves, and root bleeding sap obtained concurrently from shallow-rooted reference crop and weed species. The latter supposedly provided a bioassay for δD of plant available water in upper parts of the soil profile (< 40–50 cm deep). Using this approach, and the simple two-water-source isotopic mixing model above for *Banksia prionotes*, seasonal changes in proportional dependencies on soil and ground water were estimated as illustrated in Figure 3. The comparison involved alley crop with lupin, alley crop with weed free fallow areas between the tree rows, and tagasaste at a plantation density five times that of the two alley formats (see Lefroy and Stirzaker, this volume). The data (Figure 3) bore clear evidence of earlier and more pronounced recourse to ground water in late spring though to early summer (November/December) in the highly transpiring plantation and alley crop formats than in the alleys of tagasaste in which annual vegetation had been eliminated by chemical fallowing.

Included also in the above study were late summer measurements of δD of trunk xylem sap of two indigenous tap-rooted tree species (*Eucalyptus todtiana* and *Banksia prionotes*) at the study site, both of which were presumably exploiting the same ground water (δD of –25.4‰) as the tagasaste trees. The eucalypt and banksia sampled showed xylem sap δD values of –19.2‰ and –19.4‰ respectively, which translated to 75% dependence on ground water. Concurrently sampled plantation format tagasaste showed a xylem sap δD value of –17.0‰, indicating an almost as great seasonal peak in dependence on ground water.

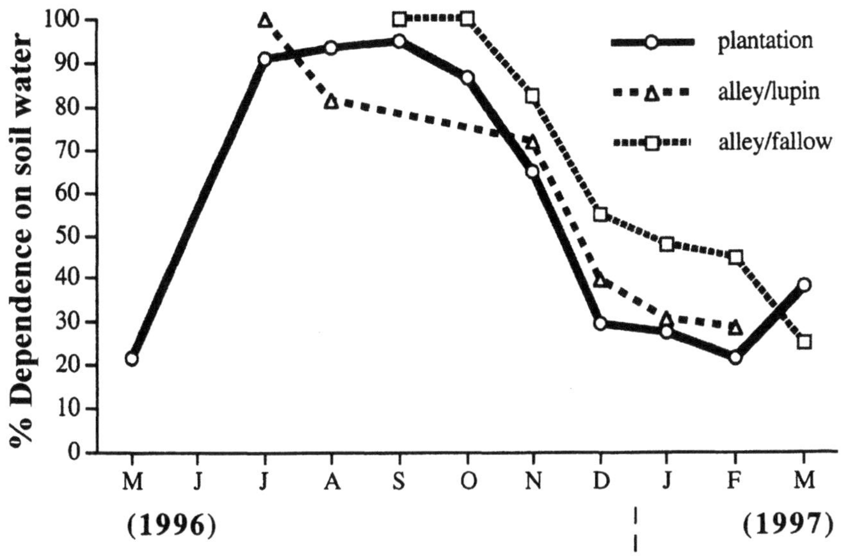

Figure 3. Comparison of the proportional dependencies of plantation, alley crop and alley fallow plantings of tagasaste on ground water during the critical soil drying out phase from late spring through to peak stress in late summer at Moora, Western Australia. Plantation trees at five times the density of those in alleys, and all plants growing on deep sand over a water table at 5 m (see Lefroy and Stirzaker, this volume). Data for dependence on ground water were derived using δD values for xylem sap of the tagasaste, reference shallow rooted species and ground water using the mixing model approach described in the text. Note the earlier and more pronounced recourse to ground water in the plantation and alley lupin treatments than in the fallowed alleys in which competition for water by annual herbaceous species had been eliminated by chemical fallowing. *Source:* E. C. Lefroy, J. S. Pate and R. J. Stirzaker (unpubl. data).

3.3. *Relationships between $\delta^{13}C$ of plant biomass and whole plant water-use efficiency*

A number of problems may arise when one attempts to relate cuvette-based exchange studies of CO_2 and H_2O of foliage to the $\delta^{13}C$ signals evident in dry matter of various components of large woody species. Firstly, one must take into account isotope discrimination effects during biochemical syntheses of various dry matter comments such as lipids, lignin and cellulose (Park and Epstein, 1961; Benner et al., 1987; Brugnoli et al., 1988; Gleixner et al., 1993). Secondly, shade, hydraulic architecture and xylem disfunction from embolisms and tyloses have been shown to influence carbon isotope ratios of leaves at different locations in a canopy, and such effects would need to be accounted for when interpreting variations in $\delta^{13}C$ values for various parts of the photosynthetic 'catchment' of a whole shrub or tree (Tyree and Sperry, 1989; Cochard and Tyree, 1990; Tyree and Ewers, 1991; Garten and Taylor, 1992; Sperry and Pockman, 1993; Sperry et al., 1993; Franks et al., 1995; Panek,

1996; Buchman et al., 1997). Thirdly, evaluations need to be undertaken of the relative extents to which the $\delta^{13}C$ values of leaves or other plant parts might have been affected through mixing of newly formed carbon with possibly differently discriminated pools of remobilised carbon produced during earlier periods of photosynthesis (Smedley et al., 1991). Fourthly, carbon dioxide gradients within the atmosphere of a forest may affect $\delta^{13}C$ signals (Medina and Minchin, 1980; Buchmann et al., 1997), as also may genetic (Geber and Dawson, 1990; Osorio and Pereira, 1994) and gender-specific differences (Dawson and Ehleringer, 1993), while the extents to which leaves of different, age, size and canopy position contribute dry matter laid down during a particular period of growth may also exercise significant effects (Schleser, 1992; Schleser and Jayasekera, 1995; Kapos et al., 1993; Hanba et al., 1997). Finally, and perhaps most importantly, assessments need to be made of the extents to which plant water-use efficiency has been compromised by respiratory losses of carbon from various part of shoots and below ground organs (Farquhar et al., 1988, 1989; Ong et al., 1996). Biomass ratios of photosynthetically-productive leaves to other shoot parts or to roots need to be considered in this connection, as also should carbon losses from root-located N_2 fixing or mycorrhizial associations, since costs of such processes are relatively expensive and likely to be energised by plant-derived carbon.

With the above complications in mind, we have adopted the relatively simple philosophy of comparing gross $\delta^{13}C$ values of the total or incremental dry matter of a plant with assessments of its total water usage over the same time frame and environmental conditions. When studying water utilisation we employ the time-honoured technique of measuring net gains in dry matter which a uniform population of pot-cultured plants makes over set intervals of growth and compare each value obtained against gravimetric assessments of cumulative transpiration losses by the same plants over the same period of time. Resulting values, defined here as in the earlier literature in terms of 'transpiration ratio' (mL H_2O transpired per unit dry matter gain), are then related to the $\delta^{13}C$ of the dry matter gained during the same study interval. By extending this approach to encompass comparisons of full seasonal ranges of $\delta^{13}C$ values likely to be encountered by a particular species in various experimental settings within the same climatic zone, it should then become possible to relate $\delta^{13}C$ values and transpiration ratios to plant growth rate and nutrient status and to seasonal changes in soil moisture which might be influencing stomatal behaviour (Condon et al., 1992; Osorio and Pereira, 1994; Ong et al., 1996).

Figure 4 illustrates a set of preliminary data for selected woody and herbaceous species. It highlights the considerable inter- and intra-specific differences likely to be encountered in transpiration ratios and attendant $\delta^{13}C$ values of well watered pot cultured species in winter, spring and summer seasons of a Mediterranean-type climate such as experienced by Perth, Western Australia. For example the data illustrated for field peas compared under identical conditions in spring, suggest lower transpiration ratios and marginally less

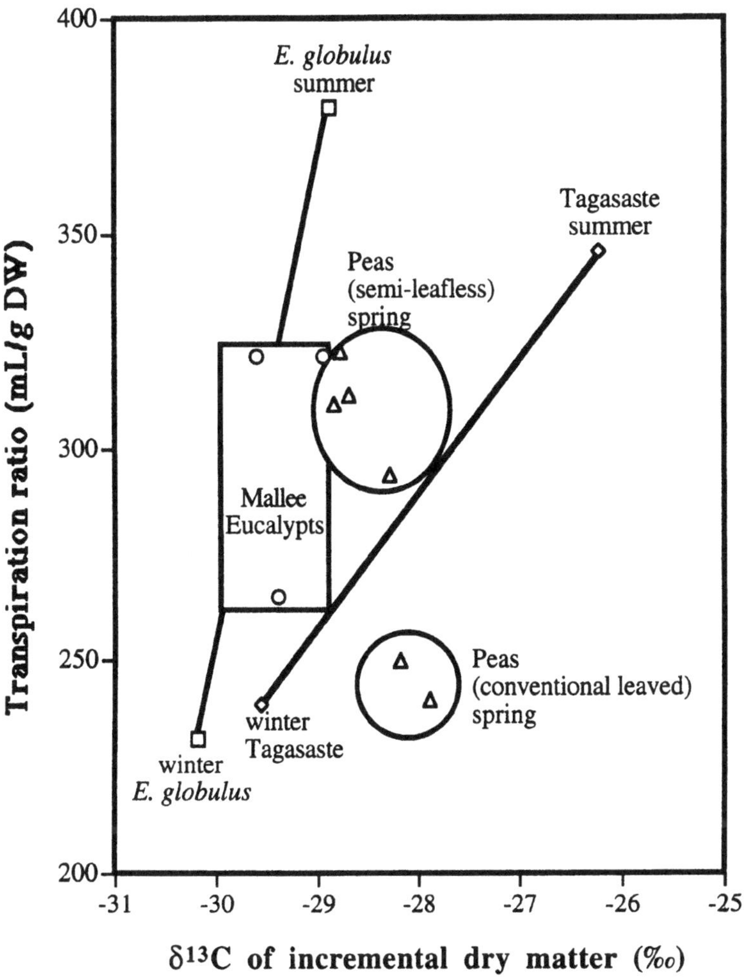

Figure 4. Data comparing $\delta^{13}C$ of recently-produced shoot dry matter and water use efficiency (transpiration ratio) of a number of herbaceous and woody species grown in glasshouse culture in different seasons of the year in Perth, Western Australia. Note the poorer overall water-use efficiencies but less negative $\delta^{13}C$ values for the tagasaste and *Eucalyptus globulus* when grown in summer than winter. The data for these two species suggest that better stomatal control under summer conditions failed to compensate for high water loss under low vapour pressure deficits and high temperatures at this time than the cool and moist winter.

negative $\delta^{13}C$ values (i.e. greater *ci/ca* ratios) in conventionally-leaved than semi-leafless genotypes. On the other hand, between season comparisons for *Eucalyptus globulus* and tagasaste demonstrate considerably more efficient use of water by both species in winter than in summer, but good evidence from respective $\delta^{13}C$ values of tighter stomatal control (higher values for *ci/ca*) under summer than winter conditions. Thus, whereas the $\delta^{13}C$ data indicate that stomatal mechanisms of both species had been highly effective in maximising leaf carbon gain per unit water loss during summer conditions, such adjustments failed by a considerable margin to offset an overriding penalty in terms of water loss under hot and dry summer conditions.

Extending the above approach it becomes instructive to match changes during a growing season in $\delta^{13}C$ of new incremental shoot dry matter produced by a species. Data of this kind are shown in Figure 5 for the same alley and plantation cultures of tagasaste considered earlier (Figure 3) in terms of δD analyses of dependence on ground water. Results proved to be consistent with progressively decreasing availability of soil water at the onset of summer driving increasingly conservative stomatal behaviour. Such effects were then faithfully reproduced in all treatments as less negative $\delta^{13}C$ values (higher *ci/ca*) for new dry matter produced over the critical soil drying period from October to December (Figure 5). Furthermore, as expected from earlier dry-down of the soil profile under the plantation than alley treatments (E. C. Lefroy and J. S. Pate, unpubl. data), the seasonal switch to less negative $\delta^{13}C$ values (tighter stomatal control) was noticeably more advanced (Figure 5) and eventually more pronounced in dense plantation culture than in other less vegetated treatments.

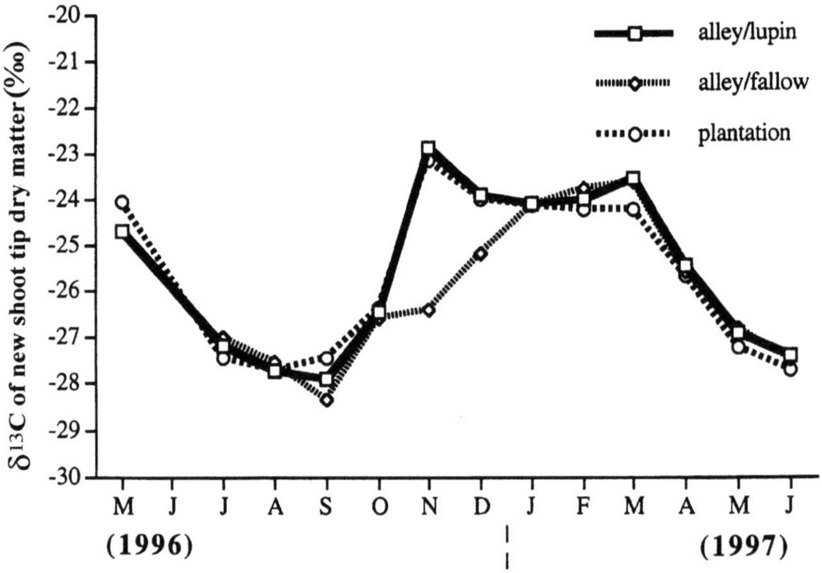

Figure 5. Comparison of a yearly changes in $\delta^{13}C$ of currently-produced terminal shoot dry matter of tagasaste grown in the same plantation, alley crop and alley fallow plot layout at Moora shown in Figure 3. Note evidence of more negative $\delta^{13}C$ values for the alley fallow trees in November/December, consistent with later drying out of the soil profile and a resulting delayed recourse to ground water (Figure 3) in this than in the alley crop or plantation layouts. *Source:* E. C. Lefroy and J. S. Pate (unpubl. data).

3.4. *Use of phloem sap analyses to assess seasonal extent of water stress and changes in plant nitrogen status*

The discovery of copious bleeding from shallow, cambial-deep, incisions in bark of *E. globulus* (Pate et al., 1998b) has permitted us to adopt measurements of $\delta^{13}C$ of translocated phloem carbon as a measure of seasonal changes in instantaneous water-use efficiency of photosynthetically active canopies of the species. Combining such data with analyses of concentrations of amino acids and sugars in the same phloem sap samples (see Pate and Arthur, 1998; Pate et al., 1998b) an integrated picture can then be obtained of seasonal interrelationships and carbon gain, water status and assimilation of nitrogen during seasonal growth of the species.

A full year comparison between a plantation of slow growing, two-year old trees under rain-fed conditions at Eulup with that of a much faster growing, unstressed plantation of the same age at Albany, south western Australia is summarised in Figure 6. Trees at the Albany site received N-rich waste water effluent from a sewage treatment facility in amounts maintaining a near fully moist upper soil profile throughout the year (see Pate and Arthur, 1998). The rain-fed trees at Eulup (Figure 6A, solid line) showed an almost 8‰ seasonal fluctuation in $\delta^{13}C$ of phloem carbon, with fluctuations closely geared to the winter wetting/summer drying cycle of the upper soil profile. Changes in phloem sap $\delta^{13}C$ values in the spring/summer adjustment period October to December (Figure 5) indicated that mean internal CO_2 concentrations (*ci*) in leaves of trees at Eulup would have changed by a massive 120 ppm, thus giving *E. globulus* possibly one of the largest gas exchange adjustment responses so far observed in a woody species. In marked contrast the essentially unstressed water status of the trees at Albany led to only slight changes in $\delta^{13}C$ values of their phloem carbon through the year of study (see Figure 6A, dashed line), indicating how little impact changes in atmospheric vapour pressure deficit and temperature had made on stomatal response of the species under situations in which soil water remained essentially non limiting.

The matching data on amino acid:sugar concentration ratios of phloem sap shown in Figure 6B indicated a nearly identical balance between transport of reduced-N and photosynthate translocation for the two plantations during winter of the study period. Maintenance of N transport relative to C transport was then maintained at more or less the same level throughout the summer in the Albany plantation, presumably due to the continuous provision of N-rich effluent water at this site. By contrast, amino acid:sugar concentration ratios in phloem sap of the rain-fed unfertilised plantation at Eulup plummeted during early summer and remained low right through to the first rains of autumn. This result was also to be expected, since mineralisation of soil N would have become greatly impeded throughout once the upper soil profiles at the site had become desiccated. Thus, the Eulup plantations were clearly experiencing both severe N and water stress during summer.

Consistent with the highly negative (unstressed) $\delta^{13}C$ values and continued

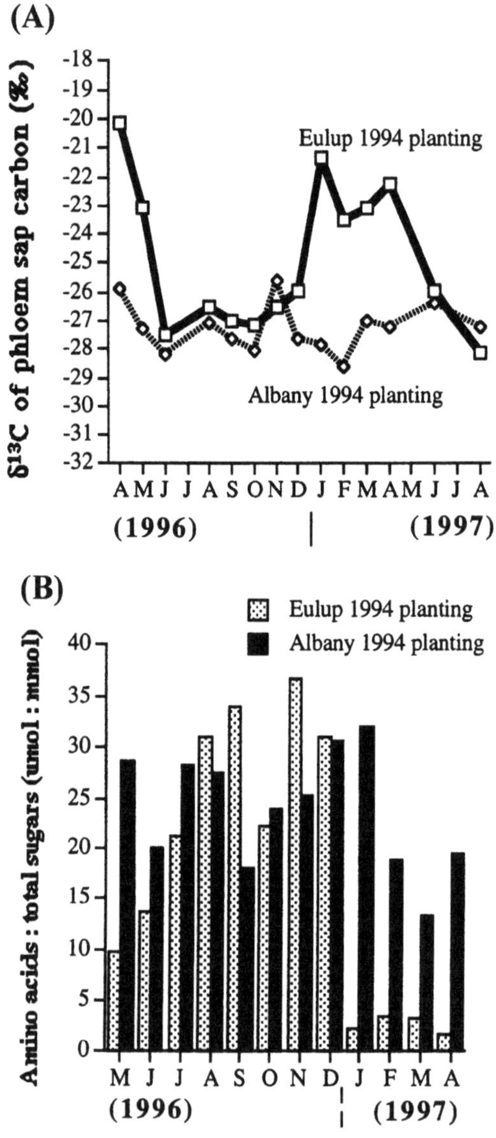

Figure 6. A) Contrasting time courses of change in $\delta^{13}C$ of phloem sap carbon collected from trunks of trees in rain-fed (Eulup) and effluent-water irrigated (Albany) plantations of *Eucalyptus globulus* in south western Australia. Note large oscillation in $\delta^{13}C$ between cool moist winter (more negative values) and hot dry summer conditions (less negative values), compared to a virtual absence of such fluctuations at Albany sites where drying of the upper soil profile was largely prevented by irrigation during summer. B) Amino acid:sugars molar ratios for the same phloem sap samples as shown in A. Note maintenance of high N (amino acid) translocation relative to sugar flow in the Albany trees fed with N-rich effluent, but a dramatic change to low amino acid:sugar ratios for trees at Eulup during summer, when desiccation of the soil profile greatly restricted mineralisation and root uptake of N. *Source:* Pate and Arthur (1998).

high N status of its phloem sap, the irrigated plantations at Albany grew unabated throughout the summer (Pate and Arthur, 1998). By contrast, the plantations at Eulup stopped growing during late summer and exhibited drought-induced deaths of trees as they entered peak stress of autumn. Again the $\delta^{13}C$ values for phloem carbon proved to be highly definitive, since any tree found to exhibit a phloem sap $\delta^{13}C$ value of –18.5‰ or less was found to succumb within a few weeks of achieving such a low value (Pate and Arthur, 1998).

3.5. *Relationships between $\delta^{13}C$ values of phloem carbon and bulk dry matter of foliage and wood*

The assumptions underlying this approach were that carbon collected in phloem sap from trunks would consist predominantly of photosynthate recently fixed by the leaf canopy, and that this translocate would accordingly be the most likely, if not the only source of carbon source for shoot extension growth and secondary (radial) thickening of trunks and branches during a particular interval of study. Seasonal changes in the $\delta^{13}C$ of phloem sap carbon would therefore provide a simple and highly definitive means of interpreting patterns of variation in carbon isotope ratios in dry matter of leaves and successive growth increments made in trees over a particular period of time.

A preliminary comparison of the above kind was undertaken during a summer of extension growth of shoots of acorn banksia in which it proved possible to match seasonal changes in $\delta^{13}C$ of trunk phloem sap carbon against $\delta^{13}C$ of dry matter of leaves produced sequentially during the growing season (Figure 7; Pate et al., 1998a). Terminal resting buds on shoots of *B. prionotes* break dormancy in early summer (October/November). Extension growth is at first very rapid (Figure 7), but then declines progressively as soil water is depleted and the tree switches to reliance on ground water (e.g. δD analyses of Figure 2A).

Changes in $\delta^{13}C$ of sequentially produced leaves and concurrently collected phloem sap turned out to be closely synchronized (Figure 7). At the beginning of the season, when shoot extension was rapid, ^{13}C values of leaf dry matter and phloem sap were low (more negative), reflecting still relatively unstressed condition. Then, at the beginning of the second half of the growing season, $\delta^{13}C$ values of leaf dry matter and phloem carbon rose concurrently, apparently signalling onset of increasingly tighter stomatal control over carbon assimilation and associated 'penalties' in terms of lowered growth rate (Figure 7). Finally, growth halted abruptly in the peak stress of late summer when the seasons resting bud was initiated. As to be expected, highest (least negative) $\delta^{13}C$ values were recorded for new leaf dry matter produced at this time (Figure 7). It can be seen that the $\delta^{13}C$ signals of phloem carbon sap obtained earlier in the study period were somewhat more negative than those recorded for current new leaf dry matter. This presumably reflected carbon

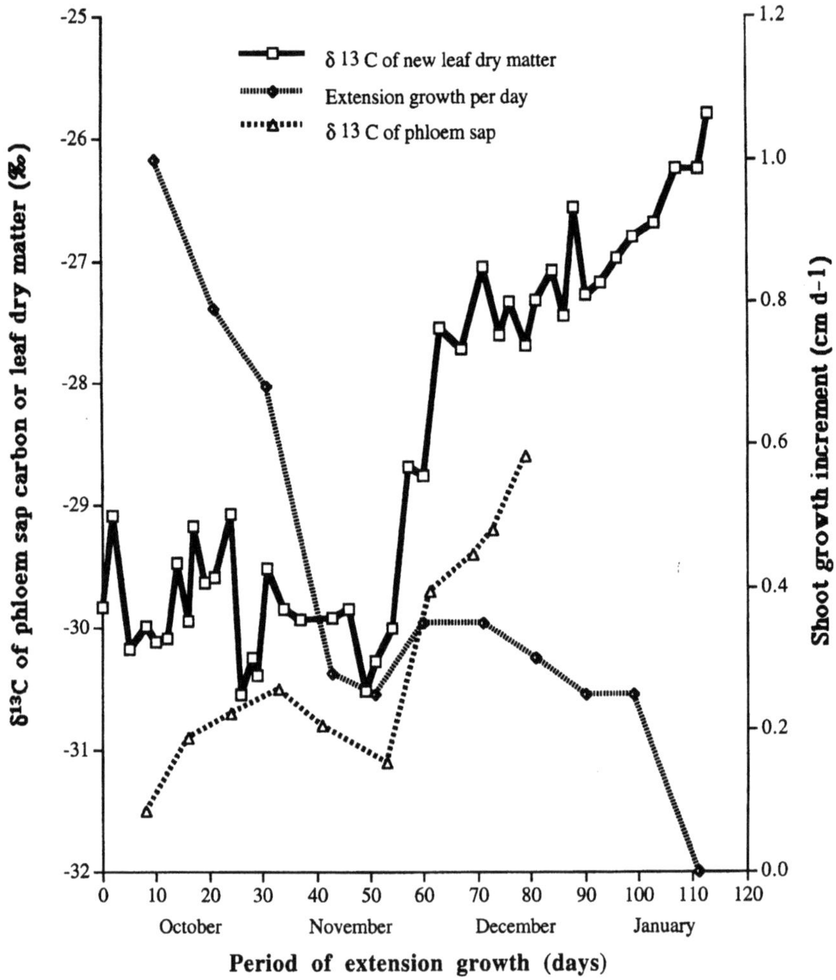

Figure 7. Time-correlated plots relating $\delta^{13}C$ of new leaf dry matter and $\delta^{13}C$ of total C of mid trunk phloem sap to main shoot extension rate in naturally growing stands of *Banksia prionotes*, at Shenton Park, Perth, Western Australia. Seasonal growth of the species commenced in October and continued until terminal resting buds formed in late summer. *Source:* Data, with addition of values for $\delta^{13}C$ phloem sap, taken from Pate et al. (1998a).

isotope discrimination during synthesis of major fabric materials of the leaf from the sugar and other assimilates which it was receiving as translocate from adjacent source foliage.

Turning to our other phloem bleeding species, *Eucalyptus globulus*, interesting relationships have been demonstrated between $\delta^{13}C$ signals of:

a) earlier-formed fully expanded leaves at different heights and aspects in a tree canopy;

b) current translocate in phloem; and
c) dry matter currently laid down in apical tissues at the top of the trunk of the same tree (Pate and Arthur, 1998).

Figure 8 summarises the situation for a contrasting pair of irrigated (Albany) and non irrigated (Eulup) two-year old trees sampled at peak water stress in April (1996). $\delta^{13}C$ values of the dry matter of earlier-formed mature leaves obtained from side branches at different heights up the trunk were in all cases more negative than that of carbon of bulk phloem translocate currently originating from the canopy. The substantial differences involved probably reflected the relatively low levels of water stress which had applied the previous winter and spring when the leaves in question were accumulating the bulk of their constituent dry matter. Then, under the highly stressed autumn conditions when phloem sap was sampled, $\delta^{13}C$ signals of the translocate produced by these same leaves provided evidence of high levels of water stress and good instantaneous water use efficiency. These $\delta^{13}C$ signals were in turn faithfully reflected in dry matter of new apical shoot tissues laid down just prior to the phloem sap collections.

There was also evidence in the above study of less negative signals for north-oriented (unshaded side) than south-oriented (shaded) branches in the tree at Albany (Figure 8A), as anticipated from supposedly more stressful conditions likely to be experienced on the sunny sides of the tree. In contrast, foliage sampled from the tree at Eulup showed closely similar $\delta^{13}C$ values (Figure 8B) for the more or less equally shaded leaves of east and west faces of its canopy. Furthermore, consistent with the much wetter soil conditions at Albany, phloem sap carbon carried $\delta^{13}C$ signals some 4‰ more negative than those recorded for the highly stressed, rain-fed site at Eulup. Nevertheless, as at Eulup, shoot apex tissue currently produced by the tree at Albany showed $\delta^{13}C$ values closely matching those of translocate currently moving in the trunk.

Using the same contrasting pair of irrigated and rain-fed plantations of blue gum, it proved possible to compare seasonal changes in $\delta^{13}C$ values of phloem carbon against respective $\delta^{13}C$ signals of the ethanol-insoluble structural xylem carbon obtained by scraping the thin veneer of gelatinous, recently differentiated xylem tissue from the innermost face of the cambium of the same trees (see Pate and Arthur, 1998). The data obtained (Figure 9), comprising successive samplings from 1993 and 1994 plantings at Eulup and at Albany over the period 1996–97, all showed close matchings between $\delta^{13}C$ values of bulk carbon of phloem and those of recent xylem. This was of course to be anticipated since phloem streams currently passing down the trunk would have comprised the most likely carbon source for secondary growth of trunk xylem.

Further inspection of the data of Fig. 9 reveals seasonal peaks and troughs in phloem $\delta^{13}C$ for the plantations at Eulup, which were reproduced one month or so later by fluctuations of comparable magnitude in $\delta^{13}C$ of new xylem carbon. The time lag involved presumably denotes the time taken from phloem

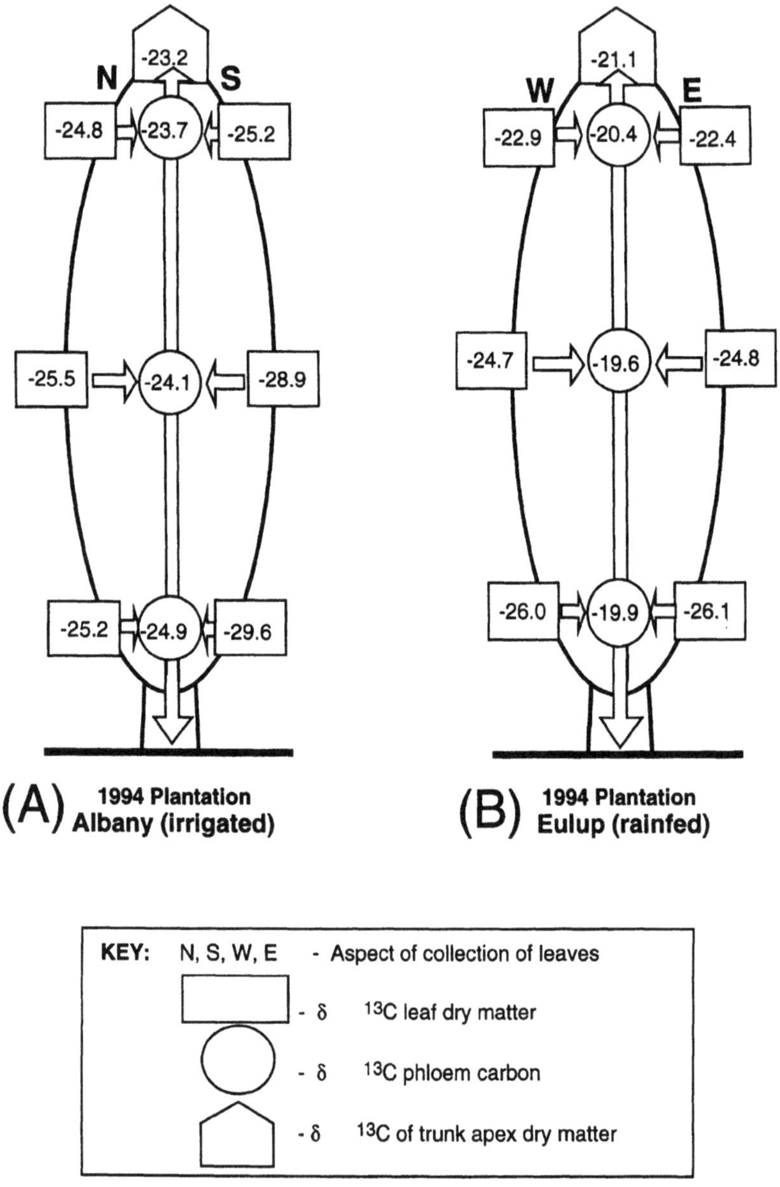

Figure 8. Schematic-type representations depicting the $\delta^{13}C$ values for total carbon of mature leaf dry matter, recently initiated apical shoot dry matter and phloem sap of trees of *E. globulus*. One tree came from a relatively unstressed, effluent irrigated site at Albany, the other from a highly stressed rain-fed site at Eulup. Relative levels up the tree at which matching samples of phloem sap and foliage were harvested are indicated and presumed directions of phloem translocation from these levels depicted by arrows. Aspect of collection of leaves from canopies are shown, viz. N canopy faces unshaded, S face densely shaded, E and W faces equally and moderately shaded. *Source:* Data from Pate and Arthur (1998).

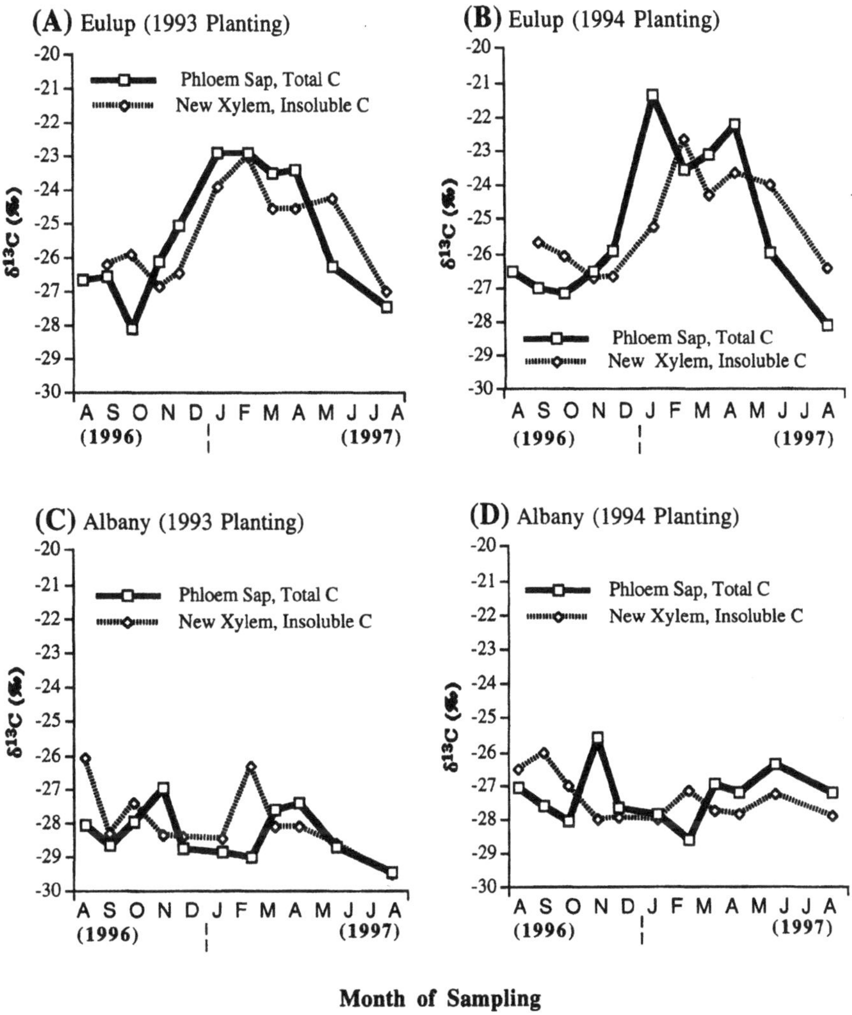

Figure 9. Matched seasonal profiles of change in $\delta^{13}C$ of total carbon of phloem sap and $\delta^{13}C$ of corresponding ethanol-insoluble (structural) carbon of newly initiated xylem tissues scraped off the inner cambial face of trunks of 1993 and 1994 plantings of *Eucalyptus globulus* at a rain-fed (Eulup) and irrigated (Albany) site in south western Australia. Note that stress-related fluctuations in of phloem carbon in the Eulup plantations are reproduced a month or so later by changes in $\delta^{13}C$ of similar magnitude in new xylem tissue. An effect of this nature is not evident in the continuously unstressed trees from Albany. *Source:* Data from Pate and Arthur (1998).

carbon being received by the cambium and becoming incorporated into recently initiated xylem tissue. Such close matchings between $\delta^{13}C$ signatures of phloem sap and new xylem tissue provide direct validation of the use $\delta^{13}C$ values of growth rings of wood as records of stress-related effects of soil and climate on carbon and water exchange by *E. globulus*. This is, of course, the very rationale underlying widespread use of growth ring analyses for $\delta^{13}C$ researchers on a wide range of woody species (Francey and Farquhar, 1982; Levitt and Long, 1982, 1986, 1988; Levitt, 1992; Dawson, 1993b; Dupouey et al., 1993; Livingston and Spittlehouse, 1996; Bert et al., 1997).

Because of the relatively minor seasonal fluctuations in $\delta^{13}C$ of phloem carbon evident in the irrigated Albany plantations, distinct temporal relationships were not evident between $\delta^{13}C$ values for phloem sap and xylem scrapings for trees at this site.

3.6. *Use of paired analyses of phloem and xylem sap to study current uptake and cycling of nutrient elements*

An earlier published account (Jeschke and Pate, 1995) using this approach for acorn banksia has already provided an interesting commentary on how this species adapts to the highly depauperate deep sands of its native habitat. Analyses of xylem bleeding sap from active clusters of proteoid roots have shown these specialized structures to act as principal absorbers of N and P during winter. Comparisons of mean concentrations of major inorganic ions of xylem and phloem of the species have also indicated unusually high concentrations of Na^+, Cl^- and SO_4^{2+} relative to K^+ and PO_4^{3-}, suggesting a substitution of the former three for the latter two ions in transport pathways of both xylem and phloem. This has been interpreted as a strategy for releasing K and P from functional roles in transport, and thereby maximising targeting of these two limited resources into vital components of growing tissues.

A recent study of a similar nature on plantation grown *Eucalyptus globulus*, growing in the highly leached and impoverished soils typical of SW Australia, has provided the data shown in Table 2. As recorded above for acorn banksia, phloem sap of *E. globulus* contains unusually high concentrations of Na^+ and Cl^- but average phosphorus and sulphur levels in its phloem and xylem streams are considerably higher than in banksia. This may well reflect a history of superphosphate application to the previously pastured sites on which blue gum plantations are typically established. No amelioration of inherent P deficiency would have occurred in the natural bushland site at which the banksia study was studied. As is generally known for other species, both blue gum and banksia show Ca^{2+} at lower level in their phloem than xylem and NO_3^- is at trace level or absent from phloem but detectable in xylem. As shown by Jeschke and Pate (1995) and in Table 2, the steepest enrichment in concentration from xylem up to phloem relates to K^+. In further studies of this type (Pate et al., 1998b), key attributes relating to nutrients in phloem or xylem (e.g. K:Na and N:P balance, amino acid:sugar ratio) have been suggested as means of assessing relative nutritional status of blue gum plantations across a wide range of edaphic conditions. Indeed, such analyses of transport fluids may well turn out to provide more effective indicators of immediate nutrient status than assessments based on conventional foliar, bark or twig analysis.

Table 2. Mean concentrations of major inorganic and organic solutes in mid trunk phloem sap and xylem sap of Tasmanian blue gum (*Eucalyptus globulus*) sampled at monthly intervals over the period May 1996 and June 1997 from a plantation of uniform two-year old trees at Eulup, near Mt Barker, Western Australia.

	Phloem Sap	Xylem Sap
Inorganic components (mg L^{-1} ± S.E.)		
K	1136 ± 15	128 ± 23
Ca	9.8 ± 0.5	43.2 ± 8.9
Mg	50.7 ± 6.2	16.5 ± 3.7
Na	241 ± 30	27.1 ± 5.5
Total S	59.3 ± 4.1	29.3 ±18.4
Total P	58.3 ±8.3	5.67 ± 1.74
Cl	76.7 ± 10.5	58.3 ± 13.6
NO_3	Trace	1.58 ± 0.27
Organic components (mmol L^{-1} ± S.E.)		
Amino acids	10.3 ± 2.7	1.77 ± 0.54
Organic acids	6.3 ± 1.4	0.3 ± 0.03
Sugars	631 ± 105	Trace (< 15)

Source: Data on inorganic constituents from Pate et al. (1998b) with additional unpublished data on amino acids, sugars and organic acids.

4. Prospective for future research

One of the principal aims of the conference on which the articles in this volume are based was to evaluate how closely particular novel agricultural systems, such as woody perennial/annual intercropped mixtures, can be matched to the self-sustaining natural ecosystems from which these mimics were designed. Central to this theme come questions as to how counterpart natural and mimic systems compare functionally in terms of utilising limiting resources of water and nutrients, and how closely any differences or similarities observed relate to growth performance and long term sustainability of the respective systems. We believe that certain of the earlier established and innnovative techniques employed in the case studies discussed in this paper should permit one to move more effectively towards this goal.

Regular monitoring of plant performance of even a few dominant 'keystone' species presents many challenges and inevitably requires deployment of a range of taxon-specific approaches and technologies. Wherever possible, the experimental strategies adopted should be not only inexpensive and easily applied but also duly informative and accountable in terms of how validly data obtained apply to the ecosystem as a whole.

Ecosystem functioning in terms of one limiting resource, water, is reviewed in some of the examples above and elsewhere in this volume (see articles by Pate and Bell, and Hatton and Nulsen). New indices for assessing how water limitation affects vegetation are exemplified in this article in sections dealing with carbon isotope discrimination of phloem sap and plant dry matter (leaves or wood), and these should be further cross related in the future to other equally instructive indices of water use, such as heat pulse measurements of sap flow, transpiration ratios, and cuvette-based assays of instantaneous water use efficiency. We would then have a powerfully integrated and informative approach for depicting performances of whole systems. Moreover, by matching changes in $\delta^{13}C$ of key species across and between seasons against simultaneous measurements of carbon gain in dry matter of the same species, it should be possible to compute with reasonable accuracy total resource utilisation and resource-use efficiency of the species concerned. Further, by possibly adding information from above ground monitoring of CO_2 and H_2O exchanges from canopies, data bases might be extended to provide overviews of short and long term performances of the system. Then, by combining the above information with studies of water source utilisation using hydrogen or oxygen isotope studies, rainfall, evaporation, evapotranspiration and deep drainage it would be possible to arrive at an entire picture of carbon and water balance of the ecosystem (Dawson et al., 1997; Dawson and Ehleringer, 1998). Such information may well be vital for proper accreditation of ecosystems in respect of projected future international trading in carbon credits.

Whether with respect to investigations of only one or a number of species, the studies using stable isotope analyses of soil water, ground water and xylem sap presented here and elsewhere (Ehleringer et al., 1991; Ehleringer and Dawson, 1992; Dawson, 1993b, 1995; Dawson et al., 1997; Dawson and Ehleringer, 1998). Dawson (1993a) and Caldwell et al. (1998) appear to offer considerable promise in future examinations of principal interactions among species of complementary water use strategies in natural and mimic systems. Both natural and agricultural ecosystems are likely to contain species of different rooting depths, morphologies, and growth/water-use phenologies, thereby allowing fully complementary co-existence in time and space. However, we are still woefully ignorant of how these interspecific differences influence the total water balance of any particular system and, how they individually and collectively impact upon long term effectiveness in terms of sustainability. This is also a highly important issue in view of current concern over rising water tables and disastrous periodic floodings in deforested agricultural land in so many parts of the globe.

The age-old dogma of rate limitation of plant growth has always assumed that only one factor, process or resource within the environment will be directly limiting to productivity at any particular time or circumstance, but pinpointing which factor applies has always proven difficult. Our present approach of analysing phloem sap and specific dry matter entities from particular species illustrates the value of a broadly based approach in which assays for $\delta^{13}C$ and certain key nutrient quantities provide an indication of likely temporal patterns in which availability of water or of specific nutrients control growth rate through a season.

Our paper attempts to illustrate the potential value of applying multifaceted comparative approaches simultaneously to natural or contrived ecosystems. While this is borne out in the species we have selected, the utility of the approach will need to be tested across a much wider array of taxa (e.g. see Brooks et al., 1997; Dawson and Fry, 1998). In any event, where woody species are being deployed in agroforestry alongside respective native counterparts, one has ideal opportunities for testing such an approach, especially if the alien species employed in the mimic turn out to be functionally very different from the natives which they replace. Such comparisons should be studied not only in terms of photosynthesis and mineral nutrition but also include temporal and spatial usage of water resources, as related to rooting morphology, seasonal water usage and water use efficiency by the players concerned. Challenges such as these present exciting prospects for the researcher bent on understanding how innovative technologies may be brought to bear on the much-vexed problems currently facing contemporary agriculture.

5. Acknowledgements

We thank our respective Institutions, The University of Western Australia, (JSP) and Cornell University (TED) for supporting a study leave opportunity to one of us (TED) during which the material for this article was assembled and some of the relevant experimental material was obtained. The Western Australian segment of the study was supported by generous funding to JSP from the Australian Research Council, the Land and Water Resources Research and Development Corporation and the Grains Research Development Corporation (Australia). The assistance of Judy Furlong and David Arthur in preparing the manuscript is gratefully acknowledged.

6. References

Benner R, Fogel ML, Sprague EK and Hodson RE (1987) Depletion of ^{13}C in lignin and its implications for stable carbon isotope studies. Nature 329: 708–710

Bert D, Leavitt SW and Dupouey JL (1997) Variations of wood $\delta^{13}C$ and water-use efficiency of *Abies alba* during the last century. Ecology 78: 1588–1596

Bollard EG (1960) Transport in the xylem. Annual Review of Plant Physiology 11: 141–166

Brooks JR, Flanagan LB, Buchmann N and Ehleringer JR (1997) Carbon isotope composition of boreal plants: functional grouping of life forms. Oecologia 110: 301–311

Brunel J-P, Walker GR and Kennett-Smith AK (1995) Field validation of isotopic procedures for determining sources of water used by plants in a semi-arid environment. Journal of Hydrology 167: 351–368

Buchmann N, Guehl JM, Barigah TS and Ehleringer JR (1997) Interseasonal comparison of CO_2 concentrations, isotopic composition, and carbon dynamics in an Amazonian rainforest (French Guiana). Oecologia 110: 120–131

Brugnoli E, Hubick KT, von Caemmerer S, Wong SC and Farquhar GD (1988) Correlation between the carbon isotope discrimination in leaf starch and sugars of C_3 plants and the ratio of intercellular and atmospheric partial pressures of carbon dioxide. Plant Physiology 88: 1418–1424

Caldwell MM, Dawson TE and Richards JH (1998) Hydraulic lift: consequences of water efflux from plant roots. Oecologia (in press)

Cochard H and Tyree MT (1990) Xylem dysfunction in *Quercus*: vessel sizes, tyloses, cavitation and seasonal changes in embolism. Tree Physiology 6: 393–407

Condon AG, Richards RA and Farquhar GD (1992) The effect of variation in soil water availability, vapour pressure deficit and nitrogen nutrition on carbon isotope discrimination in wheat. Australian Journal of Agricultural Research 43: 935–947

Cowan IR (1982) Regulation of water use in relation to carbon gain in higher plants. In: Lange OL, Nobel S, Osmond CB and Zeigler H (eds) Physiological Plant Ecology II. Encyclopedia of Plant Physiology, New Series Vol. 12B, pp 589–613. Springer-Verlag, New York

Dawson TE (1993a) Hydraulic lift and water use in plants: implications for performance, water balance and plant-plant interactions. Oecologia 95: 565–574

Dawson TE (1993b) Water sources of plants as determined from xlyem-water isotopic composition: perspectives on plant competition, distribution, and water relations. In: Ehleringer JR, Hall AE and Farquhar GD (eds) Stable Isotopes and Plant Carbon-Water Relations, pp 465–496. Academic Press, Inc., San Diego

Dawson TE (1995) L'integrazione dei metodi basati sugli isotopi stabili nello studio delle relazioni idriche e del bilancio del carbonio in specie legnose. Monti e Boschi 95: 41–49 (english title: The integration of stable isotope methods into studies of woody plant carbon and water balance. Mountains and Forests 95: 41–49)

Dawson TE (1996) Determining water use by trees and forests from isotopic, energy balance and transpiration analyses: the roles of tree size and hydraulic lift. Tree Physiology 16: 263–272

Dawson TE (1998) Water loss from tree roots influences soil water and nutrient status and plant performance. In: Flores HE, JP Lynch JP and Eissenstat DM (eds) Radical Biology: Advances and Perspectives in the Function of Plant Roots. Current Topics in Plant Physiology, Vol. 17 (special issue, Plant Physiology). American Society of Plant Physiologists, Rockville, MD, USA

Dawson TE and Ehleringer JR (1993) Gender-specific physiology, carbon isotope descrimination, and habitat distribution in Boxelder, *Acer negundo*. Ecology 74: 798–815

Dawson TE and Ehleringer JR (1998) Plants, isotopes, and water use: a catchment-level perspective. In: McDonnell JJ and Kendall C (eds) Isotope Tracers in Catchment Hydrology. Elsevier Science Publishers, Amsterdam

Dawson TE and Fry R (1998) Agriculture in nature's image. Trends in Ecology and Evolution 13(2): 50–51

Dawson TE and Pate JS (1996) Seasonal water uptake and movement in root systems of Australian phreatophytic plants of dimorphic root morphology: a stable isotope investigation. Oecologia 107: 13–20

Dawson TE, Pausch RC and Parker HM (1997) The role of hydrogen and oxygen stable isotopes in understanding water movement along the soil-plant-atmospheric continuum. In: Griffiths H, Robinson D and Van Gardingen P (eds) Stable Isotopes and the Integration of Biological, Ecological and Geochemical Processes. BIOS Scientific Publishers, London

Dupouey JL, Leavitt S, Choisnel E and Jourdain S (1993) Modelling carbon isotope fractionation in tree rings based on effective evapotranspiration and soil water status. Plant, Cell and Environment 16: 939–947

Ehleringer JR (1988) Carbon isotope ratios and physiological processes in aridland plants. In: Rundel PW, Ehleringer JR and Nagy KA (eds) Stable Isotopes in Ecological Research, pp 41–54. Springer Verlag, New York

Ehleringer JR and Osmond CB (1989) Stable isotopes. In: Pearacy RW, Ehleringer J, Mooney HA and Rundel PW (eds) Plant physiological ecology: field methods and instrumentation, pp 281–300. Chapman and Hall, New York, USA

Ehleringer JR and Rundel PW (1988) Stable isotopes: history, units and instrumentation. In: Rundel PW, Ehleringer JR and Nagy KA (eds) Stable Isotopes in Ecological Research, pp 3–13. Springer Verlag, New York

Ehleringer JR, Phillips SL, Schuster WSF and Sandquist DR (1991) Differential utilization of summer rains by desert plants. Oecologia 88: 430–434

Ehleringer JR and Dawson TE (1992) Water uptake by plants: perspectives from stable isotope composition. Plant, Cell and Environment 15: 1073–1082

Ehleringer JR, Hall AE and Farquhar GD (eds) (1993) Stable Isotopes and Plant Carbon-Water Relations. Academic Press, San Diego

Emerman SH and Dawson TE (1996) Hydraulic lift and its influence on the water content of the rhizosphere: an example from sugar maple, *Acer saccharum*. Oecologia 108: 273–278

Farquhar GD and Richards RA (1984) Isotopic composition of plant carbon correlates with water-use efficiency of wheat genotypes. Australian Journal of Plant Physiology 11: 539–552

Farquhar GD, Hubick KT, Condon AG and Richards RA (1988) Carbon isotope fractionation and plant water use efficiency. In: Rundel PW, Ehleringer JR and Nagy KA (eds) Stable Isotopes in Ecological Research, pp 21–40. Springer Verlag, New York

Farquhar GD, Ehleringer JR and Hubick KT (1989) Carbon isotope discrimination and photosynthesis. Annual Review of Plant Physiology and Plant Molecular Biology 40: 503–537

Fisher DB (1983) Year-round collection of willow sieve-tube exudate. Planta 159: 529–533

Francey RJ and Farquhar GD (1982) An explanation of $^{13}C/^{12}C$ variations in tree rings. Nature 297: 28–31

Franks PJ, Gibson A and Bachelard EP (1995) Xylem permeability and embolism susceptibility in seedlings of *Eucalyptus camaldulensis* from two different climatic zones. Australian Journal of Plant Physiology 22: 15–21

Garten Jr CT and Taylor Jr GE (1992) Foliar $\delta^{13}C$ within a temperate deciduous forest: spatial, temporal, and species sources of variation. Oecologia 90: 1–7

Gat JR (1996) Oxygen and hydrogen stable isotopes in the hydrologic cycle. Annual Review of Earth and Planetary Sciences 24: 225–320

Geber MA and Dawson TE (1990) Genetic variation in and covariation between leaf gas exchange, morphology and development in *Polygonum arenastrum*, an annual plant. Oecologia 85: 153–158

Gleixner G, Danier HJ, Werner RA and Schmidt HL (1993) Correlations between the ^{13}C content of primary and secondary plant products in different cell compartments and that in decomposing basidiomycetes. Plant Physiology 102: 1287–1290

Guy RD, Warne PG and Reid DM (1988) Stable carbon isotope ratio as an index of water-use efficiency in C^3 Halophytes – possible relationship to strategies for osmotic adjustment. In: Rundel PW, Ehleringer JR and Nagy KA (eds) Stable Isotopes in Ecological Research, pp 55–75. Springer Verlag, New York

Hanba YT, Mori S, Lei TT, Koike T and Wada E (1997) Variations in leaf $\delta^{13}C$ along a vertical profile of irradiance in a temperate Japanese forest. Oecologia 110: 253–261

Horton, JL and Hart SC (1998) Hydraulic lift: a potentially important ecosystem process. Trends in Ecology and Evolution 13: 232–235.

Jeschke WD and Pate JS (1995) Mineral nutrition and transport in xylem and phloem of *Banksia prionotes* and proteaceous trees of dimorphic root morphology. Journal of Experimental Botany 46: 895–905

Kapos V, Ganade G, Matsui E and Victoria RL (1993) $\delta^{13}C$ as an indicator of edge effects in tropical rainforest reserves. Journal of Ecology 81: 425–432

Leavitt SW (1992) Seasonal $^{13}C/^{12}C$ changes in tree rings: species and site coherence, and a possible drought influence. Canadian Journal of Forest Research 23: 210–218

Leavitt SW and Long A (1982) Evidence for 13C/12C fractionation between tree leaves and wood. Nature 298: 742–743

Leavitt SW and Long A (1986) Stable carbon isotope variability in tree foliage and wood. Ecology 67: 1002–1010

Leavitt SW and Long A (1988) Intertree variability of $\delta^{13}C$ in tree rings. In: Rundel PW, Ehleringer JR and Nagy KA (eds) Stable Isotopes in Ecological, pp 95–104. Springer Verlag, New York

Lin G and Ehleringer JR (1997) Plant Physiology 114: 391–394

Livingston NJ and Spittlehouse DL (1996) Carbon isotope fractionation in tree ring early and late wood in relation to intra-growing season water balance. Plant, Cell and Environment 19: 768–774

Medina E and Minchin P (1980) Stratification of $\delta^{13}C$ values of leaves in Amazonian rain forests. Oecologia 45: 377–378

Ong CK, Black CR, Marshall FM and Corlett JE (1996) Principles of resource capture and utilization of light and water. In: Ong CK and Huxley P (eds) Tree-crop interactions: a physiological approach, pp 73–158. Cab International

Osorio J and Pereira JS (1994) Genotypic differences in water use efficiency and ^{13}C discrimination in *Eucalyptus globulus*. Tree Physiology 14: 871–882

Panek JA (1996) Correlations between stable carbon-isotope abundance and hydraulic conductivity in Douglas-fir across a climate gradient in Oregon, USA. Tree Physiology 16: 747–755

Park R and Epstein S (1961) Metabolic fractionation of C^{13} and C^{12} in plants. Plant Physiology 36: 133–138

Pate JS (1975) Exchange of solutes between phloem and xylem and circulation in the whole plant. In: Zimmerman MH and Milburn JA (eds) Encyclopedia of Plant Physiology, New Series Volume 1. Transport in Plants. I. Phloem transport, pp 451–473. Springer-Verlag, Berlin

Pate JS (1980) Transport and partitioning of nitrogenous solutes. Annual Review of Plant Physiology 31: 313–340

Pate JS (1986) Xylem-to-phloem transfer – vital component of the nitrogen-partitioning system of a nodulated legume. In: Cronshaw J, Lucas WJ and Giaquinta RT (eds) Phloem Transport, pp 445–462. AR Liss Inc., New York

Pate JS and Arthur DJ (1998) $\delta^{13}C$ analyses of phloem sap carbon: novel means of evaluating seasonal water stress and interpreting $\delta^{13}C$ signatures of foliage and trunk wood of *Eucalyptus globulus*. Oecologia 117: 301–311

Pate JS and Jeschke WD (1993) Mineral uptake and transport in xylem and phloem of the proteaceous tree *Banksia prionotes*. Plant and Soil 155/156: 273–276

Pate JS and Jeschke WD (1995) Role of stems in transport, storage and circulation of ions and metabolites by the whole plant. In: Gartner B (ed) Stems and Trunks: their roles in plant form and function, pp 177–204. Academic Press, New York

Pate JS, Jeschke WD and Aylward MJ (1995) Hydraulic architecture and xylem structure of the dimorphic root system of S.W. Australian tree-species of Proteaceae. Journal of Experimental Botany 46: 907–915

Pate JS, Jeschke WD, Dawson TE, Raphael C, Hartung W and Bowen BJ (1998a) Growth and seasonal utilisation of water and nutrients by *Banksia prionotes*. Australian Journal of Botany 46: 511–532

Pate JS, Shedley E, Arthur DJ and Adams MA (1998b) Spatial and temporal variations in phloem sap composition of plantation – grown *Eucalyptus globulus*. Oecologia 117: 312–322

Pereira JS (1995) Gas exchange and growth. In: Schulze ED and Caldwell MM (eds) Ecophysiology of Photosynthesis, pp 147–181. Springer-Verlag, Berlin

Power, ME, Tilman D, Estes JA, Menge BA, Bond WJ, Mills LS, Daily G, Castilla JC, Lubchenco J and Paine RT (1996) Challenges in the quest for keystones. BioScience 46: 609–620

Read J and Farquhar G (1991) Comparative studies in *Nothofagus* (Fagaceae). I. Leaf carbon isotope discrimination. Functional Ecology 5: 684–695

Schleser GH (1992) $\delta^{13}C$ pattern in a forest tree as an indicator of carbon transfer in trees. Ecology 73: 1922–1925

Schleser GH and Jayasekera R (1985) $\delta^{13}C$-variations of leaves in forests as an indication of reassimilated CO_2 from the soil. Oecologia 65: 536–542

Smedley MP, Dawson TE, Comstock JP, Donovan LA, Sherrill DE, Cook CS and Ehleringer JR (1991) Seasonal carbon isotope discrimination in a grassland community. Oecologia 85: 314–320

Sperry JS, Alder NN and Eastlack SE (1993) The effect of reduced hydraulic conductance on stomatal conductance and xylem cavitation. Journal of Experimental Botany

Sperry JS and Pockman WT (1993) Limitation of transpiration by hydraulic conductance and xylem cavitation in *Betula occidentalis*. Plant, Cell and Environment 16: 279–287

Tieszen LL and Boutton TW (1988) Stable carbon isotopes in terrestrial ecosystem research. In: Rundel PW, Ehleringer JR and Nagy KA (eds) Stable Isotopes in Ecological Research, pp 169–195. Springer Verlag, New York

Thorburn PJ and Walker GR (1993) The source of water transpired by *Eucalyptus camaldulensis*: soil, groundwater, or streams? In: Ehleringer JR, Hall AE and Farquhar GD (eds) Stable Isotopes and Plant Carbon-Water Relations, pp 511–527. Academic Press, Inc., San Diego

Tyree MT and Ewers FW (1991) The hydraulic architecture of trees and other woody plants (Tansley Review No. 34). New Phytologist 119: 345–360

Tyree MT and Sperry JA (1989) Vulnerability of xylem to cavitation and embolism. Annual Review of Plant Physiology and Plant Molecular Biology 40: 19–38

Walker CD and Richardson SB. (1991) The use of stable isotopes of water in characterizing the sources of water in vegetation. Chemical Geology (Isotope Geoscience Section) 94: 145–158

Waring RH and Silvester WB (1994) Variation in foliar $\delta^{13}C$ values within the crowns of *Pinus radiata* trees. Tree Physiology 14: 1203–1213

White JWC (1988) Stable hydrogen isotope ratios in plants: a review of current theory and some potential applications. In: Rundel PW, Ehleringer JR and Nagy KA (eds) Stable Isotopes in Ecological Research. Ecological Studies, vol 68, pp 142–162. Springer-Verlag, Heidelberg

White JWC, Cook ER, Lawrence JR and Broecker WS (1985) The D/H ratios of sap in trees: implications for water sources and tree ring D/H ratios. Geochimica et Cosmochimica Acta, 49: 237–246

Agroforestry Systems **45**: 277–302, 1999.

Agroforestry for water management in the cropping zone of southern Australia

E. C. LEFROY[1,*] and R. J. STIRZAKER[2]
[1] *Centre for Legumes in Mediterranean Agriculture, University of Western Australia, Nedlands 6907, Western Australia;* [2] *CSIRO Land and Water, P.O. Pox 1600, Canberra, ACT 2601, Australia (Author for correpondence: E-mail: elefroy@cyllene.uwa.edu.au)*

Key words: alley cropping, deep drainage, revegetation, salinity, tree crop interactions, tree water use

Abstract. Agroforestry has been advocated as a means of managing excess water that has accumulated in the agricultural landscape of southern Australia since clearing of native vegetation. This article examines the feasibility and profitability of agroforestry systems designed to manage rising, saline watertables. A framework for Australian conditions is described that considers the interactions between trees, crops and their below ground environment and how they influence water use, crop yield and profitability. Data is presented from a study of a commercial scale agroforestry system under ideal conditions where trees have access to a shallow fresh water table. The discussion is then broadened to encompass soil, relief and ground water conditions more typical of the southern Australian cropping zone. The relative merits of segregating, integrating and rotating trees with crops are then examined. It is concluded that, in most cases, trees would need to be widely dispersed over a significant proportion of the landscape to manage deep drainage and salinity. Agroforestry is therefore only likely to be an effective solution to water management where trees can compete directly on commercial terms with conventional agriculture. Given the generally low rates of biomass accumulation in semi-arid woody species, this presents a significant challenge for agroforestry in the cropping zone of southern Australia.

1. Introduction

Converting the woodland, forest and heath communities of southern Australia to exotic annual grasslands of crops and pastures has severely disrupted the hydrologic cycle. The deep-rooted summer active component of the native vegetation draws on sub-soil and sometimes groundwater reserves recharged over winter. By contrast, rainfall unused by annual shallow-rooted agricultural crops and pastures accumulates in the landscape and manifests as seasonal waterlogging and rising water tables. The most dramatic consequence is that ancient stores of salt in the subsoil are carried up into the active root zone by rising watertables (Hatton and Nulsen, this volume). This article investigates the degree of intervention required if we are to manage this process through the re-introduction of trees into an agricultural landscape.

Clearing of native vegetation has proceeded largely unchallenged over most of this century due to the commitment by government and the community to expansion of agricultural development (Lines, 1991) despite early appreciation of the hydrologic changes involved (Despessis, 1902; Western Australian Government, 1917; Wood, 1924) and despite warnings of the consequences

(Burvill, 1947). The new hydrology of cleared landscapes is associated with degrading processes in addition to salinity. Soil erosion, acidity and soil structural decline, combined with pollution of groundwater and eutrophication of surface waters, are all related to changes in the way water moves through the landscape (McFarlane et al., 1993). Management of water is therefore a key element in any strategy to improve agricultural sustainability in this region. Recent predictions that salinity will affect about 30% of cleared land in south western Australia by the time a new hydraulic equilibrium is reached some time next century, (Ferdowsian et al., 1996; George et al., 1997), are driving government and farmers to find alternative systems of land use.

Four articles in this volume examine the question of designing sustainable land use systems in southern Australia from first principles using the processes occurring in pre-existing plant communities as models for theoretical agricultural mimics (Hatton and Nulsen; Pate and Bell; Hobbs and O'Connor; and Dunnin et al.). All pay particular attention to the capture and redistribution of water and all conclude that a significantly higher year-round leaf area is required for water management than could ever be found in farming systems based solely on annual plants.

A pragmatic way to achieve this objective is to evaluate farming systems that already include economically valuable perennial species in the hope that these profit-driven systems will have hydrologic side-effects of the required magnitude. The recent practice of alley cropping is an example. Alley cropping, or hedgerow intercropping, evolved as an attempt to reduce soil erosion and increase soil fertility on sloping soils in the tropics (Kang et al., 1990). A similar novel cropping system, though on a much broader scale, has been developed in southern Australia by farmers attempting to manage rising water tables and wind erosion. Trees are also planted to provide additional sources of income through their timber (*Eucalyptus* and *Pinus* spp.) or fodder value (Acacia spp. and tagasaste [*Chamaecytisus proliferus* Link, var. *palmensis*]) and typically occupy 5 to 25% of arable land (Lefroy et al., 1993). In terms of water management, the hope has been that at this density and arrangement, the trees will be able to capture and transpire sufficient water to account for the amount of excess water currently leaching below the root zone of annuals.

The alley cropping experience represents an opportunity to empirically test the effectiveness of a form of intervention that has already passed the critical tests of acceptability and practicality in the eyes of some landholders (see Pannell, this volume). The question remains whether it is capable of achieving the ecological objective of water management in its commercially acceptable form. The next section describes a theoretical framework for evaluating the trade-offs between water management and agricultural productivity of a tree-crop system. Part three describes an experiment comparing the water use and productivity of conventional cropping and a tagasaste alley cropping system on a deep infertile sand in south western Australia. The implications for agriculture are discussed in the final section.

2. Determinants of water use in a tree-crop system

The tree-crop system represented in Figure 1 can be regarded as the smallest repeating unit in an agroforestry landscape. To be acceptable to farmers we must be able to demonstrate its profitability. To be creditable to hydrologists we must quantify the average drainage through the tree-crop unit. Profitability can be summarised by the expression given by Lefroy and Scott (1994) as

Net benefit = T – Y1 + (Y3 – Y2)

where T = the value of the tree products, Y1 is the value of crop/pasture displaced, Y2 is the value of crop/pasture lost due to competition with trees and Y3 is the value of increased crop/pasture or stock yield due to shelter, reduced waterlogging, falling watertables or other effects of the trees.

Reducing drainage to an acceptable value will require a certain area under trees and the distribution of tree roots into part of the crop root zone (S2) and the zone immediately below it (S4). Factors that increase water capture

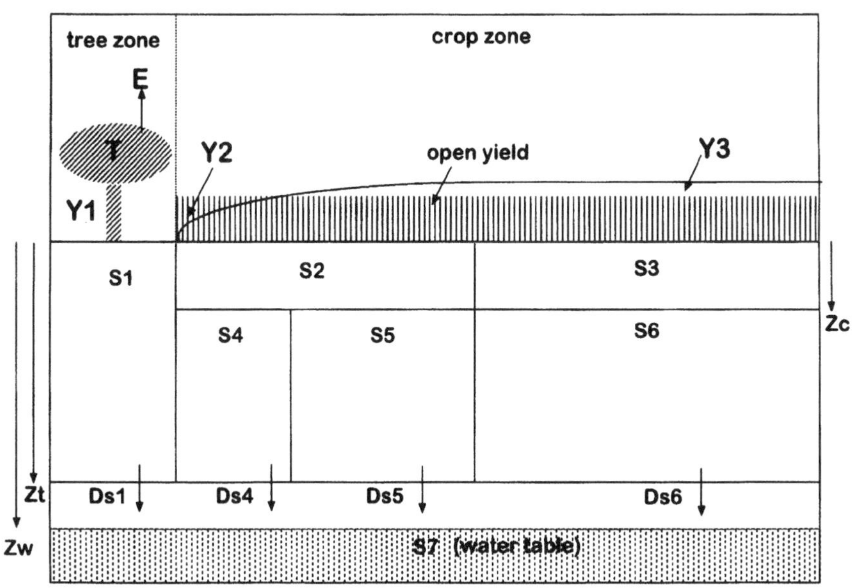

Figure 1. The interactions between trees, crops and their below ground environment in a simultaneous agroforestry system. The variables that determine the effectiveness of water management are crop rooting depth (Zc), tree rooting depth (Zt), depth to water table (Zw), size of the soil zones occupied by tree roots (S1, 2 and 4) and crop roots (S2 and 3), the transmissivity and salinity of the saturated zone (S7), tree water use (E), and drainage (Ds). The variables that determine profitability are the value of crop displaced by trees (Y1), direct value of tree products (T), yield foregone in the competition zone (Y2) and yield increase due to shelter effects outside the competition zone (Y3).

will, in most cases decrease productivity, necessitating a delicate balance between these two management objectives.

2.1. *Crop water use*

Central to the argument of reintroducing trees is the fact that the agricultural landscape leaks more water than the pre-existing vegetation. Quantification has been notoriously difficult because drainage is normally calculated as the difference between evapotranspiration and rainfall and contains the errors inherent in both these measurements. In a review of the literature, George et al. (1997) report at least a hundred fold increase in deep drainage following clearing in the lower rainfall areas; from < 0.1 mm yr^{-1} to > 10 mm yr^{-1}. Peck and Hurle (1973) used catchment scale chloride balance to estimate rates of recharge since clearing in the 500–1000 mm annual rainfall zone of south western Australia. Their estimate of recharge in seven cleared catchments ranged from 23 to 430 mm yr^{-1} (3% and 38% of rainfall respectively) with a mean of 10% of annual rainfall. Values at the high end of this range have been reported for deep sands at Moora (rainfall 460 mm), the site that is the focus for this study. Combining continuous measurement of soil water and a model using 80 years of historical weather data, Asseng et al. (1998), report the median drainage below the root zone of annual crops at Moora to be 140 mm y^{-1} or 30% of annual rainfall.

Attempts to increase evapotranspiration by manipulating crops and pastures are limited by three characteristics of the southern Australian environment. Firstly there is the asynchrony between seasonal patterns of temperature and rainfall, resulting in a surplus of water over the winter months. Secondly, many of the crop-growing soils have a low water holding capacity or subsoils hostile to root growth, so annual rainfall can rarely be stored within the rooting depth of annual plants. Thirdly the high variability in rainfall amount and distribution results in occasional large episodic events outside the winter growing season for which there is no effective avenue of transpiration in the agricultural landscape. The argument that increasing the yield of annual crops will result in significant increases in evapotranspiration and a subsequent decrease in deep drainage has been largely countered by studies showing that increases in crop transpiration are achieved at the expense of soil evaporation, not drainage, with little change in total evapotranspiration (e.g. Smith et al., 1998).

2.2. *Tree Water use*

Drainage under native vegetation in lower rainfall zones of southern Australia was negligible because, over the long term, leaf area index adjusted to match rainfall (see Hatton and Nulsen, this volume). In the < 600 mm rainfall zone, it is reasonable to assume that water use of trees in native vegetation is constrained by rainfall and not by the trees' physiological capacity to use water. However, in a spaced agroforestry system where water is left unused by crops, it is the area of land that tree roots can explore, and not rainfall, that has a greater bearing on tree water use. Theoretically, tree water use would then be limited by the energy absorbed by the canopy, and could be several times higher than annual rainfall on a projected canopy area basis. If this is the case, it may be possible to mimic the water balance of the natural system using a fraction of the land under trees with a high leaf area index and the rest under crops. Stewart (1984) used this approach to calculate the percentage of land area required under trees (A) to prevent water table rise:

$$A = (R - E_{crop})/(E_{tree} - E_{crop})$$

where R is the annual rainfall and E_{crop} and E_{tree} are the annual average water use of crops and trees respectively. Even with tree water use at only 50% of pan evaporation, the above equation gives cause for optimism. For example if annual rainfall is 500 mm, annual crop evapotranspiration 400 mm, and trees could potentially use 1200 mm, then the proportion of a catchment which must be reforested to eliminate drainage is 12%. This would seem an achievable objective.

Experience suggests however that the above example is flawed on two counts. Firstly, although rates of tree water over 1200 mm y^{-1} have been reported over short periods for irrigated plantations and sites over shallow watertables (Myers et al., 1996; Marshall et al., 1998), they seem to be the exception. Annual water use of less than 1000 mm is the norm (Raper, 1998), even when trees have apparent access to water stored either laterally or deep in the subsoil. A second more serious flaw is the assumption that trees could access and transpire all the water unused by crops. This would require absolute complementarity between tree and crop water use, a most unlikely scenario that is dealt with in the next section.

2.3. *Unsaturated soil stores (S1, S2, S3)*

The store of water under the tree canopy (S1), is determined by the rooting depth of the tree and the water holding capacity of the soil. For mature jarrah trees (*Eucalyptus marginata*) in this region, rooting depth may be 15 to 20 m (Kimber, 1974). However it is the rooting pattern beyond the tree canopy that is critical to the success of agroforestry. The ideal root morphology of species introduced for water management would be a low root length density in the topsoil with extensive lateral distribution below the rooting zone of annual plants. This would minimise competition with crops, act as a safety net for leaching nutrients and provide a buffer against recharge. Such a strategy is however at odds the characteristics of the below-ground environment of this region. Endemic and well-adapted introduced summer-active trees on sandplain soils exhibit a dimorphic structure suited to a water uptake pattern of retrieval following winter drainage rather than interception at the time of rainfall. The extensive lateral roots are probably more important for the capture of nutrients, rather than water (Pate et al., 1984; Dawson and Pate 1996; Pate and Bell, this volume) and this would explain why they tend to be confined to the topsoil.

There is little information on the dimensions of soil store S2. Zohar (1985) reported that roots from a Eucalyptus windbreak spread 20 m into an adjacent cotton field in the top 0.8 m of soil, resulting in considerable competition. Least information of all exists for soil store S4, the zone critical to the success of agroforestry as this is where trees are never in competition with crops. Data from Eastham and Rose (1990) suggest that trees at low densities develop shallower rooting patterns. This presents a challenge for agroforestry, which relies heavily on spatial complementarity between spaced tree and crop roots.

The essential factor in exploiting the unsaturated soil stores is the presence and density of tree roots in a particular zone. Unsaturated soil transmits water so slowly that water as little as 1 m from the edge of the root zone is unavailable to the tree. Stirzaker et al. (1999) describe a capture zone beyond the tree root zone where deep drainage can be controlled without competition with crops. The capture zone operates when rain falls onto a wet crop root zone (S2) adjacent to a dry tree root zone (S4). Water that has moved through the crop root zone is pulled sideways towards the dry tree root zone under the influence of capillary forces at the same time as it moves downwards under the influence of gravity. However the capture zone does not operate over a distance of more than 3 m from the extremity of the tree root zone, and would not, in most cases, increase the ability of trees to capture significant additional water in an agroforestry context.

In summary, drainage from S1 will be negligible in the < 600 mm rainfall zone, except on coarse sandy soils. We may expect drainage to be low from S4, where tree roots can capture water from both the topsoil and subsoil. Drainage from S5 would be higher than S4 (as S5 contains no tree roots) but lower than from S6, as there are no tree roots in the topsoil zone immediately above S6. Some competition will be inevitable in zone S2, but this could be minimised by selecting for maximum temporal complementarity between tree and crop (see Ong and Leakey, this volume). Questions that remain to be answered include 1) what root length densities are required in each zone to manage recharge 2) what degree of variation in root architecture can be found in well adapted tree species and 3) the effectiveness of root pruning in limiting competition in S2.

2.4. *Saturated soil store (S7)*

The saturated zone is a potentially important store of water, particularly as the area of land where watertables are within range of tree roots is increasing. When trees take up water from the watertable or the capillary fringe above the saturated zone, water will be replaced laterally at a rate proportional to the saturated hydraulic conductivity of the soil and the gradient generated by the localised drop in water table beneath the tree. The saturated conductivity of soil is orders of magnitude greater than the unsaturated conductivity so the influence of the tree on the watertable level could be seen over a considerable horizontal distance. With reference to Figure 1, if the roots in S1 reached the watertable, the tree could access water that drained below S5 and S6 via the saturated zone. Thus, unlike the unsaturated case above, trees could potentially protect a large area outside the physical extent of their own roots. For example, a belt of trees drawing on a watertable could use all the water that drains through the crop root zone for a distance of 30 m beyond the physical extent of their roots in a clay soil and 400 m in a sand. These distances can be calculated using estimates of the saturated conductivity of the soil, the depth of the saturated layer, the rooting depth of the trees and the annual drainage (Stirzaker et al., 1999).

The ability to rapidly switch between alternative water sources such as near surface soil water and groundwater is an important adaptation to difficult environments observed in endemic tree species (Thorburn et al., 1995; Mensforth and Walker, 1996; Dawson and Pate, 1996) and tagasaste (Pate and Dawson, this volume). However the ability of trees to use water from the saturated zone is severely limited by salinity. When trees use water from the watertable, salts dissolved in the groundwater will concentrate in the root zone of the trees, because roots exclude almost all the salt at the root surface during transpiration. Even if salt tolerant vegetation is used, the faster groundwater is used, the faster the salt levels will build up in the soil.

If the groundwater is already moderately saline (> 5 dS/m), then the maximum salinity in the capillary fringe at which trees use minimal water is reached within a couple of years or less (Thorburn et al., 1995; Stirzaker et al., 1999). Under such circumstances trees will use less than 0.5 mm d^{-1} from the watertable and their impact on groundwater levels will be small (George et al., 1998; Smith et al., 1998). In fact planting trees or fodder shrubs in saline discharge areas, whilst making best use of the land, cannot provide a cure for the problem of salinity unless mechanisms exist for flushing away salts that have accumulated in the soil profile.

2.5. *Tree-crop interactions*

When the tree and the crop are both drawing on limited resources from the same pool they are in competition. The skill in agroforestry design is to minimise the competition and maximise the complementarity between the two species. Complementarity occurs in a tree/crop mixture when the mixture 1) uses a greater fraction of a limiting resource and/or 2) either component makes more effective use of a limiting resource than sole crops (Ong and Black, 1994). Examples of greater resource capture include trees extracting water unavailable to crops by virtue of spatial complementarity (deeper rooting habit) or temporal complementarity (using off-season water). Examples of more efficient resource use include the greater water use efficiency in the 'quiet zone' in the lee of the windbreak (Frota et al., 1987), and the greater water use efficiency of certain C_3 plants growing in light shade. Greater efficiency of resource use may also occur through functional complementarity, e.g. using nitrogen fixing trees and trees with proteoid roots capable to extract phosphorus from low P soils.

In the early phase of agroforestry research, it was assumed that complementarity of resource capture between trees and crops would largely offset the competition (Ong and Leakey, this volume). This has not proved to be the general case in low to medium rainfall zones. Research has since suggested that competition is more likely to be encountered than complementarity, often because tree roots occupy the same soil zone as crops. This has been shown to occur as a result of physical and chemical barriers to root penetration (Ong, 1996), as a consequence of pruning regimes which can lead to a higher proportion of superficial roots (van Noordwijk and Purnomosidhi, 1995) and where soil structure is such that annual rainfall can readily be stored within the top metre or two of the profile (Dupraz et al., 1997; Dupraz et al., 1998).

The net effect of mixing trees with crops is more likely to depend on the relative value of the tree and the crop than on the interactions between. While competition close to the tree may be compensated for by positive effects at a distance, such as from shelter as in the tagasaste alley cropping experiment below, this has not been the experience with a wider range of crops over a range of sites and seasons in south western Australia. Data comparing mean crop yield in the lee of windbreaks and in the open for six crops over five years at a range of sites showed a small positive effect of shelter outside the competition zone, but no significant change when the total arable area was considered (R. Sudmeyer pers. comm.). This lends support to van Noordwijk's (1996) assertion that agroforestry systems are only viable if the direct values of tree and crop per unit area occupied are similar unless there is strong complementarity of resource use. If a tree is of lower value than the crop, it must pass significant resources on to the crop to warrant its inclusion. The onus is therefore on finding tree crops with higher annual returns than grain given the higher cost of establishment and the lag time before returns.

3. Water use and productivity of tagasaste alley cropping

This section presents a case study of a simultaneous agroforestry system on a deep sand in south western Australia where trees have access to a shallow fresh water table. This represents the ideal case in terms of tree water use.

The site at Moora (lat. 30°45′; long. 116°40′), 150 kilometres north of Perth, originally supported a woodland community dominated by *Banksia prionotes*, described by Pate and Bell (this volume). This site is typical of the sandplain regions of southern Australia that remained uncleared for agriculture until the 1950's and 60's when trace element technology enabled crop and pasture production on these very infertile soils. Following a one-off application of copper, zinc and molybdenum, these sandplain soils were typically sown to annual pastures dominated by the naturalised sandplain lupin *Lupinus cosentinii* and top-dressed each year with phosphate fertilizer. The very high level of nitrate recorded in the perched water table at this site (13 ppm N) is probably the result of thirty years of fixation and leaching in these legume-dominated pastures. Development of the sweet lupin grain crop *L. angustifolius* in the 1970's saw the rapid adoption of continuous lupin/cereal rotation as a viable cropping option (Gladstones, 1982; Marsh, 1997).

Rising water tables, herbicide resistance, wind erosion and sub-soil acidity have since raised doubts about the sustainability of annual crop and pasture systems on these soils. As described earlier, median annual drainage is about 140 mm y^{-1} in this 460 mm rainfall environment. If our objective is to prevent further rise in a water table, the question is what density and arrangement of trees would be required, and what would be the consequence for agriculture.

An 8 ha experimental site was established in 1992 to compare the water use and productivity of a tree/crop mixture with sole tree and sole crop treatments. The tree/crop mixture was a single row of the fodder tree tagasaste planted at 1 m spacing in rows 30 m apart giving a tree density of approximately 550 ha^{-1}, a design based on a commercial alley cropping system (Lefroy and Melvin, 1996). Tagasaste was introduced to Australia from the Canary Islands last century and is grown primarily in dense plantations where it is mechanically pruned to provide fodder for sheep and cattle. In the sole tree (plantation) treatment, tagasaste was planted at 1 m spacing in rows 6 m apart (2330 trees ha^{-1}). Trees were direct seeded in rows orientated north/south using a commercial tree planter and cut back from 2 m to 0.6 m high in May 1996 when measurement began. Grain lupins (*Lupinus angustifolius* var. Gungurru) were sown in the sole crop and alley crop plots in June 1996 followed by oats (*Avena sativa* var. Toodyay) in 1997. Each treatment was replicated twice in plots 50 m by 160 m.

The very low fertility and water holding capacity of the soil can be seen from the data in Table 1. Since clearing 34 years ago, a fresh (< 1 dS m^{-1}) perched water table has developed lying over a clay layer at approximately 10 m depth. For the duration of experiment it was approximately 5 m below the surface. The soil characteristics and very low relief at this site mean that the water balance is essentially a vertical exchange, and for the purposes of this study it is assumed there is no runoff and no lateral water movement above the saturated zone. The aim of the experiment was to quantify the variables described in the previous section.

Soil water content was measured to a depth of 3.7 m in sole tree and sole crop treatments using a neutron probe and TDR. Replicated transects of neutron probe access tubes were also positioned across the tree rows between the centres of adjacent alleys. Transpiration by the tagasaste was estimated by fitting stems with heat pulse sapflow sensors (Hatton et al., 1990), moved to fresh stems on a three monthly rotation. Flux per unit area conductive wood

Table 1. Description of the soil profile at the alley cropping experimental site.

	A_1 Horizon	A_2 Horizon	B Horizon
Depth (mm)	0–100	100–300	300–10,000
Colour and texture	Grey sand	Pale grey sand	Yellow sand
Bulk density	1.55	1.61	1.61
Water content at F.C. (% v/v)	12	12	12
Water content at W.P. (% v/v)	3	3	3
Sat. hydraulic. cond. (mm h^{-1})	300	300	300
pH (in 0.01 M $CaCl_2$)	5.2	4.8	4.8–5.4
Organic carbon (C %)	0.50		
Phosphorus (ppm)	13	8	2
Nitrogen (ppm NO_3)	7	2.5	< 1
Potassium (exch., soil)	0.03	0.03	0.02

was then scaled up to plot level transpiration using data from three monthly surveys of stem size distribution and conductive wood area per unit land.

3.1. *Crop water use*

The time series in Figure 2 commences in July 1996, following a very wet opening to the winter season, when the measured profile depth of 3.7 m was

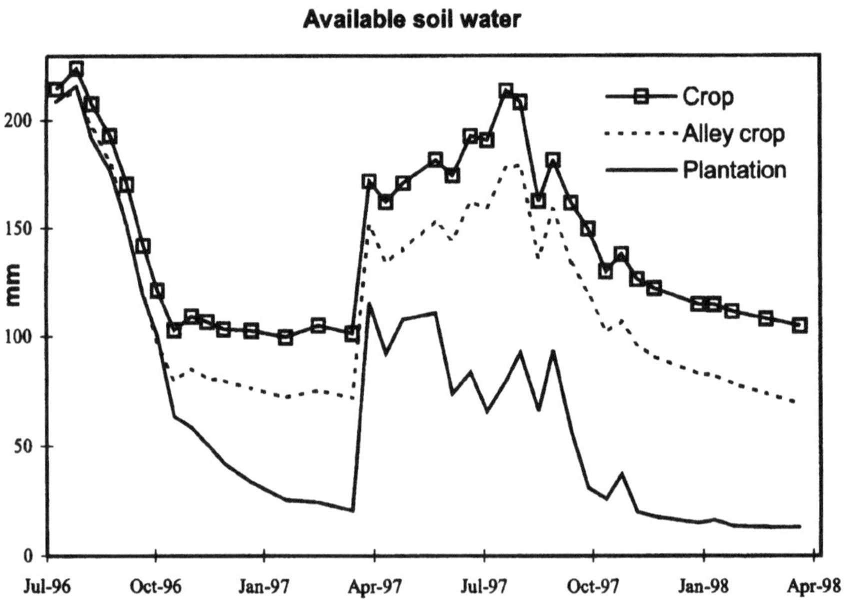

Figure 2. Change in available soil water content under crop, alley crop and tagasaste plantation at Moora, Western Australia, from July 1996 to April 1998.

at the upper drained limit in each treatment (220 mm). By the end of the lupin growing season (November 1996), available water content had been drawn down to about 102 mm. Between April 97 (break-of-season rains) and August, the soil profile under the oat crop progressively filled, with about 120 mm left in the profile at harvest. These changes in soil water content under the sole crops illustrate a pattern typical of annual plants in this environment. The profile fills from the break of season until mid to late winter followed by rapid draw down over spring when there is sufficient leaf area and radiant energy for significant transpiration. There is little change over summer and autumn when the land is fallow. Lupins extracted water from a depth of up to 3.5 m although the season terminated before they had exploited all the available water between depths of 2 and 3.5 m (Figure 3a). This illustrates the difficulty in defining the dimensions of the zone S2 and S3 of Figure 1. The oat crop did not use water below 2.2 m in the 1997 season.

3.2. *Tree water use*

At the end of the first growing season (November 1996), the soil profile to 3.7 m in the sole tree and alley crop treatments contained 64 mm (29% full) and 80 mm (37% full) respectively, compared to 102 mm (47% full) under the sole crop (Figure 2). The tagasaste had been heavily pruned to a height of 0.6 m in April 96, but was not pruned before the next rainy season. After this time, the soil profile in the sole tree treatment did not refill below 0.75 m because of the greater leaf area.

There was a marked slowing down of water extraction by the tagasaste after October 1996 coinciding with the emptying of the profile and the increasingly hotter weather. A comparison of deuterium/hydrogen ratios of groundwater, xylem sap of shallow rooted annuals (as an indicator of soil water) and xylem sap from tagasaste stems indicated that the tagasaste went through a sudden change in its relative dependence on groundwater and soil water (Pate and Dawson, this volume). Between November and December each year it switched from a 20% dependence on groundwater to 60% dependence. During this same period, wilting and desiccation of young tagasaste shoot tips was observed indicated that this mesophytic shrub, adapted to a more humid maritime environment, was experiencing heat stress. This was associated with a sharp increase in delta ^{13}C ratios of shoot tip samples over the same period (Pate and Dawson, this volume), suggesting increased periods of stomatal closure.

Figure 3 shows soil water content under sole crop, plantation and alley crop treatments for April 2 1997, at the driest time of the year. The residual water in the upper layers of the crop (Figure 3a) and alley crop treatments (Figure 3b) was a consequence of isolated storms in later summer. The most striking feature of Figure 3b is the restricted lateral extraction of water by alley cropped tagasaste trees. Soil water extraction is confined to a zone within 4 m either side of the tree rows. From Figure 3b it is evident that the apparent

a)

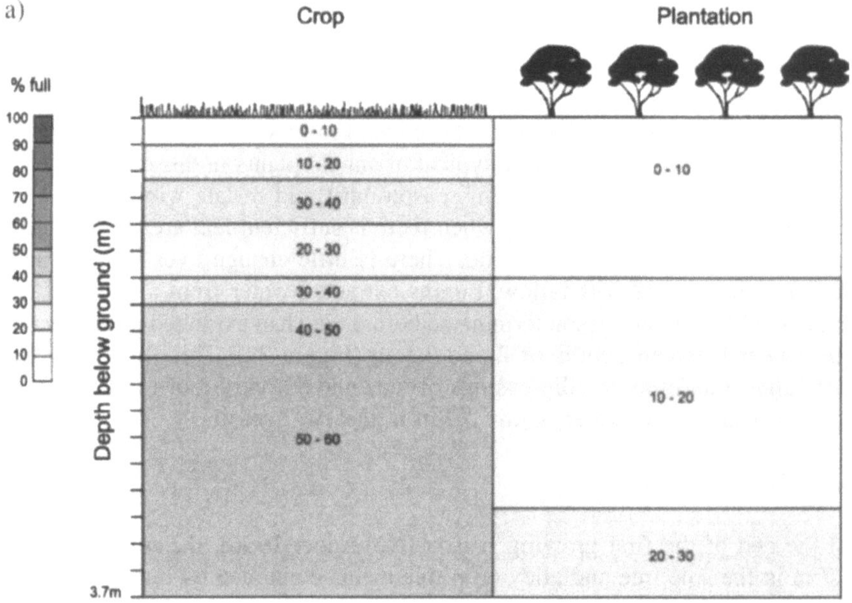

b)

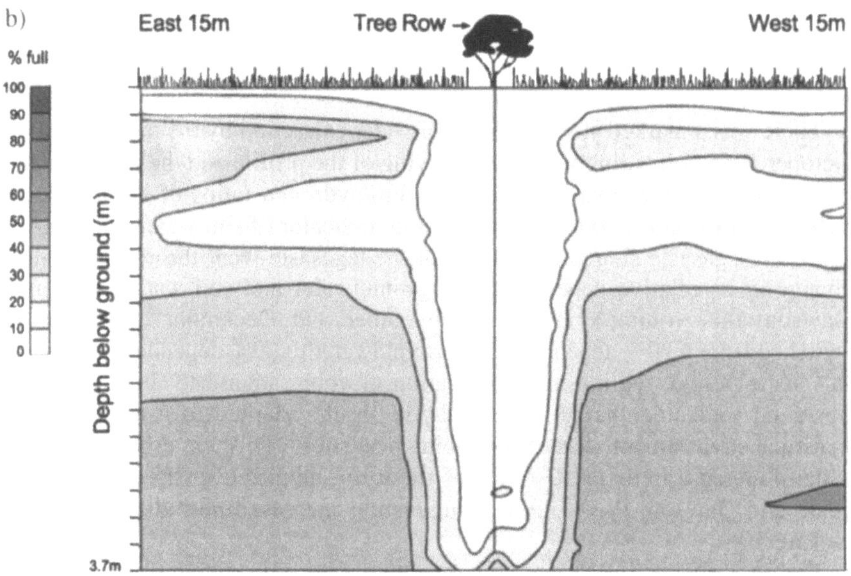

Figure 3. a). Soil water content under crop (left, mean of 12 access tube readings) and plantation (right, mean of four access tube readings) at Moora, Western Australia, on April 2, 1997. Water content expressed as % full for soil profile to a depth of 3.7 m. b). Soil water content measured in a transect across alley crop plots on April 2, 1997 showing the narrow zone of soil water extraction by rows of tagasaste trees. Water content expressed as % full for soil profile to a depth of 3.7 m (mean of two transects).

20 mm buffer of dry soil under the alley crop compared to the sole crop treatment (Figure 2) was confined to a narrow zone immediately under the trees.

Tagasaste transpiration, as measured by sap flow, was remarkably similar throughout the year (Figure 4) with a mean daily water use of 2 mm d^{-1} ranging from a low of 1.5 mm d^{-1} in winter to a high of 2.5 mm d^{-1} in summer. Despite the fact there was fresh groundwater in a transmissive aquifer available at a depth of 5 m, the tagasaste did not achieve better than 40% of potential rates over the three summer months and 48% of potential for the year. Total transpiration over 12 months from January 1997 was measured at 749 mm. Allowing for a further 150 mm as soil evaporation (estimated from changes in soil water content in the top 20 cm) and canopy interception (measured with transects of rain gauges), total evapotranspiration was in the order of 900 mm y^{-1}. Assuming for the moment no positive or negative interaction between trees and crops, a simple replacement series between sole crop with evapotranspiration of 320 mm y^{-1} (rainfall of 460 mm y^{-1} less median deep drainage of 140 mm y^{-1}) and sole tree using 900 mm y^{-1} indicates 24% tree cover at the catchment scale would be required to prevent further water table rise (Figure 5). This assumes that all trees would have access to a perched fresh water table.

3.3. *Tree-crop interactions*

Tree-crop interactions can be viewed from two perspectives – the effect the trees have on the crop, and the effect the crop has on the trees The impact of

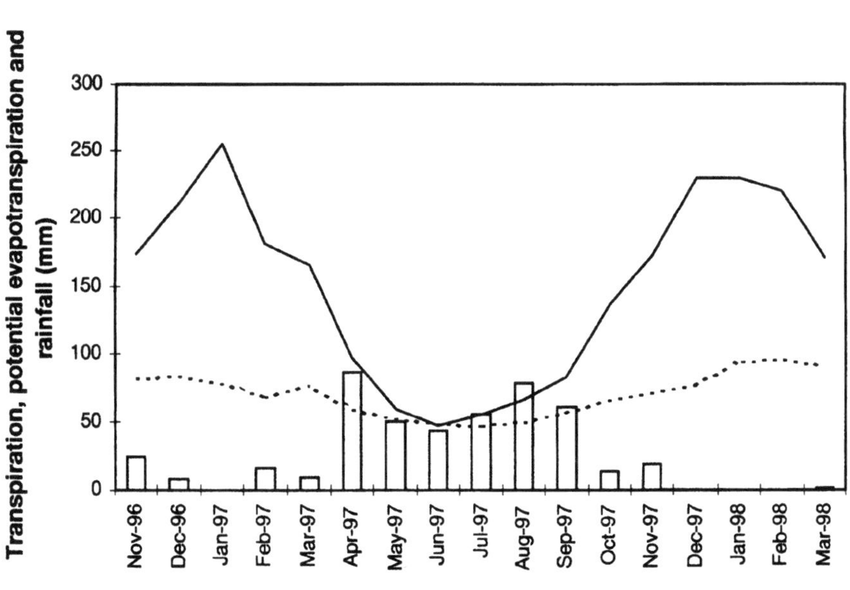

Figure 4. Transpiration by tagasaste plantation (dashed line), potential evapotranspiration (solid line) and rainfall (columns) at Moora, Western Australia.

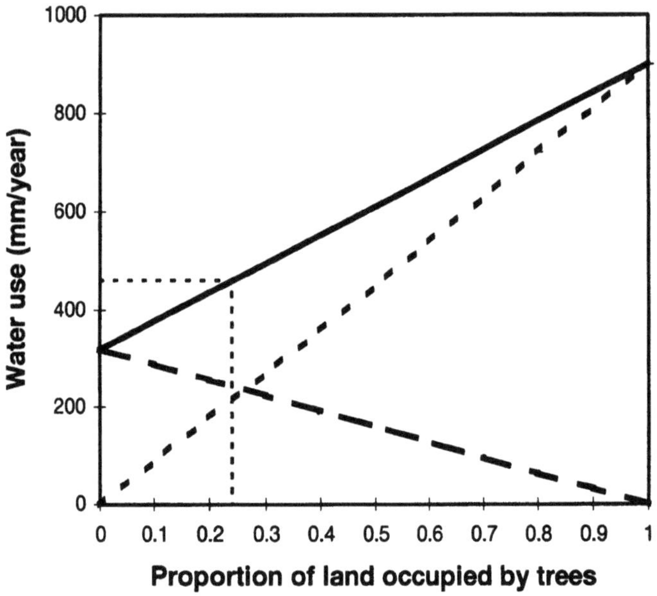

Figure 5. A replacement series involving sole crop (dashed line) with evapotranspiration of 320 mm per year (rainfall less median deep drainage) and plantation (dotted line) using 900 mm per year showing that 24% tree cover at the catchment scale would be required to account for annual rainfall of 460 mm y^{-1} and prevent further water table rise.

the trees on crop yield was measured by harvesting continuous transects across the 30 m alleys with a 1.8 m wide mechanical harvester and comparing that to the yield from a similar area in the sole crop plots (Figure 6). A positive response to shelter was measured in lupins in 1996. There was a 29% increase in grain yield over the arable area compared to the sole crop treatment and a 22% increase over the total land area, i.e. including the arable land displaced by trees (Figure 6a). The shelter effect was detected only in grain yield, with no significant difference in biomass between alley and sole crop plots measured at three stages during the vegetative phase. A pot experiment using soil taken in a transect across an alley showed no significant differences in plant growth or grain yield of lupins suggesting the yield response in the field was due to above-ground rather than below-ground influence of the trees. The yield of oats was 4% higher per unit arable land under alley cropping but not significantly different over the total area (Figure 6b). Thus the legume and cereal crops responded differently to shelter, as reported by others (Kort, 1984; Bicknell, 1991). In 1998 lupin yield was significantly lower in alleys.

Assuming yield effects similar to 1996 and 1997, and that wider tree belts would have similar effects on crop yield as single rows, increasing the area of land under trees from the present 10% to 24% (to fit the water use objective established above) would result in a 3% drop in lupin yield and a 22%

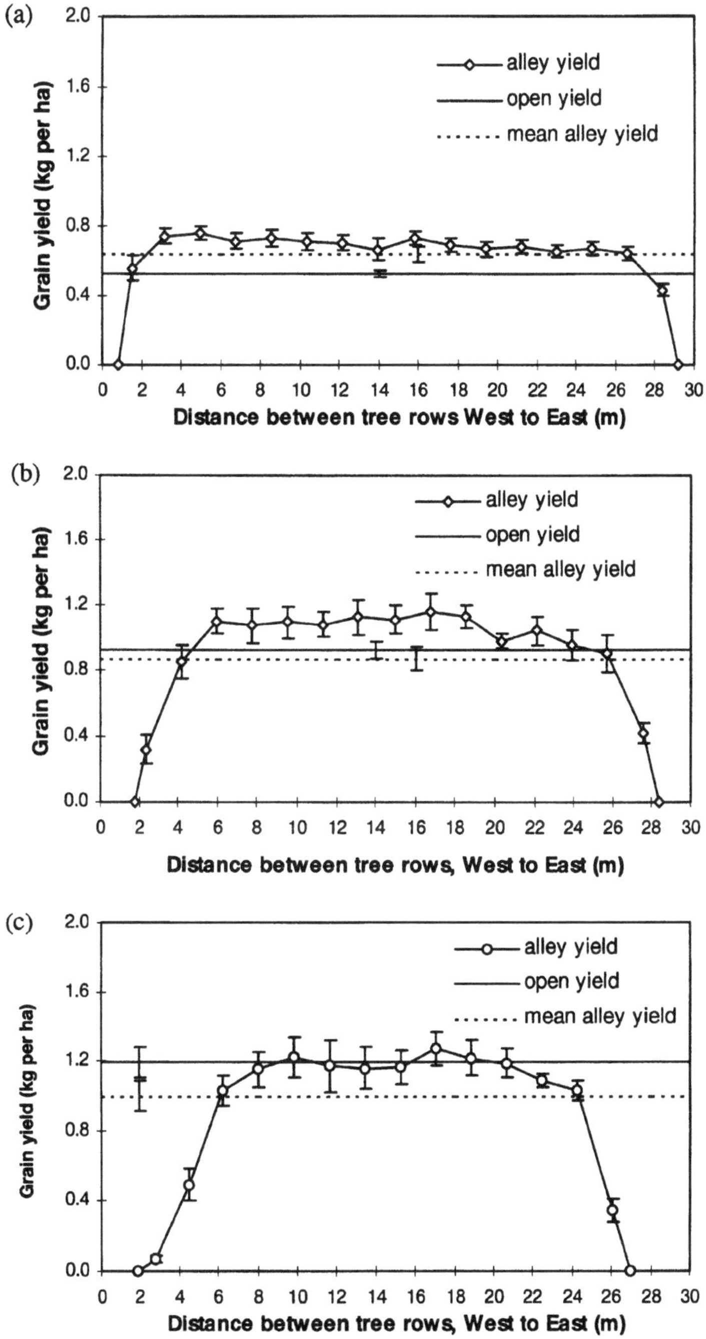

Figure 6. Yield of lupins 1996 (a), and oats 1997 (b) and lupins 1998 (c) measured in 1.8 m widths in a continuous transect across 30 m wide alleys (*n* = 5) at Moora, Western Australia.

drop in the yield of oats over the total land area under alley cropping compared to sole crop. In a lupin/cereal rotation, the cereal phase is generally the more profitable with the legume being important as a source of residual nitrogen and as a disease break. This situation would therefore be unacceptable unless the tree has a commercial product of similar or higher value than the displaced crop and its harvest does not compromise the water use or shelter roles of the tree.

Where the proportion of land required under trees for water management is low, or where shelter is the primary design criteria, yield data for the first two years suggest there may be medium to long term benefits from the response of the legume phase to shelter that could justify adoption (Figure 7). However, the initial cost of establishment and the lag between planting and yield response inevitably means a period of forgone income, in this case seven years. The amount of forgone income can be regarded as a measure of the willingness to pay for water management on the part of landholders and/or the degree of intervention required by society for such an approach to become commercially attractive.

The effect of the crop on the trees was minimal. We might expect tagasaste trees surrounded by annual crops to extract more water than those in plantation, but no difference in transpiration rate was detected. This is further evidence that tagasaste does not have the rooting pattern of an ideal agroforestry tree (Figure 3b). It may also explain why so little competition was observed in the crop zone adjacent to the trees in the first two years.

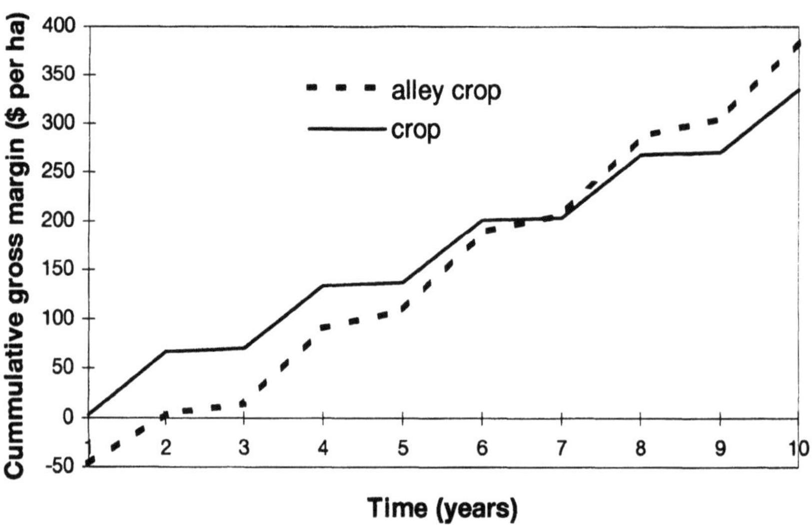

Figure 7. Cumulative gross margin for a lupin cereal rotation under conventional and alley cropping based on data collected at Moora, Western Australia. Assumes a twenty percent yield increase due to shelter effects in the lupin phase and a five percent yield increase due to shelter effects in the cereal phase. Establishment assumes direct seeded trees occupying ten percent of arable area, reaching a mature height in year five.

4. Implications for agriculture

In many parts of the world, woodland and forest have been converted to grassland without suffering the hydrologic consequences that have been experienced in southern Australia so rapidly after the introduction of agriculture. The great age and low hydraulic gradient of this landscape combined with the asynchrony between rainfall and radiant energy that typifies a semi-arid Mediterranean-type climate are the fundamental reasons. Rehabilitation strategies based on winter active plants are demand limited, ie limited by low potential evaporation at the time of major recharge events. If we are to address the problem of saline watertables there will have to be large scale introduction of deep rooted perennial vegetation into the agricultural landscape. There are potentially three ways to do this, namely to segregate, to integrate or rotate perennial vegetation with conventional agriculture (van Noordwijk et al., 1997).

4.1. *To segregate trees and crops*

Segregation involves growing and managing each species separately. Modern agriculture is based on monocultures precisely tuned to be productive when subsidised by inputs of fossil fuel energy. It may be best to continue to manage the best land to be as productive as possible in this way, and to set other land aside for environmental services (van Noordwijk, this volume). In the context of the salinity problem, a proportion of the land could be put under conventional forestry to reduce drainage, with the remainder under conventional crop and pasture monocultures. There are three problems to this approach. First, it is difficult to identify areas in a catchment where there is sufficient lateral movement so that excess water can move from cropping land to blocks of trees. Though there are some examples of strategically placed plantations intercepting groundwater (e.g. George et al., 1997), it is more likely that trees in plantations will have to survive on annual rainfall once they have mined the water left behind by annual crops or pastures. Second, hydrologists in Australia continue to revise upwards the area that needs to be revegetated as more data becomes available, to the point where annual agriculture may only be viable on just 20% of the land (Miller et al., 1981; Schofield et al., 1989; Hatton and Nulsen, this volume). Third, many regions where dryland salinity is a problem do not have sufficient rain to sustain profitable commercial forestry.

One example where segregation does have merit is if trees can access relatively fresh watertables over transmissive aquifers, as in the case study above. With drainage under crops averaged 140 mm y^{-1}, tagasaste was able to extract 420 mm y^{-1} from the watertable. Thus each hectare of tagasaste could protect some three hectares of arable land. Using this example, the area of a catchment which needs to be revegetated to prevent drainage is presented in Table 2. The proportion of trees and crops was arrived at by allo-

Table 2. The extent of the saturated zone and its effect on the proportion of a catchment requiring revegetation for catchment evapotranspiration to equal rainfall. Assumes annual median deep drainage is 140 mm, annual average rainfall is 460 mm, crop water use is 320 mm, plantations over a water table use 900 mm and plantations without access to a water table are in equilibrium with rainfall.

Extent of saturated zone (%)	Plantations + WT (%)	Annual crops and pastures (%)	Trees in alley cropping (%)	Plantations – WT (%)	Total revegetation (%)
2	2	6	1	91	94
5	5	15	2	78	85
10	10	30	3	57	70
15	15	45	5	35	55
20	20	60	6	14	40
25	25	75	0	0	25

cating all the area with access to the saturated zone to plantations of tagasaste and then assigning three times that area to annual crops or pastures and the remainder to 'dryland' plantation and trees in alley cropping (with trees occupying 10% of the arable area). The outcome is that for every 5% decrease in the area of land that is overlying a saturated zone below a 25% reforestation target, the area required under trees rises by 15%. However the assumption that all deep drainage will reach a perched aquifer is not realistic. The fact that regional groundwater systems are rising in the order of 0.2 m y^{-1} in the Moora region (R. Speed, pers. comm.) and are not responding to localised revegetation that has dried up perched systems (R. Speed et al., 1993), indicates that some recharge is bypassing these perched aquifers. Moreover, the area of land having a saturated zone within reach of trees in any given catchment is difficult to estimate without extensive drilling. Experience suggests that only a small proportion of aquifers in the cropping zone are likely to be both fresh and within the rooting depth of trees. George (1992) estimated from a drilling program in the eastern wheatbelt (annual rainfall 300–350 mm y^{-1}) that at most 20% of the sandplain soil units had an underlying shallow saturated zone that was relatively fresh (2–10 dS/m). Sandplain constitutes about 20% of soils in the crop growing region (Carder and Grasby, 1986) and the occurrence of relatively fresh watertables is rare on other soil types.

4.2. *To integrate trees and crop*

Alley cropping is an example of integration, where the domain of the tree overlaps with that of the crop so that trees and crops share or compete for the same resources. The pivotal question here is when is it better to integrate than to segregate. The rationale for growing crops and trees simultaneously is that it increases the chance of capturing all available resources. How this increase in total resource capture is partitioned between the two species will determine the productivity and economic viability of the integrated system.

A simple framework for deciding whether to segregate or integrate is illustrated in Figure 8 for the Moora region (annual rainfall and crop water use 460 and 320 mm y^{-1} respectively). For the purpose of this example it is assumed the tree component cannot access a watertable. The following assumptions are made to examine the most favourable possible outcome for integration:

1. In the segregated case (plantation) the projected area of the tree canopy and root zones are the same, so that tree water use is equal to rainfall.
2. In the integrated case (alley farming) the trees are spread out such that their roots explore the entire area under cropping.

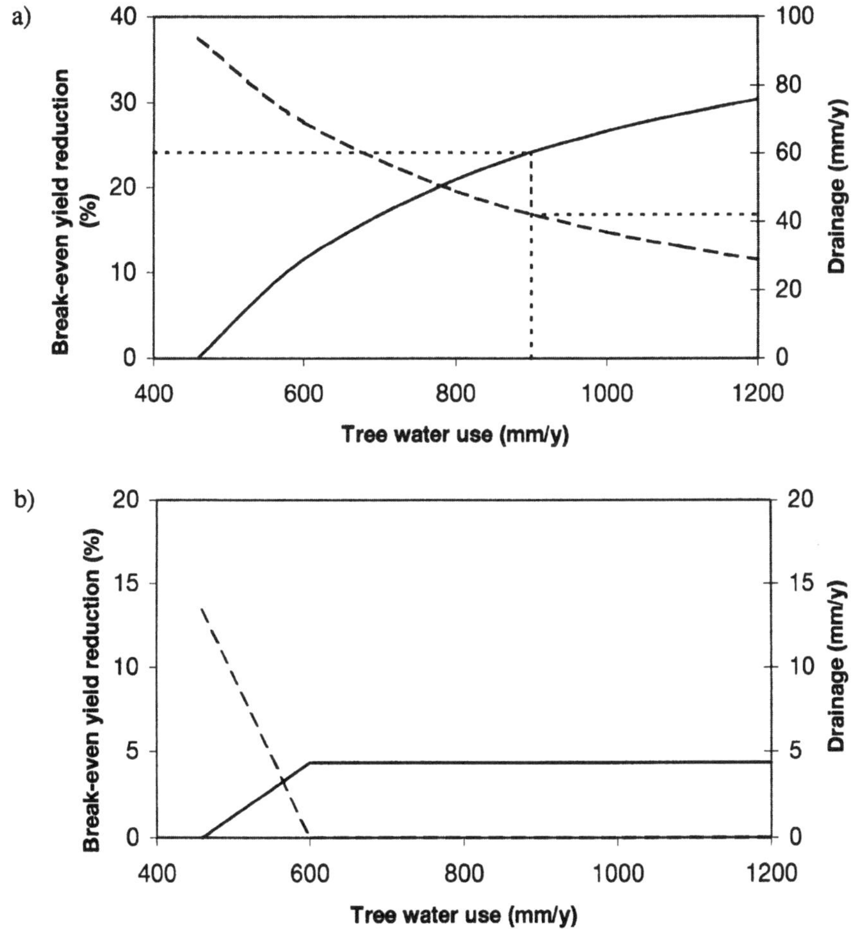

Figure 8. The break even yield reduction (solid line) and drainage (dashed line) as a function of the water use of the trees integrated with crops when drainage below the crops is a) 140 mm per year and b) 20 mm per year.

3. Annual rainfall can be contained within the depth of the soil profile occupied by tree roots (ie. the entire area of the catchment).
4. There is a linear relationship between the amount of water a crop uses and its productivity.

The trees integrated with crops will have access to water unused by the crop and will, based on assumption 2, use more water than trees in plantation. If integrated trees use more water per unit projected canopy area than segregated trees, it follows that a lower proportion of reforested land will achieve a given total tree water use. This leaves a greater area of land for crops. However, crop growth will be reduced in the integrated case because of competition from trees. Thus there is a trade-off in the integrated case between a greater area of cropped land and tree-crop competition. The following steps need to be taken to evaluate the trade-offs between greater area under crop, tree-crop competition and deep drainage:

1. Estimate the total excess water in the catchment when it is entirely planted to crops.
2. Estimate the discharge capacity of the catchment.
3. Calculate the area under crop that produces recharge equal to the discharge capacity and plant the remainder of the catchment to trees.
4. Estimate/measure the annual water use of a line of trees planted in cropping land.
5. Calculate the area of land that should be planted to lines of trees so that the sum of tree water use and crop water use is equal the discharge capacity of the catchment.

For the Moora example above, we calculate average drainage as 140 mm y^{-1}. If we wanted to reduce this figure to around 40 mm y^{-1} (i.e. by about 70%) we would need to put about 70% of the catchment under plantations of trees. However if we measured that a belt of trees in cropping land was able to use 900 mm y^{-1}, we could achieve the same objective by having 33% of the catchment planted to tree belts. Figure 8a illustrates the effect of putting 33% of the catchment under tree belts on break even yield reduction and drainage. The break even yield reduction is the percentage reduction in crop yield that can be tolerated in the alley-cropped case. For a tree water use of 900 mm y^{-1}, the break even yield reduction is 24%. At this break-even yield reduction, the total yield in the alley-crop system would equal that when trees and crops were segregated. The break even yield reduction arises because the cropping area has increased from 30% (segregated case) to 67% (integrated case). The above proportions reflect the extent of tree planting needed in the segregated and integrated cases respectively to keep drainage at 40 mm y^{-1}.

The above example shows that there may be scope for integration when the drainage is as high as occurs on the sandplain. However, it is more common in the cropping belt for drainage to be around 20 mm y^{-1}, and at these lower values the case for integration is much weaker. Once the trees use more than 600 mm y^{-1} there is no further excess water and additional water used by trees is directly at the expense of the crop. This gives a maximum break-even yield reduction of less than 5%, which would be difficult to achieve in most circumstances (Figure 8b).

It must be noted that the above examples describe the ideal case for integration where the trees capture all the water that would have drained below crops. The tree is this study (tagasaste) did not fit the assumption of wide and deep lateral rooting. Moreover the use of annual tree and crop water estimates can lead to wrong conclusions. The area of trees integrated into cropping land may be just right in an average year to soak up the water unused by crops. In a dry year there will be no excess water and the trees will always be in competition; in a wet year the area under trees will be too low to have an impact. These problems are not as severe in the segregated case, where trees will not compete in dry years and can respond by using more water in wet years. If it is essential to halt deep drainage in catchments exporting large amounts of salt, but average drainage is relatively low, it may be better to think in terms of alley forestry than alley cropping, ie wide belts of trees separated by fallow land. Trees growing in areas with insufficient rainfall to produce a commercial product could become a commercial proposition with access to water from uncropped alleys. Alley forestry would be a less risky proposition than alley cropping and would be more effective in unusually wet years.

It should be remembered, as described in section 1 and 2, that a certain amount of tree-crop integration is acceptable to farmers. In many cases it is driven by the value of the tree fodder and the protection trees afford to crops during extreme weather events. Although alley cropping in Australia is carried out with much greater tree-crop separation than normally seen in the tropics, the observation by Sanchez (1995) is pertinent: 'alley cropping is like playing with fire, because it has strongly positive and strongly negative interactions and thus requires precise fine tuning in every situation'. It remains to be seen whether we in Australia can find tree species of sufficiently high commercial value and manage to integrate them with crops with the appropriate degree of finesse to satisfy both productivity and water balance objectives.

4.3. *To rotate crops with herbaceous perennials*

Increasing awareness by agronomists of the scale of the salinity problem has fuelled a resurgence of interest in rotating lucerne with dryland crops (Ridley et al., 1998; Dunin et al., this volume). Lucerne has proven itself as a valuable plant in conventional rotations, and farmers with the skill to manage lucerne in dryland environments have found it to be both profitable and effective in de-watering subsoils. However the requirements of lucerne are exacting and it will be very hard pressed to survive in much of cropping zone of southern Australian where there are acidic or heavy clay subsoils.

Gardner (1944) identified the relative absence of herbaceous plants as a defining feature of the flora of south western Australia. This is unlike the situation in eastern Australia where extensive areas of grassland were, and still are, to be found (Moore, 1970). While early European explorers in the south west noted areas of grassland managed by aborigines with fire for hunting and ease of movement, these were, with some very minor exceptions, the grassy understorey of woodland vegetation not grasslands in their own right (Hallam, 1979). The life form analysis presented by Pate and Bell (this volume) also illustrates the characteristic dominance of woody over herbaceous perennial species. The relative absence of herbaceous perennial plants in the original vegetation does not rule out the possibility of developing well adapted cultivars of lucerne-like plants, but it does suggest the successful adaptation of this life form to a wide range of habitats in this region presents a significant challenge. The need for roots capable of penetrating physically and chemically inhospitable sub-soils and the difficulty of maintaining mesophytic foliage through hot dry summers are two challenges to the development of well adapted herbaceous perennials. The persistence of limited plantings of drought tolerant summer active perennial grasses such as *Chloris gayana*, *Erharta calycina* and *Eragrostis curvula* (all from southern Africa) in the southern and western part of the region does provide some grounds for optimism. It may be better to think in terms of very short rotation forestry.

4.4. *The challenge for agroforestry*

The foregoing discussion highlights problems with segregation, integration and rotation of perennial plants as strategies for water management. It would seem that deep rooted, summer active vegetation needs to be widely dispersed across the landscape in a manner similar to that in the original woodland. This prognosis is not new. Over 90 years ago, the Government Analyst advising a wheatbelt farmer how to prevent wells and soaks becoming saline, suggested it would require replanting trees in a manner 'practically co-extensive with the original forest' (Mann, 1907). The expectation that small localised plantings of high water use trees could manage recharge and limit the displacement of conventional agriculture is unrealistic, being limited by the area of land overlying a fresh saturated zone within reach of tree roots, the slow rate of lateral water movement through soil and the conservative rates of water use of trees. As Hatton and Nulsen (this volume) point out, little short of geological rejuvenation will alter this situation.

Those factors within our influence that could lead to more concentrated revegetation and less displacement of agriculture include identifying and mapping shallow fresh water tables, using earthworks to divert surface water as concentrated recharge within the root zone of trees (McFarlane and Cox, 1990) and using geological structures as indicators of improved soil water availability (Farrington and Salama, 1996; Clarke et al., 1998). The biological options include selection of tree species with root architecture and functional characteristics complementary to annual plants. These are likely to be secondary to the challenge of developing new tree crops that could commercially drive even a small degree of displacement of grain agriculture. Progress towards the development of viable tree crops in this region (Bartle et al., 1996) and high seed yielding herbaceous perennials in the Great Plains (Jackson, 1985) indicate that domesticating perennial species is a decades long process with a high degree of uncertainty. The low rates of biomass accumulation of semi-arid woody species (e.g. Pate and Bell, this volume), accentuate this challenge.

Whether water management involves integrating, segregating or rotating trees with crops, the transition from an exclusively annual-based agriculture to one with a significant component of woody perennials inevitably involves a lag between establishment and first returns to the landholder (e.g. Figure 7). This problem was recognised over 24 centuries ago, when the Athenian politician Peisistratus established low interest loans to farmers forced by severe soil erosion to make a transition from grain agriculture to the cultivation of vines and olives (Claiborne, 1970). Similar intervention by government or industry will be equally important if such a transition is to occur in southern Australia.

This is a region with low ecological amplitude. Unlike the young post-glacial soils of the Middle East and Europe from which grain agriculture evolved, it appears we cannot stray far from the original vegetative structure without paying a penalty. Whether viable functional mimics of that original vegetation can be developed, herbaceous or woody, remains to be seen.

5. Acknowledgements

The experimental work was part of a broader study of the deep sand ecosystems near Moora, Western Australia, carried out in collaboration with J. S. Pate and M. J. Unkovich with the assistance of C. P. Allmark, D. J. Arthur, and staff of Agriculture Western Australia. Land was generously made available by Christine Brown. The research was jointly funded by the Grains Research and Development Corporation and the Rural Industries Research and Development Corporation.

6. References

Asseng S, Fillery IRP, Anderson GC, Dolling PJ, Dunin FX and Keating BA (1998) Use of the APSIM model to predict yield, drainage and NO_3^- leaching for a deep sand. Australian Journal of Agricultural Research 49: 363–377

Bartle JR, Campbell C and White G (1996) Can trees reverse land degradation? In Proceedings, Farm Forestry and Plantations: Investing in Future Wood Supply, pp 68–75, Australian Forest Growers, Mt Gambier, South Australia

Bicknell D (1991) The role of trees in providing shelter and controlling erosion in the dry temperate southern agricultural area of Western Australia. In Proceedings, The Role of Trees in Sustainable Agriculture, pp 21–39, Rural Industries Research and Development Corporation, Canberra

Burvill GH (1947) Soil salinity in the agricultural area of Western Australia. Journal of the Institute of Agricultural Science 13: 9–19

Carder J and Grasby J (1986) A framework for regional soil conservation treatments in the medium and low rainfall agricultural districts. Western Australian Department of Agriculture Research Report 1/86

Claiborne R (1970) Climate, Man and History. Angus and Robertson, London

Clarke CJ, Mauger GW, Bell RW and Hobbs RJ (1998) Computer modelling of the effect of revegetation strategies on salinity in the western wheatbelt of Western Australia. Part 2: the interaction between revegetation strategies and major fault zones. Australian Journal of Soil Research 36: 131–142

Dawson TE and Pate JS (1996). Seasonal water uptake and movement in root systems of Australian phreatophytic plants of dimorphic root morphology: a stable isotope investigation. Oecologia 107: 13–20

Despessis A (1902) The handbook of viticulture and horticulture of Western Australia. Western Australian Government Printer 620 pp

Dupraz C, Dauzat M, Girardin N and Olivier A (1995) Root extension of young wide-spaced wild cherry trees in an agroforest as deduced from the water budget. In Ehrenreich JH, Ehrenreich DL and Lee HW (eds) Growing a sustainable Future, pp 46–50. Proceedings of the 4th North-American Agroforestry Conference, Boise, Idaho

Dupraz C, Simorte V, Dauzat M, Bertoni G, Bernadac A and Masson P (1998) Growth and nitrogen status of young walnuts as affected by intercrops in a Mediterranean climate. Accepté pour publication dans Agroforestry Systems

Eastham J, Rose CW, Cameron DM, Rance SJ and Talsma T (1990). Tree/pasture interactions at a range of tree densities in an agroforestry experiment. II Water uptake in relation to rooting patterns. Australian Journal of Agricultural Research 41: 697–707

Farrington P and Salama RB (1996) Controlling dryland salinity by planting trees in the best hydrogeological setting. Land Degradation and Development 7: 183–204

Ferdowsian R, George R, Lewis R, McFarlane D, Short R and Speed R (1996) The extent of dryland salinity in Western Australia. In Proceedings, 4th National Workshop on the Productive Use of Saline Lands pp 89–98, Promaco Conventions, Perth

Frota PCE, Ramos AD and Carrari E (1987) Wind reduction and soil water availability near a caatinga shelterbelt. In Reifsnyder WE and Darnhofer TO (eds) Meteorology and Agroforestry, pp 531–536. ICRAF, Nairobi

Gardner C (1944) The vegetation of Western Australia with special reference to the climate and soils. Journal Royal Society of Western Australia 28: 11–27

George RJ (1992) Hydraulic properties of groundwater systems in the saprolite and sediments of the wheatbelt. Journal of Hydrology 130: 250–278

George RJ, McFarlane DJ and Nulsen RA (1997) Salinity threatens the viability of agriculture and ecosystems in Western Australia. Hydrogeology Journal 5: 6–21

George RJ, Nulsen RA, Ferdowsian R and Raper GP (1998) Interactions between trees and groundwaters in recharge and discharge areas – A survey of Western Australian sites. Agricultural Water Management 1399: 1–21

Gladstones JS (1982) Breeding lupins in Western Australia. Journal of Agriculture Western Australia 23: 73–76

Hallam SJ (1979) Fire and Hearth: a study of aboriginal usage and European usurpation in south western Australia. Prehistory and Material Culture Series No. 8, Australian Institute of Aboriginal Studies, Canberra, 158 pp

Hatton TJ, Catchpole EA and Vertessy RA (1990) Integration of sapflow velocity to estimate plant water use. Tree Physiology 6: 201–209

Jackson W (1985) New Roots for Agriculture. University of Nebraska Press, Nebraska

Kang BT, Reynolds L and Atta-Krah AN (1992) Alley farming. Advances in Agronomy 43: 315–359

Kimber PC (1974) The root system of Jarrah (*Eucalyptus marginata*). Research paper No. 10, Forests Department of Western Australia, Perth, 5 pp

Kort J (1988) Benefits of windbreaks to field and forage crops. Agriculture, Ecosystems and Environment 22: 165–190

Lefroy EC, Salerian J and Hobbs RJ (1992) Integrating economic and ecological considerations; a theoretical framework. In Hobbs RJ and Saunders DA (eds) Reintegrating Fragmented Landscapes: towards sustainable agriculture and nature conservation, pp 209–244. Springer Verlag, New York

Lefroy EC and Melvin DPC (1996).The influence of alley farming and intercropping on the yield of lupins. In Proceedings, 8th Australian Agronomy Conference, Toowoomba, pp 377–380, Australian Society of Agronomy

Lefroy EC and Scot PR (1994) Alley farming: new vision for western Australian farmland. Western Australian Journal of Agriculture 35: 119–126

Lines WJ (1991) Taming the Great South Land: a history of the conquest of nature in Australia. Allen and Unwin, Sydney, 337 pp

Mann (1907) Does clearing trees increase salt in ground? Journal of the Department of Agriculture 15(3): 193

Marsh SP (1995) The adoption of lupins in Western Australia: did extension make a difference? Paper presented at the 39th Annual Conference of the Australian Agricultural Economics Society, University of Western Australia, February 14–16

Marshall JK, Morgan AL, Akilan K, Farrell RCC and Bell DT (1997) Water uptake by two river red gum (*Eucalyptus camaldulensis*) clones in a discharge site plantation in the Western Australian wheatbelt. Journal of hydrology 200: 136–148

McFarlane DJ and Cox JW (1990) Seepage interceptor drains for reducing waterlogging and salinity. Journal of Agriculture Western Australia 31: 66–69

McFarlane DJ, George RJ and Farrington P (1993) Changes in the hydrologic cycle. In Hobbs RJ and Saunders DA (eds) Reintegrating Fragmented Landscapes: Towards sustainable production and nature conservation. Springer Verlag, New York

Mensforth LJ and Walker GR (1996) Root dynamics of *Melaleuca halmaturorum* in response to fluctuating saline groundwater. Plant & Soil 184(1): 75–84

Moore RM (1970) Australian Grasslands. Australian National University Press, Canberra

Myers BJ, Theiveyanathan S, O'Brien ND and Bond WJ (1996). Growth and water use of *Eucalyptus grandis* and *Pinus radiata* plantations irrigated with effluent. Tree Physiology. 16: 211–219

Ong CK and Black CR (1994) Complementarity of resource use in intercropping and agroforestry systems. In Monteith JL and Scott RK (eds) Resource Capture by Crops, pp 255–2. Nottingham University Press, Loughborough

Ong CK and Black CR (1996). Tree crop interactions. In Ong CK and Huxley P (eds) Tree Crop Interactions; A Physiological Approach, pp 1–24. CAB International, Oxford, United Kingdom

Pate JS, Dixon KW and Orshan G (1984) Growth and life form characteristics of Kwongan species. In Pate JS and Beard JS (eds) Kwongan; Plant Life of the Sandplain, pp 84–100. University of Western Australia Press, Nedlands

Peck AJ and Hurle DH (1973) Chloride balance of some famed and forested catchments in Southwestern Australia. Water Resources Research 9: 648–657

Raper GP (1998) Agroforestry water use in Mediterranean regions of Australia. Water and salinity issues in hydrology, No. 2. Rural Industries Research and Development Corporation, Canberra

Ridley AM, Haines PJ, Wilson KF and Ellington A (1998) Phase farming using lucerne in crop rotations, pp 803–806. Proceedings, 9th Australian Agronomy Conference, Wagga Wagga

Salama RB, Bartle GA and Farrington P (1994) Water use of a plantation of eucalyptus camal-

dulensis by groundwater hydrograph separation and heat pulse method. Journal of Hydrology 156: 163–180
Sanchez PA (1995) Science in agroforestry. Agroforestry Systems 30: 5–55
Schofield NJ, Loh IC, Scott PR, Bartle JR, Ritson P, Bell RW, Borg H, Anson B and Moore R (1989) Vegetation strategies to reduce stream salinity of water resource catchments in south Western Australia. Report No. WS 33, Water Authority of Western Australia, Perth
Smith CJ, Dunin FX, Zegelin SJ and Poss R (1998) Nitrate leaching from a riverine clay soil under cereal rotation. Australian Journal of Agricultural Research 49 (In press)
Smith TA, George RJ, Scott PR, Bennett DL, Rippon R and Orr GJ (1998) Results of investigations into groundwater response and productivity for high water use agricultural systems 1990–1997, 6 Summary of all sites. Agriculture Western Australia Natural Resource Management Technical Report 179, 54 pp
Speed RJ, Wiley TL and Simmons JA (1993) Tagasaste (Chamaecytisus palmensis) is profitable in recharge control. In Proceedings, National Conference on Land Management for Dryland Salinity Control, pp 275–276. Express Printing, Bendigo, Victoria
Stewart JB (1984) Measurement and prediction of evaporation from forested and agricultural catchments. Agricultural Water Management 8: 1–28
Stirzaker RJ, Cook FJ and Knight JH (1999) Approximate solutions for the location of trees on cropping land for control of dryland salinity. Agriculture Water Management (in press)
Thorburn PJ, Walker GR and Jolly ID (1995) Uptake of saline groundwater by plants; an analytical model for semi-arid areas. Plant and Soil 175: 1–11
van Noordwijk M and Purnomosidhi P (1995) Root architecture in relation to tree-soil-crop interactions and shoot pruning in agroforestry. Agroforestry Systems 30: 161–173
van Noordwijk M (1996) Mulch and shade model for optimum alley cropping design depending on soil fertility. In Ong CK and Huxley P (eds) Tree crop interactions; A physiological approach pp 51–72. CAB International, Oxford, United Kingdom
Van Noordwijk M, Tomich TP, de Foresta H, and Michon G (1997) To segregate – or to integrate: the question of balance between production and biodiversity conservation in complex agroforestry systems. Agroforestry Today 9: 6–9
Western Australian Government (1917) Report of The Royal Commission on The Mallee Belt and Esperance Lands Together with Appendices and Reports of Evidence, Dempster C (chairman), Government Printer, Perth, 192 pp
Wood WE (1924) Increase in salt and streams following destruction of the native vegetation. Journal of the Royal Society of Western Australia 10: 35–47
Zohar Y (1985) Root distribution of a eucalypt shelterbelt. Forest Ecology and Management, 12: 305–307

Agroforestry Systems **45**: 303–341, 1999.

Application of the ecosystem mimic concept to the species-rich *Banksia* woodlands of Western Australia

J. S. PATE* and T. L. BELL
*Department of Botany, The University of Western Australia, Nedlands 6907, Australia (*Author for correspondence: darth@cyllene.uwa.edu.au)*

Key words: life form, net primary productivity, phenology, rooting patterns, trophic specialisations, water relations

Abstract. This article describes the structure and functioning of a natural *Banksia* woodland at Moora, Western Australia. Species are first grouped in terms of growth form, root morphology, phenology and nutrient acquisition strategy. Above- and belowground standing biomass of a woodland is measured and its net annual primary production per unit rainfall compared with that of adjacent crops and plantings of the tree tagasaste. Information on seasonal water use and nutrient cycling in the dominant tree species *Banksia prionotes* is used to highlight the pivotal role of deep rooted summer growing trees in the maintenance of sustainability of the system. The article then addresses how one might select species mixtures as functionally effective analogues of the woodland. Assuming the mimic system replaces cleared virgin woodland not previously subject to runoff of water and nutrients from agriculture, a selection procedure would incorporate native flora representing (a) summer-growing deep-rooted and winter-growing shallow-rooted trees and shrubs, (b) herbaceous ground cover species, (c) fire resistant and fire sensitive species, and (d) a range of complementary nutrient uptake strategies. Assuming the mimic is designed to rehabilitate agricultural land experiencing rising water tables and nitrate pollution of ground water, incorporation of fast growing deep rooted exotic trees or herbaceous perennials is recommended alongside conventional annual crops or pastures, with appropriate nutrient stripping through removal of biomass. Difficulties in this context are scale of planting required and current lack of profitable incentives for planting and maintenance of perennials.

1. Introduction

This article sets out to describe, and then suggest how one might proceed to mimic, a complex functionally sustainable natural ecosystem. The system chosen is south western Australian *Banksia* woodland, embodying some of the most oligotrophic soils and exacting climatic conditions experienced anywhere on the continent. *Banksia* woodlands, defined as containing at least one dominant species of *Banksia*, other trees at lesser density and an understorey of predominantly woody shrubs, comprise one of the major vegetational types of the deep sands of the Swan Coastal Plain and Dandaragan Plateau. They are more or less confined to a narrow strip of land 15 to 90 km wide extending for 400 km between Jurien Bay in the north to Busselton in the south (Figure 1a). Much of the biome (approximately 61%) has been cleared for agriculture and urban and semi-rural development or exploited for natural resources such as sand and limestone (Gozzard and Mouritz, 1989) and timber (Hopkins, 1989). Nevertheless, large tracts of *Banksia* woodlands

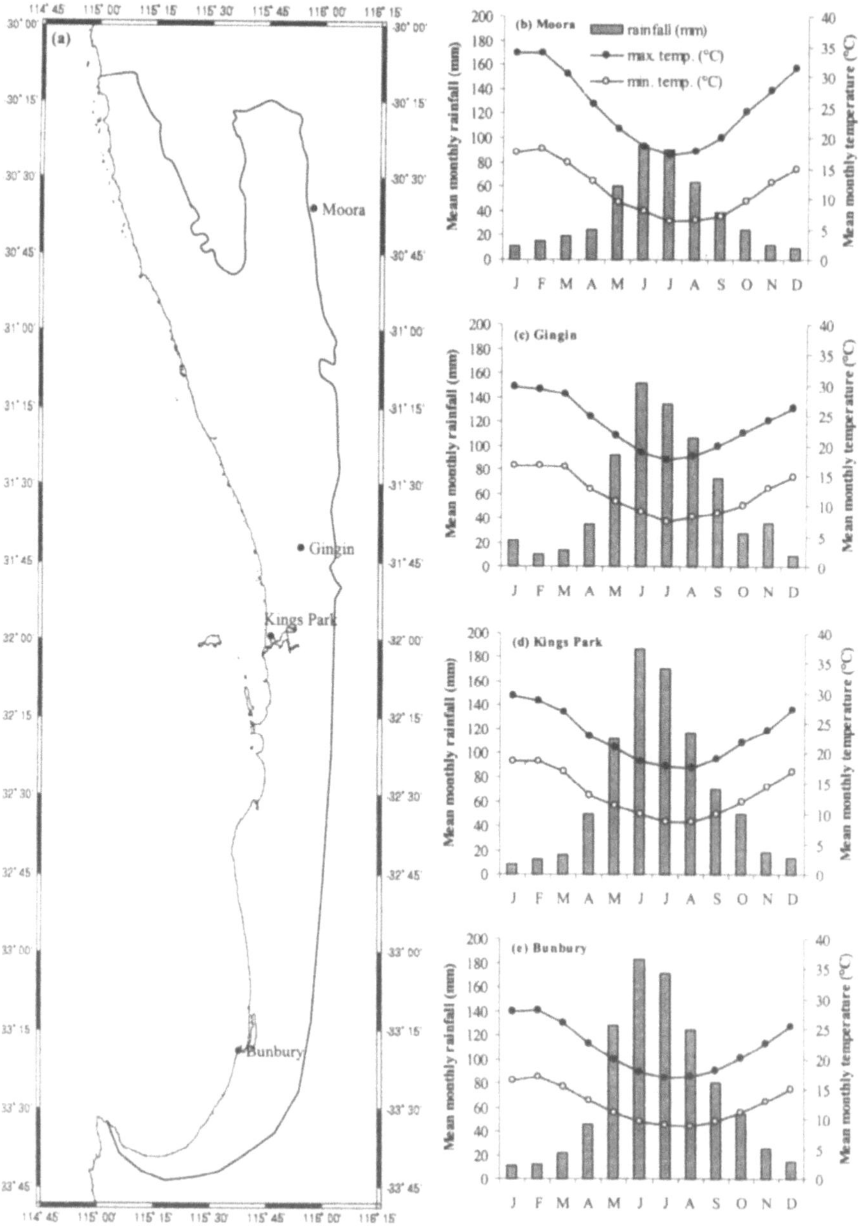

Figure 1. (a) Distribution of *Banksia* woodlands in southwest Australia with details of mean annual rainfall, maximum and minimum temperatures for four sites; (b) Moora, (c) Gingin, (d) Kings Park, Perth and (e) Bunbury. *Source:* Map (a) redrawn from Beard (1989), weather information in (b) to (e) from the Bureau of Meteorology, Perth.

and associated open heaths remain in relatively pristine state on Crown Lands or conservation reserves (Beard, 1989).

Several species of *Banksia* are characteristic dominants of the woodlands, the commonest of these being *B. attenuata*, *B. menziesii* and *B. ilicifolia* and, in low-lying wetter areas, *B. littoralis*. Accompanying tree species include *Actinostrobus arenarius*, *Allocasuarina fraseriana*, *Banksia burdettii*, *B. prionotes*, *Corymbia calophylla*, *Eucalyptus marginata*, *E. todtiana*, *Nuytsia floribunda* and *Xylomelum angustifolium* (Beard, 1989; Dodd and Griffin, 1989). The species-rich understorey is dominated by sclerophyllous woody species from the families Proteaceae, Myrtaceae, Fabaceae and Epacridaceae. The equally rich herbaceous ground flora includes members of the Restionaceae, Cyperaceae, Orchidaceae, Haemodoraceae and Anthericaceae (Dodd and Griffin, 1989). Life forms typically encompass relatively few species of tall trees (4–8 m in height), a taxonomically diverse mid-storey of woody shrubs (0.2–2.5 m) and a herbaceous ground cover of hemicryptophytes, geophytes, helophytes, and epiphytes (Pate et al., 1984). For example, in a broad survey conducted by Speck (1952) in *Banksia* woodlands of the Swan Coastal Plain near Perth, 28 species of trees and tall shrubs, 130 species of low shrubs and 182 species of herbaceous ground layer plants were listed as common components.

Soils of *Banksia* woodlands variously comprise well-drained and highly leached, yellow or white quartz sands of the inland Bassendean Dune Systems, the shallow to deep yellow, iron-coated quartz sands of the Spearwood Dune Systems or, in lower lying areas, poorly-drained organic matter-rich soils bordering interdunal swamps and lakes (Bettenay, 1984; Semeniuk and Glassford, 1989). All of these sands are exceptionally poor in nutrients and base-exchange capacity, typically in the pH range of 4.5–6.0, and highly deficient in clay components. For example, in Bassendean sands of *Banksia* woodlands near Gingin, 75 km north of Perth, Bell et al. (1994) recorded greatest concentrations of available nutrients within the top 20 cm of soil. Below this, concentrations quickly declined until, at about 50 cm depth, mineral contents were only marginally greater than those of the parent sands of the ecosystem. Soil organic matter content in the top 20 cm was only 0.7% and maximum levels of critically important macronutrients in this zone were 1 μg NO_3^-–N g dry weight $soil^{-1}$, 12 μg NH_4^+–N, 2–5 μg total P and 40 μg K.

It should be noted that similar soils, loosely termed '*Banksia* sands', occur in discrete pockets from the southwest corner of the State, eastwards along the coast to Albany and thereafter surrounding Esperance. Vegetation in these localities is also dominated by Proteaceae, but mostly in shrubland rather than woodland format. We will restrict our attention here to *Banksia* woodlands as detailed in Figure 1a.

The climate experienced by most of the Botanical Province of the southwest corner of Australia is classic 'mediterranean-type', with hot dry summers alternating with cool wet winters. The area occupied by *Banksia* woodlands is classified as moderate to severe 'dry mediterranean' (Beard, 1984, 1989), with a dry summer season spanning five to six months of the year (November to April) when no useful rain is likely to fall. Winter rainfall commences in May and peaks in June, July and early August, with rain-bearing cyclonic systems penetrating first from the northwest and then more widely from a series of wet depressions from the south west (Gardner, 1944). Effective rainfall decreases sharply during late August and September but attenuates later in the south than north of the biome. Figure 1(b) to (d) illustrates rainfall data and maximum and minimum temperatures for the last 30–100 years at four sites (i.e. Moora, Gingin, Kings Park and Bunbury) spanning the latitudinal extent of the biome.

Based on the hypothesis that soil texture and moisture relations play a more important role than nutrient availability in shaping the physiognomy, life form, and taxonomic spread within SW Australian ecosystems (Beard, 1984), *Banksia* woodlands on relatively shallow sands overlying ironstone gravel are almost invariably restricted to areas of annual rainfall of 600 to 750 mm, whereas those on deeper leached sands extend more broadly across areas of annual rainfall ranging from 500 to 900 mm (Beard, 1984). In deep sand ecosystems further south in regions of even greater rainfall (900 to 1400 mm), eucalypt-dominated forests typically replace *Banksia* woodlands, while at the other extreme, in areas of '*Banksia* sands' north and south of the State receiving less than 400 mm rainfall, low density *Banksia* woodlands intergrade with, and are finally replaced by scrub heath, albeit still dominated by proteaceous species (see Hobbs and O'Connor, this volume).

Seasonal changes in soil moisture under *Banksia* woodlands (e.g. see Dodd et al., 1984; Bell et al., 1994) show protracted minima during summer months, January to April, increasing gradually with the onset of winter rains in May and June, and remaining transiently close to or at field capacity (5% by weight) during winter months of July and August. After this, the soil dries out progressively from September to November and remains very dry until the following autumn. Large variations can occur between seasons depending on when and in what quantity rainfall events occur. For example, at our principal study site in northern *Banksia* woodlands at Moora, a whole year passed in the exceptionally dry winter of 1997 without the soil profile becoming fully wetted from the surface down to the groundwater at 5 m depth. Not surprisingly, seasonal shoot extension of trees during the following summer proved to be minimal (A. M. Grigg, M. J. Unkovich and J. S. Pate, unpubl. data).

2. Morphological and functional characteristics of *Banksia* woodland

In the following sections we assemble information on the phenologies and morphological and physiological characteristics exhibited by floristic elements of typical *Banksia* woodlands, using as examples data from Melaleuca Park and Gnangara, 40–50 km north of Perth (Dodd et al., 1984; Farrington et al.,

1989; Farrington and Bartle, 1991), Kings Park, Perth (Stewart et al., 1993; Pate et al., 1993) and Moora, 250 km north of Perth (T. E. Dawson, T. J. Hatton, D. J. Arthur, M. J. Unkovich, A. M. Grigg, T. L. Bell and J. S. Pate, unpubl. data). *Banksia* spp. and other proteaceous species collectively make up to 70% or more of the ground cover at these sites, so special attention will be directed at water and nutrient relationships of members of this family. A later section of the article presents information relating to one particularly well researched species, *Banksia prionotes* (see review of Pate et al., 1998).

2.1. *Growth and life forms, phenologies of growth and flowering*

Table 1 lists the common species encountered at the Moora and Kings Park sites, according to growth and life form and the times of the year when shoot extension and flowering would normally occur. The majority of understorey species at Moora are woody shrub species (57% of species) alongside fewer species of herbaceous perennials (24% of species). At Kings Park, the proportion of herbaceous perennial species (43%) is slightly greater than woody shrubs (37%). Both sites contain relatively small percentages of their floras as ephemerals (8% of species at Moora and 9% of species at Kings Park) and trees (6% of species at Moora and 7% of species at Kings Park). Dominance of the families Proteaceae (24% of species at Moora and 13% of species at Kings Park), Myrtaceae (19% of species at Moora and 4% of species at Kings Park) and Fabaceae (5% of species at Moora and 9% of species at Kings Park) is clearly evident amongst shrub and tree components. A phenology involving shoot extension in winter accompanied by flowering in winter or early spring (hereafter referred to as winter-active species) is exhibited by a preponderance of species (71% at Moora and 81% at Kings Park) compared with those showing summer shoot extension and summer or early autumn flowering (hereafter termed summer-active species), for which the relevant proportions are 29% at Moora and 19% at Kings Park. It must be remembered that the term 'summer-active' refers solely to timing of shoot extension and flowering and that species of this ilk are likely to be 'winter-active' in terms of nutrient uptake in having their feeding roots located mostly in top layers of soil.

2.2. *Responses to fire*

Banksia woodlands are frequently subjected to fire and, in common with other south western Australian ecosystems, contain widely ranging mixtures of 'resprouter' species surviving fire by sprouting from root stocks or trunks, and 'obligate seeders' or 'non-sprouter' species succumbing to fire and recruiting thereafter exclusively from seed (Bell et al., 1984). As shown by Pate et al. (1990) and Bell and Pate (1996), resprouters are typically long-lived, poorly active in reproduction, of slow growth during seedling stages, late in achievement of adulthood and exhibit shoot:root dry weight ratios in the range of 0.6 to 2.5. Seeders by contrast are short-lived, fast growing, highly

Table 1. Life and growth forms, shoot and flowering phenology, fire response and rooting type of common species encountered in two *Banksia* woodland sites at (a) Moora and (b) Kings Park.

Life and growth form	Family	Species	Shoot extension	Flowering phenology	Fire response	Rooting type
(a) MOORA						
Ephemerals	Asteraceae	**Hypochaeris glabra* L.	W	W/Sp	S	1
		Podotheca gnaphalioides Grah.	W	Sp	S	1
		**Ursinia anthemoides* (L.) Gaertn.	W	Sp	S	1
		Waitzia acuminata Steetz in Lehm.	W	Sp	S	1
	Geraniaceae	*Erodium cygnorum* Nees	W	W	S	1
	Gyrostemonaceae	*Gyrostemon ramulosus* Desf.	W	W/Sp	S	2
Herbaceous perennials	Chloanthaceae	*Pityrodia bartlingii* (Lehm.) Benth.	W	Sp/S	R	3
		Pityrodia verbascina (F. Muell.) Benth.	W	W	R	3
	Cyperaceae	*Caustis dioica* R. Br.	W	Sp	R	1
		Mesomelaena stygia (R. Br.) Nees	W	W	R	1
	Dasypogonaceae	*Lomandra* sp.	W	Sp	R	1
	Droseraceae	*Drosera erythrorhiza* Lindl.	W	W/Sp	R	1
		Drosera sp. (pygmy)	W	W	S	1
	Goodeniaceae	*Verrauxia reinwardtii* (Vriese) Benth.	Sp	Sp/S	R	3
	Haemodoraceae	*Anigozanthos humilis* Lindl.	W	W/Sp	R	1
		Conostylis setigera R. Br.	W	Sp	R	1
	Orchidaceae	*Caladenia flava* R. Br.	W	Sp	R	1
		Lyperanthus nigricans R. Br.	W	Sp	R	1
	Phormiaceae	*Dianella revoluta* R. Br.	S	Sp	R	1
	Poaceae	*Neurachne alopecuroidea* R. Br.	W	W/Sp	R	1
	Restionaceae	*Lepidobolus preissianus* Nees	W	W/Sp	R	1
	Rubiaceae	*Opercularia vaginata* Labill.	S	Sp	S	2
	Rutaceae	*Boronia* sp.	W	W	S	2
	Stylidiaceae	*Stylidium aff. repens* R. Br.	W	W/Sp	S	1
		Stylidium brunonianum ssp. *brunonianum* Benth.	W/Sp	Sp	S	1

Parasites	Lauraceae	*Cassytha* sp.				–
	Loranthaceae	*Nuytsia floribunda* (Labill.) R. Br. Ex Fenzl	S	S	R	5
	Orobanchaceae	**Orobanche minor* Sm.	W	Sp	S	–
Woody shrubs	Casuarinaceae	*Allocasuarina humilis* (Otto & Dietr.) L. Johnson	S	Sp/S	R	5
	Dilleniaceae	*Hibbertia acerosa* (R. Br. Ex DC.) Benth.	W	Sp	R	2
		Hibbertia pachyrrhiza Steudel	Sp	Sp/S	S	2
		Hibbertia sp.	W		S	2
	Epacridaceae	*Astroloma xerophyllum* (DC.) Sonder	W	W	S	2
		Leucopogon polymorphus Sonder	W	W	S	2
		Lysinema ciliatum R. Br.	W	W	S	2
	Mimosaceae	*Acacia pulchella* ssp. *glaberrima* Meissner	S	W	S	4
	Myrtaceae	*Baeckea camphorosmae* Endl.	S	Sp/S	S	4
		Beaufortia elegans Schauer	S	S	R	4
		Calothamnus sanguineus Labill.	W	Sp	R	4
		Calytrix fraseri Cunn.	Sp	W/Sp	R	2
		Calytrix sp.	Sp	Sp/S	S	2
		Eremaea beaufortioides Benth.	S	Sp	R	5
		Leptospermum erubescens Schauer	S	W/Sp	R	5
		Leptospermum spinescens Endl.	W	Sp	R	4
		Melaleuca scabra R. Br.	W	Sp/S	R	5
		Melaleuca sp. (round fruit)	W	Sp/S	R	4
		Pileanthus filifolius Meisn.	W	Sp/S	S	1
		Verticordia acerosa Lindl.	W	Sp	S	2
		Verticordia densiflora Lindl.	W	Sp	S	2
		Verticordia sp. (large pink)	W	Sp	S	2
	Fabaceae	*Gompholobium* sp. (yellow flower)	S	Sp	S	2
		Jacksonia aff. *stricta*	S	Sp	R	5
		Jacksonia ulicifolia	S	Sp	R	5
		Mirbelia spinosa Benth.	W/Sp	W	S	2
	Proteaceae	*Adenanthos cygnorum* Diels	W	Sp/S	S	5
		Banksia leptophylla A. S. George	S	S	S	5
		Conospermum incurvum Lindl.	W	W/Sp	S	3

Table 1. (*Continued*)

Life and growth form	Family	Species	Shoot extension	Flowering phenology	Fire response	Rooting type
		Conospermum stoechadis Endl.	W	W/Sp	R	5
		Dryandra nivea (Labill.) R. Br.	W	W/Sp	R	5
		Dryandra sessilis (Knight) Domin	W	W/Sp	S	5
		Grevillea eriostachya ssp. *eriostachya* McGillivray	W	W/Sp	S	5
		Hakea obliqua R. Br.	S	W/Sp	S	5
		Hakea platysperma Hook.	S	W/Sp	S	5
		Hakea trifurcata (Smith) R. Br.	W/Sp	Sp	R	5
		Isopogon dubius (R. Br.) Druce	S	Sp	R	5
		Isopogon scabriusculus Meissn.	W/Sp	Sp	R	5
		Petrophile brevifolia Lindl.	S	Sp	R	5
		Stirlingia latifolia (R. Br.) Steudel	Sp/S	Fire	R	5
		Synaphea spinulosa ssp. *spinulosa* (Burm. F.) Merr.	W	W/Sp	R	5
	Rhamnaceae	*Cryptandra pungens* Steudel	W	W	S	2
	Thymelaeaceae	*Pimelea leucantha* Diels	W/Sp	W/Sp	S	2
		Pimelea spectabilis Lindl.	W	Sp	S	2
Trees	Cupressaceae	*Actinostrobus arenarius*	W		S	5
	Myrtaceae	*Eucalyptus todtiana* F. Muell.	S	S	R	5
	Proteaceae	*Banksia attenuata* R. Br.	S	Sp/S	R	5
		Banksia menziesii R. Br.	S	S	R	5
		Banksia prionotes Lindl.	S	S	S	5
		Xylomelum angustifolium Kipp. et Meissn.	S	S	R	5
(b) KINGS PARK						
Ephemerals	Apiaceae	*Eryngium rostratum* Cav.	W/Sp	W/Sp	S	2
	Asteraceae	*Podolepis gracilis* (Lehm.) Grah.	W	W/Sp	S	1
		Podotheca angustifolia (Labill.) Less.	W	Sp	S	1
		Podotheca chrysantha (Steetz) Benth.	W	Sp/S	S	1

		**Ursinia anthemoides* (L.) Gaertn.	W	Sp	S	1
		Waitzia suaveolens (Benth.) Druce	W	Sp/S	S	1
	Cyperaceae	*Isolepis marginata* (Thunb.) A. Dietr.	W/Sp	Sp	S	1
	Euphorbiaceae	*Poranthera microphylla* Brongn.	W	Sp	S	1
	Lobeliceae	*Lobelia gibbosa* Labill.	W	Sp/S	S	1
		Lobelia tenuior R. Br.	W	Sp/S	S	1
Herbaceous perennials	Amaranthaceae	*Ptilotus polystachyus* (Gaudich.) F. Muell.	Sp	W/Sp	R	3
	Anthericaceae	*Arthropodium preissii* Endl.	W/Sp	Sp/S	R	2
		Corynotheca micrantha (Lindl.) J. F. Mcbride	W/Sp	S	R	
		Sowerbaea laxiflora Lindl.	W/Sp	W/Sp	R	1
		Thysanotus manglesianus Kunth	W/Sp	Sp	R	1
	Asteraceae	*Helichrysum cordatum* DC.	Sp/S	Sp/S	R	1
	Colchicaceae	*Burchardia umbellata* R. Br.	W	Sp	R	1
	Cyperaceae	*Lepidosperma angustatum* R. Br.	W/Sp	W/Sp	R	1
		Lepidosperma leptostachyum Benth.	W/Sp	W/Sp	R	1
		Lepidosperma scabrum Nees	W/Sp	W	R	1
		Mesomelaena stygia (R. Br.) Nees	W/Sp	S	R	1
		Schoenus curvifolius (R. Br.) Benth.	W/Sp	W/Sp	S	1
		Schoenus grandiflorus (Nees) F. Muell.	W/Sp	W	R	1
		Tetraria octandra (Nees) Kuek.	W/Sp	W/Sp		1
	Dasypogonaceae	*Calectasia cyanea* R. Br.	W/Sp	W/Sp	R	2
		Dasypogon bromeliifolius R. B.	Sp/S	Sp/S	R	1
		Lomandra caespitosa (Benth.) Ewart	W/Sp	W/Sp		1
		Lomandra nigricans T. D. Macfarlane	W/Sp	W		1
		Lomandra preissii (Endl.) Ewart	W/Sp	W		1
	Droseraceae	*Drosera pallida* Lindl.	W	W/Sp		1
	Fabaceae	*Hardenbergia comptoniana* (Andrews) Benth.	Sp	W/Sp	R	5
		Kennedia prostrata R. Br.	Sp	Sp	S	5
	Haemodoraceae	*Anigozanthos humilis* Lindl.	W	W/Sp	S	1
		Anigozanthos manglesii D. Don	W/Sp	Sp	S	1
		Conostylis aculeata R. Br.	W/Sp	Sp		1
		Haemodorum laxum R. Br.	W/Sp	Sp	R	1

Table 1. (Continued)

Life and growth form	Family	Species	Shoot extension	Flowering phenology	Fire response	Rooting type
		Haemodorum paniculatum Lindl.	W/Sp	Sp/S	R	1
		Haemodorum spicatum R. Br.	W/Sp	Sp/S	R	1
	Iridaceae	*Patersonia occidentalis* R. Br.	Sp	Sp	R	1
	Orchidaceae	*Caladenia deformis* R. Br.	W	W	R	1
		Caladenia flava R. Br.	W	Sp	R	1
		Caladenia latifolia R. Br.	W	W/Sp	R	1
		Caladenia longicauda Lindl.	W	W/Sp	R	1
		Diuris longifolia R. Br.	W	Sp	R	1
		Elythranthera brunonis (Endl.) A. S. George	W	Sp	R	1
		Lyperanthus nigricans R. Br.	W	Sp	R	1
		Pterostylis scabra var. *robusta* (Rogers) A. S. George	W	W	R	1
	Phormiaceae	*Dianella divaricata* R. Br.	W/Sp	Sp	R	1
	Restionaceae	*Alexgeorgea nitens*	Sp	W	R	1
		Desmocladus fasciculatus	Sp	W/Sp	R	1
		Desmocladus flexuosus	Sp	Sp	R	1
		Hypolaena exsulca R. Br.	Sp	Sp/S	R	1
		Lyginia barbata R. Br.	Sp	W/Sp	R	1
	Rubiaceae	*Opercularia vaginata* Labill.	S	Sp	S	2
	Stackhousiaceae	*Tripterococcus brunonis* Endl.	W/Sp	W/Sp	S	1
	Stylidiaceae	*Stylidium brunonianum* Benth.	W/Sp	Sp	S	1
		Stylidium calcaratum R. Br.	W	Sp	S	1
	Violaceae	*Hybanthus calycinus* (DC. ex Ging.) F. Muell.	W/Sp	Sp	S	2
Parasites	Lauraceae	*Cassytha glabella* R. Br.			S	–
	Loranthaceae	*Amyema miquelii* (Lehm. ex Miq.) Tieghem		W/Sp	S	–
		Nuytsia floribunda (Labill.) R. Br. Ex Fenzl	S	S	R	5
	Orobanchaceae	**Orobanche minor* Sm.	W	Sp	S	1

Woody shrubs	Casuarinaceae	*Allocasuarina humilis* (Otto & Dietr.) L. Johnson	S	Sp/S	R	5
	Dilleniaceae	*Hibbertia huegelii* (Endl.) F. Muell.		Sp	S	2
		Hibbertia hypericoides (DC.) Benth		Sp	R	2
	Epacridaceae	*Astroloma macrocalyx* Sond.	W	W	R	2
		Astroloma pallidum R. Br.	W	W	R	2
		Conostephium pendulum Benth.	W	W	R	2
		Conostephuim preissii Sond.	W	W	R	2
		Leucopogon propinquus R. Br.	W	W	R	2
	Euphorbiaceae	*Phyllanthus calycinus* Labill.	W/Sp	Sp	R	2
	Fabaceae	*Bossiaea eriocarpa* Benth.		W/Sp	R	3
		Daviesia divaricata Benth.		W/Sp	R	3
		Daviesia nudiflora Meiss.		W	R	3
		Daviesia triflora M. D. Crisp		W/Sp	R	3
		Gompholobium tomentosum Labill.	Sp/S	Sp/S	S	2
		Hovea trisperma Benth.	W/Sp	Sp	S	2
		Isotropis cuneifolia (Smith) Benth. ex B. D. Jackson	W/Sp	Sp	R	2
	Fabaceae	*Jacksonia furcellata* (Bonpl.) DC.	S	Sp/S	R	5
		Jacksonia sericea Benth.	S	S	R	5
		Oxylobium capitatum Benth.	W/Sp	W/Sp	R	2
	Goodeniaceae	*Dampiera linearis* R. Br.		W/Sp	R	2
		Scaevola canescens Benth.		W/Sp	R	2
		Scaevola holosericea De Vriese		Sp	R	2
		Scaevola paludosa R. Br.		Sp	R	2
	Mimosaceae	*Acacia cochlearis* (Labill.) H. L. Wendl.		Sp		5
		Acacia pulchella R. Br.	S	W	S	4
		Acacia stenoptera Benth.		W/Sp	R	
	Myrtaceae	*Calytrix fraseri* A. Cunn.	W	S	R	2
		Hypocalymma robustum Endl.	W	Sp	R	2
		Melaleuca acerosa Schau.	W	Sp	R	2
	Proteaceae	*Adenanthos cygnorum* Diels	W	Sp/S	S	4
		Dryandra nivea (Labill.) R. Br.		W/Sp	R	5
		Dryandra sessilis (Knight) Domin		W/Sp	S	5

Table 1. (Continued)

Life and growth form	Family	Species	Shoot extension	Flowering phenology	Fire response	Rooting type
		Grevillea vestita Endl.	Sp/S		W	5
		Hakea prostrata R. Br.	S	Sp	R	5
		Persoonia saccata R. Br.		W/Sp	S	
		Petrophile linearis R. Br.	Sp/S	Sp	R	4
		Petrophile macrostachya R. Br.	Sp/S	Sp		4
		Stirlingia latifolia Endl.	Sp/S	Fire	R	4
	Rhamnaceae	*Spyridium tridentatum* (Steudel) Benth.		S	S	2
	Rutaceae	*Eriostemon spicatus* A. Rich.	W/Sp	W/Sp		2
	Xanthorrhoeaceae	*Xanthorrhoea preissii* Endl.	Sp/S	Sp	R	1
	Zamiaceae	*Macrozamia riedlei* (Fischer ex Gaudich.) C. Gardner	Sp/S	W/Sp	R	5
Trees	Casuarinaceae	*Allocasuarina fraseriana* (Miq.) L. Johnson	S	W/Sp	R	5
	Myrtaceae	*Eucalyptus calophylla* Lindl.	S	S	R	5
		Eucalyptus marginata Donn ex Sm.	S	S	R	5
	Proteaceae	*Banksia attenuata* R. Br.	S	S	R	5
		Banksia grandis Willd.	S	S	R	5
		Banksia ilicifolia R. Br.	S	S	R	5
		Banksia menziesii R. Br.	S	S	R	5
		Banksia prionotes Lindl.	S	S	S	5

Key and summary		Number of species	
		Moora	Kings Park
* – Introduced species		3	2
Shoot extension	W– winter (June to August)	45	30
	W/Sp – winter/spring (July to October)	5	31
	Sp – spring (September to November)	4	9
	Sp/S – spring/summer (October to December)	1	9
	S – summer (December to February)	22	15
Flowering phenology	W – winter (June to August)	11	15
	W/Sp – winter/spring (July to October)	18	32
	Sp – spring (September to November)	27	36
	Sp/S – spring/summer (October to December)	11	14
	S – summer (December to February)	7	13
	Fire – fire stimulated flowering	1	1
Fire response	R – resprouter	39	31
	S – obligate seeder	38	71
Rooting types (see Figure 2)	1 – no obvious main root, shallow laterals	20	51
	2 – shallow single main root, shallow laterals	19	26
	3 – deep tap root, few laterals	4	5
	4 – deep tap root, laterals throughout profile	6	5
	5 – one or more deep tap roots, shallow laterals	27	20

reproductive and exhibit high shoot:root ratios in the range of four to 10. Table 1 provides information on the respective fire response status of the common species listed for the two *Banksia* woodland sites. Note the predominance of resprouters (70% of species) at the Kings Park site and better representation of seeders (50% of species) at the Moora site.

Hot fires, which periodically consume all flammable biomass of *Banksia* woodlands, have very dramatic effects on vegetation. Seeds of seeders and some resprouters, and particularly of short-lived fire ephemerals, germinate prolifically and exhibit rapid initial growth rates in post-fire environments (see Pate et al., 1985). Fast growth is possible since water, light and limiting nutrients such as phosphorus are at much greater availability after fire than in fully vegetated long unburnt communities. Aboveground living biomass is restored to pre-fire levels in the order of five to ten years, being at first composed principally of replacement shoots of resprouters and then increasingly of new shoots of recruiting seeders (Bell et al., 1984; Hobbs and Atkins, 1990). Short-lived understorey legumes provide substantial amounts of fixed N at this stage (Pate and Unkovich, in press) and these inputs are highly significant in replacing N losses to the atmosphere during fire. This fixed N is first released from decomposing legume biomass and then cycles progressively through that of other community members. However, with increasing time after fire, the woodland returns to its supposedly basic P-limited condition (see Pate and Dell, 1984), whereupon success of mycorrhizal and other agencies concerned principally with P acquisition becomes paramount to growth and survival of species.

2.3. *Rooting morphologies*

Using the simplified scheme for categorising rooting morphologies shown in Figure 2, the Moora and Kings Park sites provide examples of all categories depicted (see entries of Table 1), but with all of their large tree and shrub species possessing deeply penetrating tap roots (Root types 3–5). Such phreatophytic species access deeply located soil water resources and, as seen in Table 1, tend to be mostly summer-active in their shoot growth and flowering. Conversely, shallow-rooted shrubs, herbaceous perennials and ephemerals, geophytes and hemicryptophytes conform mostly to rooting patterns 1 and 2, thus constraining their growth and reproduction to winter and early spring. They survive the summer in a dormant drought-tolerant state, induced rapidly once the upper soil profile in which their roots are located has dried out to critical levels (see also Dodd et al., 1984; Dodd and Heddle, 1989; Specht et al., 1991).

Woody species of all categories develop crops of fine multi-branched roots on the perennial parts of their root systems each autumn and these engage in intense uptake of nutrients through winter and spring until the upper soil dries in early summer (Pate and Jeschke, 1993). Feeding roots of this nature are confined mostly to the upper 50 cm or so of the soil profile, where the bulk

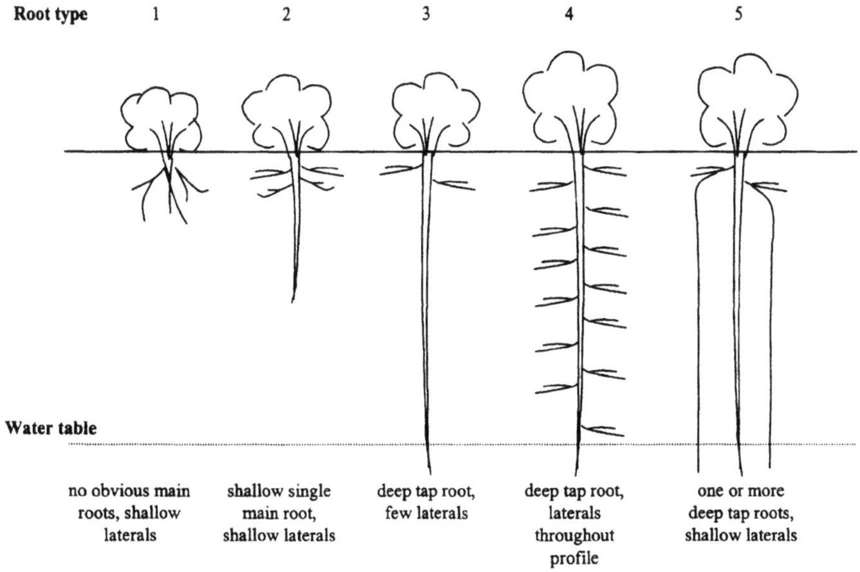

Figure 2. Schematic representation of five basic rooting patterns encountered in deep sands of *Banksia* woodlands in south western Australia. Types 1 and 2 are typically shallow, reaching depths of only 50 to 70 cm, whereas Types 3 to 5 invariably attain greater depths and reach the water table if at accessible depth (dashed line).

of the predominantly organically-bound reserves of nutrients are located. Interspecific competition for nutrients through various mycorrhizal-mediated and non-mycorrhizal strategies is particularly intense at times when these upper layers of the soils are suitably moist (Bell, 1995). According to studies on seedling Proteaceae by Bowen and Pate (1991), seeder species typically make unusually greater proportional investments into biomass of lateral feeding roots, presumably to accommodate the large nutrient demands associated with the fast growth rates and high reproductive outputs typical of seeders. The corresponding situation for resprouters shows accentuated biomass allocation into enlarged tap roots and lignotubers, and deposition within these organs of substantial carbohydrate reserves. The latter are utilised during resprouting after fire (Bowen and Pate, 1993; Bell et al., 1994).

2.4. *Water relations of ecosystem and species*

Studies at Gnangara by Dodd et al. (1984), Farrington et al. (1989) and Farrington and Bartle (1991) provide evidence of annual fluctuations of the order of 500–800 mm in water table depths of unconfined groundwater systems typical of much of the higher rainfall regions of the *Banksia* woodland biome. Minimum depth of water table is normally reached in September or October, following percolation of spring rain through a soil profile already fully recharged by the peak winter rains of June, July and August (see Figure 3). The corresponding maximum then occurs in late autumn, when sand several metres down the profile is still fully dried out and deep-rooted summer-active species are drawing extensively from the capillary fringe of a retreating

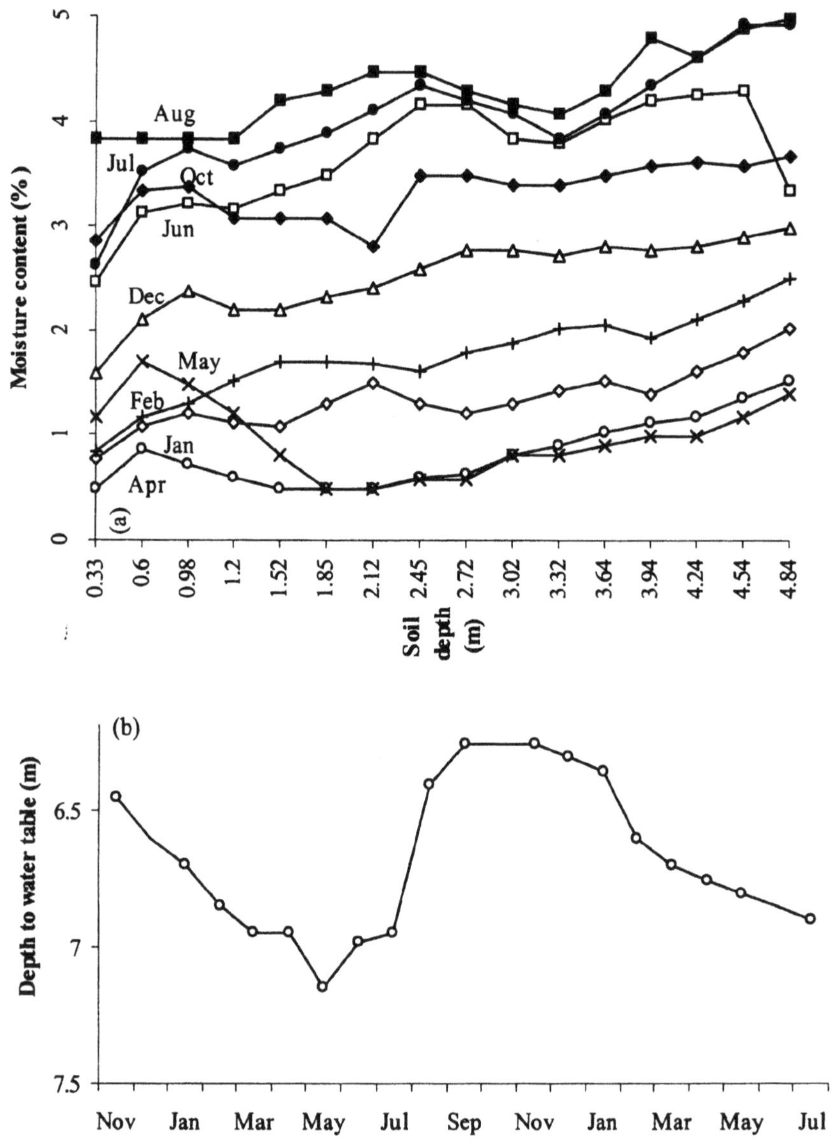

Figure 3. (a) Seasonal variation in soil moisture content at Melaleuca Park, Western Australia during drying (August to April) and recharge by winter rainfall (May to July), and (b) depth to water table during the same period (August 1981 to September 1982). *Source:* Redrawn from Dodd et al. (1984).

water table. Judging from seasonal patterns of evapotranspiration of shrubs such as *Adenanthos cygnorum*, *Eremaea pauciflora* and *Stirlingia latifolia* studied by Dodd et al. (1984), and similar information for tree banksias (*B. attenuata* and *B. menziesii*), *A. cygnorum* and groundflora components of *Banksia* woodlands provided by Farrington et al. (1989), peak water loss to the atmosphere typically occurs in late spring and early summer and corresponding minima in the cooler and wetter months of June and July (Figure 4a). The study of Farrington et al. (1989), comparing yearly losses as total evapotranspiration (transpiration and evaporation from soil and foliage surfaces) for the vegetation (Figure 4b) against rainfall received indicated that 60% of the annual evaporation took place during a six month dry period when virtually no rain was recorded. Since evaporation losses from dry soil and shallow-rooted understorey species were minimal at this time, most of the evaporative loss would have comprised transpiration of ground water by actively growing phreatophytic tree and shrub species. Cumulative evaporation loss for this six-month dry period amounted to approximately 400 mm, or the equivalent of 47% of the total rainfall for the year. Conversely, total evaporation for the other wet and cool six months of the year was only 270 mm and, with 85% of the year's rain falling at this time, substantial recharge of soil and water table would be expected, as was indeed observed during this period.

In a somewhat similar study following mean daily rates of transpiration of *Banksia attenuata* and *B. menziesii* at Melaleuca Park, 35 km north of Perth, Dodd and Bell (1993) showed that tree water use was greatest in late summer (rates up to 1.9 mm day^{-1} for *B. attenuata* and 0.3 mm day^{-1} for *B. menziesii*), compared to only 0.3 mm day^{-1} and 0.05 mm day^{-1} for the same species in winter. The bulk of the yearly transpiration from the tree storey again occurred during months of little or no rain, a finding reinforcing the efficacy of phreatophytic woody species in drying out soil profiles during summer and autumn and in progressively lowering water tables back to pre-rain levels of the autumn of the previous year. Data illustrating these principles (Dodd and Bell, 1993) are detailed for *B. attenuata* in Figure 4(c) and (d), where it can be seen how closely tree water use follows total global radiation over the two years of study (1981 and 1982), and the marked extent to which tree transpiration is out of phase with rainfall. Changes in levels of the water table recorded at their site again reflect deep drainage in late winter and utilisation of this water by phreatophytic trees the following summer.

From the limited data sets detailed above, it seems likely that rain considerably in excess of transpirational demand will be received by *Banksia* woodlands during an average winter season, and that this will promote deep drainage to ground water (see also the article on similar phenomena in other natural and agricultural systems by Hatton and Nulsen, this volume). Deep-rooted components of the woodlands are thus of vital importance in drawing on the winter surplus of water, firstly by drying out the soil profile, and then by increasingly using ground water as the summer progresses. Where such

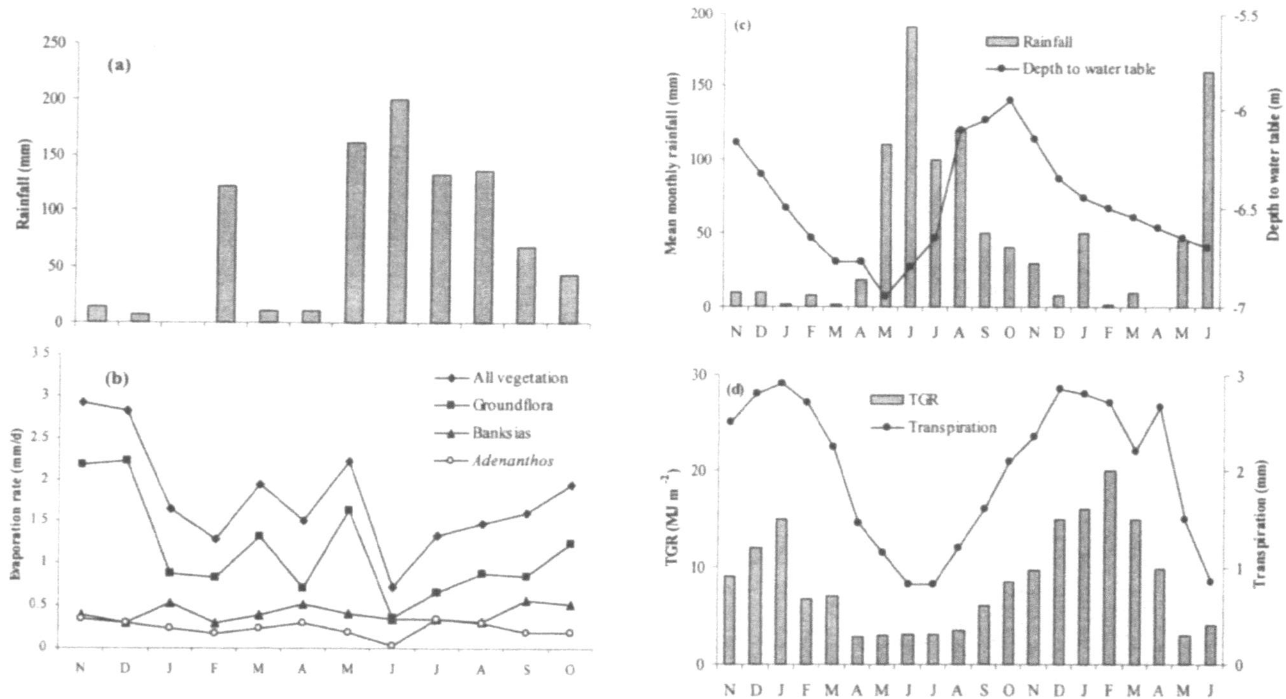

Figure 4. (a) Monthly rainfall totals and (b) rates of evaporation for total vegetation and the three main vegetation components for *Banksia* woodland near Gnangara, Western Australia during 1985 and 1986. (c) Depth to water table (line) and monthly rainfall totals (bars) and (d) mean daily values of transpiration in *Banksia attenuata* (line) and total global radiation (TGR, bars) are also given for each month of the study period for *Banksia* woodland of Melaluca Park, Western Australia during 1980 to 1982. *Source:* (a) and (b) redrawn from Farrington et al. (1989), (c) and (d) redrawn from Dodd and Bell (1993).

species are removed in land clearing and replaced by shallow-rooted annual species, water tables inevitably rise dramatically across a landscape especially in areas of low relief. For example, ground water levels under cleared land at Moora are estimated to be increasing, over long term, at an average extent of 10–20 cm per year (R. Speed, pers. comm.).

2.5. *Trophic specialisations*

Banksia woodlands occupy highly oligotrophic soils limited in readily available key elements, especially P and N. Tight nutrient cycling is evident from the high efficiency of major and minor components of the flora in uptake and usage of potentially limiting nutrients during growth, and in retrieval of the same key elements prior to senescence of foliage (Lamont, 1984; Pate and Dell, 1984; Pate, 1994). Such effectiveness leads to extremely low rates of leaching of essential nutrients to ground water. At the Moora site, for instance, nitrate and phosphate are barely detectable in ground water under pristine woodland and seasonal peaks in mineralised N in the top 10 cm of soil of the woodlands are very low, *viz.* 0.4 mg NO_3^-–N and 0.6 mg NH_4^+–N kg^{-1} of soil (Pate and Unkovich, in press). Not surprisingly, virtually all major components of the flora show nutrient acquisition strategies of one form or another in respect of P and N and probably also for other sparingly available nutrients.

Possible trophic specialisations fostering effective nutrient uptake relating generally to south western Australian heathlands and open woodlands have been listed by Pate (1994). These include a variety of mycorrhizal and non-mycorrhizal associations, symbiotic nitrogen fixation, and the ability to effect heterotrophic inputs of organic and inorganic solutes through development of carnivorous or parasitic habits. Families and species of plants subscribing to these contrasting trophic categories are listed in Table 2 for the woodland at the Moora and Kings Park sites. Virtually all nutrient acquisition strategies are represented at one or both of the sites, but particularly common are vesicular-arbuscular (VA) mycorrhizal woody shrubs (20% at Kings Park and 23% at Moora), woody species with non-mycorrhizal proteoid roots (16% at Kings Park and 33% at Moora), VA- and orchid-type mycorrhizal herbaceous species (16% at Kings Park and 11% at Moora) and various non-mycorrhizal ephemeral and herbaceous taxa (23% at Kings Park and 14% at Moora). Eighteen putative N_2-fixing species were present at Kings Park and six such species at Moora.

It is highly probable that competition for nutrients is intense between species possessing the same acquisition strategies, especially in view of the extremely low nutrient levels in the rooting substrate and the short season during which soils are sufficiently moist for various uptake systems to operate effectively. On the other hand, complementarity between different trophic categories is equally likely, especially where a critically important nutrient is present in several different forms in the soil. Mutual benefits in terms of

Table 2. Classification of trophic types encountered in *Banksia* woodlands examined at Moora and Kings Park amongst herbaceous (A) and woody (B) species.

		Moora	Kings Park
(A) HERBACEOUS HABIT			
Autotrophy	Non-mycorrhizal		
	VA mycorrhizal	e.g. *Conostylis*, *Lomandra*, *Pityrodia* (18 species)	e.g. *Anigozanthos*, *Haemodorum* (24 species)
		Neurachne (1 species)	*Helichrysum*, *Stylidium* (3 species)
	Orchid-type mycorrhizal	*Calaldenia* (1 species)	e.g. *Diuris*, *Lyperanthus* (8 species)
Nitrogen-fixing	VA mycorrhizal	None present	*Hardenbergia*, *Kennedia* (2 species)
Carnivory	Non-mycorrhizal	*Drosera* (2 species)	*Drosera* (1 species)
Holoparasitism	Non-mycorrhizal	*Cassytha*, *Orobanche* (2 species)	*Cassytha*, *Orobanche* (2 species)
Heterotrophic epiparasitism	Orchid-type mycorrhiza	None present	None present
Mycoautotrophy	Ectomycorrhizal	None present	*Lobelia* (2 species)
(B) WOODY HABIT			
Autotrophy	Non-mycorrhizal (proteoid roots)	e.g. *Banksia*, *Hakea*, *Xylomelum* (16 species)	e.g. *Adenanthos*, *Grevillea* (13 species)
	VA mycorrhizal	e.g. *Boronia*, *Hibbertia*, *Pimelea* (10 species)	e.g. *Hibbertia*, *Macrozamia*, *Scaevola* (7 species)
	Ectomycorrhizal	e.g. *Actinostrobus*, *Eucalyptus* (4 species)	e.g. *Eucalyptus*, *Hypocalymma* (7 species)
	Ericoid mycorrhizal	e.g. *Astroloma*, *Lysinema* (3 species)	e.g. *Conostephium*, *Leucopogon* (5 species)
Nitrogen-fixing	VA mycorrhizal	*Acacia*, *Gompholobium* (4 species)	*Acacia* (3 species)
	Ecto + VA mycorrhizal	*Allocasuarina* (1 species)	e.g. *Gompholobium*, *Hovea* (5 species)
	Cluster roots (non-mycorrhizal)	*Jacksonia* (2 species)	*Davesia*, *Jacksonia* (4 species)
Root hemiparasitism	Non-mycorrhizal	*Nuytsia* (1 species)	*Nuytsia* (1 species)
Aerial hemiparasitism	Non-mycorrhizal	None present	*Amyema* (1 species)

efficiencies of nutrient cycling may also apply where root distributions of different species are mostly non-overlapping down a rooting profile (see Bell, 1995 for examples of such situations between epacridaceous, myrtaceous and proteaceous species). The case studies from Kings Park (Pate et al., 1993; Stewart et al., 1993) provide a good example of complementary patterns of utilisation in respect of N resources, in demonstrating apparently mutualistic interactions of (a) non-mycorrhizal herbaceous species relying principally on nitrate, (b) N_2 fixers subsisting on atmospheric N and (c) mycorrhizal woody species presumably reliant on highly intractable organic forms of N. The latter trophic specialisation is well exemplified by ericoid-type mycorrhizal associations of Epacridaceae (Bell and Pate, 1996), whereas non-mycorrhizal nitrate users are especially prominent after fire, when mineralisation of N of belowground biomass of species killed in the fire provides transient large resources of N (see Pate and Unkovich, in press).

2.6. *Estimating proportional amounts of total biomass of major components of a* Banksia *woodland ecosystem*

Using mechanically-aided excavations of 3 m diameter × 3 m deep soil cores, root biomass was exhaustively collected from three to ten replicate plants of all common woody and large herbaceous species at the study site at Moora. By comparing dry weights of total roots recovered with aboveground biomass, shoot:root dry weight ratios were computed for all major species (Table 3, column 2). The last fire at the site occurred 35 years previously (March 1962) so the woodland was rated to be at or near post-fire maximum biomass accumulation when sampled in 1997–98. Growth ring counts in trunks of major seeder tree and shrub species indicated that virtually all major recruitment occurred immediately after that fire (A. M. Grigg, J. S. Pate and M. J. Unkovich (unpublished data)).

At Moora mean shoot dry weights of these same species were then assessed by random sampling of 30 individuals of each species, and using the available data on shoot:root ratios (Table 3, column 2), values for mean standing above- and belowground biomass (DM) per plant were computed for each species (Table 3, columns 3 and 4). Density of each species per square metre was assessed by random quadrat sampling throughout the bushland (Table 3, column 5), and these values combined with previous biomass determinations to estimate total standing biomass of each species per hectare (Table 3, column 6). The resulting data reinforce the prominence of biomass as large trees and shrubs, and highlight the major impact that this is likely to have on nutrient and water relationships at the site. It should be noted that it was not feasible to recover biomass of seasonally-produced fine feeding roots of large woody species, so estimates of belowground biomass of these are likely to be highly conservative. This issue is referred to later when dealing specifically with *Banksia prionotes*.

The total biomass of all species in the undisturbed woodland turned out to

Table 3. Estimates of above and belowground biomass of 45 common woody species and the herbaceous perennial layer found in *Banksia* woodland at Moora, Western Australia.

Species	Shoot:root dry weight ratio	Biomass (kg ha^{-1})			Density (no. plants ha^{-1})
		Aboveground	Belowground	Total	
Acacia pulchella	5.1	7	1	9	60
Actinostrobus arenarius	9.0	3290	366	3660	2510
Adenanthos cygnorum	4.1	1290	316	1610	461
Allocasuarina humilis	1.1	238	217	455	165
Astroloma xerophyllum	4.7	1	1	1	30
Banksia attenuata	1.5	454	309	763	185
Banksia leptophylla	4.0	685	171	857	296
Banksia prionotes	4.1	13100	2610	17800	1450
Beaufortia elegans	3.3	72	22	94	30
Boronia gracilipes	9.0	1	1	1	50
Calytrix sp.	1.2	23	19	42	687
Calytrix fraserii	2.5	34	14	48	687
Conospermum stochaedis	1.0	1	1	1	10
Cryptandra pungens	2.7	1	1	1	110
Dryandra nivea	1.0	8	8	16	60
Eremaea beaufortioides	3.5	1230	354	1580	1500
Eucalyptus todtiana	1.0	100	100	200	5
Grevillea eriostachya	4.6	1	1	1	20
Hakea obliqua	4.0	20	5	25	20
Hakea platysperma	4.0	20	5	25	20
Hakea trifurcata	3.8	23	6	29	10
Hibbertia acerosa	1.5	7	4	11	100
Hibbertia pachyrrhiza	2.8	2	1	3	360

Isopogon dubius	1.0	6	6	12	30
Isopogon scabriuscula	4.5	155	34	190	200
Jacksonia aff stricta	2.5	4	2	6	20
Jacksonia ulicifolia	1.8	63	36	100	190
Leptospermum erubescens	1.0	125	125	250	950
Leptospermum spinecens	0.7	8	13	21	30
Leucopogon polymorphus	4.3	14	3	17	650
Melaleuca scabra	1.5	1080	720	1800	2740
Melaleuca sp.	3.5	14	4	18	40
Mirbelia spinosa	4.1	7	2	10	170
Nuytsia floribunda	1.0	100	100	200	10
Petrophile brevifolia	0.6	15	26	42	50
Pileanthus sp.	3.5	65	19	83	687
Pimelea sp.	3.3	1	1	1	10
Pimelea spectabilis	1.6	1	1	1	10
Pityrodia bartlingii	1.3	56	42	97	680
Stirlingia latifolia	1.0	10	10	19	30
Synaphea spinosa	3.6	3	1	4	10
Verrauxia reinwardtii	1.0	1	1	2	10
Verticordia sp. A	3.2	2	1	3	91
Verticordia sp. B	4.4	5	1	7	91
Xylomelum angustifolium	1.7	306	182	489	120
Herbaceous layer[1]	(1.0)	995	995	1990	n.a.[2]
Mean ± SE	2.9 ± 0.0				
Total (t ha^{-1})		24.5	7.8	34.5	

[1] Herbaceous biomass estimated as total dry weight of fully excavated plants per square meter. Data for aboveground biomass of herbaceous layer estimated on basis of its shoot:root dry weight ratio being 1.0, in view of the extremely high proportion of cryptophytes and resprouter species.
[2] Not measured.

be 32.5 t DM ha^{-1} (Table 3), with proportions of this as trees, woody shrubs, herbaceous perennials and winter ephemerals (as shown in Figure 5A). Viewed in terms of trophic specialisations, about two-thirds of the biomass was associated with proteoid root-forming species, about one-seventh was variously mycorrhizal and only a few percent comprised of potential nitrogen-fixing species (Figure 5B). Analysis in respect of seasonality of growth showed most of the standing biomass to be associated with a summer pattern of shoot extension, the remainder being winter-active. Virtually all of the

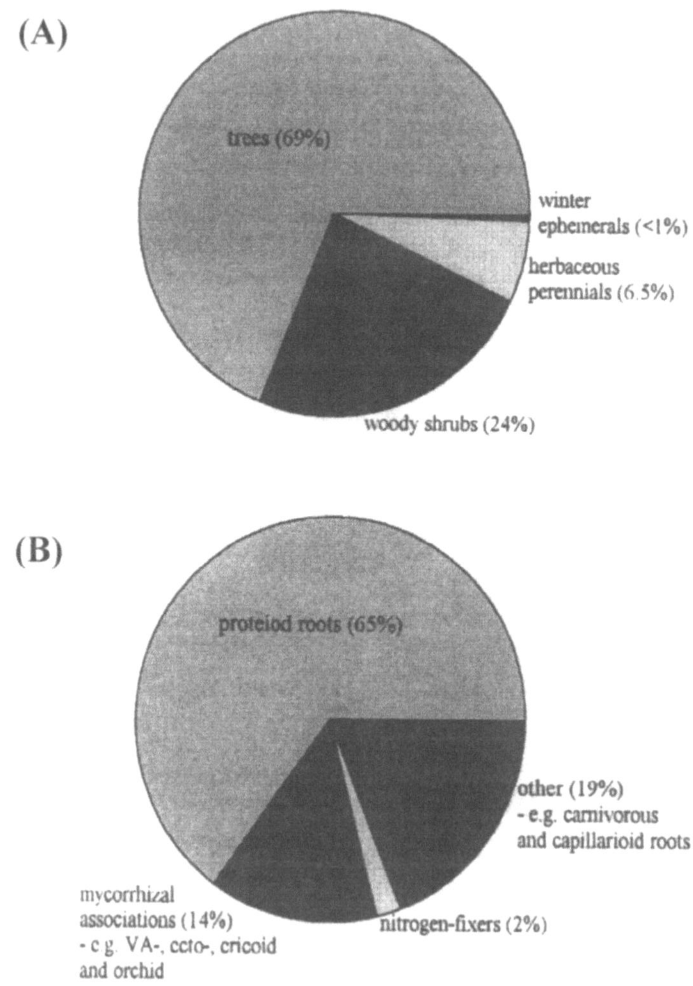

Figure 5. Relative proportions of biomass of species from the *Banksia* woodland at Moora, Western Australia in terms of (a) life form as trees, woody shrubs, herbaceous perennials and winter ephemerals, and (b) trophic specialisations including proteiod roots, mycorrhizal associations and nitrogen-fixers.

summer-active species were phreatophytic trees or shrubs, whereas most of winter-active species were shallow-rooted (maximum depth of penetration 1.5 m) and therefore unable to access ground water at 5–6 m depth at the sites concerned.

3. *Banksia prionotes* – major player of the woodlands

In this section we summarise the adaptive features exhibited by *Banksia prionotes* – the arboreal dominant at our woodland site at Moora. The aim is to provide a profile of the cardinal features deemed most desirable when selecting a deep-rooted mimic species for the bushland.

Banksia prionotes is a fire-sensitive species recruiting most prolifically after fire when dense, self-thinning stands quickly dominate post-fire vegetation of regions where it is a major arboreal component. Further recruitment after this post-fire germination response is generally insignificant. For example, counts of trunk growth rings of *B. prionotes* in the woodland at Moora indicated that virtually all trees currently present had recruited from a hot summer fire in the area 35 years previously. According to studies of populations at Yanchep, growth of individuals in post-fire stands of *B. prionotes* increases exponentially over the first eight or so years after germination, with shoot:root dry weight ratios increasing from two to three in young plants to a plateau value of 3.8–4.5 for plants of four years of age or older (Bowen and Pate, 1991; Pate et al., 1998). Distribution of dry matter between plant parts shows fastest relative growth rates to be accompanied by peak ratios of leaf area:total plant weight in plants of three to four years of age and greatest rates of accumulation of dry matter per unit leaf area at the same time.

The dimorphic rooting character of *B. prionotes*, alluded to in this article and that of Pate and Dawson (this volume) becomes evident early after germination and tap (sinker) roots of surviving seedlings would typically have already penetrated to 2 m depth by the end of their first summer. Tap roots of the species eventually descend to 15 m or more if a more elevated water table is absent, or upon meeting the capillary fringe of a water table, will branch out in a complex network of fine water-absorbing roots. Seedlings show an initial complement of three to five lateral roots radiating from the top of the root stock and before a tree reaches ten years of age, eight to ten lateral roots should have developed. Mapping of locations and branching patterns of these roots (Pate et al., 1998) reveals effective exploitation of space, to the extent that rooting catchments of mature trees will generally occupy ground areas four to six times greater than that shaded by their crowns. Roots of older trees at typical woodland densities are thus likely to trespass to a great extent into rooting zones of neighbours. Competition for water and nutrients is therefore viewed as a major contributing factor to the progressive thinning of stands and slowing down of growth rates in middle-aged populations of the species. Most trees at Moora show 'secondary' tap roots descending from mid regions of one or more of their major lateral roots (J. S. Pate and S. S. O. Burgess, unpubl. data). These auxilliary tap roots eventually access ground water just as effectively as does the primary tap root.

Seasonal production of proteoid roots commences each autumn following outgrowth of new initials from periderm-invested perennating parts of the lateral root system. There is evidence from deuterium:hydrogen (D/H) studies and heat-pulse experiments (Dawson and Pate, 1996; Pate and Dawson, this volume) that water hydraulically lifted by tap roots supports early growth of feeding roots in autumn, maintains integrity of proteoid root clusters through transient drought in winter and spring, and extends activity of such roots thereafter well into a summer season. This facility would clearly ameliorate the vulnerability of superficially (5–15 cm) located feeding roots to sudden change in soil water status. It is not known whether shallow-rooted species benefit appreciably from hydraulically-lifted water.

Study of seasonal changes in xylem sap composition of lateral roots, sinker roots, trunks and proteoid root clusters of *B. prionotes*, taken together with corresponding analyses of seasonal filling and emptying of storage pools of minerals in trunks and leaves, have provided an overall picture of mineral resource partitioning (Pate and Jeschke, 1993). Throughout their peak activity in late winter and spring proteoid roots comprise principal contributors of PO_4^{3-}, amino acids and K^+ to the xylem stream and may also function significantly in reducing nitrate released following sporadic nitrification events in the upper soil profile (Jesckhe and Pate, 1995). Xylem sap of laterals in receipt of minerals from proteoid roots is therefore consistently more concentrated in a range of essential nutrients, such as N and P, during the wet season than is that of the tap root. Xylem sap collected simultaneously from the trunk base shows mineral concentrations intermediate between those of lateral and tap roots, implying that the shoot concerned is being supplied with a mixture of both xylem streams, more or less as predicted from the D/H studies detailed by Dawson and Pate (1996) Pate and Dawson (this volume). Analyses of gradients in xylem sap concentrations of P, N and other nutrients from base to top of trunks of trees harvested throughout a year provide evidence of lower parts of stems withdrawing nutrients laterally from the xylem during winter and accommodating these nutrients in storage parenchyma of xylem, especially that of rays (Pate and Jeschke, 1995). The following summer through to autumn, nutrients are released back to the xylem, and together with nutrients mobilised from oldest leaves, provide a major resource base for nutrition of the rapidly extending shoot. Further uptake of minerals from the root is of course limited during summer, due to (a) loss of proteoid roots with drying out of the upper soil, and (b) general deficiencies of macronutrients in the ground water source accessed by the tap root. However, large quantities of water from deep sources remain vital for maintenance of plant water status and, more importantly, for enabling high rates of photosynthesis to implement shoot growth in summer. These features remain contingent upon tap roots drawing successfully from the retreating water table. Indeed, where this does not happen, sudden death of trees is observed, especially in late autumn following a previous winter of poor rainfall. Presumably, the more vigorous surviving trees had lowered ground water to levels beyond the reach of inferiors which have succumbed.

The complex hydraulic architecture of xylem of trunk and roots of *B. prionotes* and other dimorphic-rooted proteaceous species (Pate et al., 1995) is viewed as an essential ingredient for maintenance of integrity of vessel functioning through seasonal conditions in which large scale changes occur in water status of upper soil and atmospheric water deficit. The very long and wide vessels of the tap root, together with increases in the specific hydraulic conductivity of xylem with downward attenuation in root diameter, result in the xylem pathway from ground water to soil level being of uniformly very low resistance. In marked contrast, the much shorter and narrower vessels of the trunk give high resistances to flow, thus comprising a large buffer resistance against which stomata can respond to transient or long term water stresses. It remains to be seen whether some of the exotic species of trees which we suggest later as possible mimics of *Banksia* woodlands possess the same adaptive qualities.

Particularly relevant to replacement of natural vegetation with a functional equivalent mimic is the requirement to evaluate the net primary productivity of biomass of the original bushland and relate this to annually available resources of rainfall and nutrients. We have already shown that standing living above- and belowground biomass of the Moora woodland was rated at approximately 32.5 t DM ha^{-1} and that 55% (17.8 t DM ha^{-1}) of this consisted of the single dominant *B. prionotes*, and a further 16% (5.1 t DM ha^{-1}) comprised of other woody trees or shrubs of 2 m or more in height. Our protocol for estimating net annual primary production of these woody elements was as follows.

Due to a strictly summer pattern of growth in which seasonal flushes of new shoots and foliage persist on the plant for three seasons, this last three years worth of extension growth to main and side branches of tree or shrub species of genera such as *Banksia*, *Xylomelum*, *Adenanthos* or *Hakea* can be readily identified and collected separately from older biomass. Using this protocol the last three years of shoot growth of *B. prionotes* at Moora was estimated as equivalent to 21.1% (or 2.7 t DM ha^{-1}) of the current aboveground biomass of the trees (see Figure 6). By cutting out a segment from the trunk base at 10 cm or so aboveground level and examining its annual growth rings, the last three years of secondary growth can be easily chiselled off from a circumferential sector of the outer trunk and the dry weight of this off cut compared with that of the remaining earlier formed wood of the sector. In this way net productivity in terms of secondary thickening was assessed for *B. prionotes* as 8.8% (or 1.0 t DM ha^{-1}) of the wood biomass currently present in the trunk (Figure 6). Then, assuming that secondary thickening of the perennial superstructure of the root increased similarly (see Figure 6 roots),

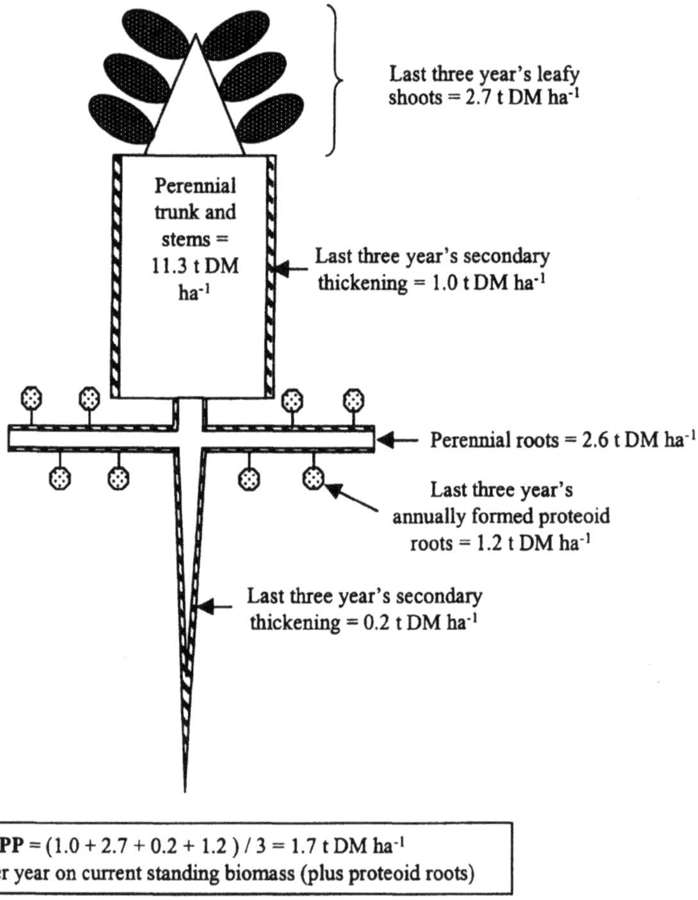

Figure 6. Model depicting dry matter in total standing biomass and biomass components relating to three years leafy growth, annual proteoid root production and secondary thickening of *Banksia prionotes* (1995–1997) in pristine bush at Moora, Western Australia. Annual net primary production (ANPP) was determined on the basis of total current standing biomass including assumed proteoid root production averaged for the three year period of the study.

our earlier estimate of 17.8 t DM ha^{-1} for *B. prionotes* at the site would have effected a net biomass increment of 3.9 t DM over the past three years, made up of 2.7 t of new shoot extensions and 1.2 t DM of growth increase in trunk and perennial parts of the root system. We then need to add to this a further estimate of the new biomass laid down each year in annual investment of fine feeding roots, predominantly composed of proteoid root biomass (Figure 6). As suggested from previous evaluations (Pate et al., 1998) we tentatively assess this as representing the equivalent of a total investment of 46% of the current biomass of perennial parts of the root system over the three year period. Rated as such, it is given a value of 1.2 t DM ha^{-1} for the three year study period depicted in Figure 6. Using the data represented pictorially in Figure 6, our model accordingly suggests a net above- plus belowground productivity of 1.7 t DM ha^{-1} $year^{-1}$ for the population of *B. prionotes* at the site, and a yearly net increment equivalent to 10.6% of the present standing biomass.

The above approach (Figure 6), involving a three year rather than a single year estimate of net primary production of biomass, carried the advantage of averaging tree response at Moora over a first year of well above average (1995 – 703 mm) rainfall, a second one of near average (1996 – 448 mm) rainfall and a third (1997 – 431 mm) of marginally less, most of which fell in summer at times of high evaporation from the soil surface. Leaf weights on each summer's annual shoot extension of this chronosequence varied markedly, *viz.* in a weight ratio of 11 (1995):7 (1996):3 (1997), i.e. more or less reflecting the effectiveness of rainfall of the previous winter in replenishing water in soil and ground water and mobilising soil nutrients from soil to standing biomass.

Our approach to measuring net primary productivity proved to be applicable to any summer-active woody species that showed tell-tale morphological evidence of seasonal growth increments. It was encouraging in this connection to find similar proportional increases in shoot biomass over the three year period to that recorded above when this was applied to other species at the Moora site (*viz.* 24% for *B. leptophylla*, 26% for *B. attenuata* and 24% for *Adenanthos cygnorum versus* the already mentioned value of 21% for *B. prionotes*).

Assuming the above approach to apply generally to all biomass at the site we estimated total net primary productivity for the *Banksia* woodland system as 3.2 t DM ha^{-1} y^{-1} (i.e. 10.6% of the 32.5 t DM ha^{-1} of total biomass shown in Table 3). With a mean annual rainfall of 527 mm for the three years studied, net productivity of the ecosystem then translates to 6.1 kg of total biomass ha^{-1} mm^{-1} of annual rainfall, or 4.5 kg DM ha^{-1} mm^{-1} if estimated solely in terms of aboveground biomass.

Matching data on the productivity of total aboveground biomass for pasture, lupin and wheat for the period 1994–96 at a location a few kilometres from the *Banksia* woodland study site (see data of Anderson et al., 1998a and the unpublished observations of E. C. Lefroy and J. S. Pate for a tagasaste trial adjacent to the Moora bushland site) are given alongside those for the *Banksia* woodland in Table 4. A somewhat surprising feature of the comparison is that annual productivity per unit rainfall of the bush (4.5 kg DM mm^{-1}) is least of all values recorded, with the exception of one particularly poor wheat crop (4.0 kg DM mm^{-1}, 1995 season). The border of the pristine bush at the study site receives run-off from nutrient-rich water of the elevated water table on adjacent agricultural land, and trees of *B. prionotes* and other nutrient-tolerant native species in this 'hedge' of vegetation have achieved gigantic proportions embodying productivity rates of up to 15.9 kg DM mm^{-1} rainfall (see Table 4). However, even in this cossetted situation of much freer access to

Table 4. Net annual primary productivity of aboveground biomass of annual pasture and crops, a plantation of tagasaste (*Chamaecytisus proliferus*) and neighbouring *Banksia* woodland in the Moora region of Western Australia.

	Net annual productivity of aboveground biomass ($t\ ha^{-1}$)	Biomass production per unit rainfall ($kg\ mm^{-1}$)
1994: 294 mm annual rainfall		
Lupin[a]	5.3	18.0
Wheat[a]	3.5	11.9
1995: 703 mm annual rainfall		
Lupin[a]	6.5	9.2
Wheat[a]	3.7, 2.8	5.3, 4.0
Pasture[a]	7.3, 7.6	10.3, 10.8
1996: 438 mm annual rainfall		
Lupin[a]	5.8	13.2
Lupin[b]	4.2	9.6
Wheat[a]	7.5, 6.8	17.1, 15.5
Pasture[a]	4.2, 5.0	9.6, 11.4
Pasture[b]	3.4	7.8
Tagasaste[b, e]	4.4	10.0
1997: 412 mm annual rainfall		
Oats[b]	3.1	7.5
Tagasaste[b]	20.1	48.7
1995 – 1997: 527 mm mean annual rainfall		
Banksia woodland[c]	2.4	4.5
Hedge of Banksia woodland bordering agricultural land[d]	8.4	15.9

Source and notes: [a] Data from Anderson et al. (1998a); [b] Data from E. C. Lefroy and J. S. Pate (unpubl. data); [c] Data derived from pristine bush unaffected by agricultural runoff as shown in text and Table 3; [d] Receives subterranean runoff of nitrate-polluted water throughout the year from elevated water table in agricultural land (A. Grigg, D. J. Arthur, T. E. Dawson, T. J. Hatton, M. J. Unkovich and J. S. Pate, unpubl. data); [e] Tagasaste cut in May 1996 and not cut in 1997.

water and nutrients, the *Banksia* and other trees still fail to out-produce the best wheat crops shown in the data set, and achieve a net annual production of aboveground biomass equivalent to only 40% of that of a tagasaste (*Chamaecytisus proliferus* Link ssp. *palmensis*) plantation in its second year after cutting (Table 4; see also Lefroy and Stirtzaker, this volume).

The apparently greater productivity of the annual crops than *Banksia* in terms of rainfall received becomes even more surprising when one considers that deep drainage losses of unused water occur during growth of the former (see data for deep sands of the Moora region by Anderson et al., 1998b; Asseng et al., 1998), whereas, seasonal rises and falls in water table in the woodland area indicate long-term balance of rainfall utilisation and therefore a more meaningful relationship between net primary productivity and rainfall. We conclude that this alleged 'ineffectiveness' of the native bush in dry matter production relates to the perennial character of the vegetation and the major constraints on growth likely to be imposed by the highly oligotrophic nature of the rooting substrate. In relation to perenniality, the model of Figure 5 shows relatively large annual investments of dry matter in secondary thickening of root and trunk and feeding roots essential for acquisition of nutrients in the wet season. We have little idea of respiratory costs in construction, maintenance and functioning of these heterotrophic parts of the plant, but judging from budgets of allocation of photosynthate for proteaceous seedlings including *B. prionotes* (see Bowen and Pate, 1991), respiration losses of carbon are at least equal to that allocated to dry matter. It is reasonable to assume that maintenance costs would be much less in annual crop plants, given the ephemeral character of their shoots and roots and the much higher proportions of their biomass as photosynthesising leaves as opposed to stems and roots in comparison to life forms such as large and old woody trees. Furthermore, costs of nutrient acquisition are likely to be much less in fertile agricultural land due to fertiliser inputs and greater availability of recycled nutrients. Also, where winter annual crops are deployed there is no requirement for deep long-lived roots to be developed to combat summer heat and drought and access ground water.

Notwithstanding the above, it is still difficult to explain the marked discrepancy in productivity between the 'hedge' of *Banksia* receiving nutrient run off from the surrounding cleared land, and performance in the second year after cutting of tagasaste plantations which benefited similarly as sumps for water and nutrients from a hinterland of pasture and annually cropped areas. Cutting of the tagasaste the previous year, and the relative youth of the trees concerned relative to the near climax community of *Banksia* trees of the bush, might be identified as the critical ingredient for much higher productivity of the former than latter system. Thus, as shown for a number of native woody species of similar deep-sand habitats (e.g. Hansen et al., 1991; Bowen and Pate, 1991, 1993; Bell and Pate, 1996), destruction of shoots by fire or cutting can elicit extremely fast rates of shoot growth as shoot:root biomass ratios are restored to pre-disturbance levels. With relatively small stem biomass and high ratios of leaf mass to total plant mass at such times, maintenance costs should be relatively low leading to high effectiveness in dry matter production, even possibly equalling that of juvenile plants. We would suggest that the cut plantation of tagasaste was in such a highly productive, yet water expensive, state when sampled in the 1997 season at Moora.

4. Towards a mimic system for *Banksia* woodlands

4.1. *Radical changes following clearing for agriculture*

Banksia woodlands, or in more general terms, *Banksia* sand ecosystems of Western Australia were initially avoided by farmers in view of their highly oligotrophic character, and did not become amenable to economically-viable cropping or pasture production until specific trace element additions together with superphosphate were applied to boost yields. Such areas are now mostly devoted to N_2 fixing pasture and crop legumes in rotation with cereals. As alluded to earlier, annual crops on these deep sands fail to utilise seasonal rainfall, leading to particularly severe examples of rising ground water levels. Although not usually affected by salinity, cleared deep sands also provide examples of alarmingly high levels of nitrate pollution of ground water. For example, Pate and Unkovich (in press) in their comparison of N_2 fixing inputs in the natural bushland and adjacent ecosystems at Moora, found groundwater under pasture or cereals to contain 6–12 mg NO_3^-–N L^{-1} *versus* only 0.2 mg NO_3^-–N L^{-1} under neighbouring undisturbed native bush. This nitrate in cleared land has clearly been generated by agricultural practices and leached progressively to ground water since first clearing of the bush in the region in the late 1950's and 1960's. Furthermore, since nitrogen fertilisers are used only moderately for cereal production in the region, the N pollution currently present is mostly attributable to mineralisation of the easily decomposed biomass residues of N_2 fixing leguminous crops and pasture components. With virtually zero anion-retaining capacity in the rooting substrate, and significant episodes of a season's mineralisation occurring out of phase with principal periods of crop and pasture growth, nitrate leaching may be substantial and more pronounced under certain crops such as lupins than under pastures (Anderson et al., 1998b). In any event, with additions of P and trace elements, unbridled inputs of fixed N, elevated water tables and nitrate leaching, the agricultural scenario features nutritional and hydrological states which are grossly perverted from the balanced and self-sustaining natural ecosystems originally present in the area. Herein lies the crux of the problem when designing 'alternative' farming systems suited for rehabilitation of such areas.

4.2. *Two contrasting scenarios for application of mimic principles*

The above issues raise the question of what systems one should address when developing the mimic principal. One strictly limited scenario would be, say, where one wishes to revegetate a *Banksia* sand ecosystem which had been essentially denuded of biomass in timber harvesting, sand mining or some form of clearing, but where no fertiliser inputs had been involved and no eutrophication incurred through run-off from adjacent areas. Under such circumstances, there would surely be no better solution than to replant with a selection of major indigenous species, and to do so with judicious consideration of proper balance between life forms, rooting morphologies, phenologies of growth and trophic specialisations present at the site prior to disturbance.

Judging from the universally high prevalence of deep-rooted phreatophytic shrubs and trees in *Banksia* sand ecosystems, our first priority would be to select these as principal players. Planting would then be designed to achieve effective balance between summer-active deep-rooted *versus* winter-active shallow-rooted types, and hence, maintain an overall balance in seasonal water utilisation. It remains to be seen how critical the planting ratio between these two types is, or whether, through thinning and attrition, the same balance would result more or less regardless of proportional representations at the time of planting. It would also be of value to follow short- and long-term implications on water table levels and nutrient balances in systems where bias was artificially maintained towards shallow-rooted winter-active species over phreatophytic summer-growing species, and *vice versa*. In essence, we suggest two perennial functional types; summer-active deep-rooted and winter-active shallow-rooted (see Character option 1, Figure 7), with winter-active deep-

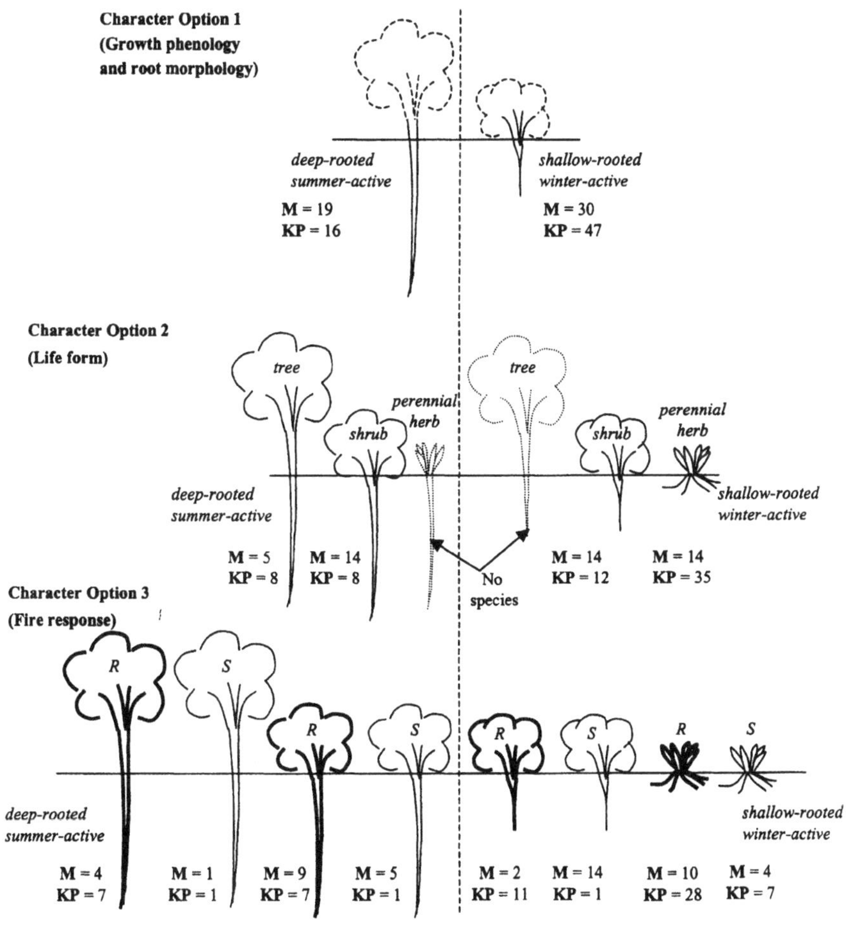

Figure 7. A four stage species selection suggested for a mimic system to replace vegetation of cleared *Banksia* woodlands in south western Australia. Numbers of species found in Moora (M) and Kings Park (KP) fitting the particular character states are indicated. Fire response is indicated as resprouter (R) or seeder (s).

rooted and summer-active shallow-rooted species essentially not represented in the selection, or indeed to any extent in the original vegetation. Numbers of species available in the natural ecosystems to cover these two types are shown in Figure 7.

A next consideration (Character option 2, Figure 7) would be to include under each of the above character states, non-woody herbaceous species and small shrubs along with trees to provide a more or less continuous ground cover. Most *Banksia* woodlands have good representation of such smaller species. However, it turns out that the floras of Moora and Kings Park do not include species which are deep-rooted summer-active perennial herbs or shallow-rooted winter-active trees. These types, indicated by dotted lines in Figure 7, are accordingly excluded from further consideration.

Thirdly, the menu for species selection proposed incorporates resilience to fire by balanced representation of fire-sensitive (seeder) and fire-resistant (resprouter) species (Character option 3, Figure 7). Large trees and shrubs come in both categories and preponderance of one over the other varies with site, as shown in the comparison between Moora and Kings Park floras in this article. However, herbaceous perennial species are dominantly resprouters and their value in holding topsoil together and their activity, and that of herbaceous ephemerals, in reabsorbing minerals released after fire cannot be overestimated. Slow growth of resprouters as seedlings, scarcity of seed stocks and poor capacity to compete initially with faster growing seeders, can cause problems when trying to achieve adequate stocking of fire-resistant species on cleared land. Again, experimentation within this area is clearly called for. At this stage our choice portfolio (Figure 7) contains eight sets of variously combined character states and we would regard this as possibly a basic minimum to successful functioning of the replacement system. As can be see from Figure 7, the floras of Moora and Kings Park contain at least one or more representations in each of the eight categories and each could, therefore, be used feasibly in revegetation programs.

Finally, we would envisage a further sophisticated series of possible options to amplify species selection and that these should be based on major trophic specialisations within the original floras (i.e. proteoid roots, various mycorrhizal types, symbiotic N_2 fixation, carnivorous and parasitic habit). With this further set of options (Table 5), the species listing escalates to 48 possible options, a number of which are minimally represented or not represented at all in the floras of the study sites.

There are two obvious dangers in the above approach. Firstly, it would be unwise to include only one representative of each set of character states, simply because it might either fail to establish to the desired density or because it was a poor choice from among the replicate candidates available carrying that set of options. Secondly, the scheme presented assumes that the selected sets of character options are the only ones necessary for the mimic to function as a fully sustainable equivalent to the indigenous system. In the absence of information on levels of other possibly equally essential species interactions

Table 5. Possible options for species selection for replacement model based on trophic specialisations after previous selection of species from Character options 1–3 outlined in Figure 7.

Trophic type		Deep-rooted summer-active				Shallow-rooted winter-active			
		Tree		Shrub		Shrub		Perennial herb	
		S[1]	R[1]	S	R	S	R	S	R
VA mycorrhiza	M	1	3	3	3	No sp.	No sp.	No sp.	No sp.
	KP	1	4	0	3				
Other mycorrhiza[2]	M	No sp.[3]	0	1	3	7	1	3	3
	KP		3	1	3	1	7	2	10
Proteoid roots	M	No sp.	1	No sp.	1	0	0	No sp.	No sp.
	KP		3		1	1	2		
N_2 fixation	M	No sp.	0	1	3	0	0	No sp.	No sp.
	KP		1	1	3	1	1		
Carnivory	M	No sp.	No sp.	No sp.	No sp.	No sp.	No sp.	1	1
	KP							0	1
Parasitism	M	No sp.	1	No sp.	No sp.	0	No sp.	1	No sp.
	KP		1			1	2		

[1] S = seeder; R = resprouter.
[2] Includes ectomycorrhiza, ericoid-type mycorrhiza, orchid mycorrhiza.
[3] No sp. = no species were recorded for this choice.

in the bushland, the protocol of choice of players indicated in Figure 7 and extended into Table 5, might well lead to dysfunctionality, especially if the final biomass balance between physiognomic groupings turned out to be very different to that of the original bush.

A second entirely different scenario concerns species selection and planting designs directed at amelioration of the substantial hydrological and nutritional balances ‘within an agricultural setting’ or ‘under agricultural use’ or ‘commercial use’. We are hesitant to use the term ‘mimic’ in such a case, since the desired objectives of the proposed system in the first instance would be to make progressive inroads into elevated water tables and excessive nitrate and other nutrient loadings in the system. This might well require use of summer-active deep-rooted tree or shrub species which are much faster growing than those of native bushlands, and species which would also prove amenable to planting in alley or plantation formats, where they would supposedly achieve greater utilisation of ground water and nutrients per unit planted area than is currently experienced in annual crop species, or indeed in native bush. A number of exotic tree species, such as *Pinus pinaster*, *P. radiata*, *Paulownia* sp., various mallee eucalypts, *Acacia saligna* (for relatively saline areas) and the use of tagasaste over the past decade or so come to mind in this connection. Were row plantings of such species to be accompanied by regular cropping with winter-active cereals and legumes, as currently practised with tagasaste, desirable outcomes might well be achieved in both short and long term. However, two important caveats would apply. Firstly, we would view with mistrust continued use of shrub legumes such as tagasaste, since when these are cut or grazed annually for fodder they typically make a greater net contribution of fixed N_2 to the system than they utilise from soil and groundwater in the form of nitrate-nitrogen (see case studies of Pate and Unkovich (in press) for tagasaste at Moora). Indeed, unless one were to select a non-nodulating genotype of tagasaste, or to genetically engineer ineffective nodulation of the species, its use, while clearly desirable in terms of hydrological balance, might exacerbate rather than ameliorate current nitrate pollution problems. Secondly, use of shallow- and deep-rooted species for nutrient stripping in terms of nitrate and possibly also excess P, would be accomplished only if the nutrients in question were removed from the site as, say, foliage, grain, animal products, or similar. This is unlikely to eventuate in the case of unpalatable shrubs and trees unless a commercial incentive were to accrue in the form of harvestable timber, biomass of wood chipping, or possible specific highly valuable products such as oil from leaves, as in the case of mallee eucalypts. As long as government agencies fail to subsidise environmental rehabilitation on a scale attractive to landowners, options available to farmers remain strictly limited, especially in current gloomy economic times for the agricultural industry. However, current discussions on trading of carbon credits might well ultimately provide extra profit incentive in this connection. In any event, vast scales of operation, enormous investments, and an unprecedented degree of cooperation between landowners will be required to plant whole catchments with sufficient deep-rooted woody species to redress the crucial problems on the deep sand ecosystems of Western Australia.

Finally, another option to be considered is that of utilising solely herbaceous species for rehabilitation programmes, in this case using annual winter crops and pastures as currently employed, alongside novel plantings of deep-rooted summer-active herbaceous perennials. Lucerne (*Medicago sativa*), despite possible disadvantages in view of symbiotic inputs of N, and summer-active C_4-type grasses such as Rhodes grass (*Chloris guyana*) would be obvious candidates for summer usage of water and nutrients, but their success would obviously depend on capacity to access underlying ground water. Both species root to at least 4–5 m in depth in deep sands, but their trialling is still very limited in the highly depauperate acidic sands of Western Australia. Despite obvious non-comparability in every respect against the sclerophyllous woodlands which they would replace, such all-herbaceous mimics might offer distinct advantages in terms of ease of management and profitability relative to alley or plantation tree-based systems. Indeed, if water table rise continues unabated in certain areas, relatively shallow-rooted herbaceous perennials might offer the only viable option for dealing with near surface ground water tables, where flooding-intolerant woody species would be unlikely to survive.

5. Acknowledgements

The authors wish to acknowledge the use of a great deal of as yet, unpublished data generously supplied by our various collaborators in the Moora project: T. E. Dawson, T. J. Hatton, S. S. O. Burgess, E. C. Griffin, D. J. Arthur, M. J. Unkovich, A. M. Grigg and E. C. Lefroy. Thanks are also due to A. J. Smit for assistance in preparation of the figures. Funding of research by grants from the Land and Water Resources Research and Development Corporation, the Australian Research Council and the Department of Conservation and Land Management, Western Australia is gratefully acknowledged.

6. References

Anderson GC, Fillery IRP, Dolling PJ and Asseng S (1998a) Nitrogen and water flows under pasture-wheat and lupin-wheat rotations in deep sands in Western Australia. 1. Nitrogen fixation in legumes, net N mineralisation, and utilisation of soil-derived nitrogen. Australian Journal of Agricultural Research 49: 329–343

Anderson GC, Fillery IRP, Dunin FX, Dolling PJ and Asseng S (1998b) Nitrogen and water flows under pasture-wheat and lupin-wheat rotations in deep sands in Western Australia. 2. Drainage and nitrate leaching. Australian Journal of Agricultural Research 49: 345–361

Asseng S, Fillery IRP, Anderson GC, Dolling PJ, Dunin FX and Keating BA (1998) Use of the APISM wheat model to predict yield, drainage, and NO_3^- leaching for a deep sand. Australian Journal of Agricultural Research 49: 363–377

Beard JS (1984) Biogeography of the kwongan. In: Pate JS and Beard JS (eds) Kwongan, Plant Life of the Sandplain, pp 1–26. University of Western Australia Press, Nedlands

Beard JS (1989) Definition and location of the *Banksia* woodlands. Journal of the Royal Society of Western Australia 71: 85–86

Bell DT, Hopkins AJM and Pate JS (1984) Fire in the kwongan. In: Pate JS and Beard JS (eds) Kwongan, Plant Life of the Sandplain, pp 178–204. University of Western Australia Press, Nedlands

Bell TL (1995) Biology of Australian Epacridaceae: With Special Reference to Growth, Fire Response and Mycorrhizal Nutrition. PhD Thesis, The University of Western Australia

Bell TL and Pate JS (1996) Nitrogen and phosphorus nutrition in mycorrhizal Epacridaceae of south-west Australia. Annals of Botany 77: 389–397

Bell TL, Pate JS and Dixon KW (1994) Response of mycorrhizal seedlings of SW Australian sandplain Epacridaceae to added nitrogen and phosphorus. Journal of Experimental Botany 45: 779–790

Bell TL, Pate JS and Dixon KW (1996) Relationships between fire response, morphology, root anatomy and starch distribution in South-west Australian Epacridaceae. Annals of Botany 77: 357–364

Bettenay E (1984) Origin and nature of the sandplains. In: Pate JS and Beard JS (eds) Kwongan, Plant Life of the Sandplain, pp 51–68. University of Western Australia Press, Nedlands

Bowen BJ and Pate JS (1991) Adaptations of SW Australian members of the Proteaceae: allocation of resources during early growth. Proceedings of the International Protea Association, Sixth Biennial Conference, Perth, Western Australia, pp 347–356. Promaco Conventions Pty Ltd

Bowen BJ and Pate JS (1993) The significance of root starch in post-fire shoot recovery of the resprouter *Stirlingia latifolia* R. Br. (Proteaceae). Annals of Botany 72: 7–16

Dawson TE and Pate JS (1996) Seasonal water uptake and movement in root systems of Australian phreatophytic plants of dimorphic root morphology: a stable isotope investigation. Oecologia 107: 13–20

Dodd J and Bell DT (1993) Water relations of the canopy species in a *Banksia* woodland, Swan Coastal Plain, Western Australia. Australian Journal of Ecology 18: 281–293

Dodd J and Griffin EA (1989) Floristics of the *Banksia* woodlands. Journal of the Royal Society of Western Australia 71: 89–90

Dodd J and Heddle EM (1989) Water relations of *Banksia* woodlands. Journal of the Royal Society of Western Australia 71: 91–92

Dodd J, Heddle EM, Pate JS and Dixon KW (1984) Rooting patterns of sandplain plants and their functional significance. In: Pate JS and Beard JS (eds) Kwongan, Plant Life of the Sandplain, pp 146–177. University of Western Australia Press, Nedlands

Farrington P, Greenwood EAN, Bartle GA, Beresford JD and Watson GD (1989) Evaporation from *Banksia* woodland on a groundwater mound. Journal of Hydrology 105: 173–186

Farrington P and Bartle GA (1991) Recharge beneath a *Banksia* woodland and a Pinus pinaster plantation on coastal deep sands in south Western Australia. Forest Ecology and Management 40: 101–118

Gardner CA (1944) The vegetation of Western Australia with special reference to the climate and soils. Journal of the Royal Society of Western Australia 28: 11–87

Gozzard JR and Mouritz MJ (1989) Mineral resources and mining of the Spearwood and Bassendean Dune Systems. Journal of the Royal Society of Western Australia 71: 109–110

Hansen A, Pate JS and Hansen AP (1991) Growth and reproductive performance of a seeder and a resprouter species of Bossiaea as a function of plant age after fire. Annals of Botany 67: 497–509

Hobbs RJ and Atkins L (1990) Fire-related dynamics of a *Banksia* woodland in south-western Western Australia. Australian Journal of Botany 38: 97–110

Hopkins ER (1989) Forestry and *Banksia* woodlands on the Swan Coastal Plain. Journal of the Royal Society of Western Australia 71: 107–108

Jeschke WD and Pate JS (1995) Mineral nutrition and transport in xylem and phloem of *Banksia prionotes* and proteaceous trees of dimorphic root morphology. Journal of Experimental Botany 46: 895–905

Lamont BB (1984) Specialised modes of nutrition. In: Pate JS and Beard JS (eds) Kwongan, Plant Life of the Sandplain, pp 236–245. University of Western Australia Press, Nedlands

Pate JS (1994) The mycorrhizal association: just one of many nutrient acquiring specialisations in natural ecosystems. In: Robson AD, Abbott LK and Malajczuk N (eds) Management of Mycorrhizas in Agriculture, Horticulture and Forestry, pp 1–10. Kluwer Academic Press, Netherlands

Pate JS, Casson NE, Rullo JC and Kuo J (1985) Biology of fire ephemerals of the sandplains of the kwongan of south-western Australia. Australian Journal of Plant Physiology 12: 641–655

Pate JS and Dell B (1984) Economy of mineral nutrients in sandplain species. In: Pate JS and Beard JS (eds) Kwongan, Plant Life of the Sandplain, pp 236–245. University of Western Australia Press, Nedlands

Pate JS, Dixon KW and Orshan G (1984) Growth and life form characteristics of kwongan species. In: Pate JS and Beard JS (eds) Kwongan, Plant Life of the Sandplain, pp 84–100. University of Western Australia Press, Nedlands

Pate JS, Froend RH, Bowen BJ, Hansen A and Kuo J (1990) Seedling growth and storage characteristics of seeder and resprouter species of Mediterranean-type ecosytems of S.W. Australia. Annals of Botany 65: 585–601

Pate JS and Jeschke WD (1993) Mineral uptake and transport in xylem and phloem of the proteaceous tree *Banksia prionotes*. Plant and Soil 155: 273–276

Pate JS and Jeschke WD (1995) Role of stems in transport, storage and circulation of ions and metabolites by the whole plant. In: Gartner B (ed) Stems and Trunks: Their Roles in Plant Form and Function, pp 177–204. Academic Press, New York

Pate JS, Jeschke WD and Aylward MJ (1995) Hydraulic architecture and xylem structure of the dimorphic root systems of South-West Australian species of Proteaceae. Journal of Experimental Botany 46: 907–915

Pate JS, Jeschke WD, Dawson TE, Raphael C, Hartung W and Bowen BJ (1998) Growth and seasonal utilisation of water and nutrients by *Banksia prionotes*. Australian Journal of Botany 46: 511–532

Pate JS, Raisins E, Rullo J and Kuo J (1986) Seed nutrient reserves of Proteaceae with special reference to protein bodies and their inclusions. Annals of Botany 57: 747–770

Pate JS, Stewart GR and Unkovich MJ (1993) ^{15}N natural abundance of plant and soil components of a *Banksia* woodland ecosystem in relation to nitrate utilization, life form, mycorrhizal status and N_2-fixing abilities of component species. Plant, Cell and Environment 16: 365–373

Pate JS and Unkovich MJ (in press) Measuring symbiotic nitrogen fixation – Case studies of natural and agricultural ecosystems in a Western Australian setting. In: Advances in Physiological Plant Ecology, Proceedings of the 1998 British Ecological Society Symposium, York, September 1998

Semeniuk V and Glassford DK (1989) Bassendean and Spearwood Dunes: their geomorphology, stratigraphy and soils as a basis for habitats of *Banksia* woodlands. Journal of the Royal Society of Western Australia 71: 87–88

Specht RL, Yates DJ, Sommerville JEM and Moll EJ (1991) Foliage structure and shoot growth in heathlands in the mediterranean-type climate of southern Australia and South Africa. Ecologia Mediterranea 16: 195–207

Speck NH (1952) Plant Ecology of the Metropolitan Sector of the Swan Coastal Plain. MSc Thesis, The University of Western Australia

Stewart GR, Pate JS and Unkovich MJ (1993) Characteristics of inorganic nitrogen assimilation of plants in fire-prone Mediterranean-type vegetation. Plant, Cell and Environment 16: 351–363

Agroforestry Systems **45**: 343–364, 1999.

Can agricultural management emulate natural ecosystems in recharge control in south eastern Australia?

F. X. DUNIN[1,*], J. WILLIAMS[2], K. VERBURG[2] and B. A. KEATING[3]
[1] *CSIRO Plant Industries, Private Bag P.O. Wembley, 6014, Australia;* [2] *CSIRO Land and Water, G.P.O. Box 1666, Canberra, 2601, Australia;* [3] *CSIRO Tropical Agriculture, 306 Carmody Road, St. Lucia, 4067, Australia (*Author for correspondence: f.dunin@ccmar.csiro.au)*

Key words: annual crops, drainage, evapotranspiration, lucerne, uptake, water balance

Abstract. Water balance information at locations across southern Australia is analysed to identify mechanistic causes for higher water tables under agricultural systems compared to natural vegetation. Contrasting patterns of water use indicate pronounced physiological activity in summer by natural ecosystems ensuring the persistence of the dominant perennial species. A strategy of tempered water use during periods of rainfall excess in winter/spring enables seasonal carryover of soil water to be withdrawn by deep roots for summer functioning. Characteristic patterns of water use by agricultural systems, comprised mainly of determinant annual species, are truncated in time but feature elevated peak rates coinciding with periods of maximum soil water availability during the cooler months. Low seasonal vapour pressure deficit explains the observed benefits of enhanced biomass production due to high water use efficiency. The associated limited scope for soil water uptake carries the penalty of increased frequency of drainage and the subsequent resource degradation currently associated with farmland. Inclusion of a perennial lucerne phase in rotation with crops confers environmental benefit through summer uptake from a rooting depth double that of crops. An effective buffer is thus provided to restrict drainage under an ensuing cropping phase in a manner that more closely emulates natural systems. Simulations of phase farming point to a halving in long term drainage. They suggest adoption over a range of arable land units could cause local retreat of water tables by 0.3 m y^{-1} initially, leading to a new equilibrium level which would still remain elevated with respect to that under natural vegetation.

1. Introduction

Rural production has played a key role in Australia's economic development; however, it has also had a profoundly detrimental impact on land, water and vegetation resources, and on the environment in general (Lovering and Crabb, 1997; Hamblin and Williams, 1995). Australian rural productions systems have been built by drastically changing the nature and seasonal patterns in the hydrological and nutrient cycles of native ecosystems; tropical rainforest was replaced with sugarcane monoculture; semi-arid clay plains became irrigated crop lands; and heathlands on sandplains were converted to wheat and lupin fields. Consequently, the diverse production systems of Australia's rural industries all face a common core of resource and environmental problems which settle about the management of soil and water. The signs of depletion and degradation of our natural resources are becoming ever more obvious.

We have developed Australian agriculture largely by importing plants, animals and production systems from the northern hemisphere, and it is becoming apparent that the cost of enhanced production is being expressed with deleterious edaphic impacts on Australia's distinctive agricultural environments. In view of the measured depletion of many of our natural resources (Williams et al., 1998; Reeves et al., 1997; Commonwealth of Australia, 1996), there must now be questions about the extent to which Australia can continue to be competitive in international market places through sole reliance on our current production systems. It is necessary to determine whether alternative agricultural systems that are more suited to the requirements of the Australian environment might be developed. Taking the Australian environment and its natural ecosystems as a starting point, there may be an opportunity to redesign agricultural systems that emulate the hydrological and nutrient cycles that have evolved in the Australian landscape.

Many of our current production systems are not ecologically sustainable because they leak water and nutrients which, in turn, may result in waterlogging, salinity and soil acidification (Lovering and Crabb, 1997; Hamblin and Williams, 1995; Williams, 1991) This characterisation of the problem raises the potential that if we can develop systems that make full use of available water and nutrients, they may be both more productive and also more ecologically sustainable. These issues of sustainability motivated us to review the function of natural ecosystems with the aim of identifying key attributes, which, if incorporated into land management, will ensure landscape health, even under a management that is predominantly geared to agricultural production.

This review focuses on the water balance with emphasis on control of water outflow, lateral and vertical, exercised by natural and exotic communities. Contrasting the function of natural and introduced agricultural systems provides baseline information for the design of a system of agricultural management that overcomes contemporary degradation without compromising productivity. We provide details of the control imposed by a cereal rotation incorporating lucerne as the ley phase and explore the generality of the experimental findings through a series of simulations of potential recharge to groundwater.

1.1. *Scope of the review*

Temperate Australia spanning a latitude range across 32°–37° S is surveyed for aspects of the water balance that determine those factors that have caused the current hydrologic imbalance, which has led to rising water tables and associated outbreaks of land degradation. We start with available data on natural ecosystems to draw out pertinent features of their function. In particular, we analyse how they use annual rainfall to yield smaller amounts of deep drainage beneath the root zone, which maintains water tables much lower than under current agricultural ecosystems. The lower leaching rates and better use of nitrate will be examined wherever possible. We contrast these features with contemporary agriculture based on annual species in order to elucidate those shifts in the water balance which give rise to elevated water tables.

We then report how the reinstatement of a perennial legume and other perennial options in temporal and spatial configurations into a cereal rotation is efficacious in reducing deep drainage from the root zone to effect lowering of the water table. We seek to design new agricultural ecosystems that have deep drainage and uptake of nitrogen that functionally emulate as closely as possible the natural ecosystem and therefore minimises the disturbance to hydrological and nutrient balances under which the Australian landscape has evolved.

2. Water balance of natural ecosystems

2.1. *Seasonal patterns of water use*

Patterns of evapotranspiration rates obtained from lysimeter observations over several years show time courses that tend to follow inputs of available radiant energy (R) or its water equivalent (R/λ) where λ is latent heat of uptake water (Figures 1 and 2). Summer peaks and winter minima are observed for both a forest and a grassland in the humid environment (average annual rainfall > 750 mm) adjacent to the South Coast of NSW. Significantly, these rates fall short of energy supply throughout each of the two series with the upper bound for the forest indicated as equilibrium evaporation (Eq), being ~0.75 Ep, the Penman estimate of potential evapotranspiration by short green grass. $Eq = (\Delta/(\Delta + \lambda))\ (R\text{-}G)$ with Δ as the slope of the saturation vapour pressure curve with temperature and λ is the psychrometric constant. These lowered rates can be attributed to high canopy resistance to limited vapour transport. In the case of the eucalypt forest with a leaf area index (LAI) > 2.5 for most of the time, intrinsically high leaf resistance would appear to be a major contributor to rate reduction (Dunin and Aston, 1984). The C_4 *Themeda* grassland experiences seasonal variation in LAI between 0.2 (winter) and 1.5 (summer) producing significant canopy control on the process transpiration (Dunin and Reyenga, 1978). Two important hydro-ecological outcomes result from these regimes with lowered rates of evaporative loss. Conserved soil water from winter/spring seasons enables the critical physiological functioning during summer that ensures the persistence of dominant perennial species in both communities. Water redistribution within the root zone is enhanced in periods of low evaporative demand enabling root function to be either maintained or restored for this critical summer uptake. The redistribution leads to a leaching fraction which removes rain-borne salt from the profile. Salt export overcomes potential limitations to plant water uptake. High frequency of this transport poses little threat to the environment when associated with aquifers of high transmission. This regime is common in

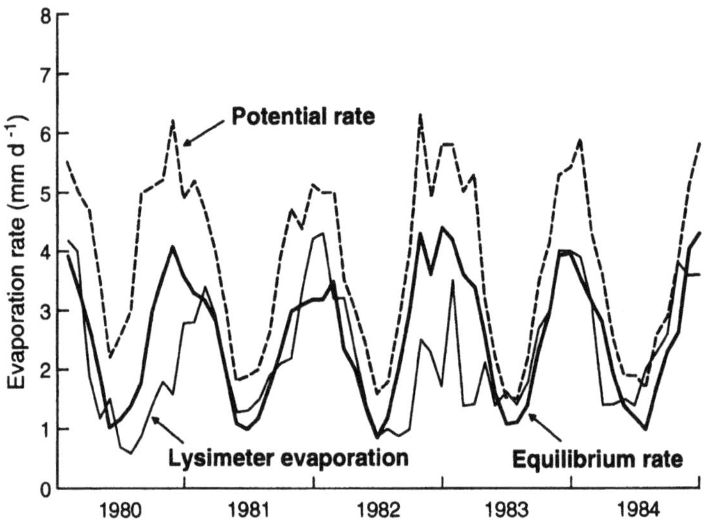

Figure 1. Time series of forest evapotranspiration rates (~ monthly lysimeter averages at Kioloa, New South Wales) indicating the relationships of actual evaporation to averaged reference rates for Penman potential, $\%_p$, and equilibrium, $\%_q$. *Source:* Adapted from Dunin and Aston (1984).

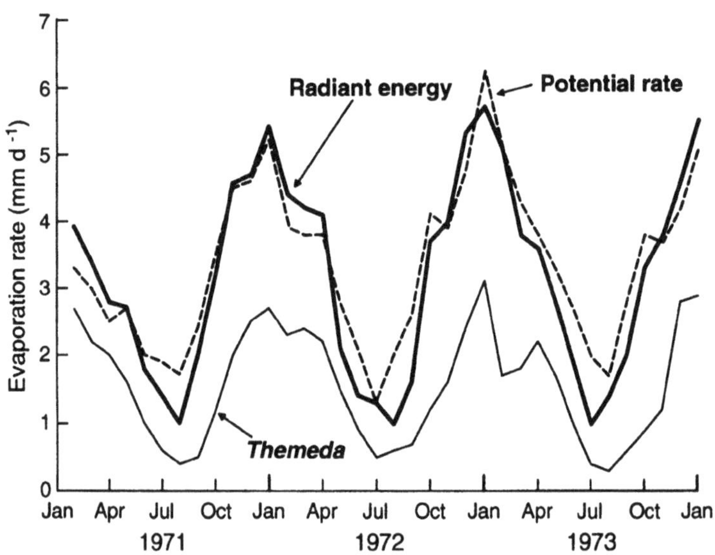

Figure 2. Time series of Themeda grassland evapotranspiration rate (~ monthly averages at Krawarree, New South Wales, Australia) indicating the relationship to averaged reference rates for Penman potential evaporation, $\%_p$, and the water equivalent of available radiant energy, $2/\lambda$. *Source:* Adapted from Dunin and Reyenga (1978).

forested catchments under high rainfall. Similarly, sandy profiles are purged of salt, over a wide rainfall range. Infrequent leaching associated with groundwater systems of low transmission are characterised by salt build-up in the regolith. Perturbing these systems and causing increased drainage frequency produces more salt in transition leading to land salinisation and diminished quality of streamflow.

2.2. *Diurnal responses of woodland communities in summer*

A Eucalypt woodland displays diurnal progressions of latent heat flux that generally lie well below inputs of available radiant energy for summer conditions both immediately after rainfall and also after prolonged drying (Figure 3). The departure from energy input is consistent with the low LAI of 1.5 and 0.3 for the dry sclerophyll forest and mallee woodland respectively.

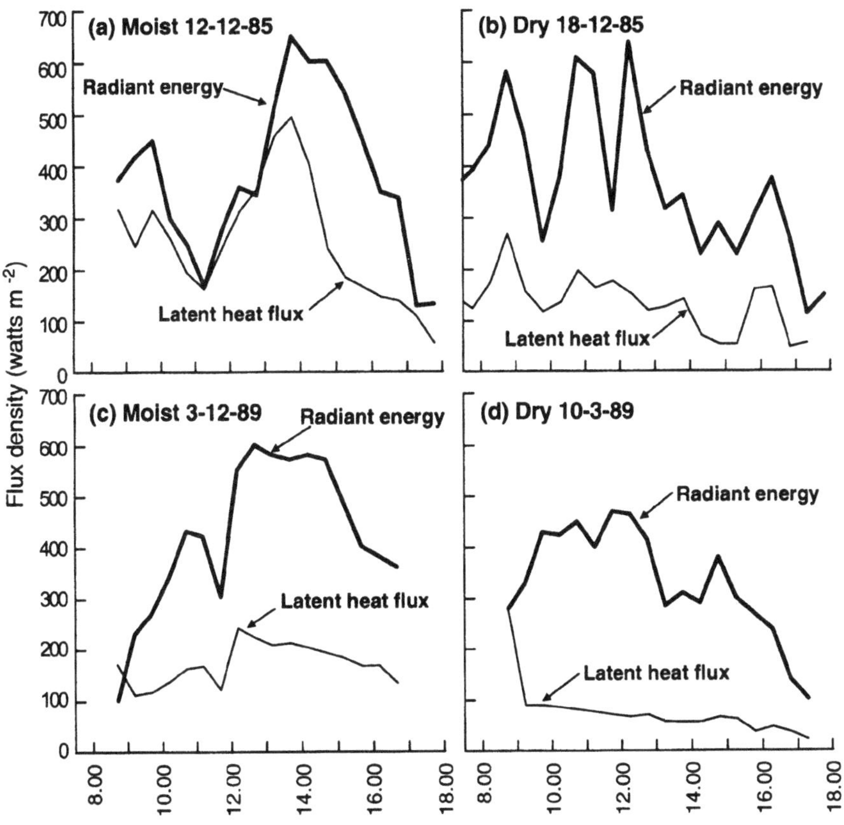

Figure 3. Diurnal responses of latent heat flux in relation to available radiant energy for a eucalypt woodland community at Mogo, New South Wales, Australia [(a) and (b)] compared to those for a mallee community at Hinks, South Australia [(c) and (d)] under moist and dry summer conditions respectively.

It is typical of climax ecosystems subject to low annual rainfall and erratic summer soil water supply. Morning conformity of latent heat loss with available radiant energy reflected substantial soil evaporation under the sparse canopies as the result of surface soil wetting from the previous day's rainfall (Figure 3a,c). During the afternoon, latent heat loss had retreated to < 40% of the radiant energy load, indicative of transpiration dominance with limited LAI following drying of the soil surface. On a daytime basis, evapotranspiration was 40% or less than the water equivalent of the radiant energy input and could be rated as ~0.5 *Eq* or < 3 mm d^{-1}. Peak daily values of the evapotranspiration regime by these sparse woodland communities may be adjudged as the equivalent of Ep that would occur during early spring. For the more usual summer condition with depleted soil water reserves (Figure 3b,d), evaporation rates are low at 1–2 mm d^{-1}. The transient morning agreement between λE and R-G in Figure 3d is explained as the evaporation of dew. Sustaining these rates for extended seasonal conditions of summer/autumn without rain can represent the removal of the rainfall excesses that occur during the previous winter/spring and which accrue deep in the profile. This implies that these woodland species operate to store excess winter/spring rainfall within a deep root zone to be used at sparing rates over summer and autumn.

These strategies of deep emplacement of soil water under sparse woodland have a variable recharge response depending on climate and soils. The dry sclerophyll woodland at Mogo with 800 mm y^{-1} rainfall on a coarse deep sand returns ~25% of rainfall to discharge of the Shoalhaven River via a relatively shallow aquifer of high transmission and with low storage capacity. In contrast, mallee woodlands occupy land with 300–400 mm y^{-1} average rainfall and characteristically release < 1% of rainfall for recharge to groundwater of low transmission but high storage capacity. Estimates of deep drainage from woodland in the Murray Darling Basin appear to be in accord with the range of response reflected in these examples. At rainfall values of 1000 mm y^{-1}, 130 mm of deep drainage has been estimated while for areas with less than 600 mm y^{-1}, drainage is rated as < 5 mm (Allison et al., 1990; Dawes et al., 1997; Hume, 1997; Williamson, 1998).

3. Water use by agricultural communities

Annual patterns for a wheat crop and a first year lucerne pasture show evapotranspiration rates (averaged weekly), starting at sowing following the break of season and progressing through the growing season and the off-season period till the next break of season (Figure 4). Peak daily rates of evapotranspiration were as high as 6 mm d^{-1} in both these time series but these occurred in mid-spring as opposed to natural ecosystems whose peak rates occur during summer.

While growing season total evapotranspiration (Table 1) (Smith et al., 1998;

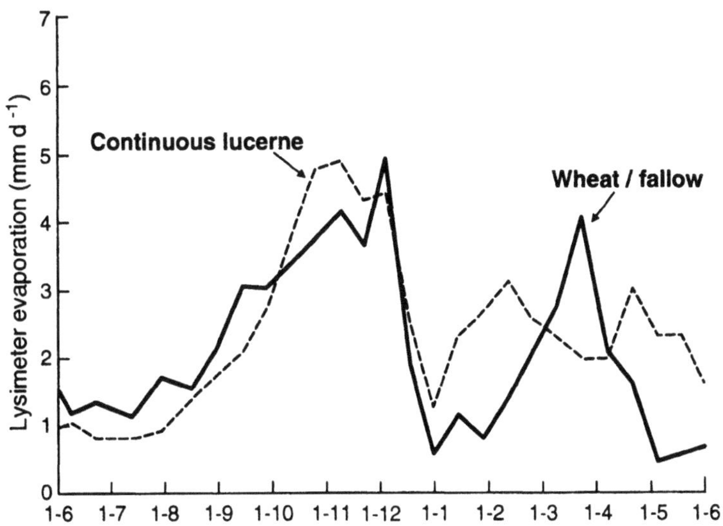

Figure 4. Annual patterns of evapotranspiration at Wagga Wagga, New South Wales, Australia (two to four weekly averages) for wheat fallow rotation 1993/94 compared with first year lucerne 1995/96. *Source:* Adapted from Smith et al. (1998).

Condon and Richards, 1993; Poss et al., 1995) can be similar between annual and perennial agricultural communities, continuing uptake from subsoil by the perennial community in summer can result in a seasonal total of evapotranspiration being in excess of seasonal rainfall by 100 mm (cf. Figure 4). Seasonal evapotranspiration totals for the short fallow phase or for sparse pasture are approximated by the off-season rainfall.

Total evapotranspiration for the growing season of the crop appears to be limited to about 400 mm and growth-enhancing agronomic treatments produce only minor perturbations to these totals. In the progression through the growing season, the upper bound of evaporation rate is equivalent to *Ep*. During the early stages of crop growth with low energy inputs and periods of wet surface soil, conformity to *Ep* is frequent, irrespective of LAI, indicating that transpiration is complemented by soil evaporation to attain the potential rate. Differences in plant density later in the season can cause departures from *Ep* that can produce minor treatment effects on the cumulative evapotranspiration for growing season.

4. Some concurrent hydrologic comparisons between native forest and exotic perennial pasture

A comparison of natural forest with an exotic perennial pasture (Demeter fescue) revealed rankings in evapotranspiration rate which changed with time

Table 1. Seasonal water use and productivity for agricultural communities at Wagga Wagga, New South Wales, Australia. *P* is rainfall, E_T is evapotranspiration, *D* (0.9) is drainage at 0.9 m, *D*(ex) is drainage from the root zone, *–D* is soil water uptake from depth > 0.9m, *M* is biomass production, ( ) is grain yield, No is Nil N supplement prior to tillering, and N140 is 140 kg N ha^{-1} prior to tillering.

Interval	Treatment	P (mm)	ET (mm)	D (0.9) (mm)	D (ex) (mm)	M (t ha^{-1})
Growing season						
18/5/89–20/11/89	Wheat cv Matong	321	402	22	22	15.1 (5.9)
18/5/89–20/11/89	Wheat cv Quarrion	321	381	22	22	14.1 (5.5)
26/5/89-5/12/89	Wheat cv Matong	329	323	0	0	12.8 (4.8)
6/6/93–15/12/93	Wheat cv Matong (No)	402	414	97	97	10.4 (3.4)
6/6/93–15/12/93	Wheat cv Matong (N140)	402	379	89	89	13.6 (6.2)
1/6/94–30/11/94	1st year lucerne	244	318	0	0	~1
8/5/95–25/11/95	1st year lucerne	565	401	112	112	~4
16/5/96–15/11/96	2nd year lucerne	327	301	18	0	~2.5
Off season						
15/12/93–1/6/94	Fallow	226	211	0	0	0
30/11/94–8/5/94	Sparse lucerne	163	179	0	0	< 1.0
26/11/95–16/5/96	1st year lucerne	196	289	–73	0	1.0
15/11/96–8/5/97	2nd year lucerne	189	221	–15	0	1.0

scale of averaging (Figure 5). The perennial pasture is capable of higher rates than the forest in the short term but cannot sustain these rates due to lesser soil water availability as the shallower root zone dries. Monthly rates from forest are systematically greater than the perennial pasture. This explains how forests can use a greater fraction of annual rainfall than corresponding agricultural communities in the same environment despite a lower capacity for short-term water use.

Differing patterns of daily evapotranspiration are apparent in the comparison of two perennial pastures, one a natural grassland dominated by Kangaroo grass (*Themeda* sp.), the other by the exotic *Festuca arundinacea* var. Demeter, over a two month drying phase in spring/summer at Krawaree (Figure 6a) (Dunin, 1991). Early in the progression, the greater rates of evapotranspiration by the exotic pasture are approximated by *Ep* but midway in the progression, the greater rates by the natural grassland tend towards *Eq*. Towards the end of the progression, rates converge as soil water becomes limiting in both cases (Figure 6b). Overall, evapotranspiration totals are comparable for the study period and are consistent with drying patterns, which tend towards convergence at the wilting point of the soil. Phasic differences in short term evaporation response are in part attributable to phenology, but these can become masked in the longer term due to effects of soil water availability causing parity in seasonal total evapotranspiration. In the case of

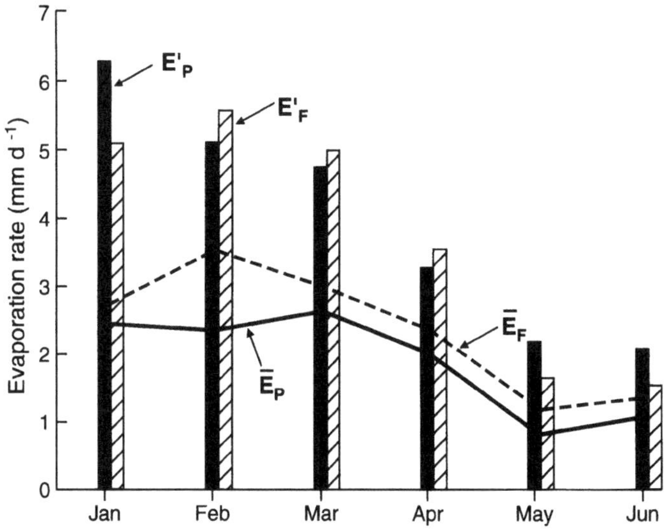

Figure 5. Comparison of evapotranspiration rates by forest and fertilised perennial pasture at adjoining sites in the coastal plain of southern NSW. Monthly averaged rates are consistently greater from forest (E_F) than pasture (E_P). Peak daily values in each month rate different, with pasture values (E'_P) being either comparable to or greater than forest values (E'_P). *Source:* Adapted from Dunin et al. (1985).

Themeda grassland, deferred water use is achieved as a consequence of C_4 pathway metabolism compared with the C_3 fescue. This delays canopy development until the required temperature thresholds are exceeded.

For a paired catchment experiment in a Mediterranean type environment at Bacchus Marsh, Victoria, (Dunin and Downes, 1962) reduction in annual volumes of runoff resulted from the conversion of a natural grassland dominated by *Themeda* to annual fertilised pasture (Table 2).

A 10% difference in the upper bound of evapotranspiration between the pasture communities accounted for changed runoff rather than any cover effect on infiltration characteristics of the shallow duplex soil. These differences in evapotranspiration had their impact during winter/spring seasons when > 75% of the annual discharge occurred. By contrast, peak flows and erosion hazard were increased following the introduction of the annual pasture; major peak discharge events were inevitably associated with summer thunderstorms with only small contributions to annual discharge. Summer infiltration under natural pasture was enhanced due to deferred water use by *Themeda* creating larger soil water deficits in the subsoil than occurred under the introduced annual pasture (Figure 7; Dunin, 1976).

This subsoil deficit could be maintained well into the growing season to serve as buffer against deep drainage. Thus, the natural grassland with lowered evaporative rate in the growing season tended to partition winter excesses into surface runoff whilst the introduced shallower rooted exotic disposed of

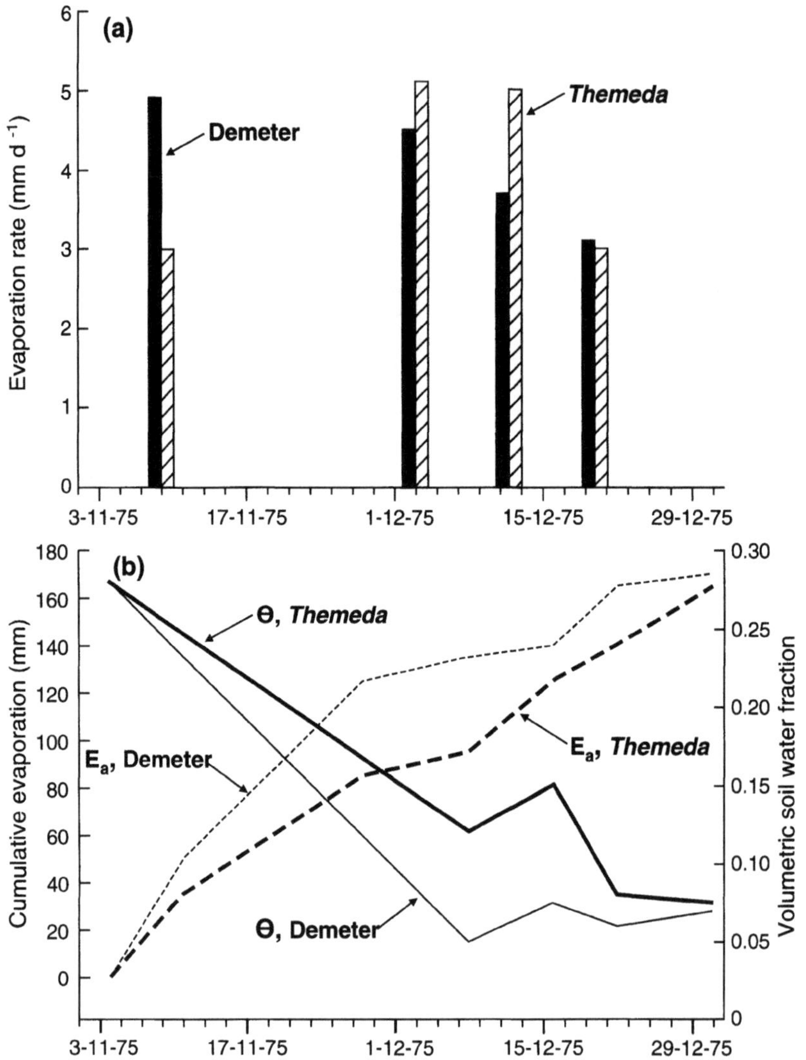

Figure 6. Two monthly comparison of natural grassland (*Themeda* sp.) and fertilised perennial pasture (Demeter fescue) at Krawarree, NSW (a) change in ranking between pastures of daily rates of evapotranspiration and (b) convergence in cumulative evapotranspiration, E_a, and soil water content θ between pastures. *Source:* Adapted from Dunin (1991).

excesses as deep drainage. This difference in water disposition between communities explains how salinity outbreaks have been increasing with the conversion of natural grassland to introduced annual pasture.

Table 2. Four year comparison (1961–1964) of hydrologic response of adjoining 2 ha experimental catchments. S is the sorptivity at $\theta_i = 0.20$, E_a is the actual evapotranspiration at Bacchus Marsh, Victoria, and E_o is the evaporation from an Australia sunken pan evaporimeter.

	Improved pasture	Natural grassland
Average annual yield of runoff (mm)	3	26
Peak flow (cumec X10^{-3})	47	11
Infiltration characteristic S (cm $min^{-0.5}$)	0.24	0.21
Evaporation characteristic E_a/E_o	1.07	0.98

Source: Dunin (1976).

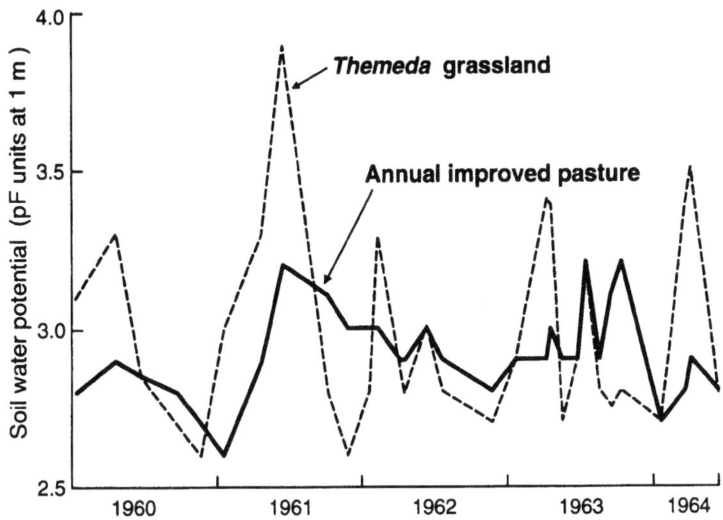

Figure 7. Annual patterns of catchment soil water status (in pF units) comparing *Themeda* with annual introduced pasture at Bacchus Marsh, Victoria. *Source:* Adapted from Dunin (1976).

5. Can a deep rooted perennial fodder plant emulate the function of native woodland to mitigate water table rise?

5.1. *Some experimental observations*

The impact of including lucerne in a cereal rotation on the patterns of plant water uptake are shown in Figure 8 for an experimental site near Wagga Wagga (Smith et al., 1998). The data are based on evaporation records from undisturbed lysimeters (1993–1996) and regular core measurements of soil water content. The season 1994/95 is not included, because a severe drought caused lucerne failure. Lucerne was re-established in 1995.

The cropping phase of 1993/94 was characterised by approximately 400 mm of evapotranspiration, which comprised of 200 mm soil evaporation and

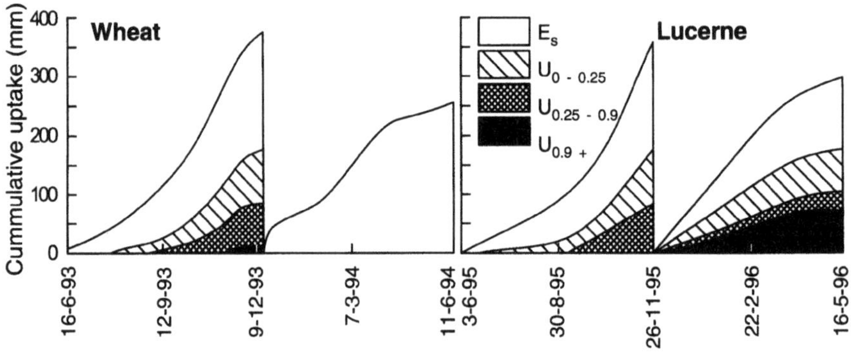

Figure 8. Patterns of water uptake from soil depth intervals (Un) and soil evaporation (Es) during a crop rotation involving lucerne as the pasture phase. The growing season for the wheat crop (1993) and first year lucerne (1995) is distinguished from the off-season phase as fallow after crop (1993/94) and continuing lucerne growth in summer (1995/96). *Source:* Adapted from Smith et al. (1998).

a similar quantity of plant water uptake. About 100 mm of plant water uptake came from the 0–0.25 m layer and another 100 mm from the 0.25–0.90 m layer. Additional uptake beyond 0.90 m was only 10 mm. During the fallow off-season, the seasonal rainfall of 226 mm was effectively disposed as soil evaporation with only a measured 9 mm conserved for potential plant uptake in the next phase.

First year lucerne for the 1995/96 season performed in a similar manner to the wheat crop of 1993/94. During the nominal growing season, which was based on the district experience for sowing and harvest in the year, the total seasonal evapotranspiration of 400 mm as well as the partition between soil evaporation and uptake from different depths were comparable in magnitude to that of the wheat crop. For the lucerne off-season (26 November–16 May), total evapotranspiration (289 mm) exceeded seasonal rainfall (196 mm) by 93 mm. The additional uptake was accounted for as 20 mm from the 0.25–0.90 m soil layer and 73 mm from 0.90–1.80 m. The dry buffer of 73 mm, created below the nominal crop rooting depth of 0.90 m in this soil, provides storage opportunity to divert deep drainage (in subsequent cropping years) from groundwater recharge.

At this Wagga Wagga site, the dry buffer may be able to store three years of deep drainage from the cropping phase before recharge is recommenced. This claim is based on an average value of deep drainage of 25 mm y^{-1}, derived from an 11 year record from the lysimeters. Control of deep drainage by 25 mm y^{-1} could cause the local water table to retreat by 0.5 m y^{-1}. Simulation provides the means to generalise from this simplistic calculation and explore a range of weather conditions to compare various sequences of crop with annual and perennial pasture.

5.2. *Simulation study of the water balance under three agricultural management systems*

A preliminary analysis of such a simulation study is presented here. Using a comprehensive farming system simulation model and historical weather data, the impacts of incorporating lucerne in a wheat rotation are examined. While it is difficult to formally validate such complex models that include many different processes, the model has been shown to simulate the key features of the water balance of lysimeter studies (Smith et al., 1998) in the region of application (K Verburg, unpubl. data). The analysis illustrates how simulation models can be used to study alternative management practices, in this case the introduction of lucerne, and their ability to emulate the water balance of the native ecosystem. The inclusion of long-term weather variability allows the assessment of the likely success or otherwise of various management practices in controlling deep drainage and water table rise. Three agricultural management systems were simulated in this study:

1. *Wheat only.* Wheat was sown upon sufficient opening rain (25 mm in a 10 day period) in a window from 1 May until 30 August. It was harvested in December, 14 days after maturity. Trash was retained and incorporated at planting. 17 kg N ha^{-1} of ammonium fertiliser was applied at sowing, with another 50 kg N ha^{-1} of urea applied 80 days later.
2. *Lucerne fodder crop only.* Lucerne was grown for six years, then re-sown in the same sowing window as for wheat.
3. *Wheat/Lucerne rotation.* A rotation consisting of three years wheat, followed by 2.5 years lucerne, and half a year of fallow was chosen. Wheat management was as for the continuous wheat scenario.

The simulations were conducted using the APSIM (Agricultural Production Systems Simulator) framework (McCown et al., 1996; Hook, 1997). This model is structured to deal with crops and major soil processes in separate modules. Here we used a wheat module (NWHEAT, Probert et al., 1998c), a lucerne module (Probert et al., 1998b), a water balance module (SWIM, Verburg et al., 1996), a surface residue module and a soil nitrogen module (Probert et al., 1998a). The soil properties used were those of a Red Kandosol (Isbell, 1996) reported by Bond et al. (1998) for their Wagga Wagga site. This soil is better drained than that in the above field experiment of Smith et al. (1998). In the simulations, roots of wheat (var. Janz) were allowed to penetrate to a maximum depth of 2 m, whilst those of lucerne penetrated to a depth of 3 m. These rooting depths are only approximate and the predicted deep drainage term of the water balance will be sensitive to this information. Results presented here should, therefore, only be interpreted in a relative sense. Instead of grazing, the lucerne was cut frequently. The impact of two cutting regimes on lucerne were examined and shown to have only a relatively small impact on the water balance. The simulations used historical weather data (1972–1996) consisting of measured daily rainfall, and maximum and minimum temperature data from the Wagga Wagga Meteorological Office, as well as estimated radiation data (Meinke et al., 1995). The results of the simulations are set out in Figures 9 and 10, and summarised in Table 3.

The perennial deep rooted lucerne grown as a fodder crop dries the profile most, continuous wheat least, and the results for the rotation are intermediate (Figure 9a). The biomass production of wheat crops which follow the lucerne phase depends on the relative importance of water and nitrogen stress in that year. Thus there is significant variation (Figure 9b). In general, wheat crops grown in rotation performed better than those in the continuous cropping scenario, an effect believed to be related to improved N supply in the lucerne rotation system.

Figure 10 illustrates the large annual variation in deep drainage in all three systems. The continuous wheat cropping system generally resulted in significant amounts of drainage, with a long term average of 185 mm y^{-1} at 4 m depth (Table 3). The lucerne fodder crop was effective in reducing the deep drainage by a factor of 7–8. As expected the lucerne/wheat rotation fell between these two and was able to reduce by half the deep drainage under wheat.

The absolute quantities of drainage predicted in these simulations are high, relative to the rates generally acknowledged and the rates associated with the lysimeter study at Wagga Wagga (Smith et al., 1998). We have investigated the reasons for this, and can demonstrate how the magnitude of the water balance terms, and in particular ET and drainage, are site specific and a

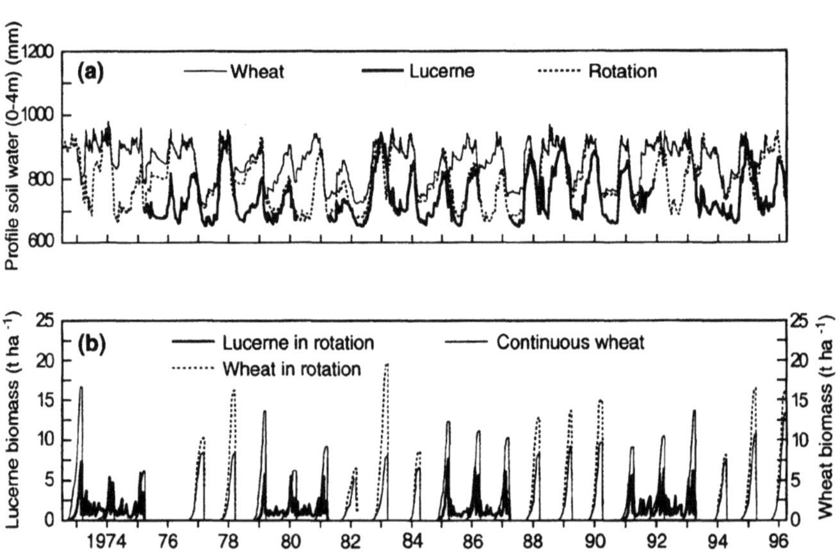

Figure 9. Comparison of (a) profile soil water (0–400 cm depth) and (b) biomass yields simulated for the scenarios of continuous wheat, lucerne fodder crop, a three year lucerne/wheat rotation in a Red Kandosol at Wagga Wagga, New South Wales, Australia (1973–1996).

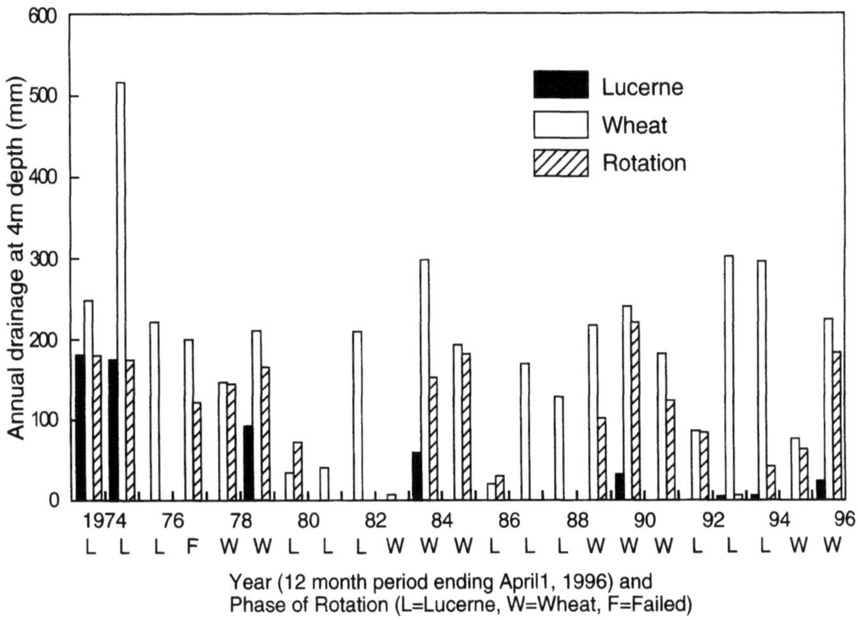

Figure 10. Comparison of simulated annual deep drainage (mm) at 4 m in a Red Kandosol at Wagga Wagga, New South Wales, Australia (1973–1996) for the scenarios of continuous wheat, lucerne fodder crop, and a three year lucerne/wheat rotation.

Table 3. Comparison of simulated average annual water balances in a Red Kandosol at Wagga Wagga, New South Wales, Australia (1973–1996) for the scenarios of continuous wheat, lucerne fodder crop, and a three year lucerne/wheat rotation.

System	Rain (mm)	Runoff (mm)	ET (mm)	Drainage at 4 m (mm)	Drainage at 1 m (mm)
Wheat	611	15	411	185	223
Rotate	611	15	507	89	181
Lucerne	611	15	579	25	134

function of soil type, average rainfall, vegetation type and growth patterns and rooting depth. Both soil properties and root system depth strongly affect the relative effectiveness of a lucerne phase in reducing deep drainage. To illustrate the soil type effect, we have shown that the 185 mm drainage term for a continuous wheat system on the highly permeable soil examined here, is reduced to 84 mm on a less permeable soil of the type similar to the lysimeter study by Smith et al. (1998). The equivalent shifts in the drainage estimates were from 89 mm to 15 mm for the phase farming system and from 25 mm to 2 mm for the continuous lucerne fodder crop system.

Data on deep drainage under natural ecosystems in Australia are scarce.

For cypress pine/eucalypt woodland in southern Australia deep drainage is estimated to be 5–15 mm y^{-1} (Dawes et al., 1995; Salama and Hatton, 1998, Allison et al., 1990). Depending on climate and soil type, this is comparable with the estimates for the lucerne fodder crop, while predicted drainage under the lucerne/wheat rotation is considerably higher (Table 3).

In addition to the estimates of annual drainage amounts, the simulations also portray the creation of a dry subsoil buffer during the lucerne phase (Figure 11a,b). This buffer is predicted to be gradually refilled during the subsequent wheat phase, as shown by the water content profiles at the end of next three wheat crops (Figure 11c–e). Water extraction from the top soil is primarily a function of crop performance, for example with the 1983 rotation wheat crop yielding higher and extracting more water than the crop in the continuous wheat scenario (compare Figures 9, 11d).

The time required for creation of the dry buffer under lucerne as well as the rate of refilling of the profile will also vary with soil type and rooting depths. To evaluate the benefits and penalties of a lucerne phase for a par-

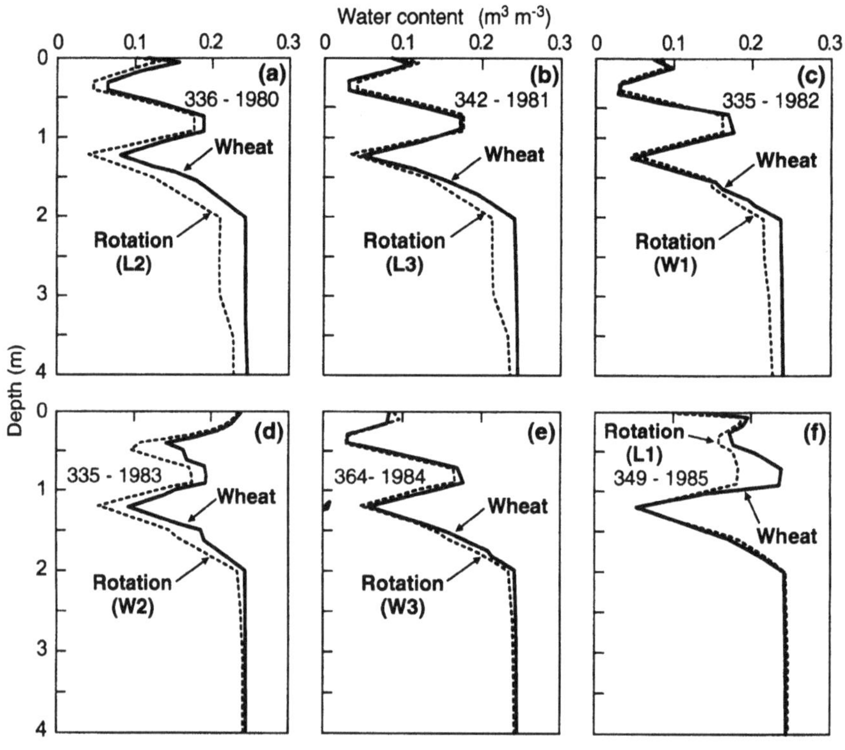

Figure 11. Simulated water content profiles at the end of six consecutive cropping seasons (mid to late December) for the continuous wheat and lucerne/wheat rotation scenarios. Stage of the rotation is indicated by L1, L2, L3 (lucerne) and W1, W2, W3 (wheat). Figures on the graphs indicate day of year and year.

ticular soil type and location, site specific parameterisation is, therefore, required. The simulations nevertheless serve to illustrate that emulating the deep drainage under native ecosystems with currently available deep rooted perennials such as lucerne is difficult when it is part of a rotation with an annual cereal like wheat. It appears from our simulations that, at least on some soils, the storage buffer provided by lucerne may not be able to store the drainage from more than two or three crops of wheat. The deep rooted perennial must be able to use water deep in the profile during summer so that there is sufficient buffer in which to store the winter rainfall and thus minimise deep drainage. In fact in some soils and seasons, this buffer could be filled by one wet year in the cereal phase of the rotation. Comparisons of the cropping systems simulations with the limited existing data and estimates for deep drainage in natural ecosystems are made difficult by the site and season specificity of these data. The strong dependence of the simulation results on soil type, rainfall record and rooting depth suggests that more extensive simulation studies coupled with comparative experimental studies are needed before the deep drainage terms in these phase farming systems can be fully assessed.

6. Agricultural management to emulate the water balance of native ecosystems

6.1. *Phase differences in patterns of water uptake*

The distinctive feature of natural ecosystems in this survey within temperate south eastern Australia is the attainment of peak values of water use in summer irrespective of the rainfall amount and its distribution. This has been suggested as a strategy that permits biological functions that enable the dominant perennial species in these ecosystems to persist. Water use in summer with limited soil water does not reap large returns in terms of biomass production because transpiration efficiency is reduced due to the seasonally high values of vapour pressure deficit in the atmosphere. Thus, we note the principle that seasonal phase shifts in physiological functioning have important consequences for persistence at the expense of productivity, an observation made by Fischer and Turner (1978).

The carryover of soil water from its storage in periods of low evaporative demand to its use in summer is important insurance for critical functioning and survival of natural ecosystems. Maintaining an adequate soil water store over several months not only facilitates critical carbon gain at temperatures optimal for metabolism but also enables root function for deep uptake by maintaining root turgor. It seems a paradox in the current debate about degradation caused by deep drainage that the drainage flux profile under natural ecosystems will be greater than under agricultural systems over the depth range that corresponds to the ultimate rooting depth of these annuals. Below this depth, drainage rates become attenuated under natural ecosystems due to root water uptake extended into this zone. Drainage export from the root zone will tend to be greater from the agricultural system than from the native system. However, it does not necessarily follow that an extended root zone prevents drainage export. Some of the factors that contribute to water conservation strategies are of interest in determining the partition of deep soil water between its uptake and its export to provide a leaching fraction.

Strategies of water conservation for summer uptake by natural ecosystems are varied but are worthy of note for the design of agricultural systems that can avoid the accelerated outflow of water and nutrients that has led to degradation. Conservative rates of evapotranspiration in the cooler, wetter months are apparent in all systems that have been measured and reasons for conservatism have been outlined in terms of stomatal control. Equally important for conservation is the attenuation of direct evaporation from soil. Noteworthy in this respect is the loss of 380 mm of soil water by direct evaporation in a wheat/fallow rotation over a 12 month period (Figure 8). This loss is equivalent to 60% of the rainfall over this period. Maintaining litter on the soil surface can reduce this fraction substantially (Denmead et al., 1997), as is apparent in the evaporation regime of *Themeda*. With a dead litter component of 5 tonne ha^{-1} in the sward (Groves et al., 1973), some 50–80 mm may have been saved from soil evaporation between June and December.

Redistribution effects on infiltration are notable in some woodlands. It protects soil water from significant losses by direct evaporation as well as providing deep emplacement of soil water to maintain root turgor. Under the multi-stem mallee canopy, infiltration of rainfall is concentrated by stemflow around the tree butt. This confines the wetting front within a wetted cylinder causing deeper penetration than would be attained with a really uniform infiltration (Allison and Hughes, 1983). Runoff generated over the intervening space between the widely distributed trees augments the concentrated infiltration process under a mallee canopy (Nulsen et al., 1986). Even more intriguing is the recent discovery of hydraulic redistribution of soil water via root tissue in both directions to maintain turgor within the root zone (Burgess et al., 1998). 'Hydraulic lift' is now a widely accepted phenomenon whereby deep soil water is uplifted along the translocation pathway during summer stress to be diverted into shallow roots to maintain turgor and function (Caldwell and Richards, 1989). The reverse transfer, immediately after 'break-of-season', can be viewed as insurance for deep roots against prolonged desiccation in seasons where the seasonal wetting front fails to reach their depth.

6.2. *Ecological stability and resource management*

Stable forest ecosystems of low to moderate biomass productivity are important for urban water supply. These issue clean, perennial streamflow that arises from frequent drainage episodes in these higher rainfall regions. Modifications to these forest ecosystems have compromised their value for water harvest, in terms of quantity and quality. Timber logging has caused point sources of sediment to cause periodic deterioration in water quality. Clearing and fire have had adverse effects on both quantity and quality of streamflow. Forest clearing for permanent pasture has induced reduction in subsoil hydraulic conductivity following the loss of tree roots (Talsma and Hallam, 1979). This has changed the streamflow regime and caused a greater proportion of overland flow. Higher flood peaks in streamflow discharge pose threats to water quality and a changed flow distribution over time causes problems in reservoir management in meeting community demands for an assured yield of water.

The question of sustaining water harvest from a recharge basin and the ecological stability of the resident climax plant community is recurrent globally. Recently, this is being addressed at a number of locations in Australia. On the coastal plain north of Perth, groundwater from the Gnangara Mound has been drawn into the reticulation network for municipal water supply. Currently, there are indications of dieback in the Banksia woodland on deep sand overlying the groundwater lens. This suggests that the rate of groundwater withdrawal is impacting on the health of the woodland; the implication is that conservative winter activity by the woodland permits seasonal recharge to groundwater but there is a requirement for summer uptake from the capillary fringe to ensure community survival during the heat stress of a hot dry summer. Apart from a diminished conservation value following extensive dieback, a likely shift in the water balance that could raise the water table may well prove to be deleterious to the water resource. Alienating *Banskia* woodland for agriculture nearby at Moora has been accompanied by accelerated nitrate leaching that has caused acidification of the deep sand, together with serious waterlogging and eutrophication of groundwater.

The case of kangaroo grass (*Themeda* spp.) warrants special mention. Despite its tropical origins, it was ubiquitous in south eastern Australia both in association with trees in woodland and as a grassland; the Western District of Victoria was devoid of trees across the basalt plain and dominated by this C_4 plant. Only following its demise due to excessive grazing pressure was there any sign of secondary salinisation in the plain. Clearly deep drainage was not a major issue in the natural state. Summer activity by the grassland created a sufficient subsoil buffer against excessive deep drainage in the cooler months when the C_4 species was quiescent.

6.3. *Reclamation involving perennial species*

The plea for re-instating perennials in agricultural land has been longstanding (Downes, 1959). This aimed at averting the degradation that has been an outcome of a shift in the water balance. The foregoing example where lucerne was used in rotation with wheat, indicated a reduction in the local water table. This was achieved by extending the period of plant water uptake in summer in a way that parallels the prior activity under natural communities. However, where growing season water use is greater than natural systems, this can place the deep roots at risk of desiccation for extended periods. This raises the issue of persistence and explains how dryland lucerne loses vigour in the long term with a progressive decline in plant numbers. The strategy of using lucerne in rotation with crops calls for limited persistence to redress recharge and its adoption can be viewed as a system with promise for sustainability of arable land.

Woody perennials offer a progressive 'mining' of deep soil water over time to generate buffer sizes that may exceed lucerne by as much as 100 mm at a specific site. However, if this progression is rapid and is accompanied by growing season rates of *Et* at *Ep*, then the community is at risk of summer desiccation early in the progression. The case of blue gum plantations in south western Australia is worthy of note in this respect. Loss of growth rate mid-term in the rotation and ultimate death of the plantation is common, raising questions of sustainability and management options to enable drought tolerance. These issues can be traced directly to the need for carryover of soil water to maintain root infrastructure.

These discussion points form the basis of our conclusion that we might emulate natural ecosystems in their function of summer water uptake but there is need for caution in how this might be achieved. Too much water use too quickly can jeopardise persistence. This matters little in the case of the lucerne rotation but may prove critical in tree plantations where mid-term or second rotation productivity is compromised and even death of trees can follow.

7. Acknowledgements

One of us (FXD) has benefited greatly in this foray into comparative physical ecology from two critical associations. The late RG Downes had a profound influence in directing study towards issues of sustainability. Mr W Reyenga was indispensable in micrometeorological campaigns that gained the insights into physiological function. The financial support and encouragement of Dr Phil Price of the Land and Water Resources Research and Development Corporation through the program 'Redesign of Australian Agricultural Plant Production Systems' has been central to the development of concepts explored in this paper. We wish to acknowledge the helpful discussion and access to data provided by Drs Chris Smith and Warren Bond of CSIRO Land and Water, whose work is also supported by Grains Research and Development Corporation, and the helpful comments on the APSIM simulations by Mr Neil Huth and Dr Mervyn Probert. Long term weather data for the APSIM runs were supplied by the Queensland Department of Natural Resources and the Bureau of Meteorology. The assistance of Mr Tony Tatarow of CSIRO Land and Water in the preparation of the figures is gratefully acknowledged.

8. References

Allison GB and Hughes MW (1983) The use of natural traces as indicators of soil-water movement in a temperature semi-arid region. Journal of Hydrology 60: 157–173

Allison GB, Cook PG, Barnett SR, Walker GR, Jolly ID and Hughes MW (1990) Land clearance and river salinisation in the western Murray Basin, Australia. Journal of Hydrology 119: 1–20

Bond WJ, Smith CJ and Ross PJ (1998) Field validation of a water and solute transport model for the unsaturated zone. In: Dillon P and Simmers I (eds) International Contributions to Hydrogeology. 18. Shallow groundwater systems, pp 63–73. A.A Balkema Publishers, Rotterdam

Burgess SSO, Adams MA, Turner NC and Ong CK (1998) The distribution of soil water by tree root systems. Oecologia 115: 306–311

Caldwell MM and Richards JH (1989) Hydraulic lift water efflux from upper roots improves effectiveness of water uptake by deep roots. Oecologia 79: 1–5

Commonwealth of Australia (1996) State of the Environment, Australia 1996. CSIRO Publishing, Melbourne

Condon AG and Richards RA (1993) Challenges ahead in using carbon isotope discrimination in plant-breeding programs. In: Ehleringer JR, Hall AE and Farquhar GD (eds) Stable Isotopes and Plant Carbon-Water Relations, pp 451–462. Academic Press

Dawes WR, Hatton TJ and Zhang L (1995) Linking land use with regional scale flownet modelling: Applications of the WAVES Model to the Loddon-Campaspe. Proceedings of the Murray-Darling Workshop, Wagga Wagga, 23–29 September 1995

Denmead OT, Mayocchi CL and Dunin FX (1997) Does green cane harvesting conserve soil water? Proceedings of Australian Society of Sugar Cane Technologists 19: 139–146

Downes RG (1959) The ecology and prevention of soil erosion. In: Biogeography and Ecology in Australia, pp 472–486. Junk, den Haag, The Netherlands

Dunin FX and Downes RG (1962) The effect of subterranean clover and Wimmera ryegrass in controlling surface runoff from four-acre catchments near Bacchus Marsh, Victoria. Australian Journal of Experimental Agriculture and Animal Husbandry 2: 148–152

Dunin FX (1976) Changes in the water balance with land modification in southern Australia. In: Heady HF, Falkenborg DH and Riley JP (eds) Watershed Management on Range and Forest Lands, pp 157–163. Proceedings 5th Workshop US-Australia Rangelands Panel, Boise, Idaho 1975

Dunin FX and Reyenga W (1978) Evaporation from a themeda grassland. 1. Controls imposed on the process in a sub-humid environment. Journal of Applied Ecology 15: 317–325

Dunin FX and Aston AR (1984) The development and proving of models of large scale evapotranspiration: an Australian study. Agricultural Water Management 8: 305–323

Dunin FX, McIlroy IC and O'Loughlin EM (1985) A lysimeter characterisation of evaporation by eucalypt forest and its representativeness for the local environment. The Forest Atmosphere Interaction, Proceedings Forest Environmental Measurements Conference, Oak Ridge, Tennessee, 1983, pp 271–291

Dunin FX (1991) Extrapolation of 'point' measurements of evaporation: some issues of scale. Vegetation 91: 39–47

Fischer RA and Turner NC (1978) Plant productivity in the arid and semi-arid zones. Annual Review of Plant Physiology 29: 277–317

Groves GH, Keraitis K and Moore CWE (1973) Relative Growth of *Themeda australis* and *Poa labillardieri* in pots in response to phosphorus and nitrogen. Australian Journal of Botany 21: 1–11

Hamblin AP and Williams J (1995) Soils: key issues in sustaining the land resource base. In: Sustaining the Agricultural Resource Base, Prime Minister's Science and Engineering Council, 12th Meeting, Canberra, ACT, 23 June 1995, pp 31–39

Hook RA (1997) Predicting Farm Production and Catchment Processes: A Directory of

Australian Models and Modelling Groups. CSIRO Publishing, Collingwood, Victoria, Australia
Hume I (1997) Episodic deep drainage under crops and shrubs in the mallee zone. In: Wilson S and Lawry T (eds) Proceedings Dryland Forum, 19–26 October 1997, North Adelaide, SA, pp 19–26. Murray-Darling Basin Commission
Isbell RF (1996) The Australian Soil Classification. Australian soil and land survey handbook series; Vol. 4. CSIRO, Melbourne, Australia
Lovering JF and Crabb P (1997) A fundamental necessity for sustainable agricultural development: an historical perspective for Australia. Proceedings of Property and Catchment Planning Symposium, Australian Institute of Agriculture, Science and Technology, Waite Institute, Adelaide, SA
McCown RL, Hammer GL, Hargreaves JNG, Holzworth DL and Freebairn DM (1996) APSIM: A novel software system for model development, model testing and simulation in agricultural systems research. Agricultural Systems 50: 255–271
Meinke H, Carberry PS, McCaskill MR, Hills MA and McLeod I (1995) Evaluation of radiation and temperature data generators in the Australian tropics and sub-tropics using crop simulation models. Agricultural and Forest Meteorology 72: 295–316
Nulsen RA, Bligh KJ, Baxter IN, Solin EJ and Imrie DH (1986) The fate of rainfall in a mallee and health vegetated catchment in southern Western Australia. Australian Journal of Ecology 11: 361–371
Poss R, Smith CJ, Dunin FX and Angus JF (1995) Rate of soil acidification under wheat in a semi-arid environment. Plant and Soil 177: 85–100
Probert ME, Dimes JP, Keating BA, Dalal RC and Strong WM (1998a) APSIM's water and nitrogen modules and simulation of the dynamics of water and nitrogen in fallow systems. Agricultural Systems 56: 1–28
Probert ME, Robertson MJ, Poulton PL, Carberry PS, Weston EJ and Lehane KJ (1998b) Modelling lucerne growth using APSIM. Proceedings 9th Australian Agronomy Conference, Wagga Wagga, July 1998, pp 247–250
Probert ME, Carberry PS, McCown RL and Turpin JE (1998c) Simulation of legume-cereal systems using APSIM. Australian Journal of Agricultural Research 49: 317–327
Reeves G, Chartres CJ and Breckwoldt R (1997) Does the answer lie in the soil? A national review of soil health issues. Centre for International Economics, Canberra and Sydney, July 1997
Salama RB and Hatton TJ (1998) Predicting land use impacts on regional scale groundwater recharge and saline discharge. Journal of Environmental Quality (in press)
Smith CJ, Dunin FX, Zegelin SJ and Poss R (1998) Nitrate leaching from a riverine clay soil under cereal rotation. Australian Journal of Agricultural Research (in press)
Talsma T and Hallam PM (1980) Hydraulic conductivity measurements of forest catchments. Australian Journal of Soil Research 18: 139–148
Verburg K, Ross PJ and Bristow KL (1996) SWIMv2.1 User Manual. Divisional Report 130. CSIRO Division of Soils, Australia
Williams J (1991) Search for sustainability: agriculture and its place in the natural ecosystem. Agricultural Science 4: 32–39
Williams J, Hook RA and Gascoigne HL (1998) Farming Action: Catchment Reaction – Impacts of dryland farming on the natural environment. CSIRO Publishing, Collingwood, Victoria, Australia
Williamson DR (1998) Land degradation processes and water quality: waterlogging and salinisation. In: Williams J (ed) Farming Action: Catchment Reaction – Impacts of dryland farming on the natural environment, pp 173–202. CSIRO Publishing, Collingwood, Victoria, Australia

Agroforestry Systems **45**: 365–392, 1999.

Designing mimics from incomplete data sets: salmon gum woodland and heathland ecosystems in South West Australia

R. J. HOBBS[1,]* and M. H. O'CONNOR[2]
[1] *CSIRO Wildlife and Ecology, Wembley 6014, Australia;* [2] *Muresk Institute of Agriculture, Curtin University of Technology, Bentley 6102, Australia (*Author for correspondence; E-mail: richard.hobbs@per.dwe.csiro.au)*

Key words: agricultural mimics, ecosystem function, eucalypt woodland, kwongan

Abstract. In this paper we assess the potential for developing mimics in situations where detailed data on the ecophysiology and ecosystem function of the native ecosystem are not available. We examine two ecosystems in south western Australia, and draw together available information on these. We discuss woodlands dominated by *Eucalyptus salmonophloia*, which formerly covered much of the valley bottom land in the Western Australian wheatbelt, and kwongan or heath which occurs on sandplain soils at the top of the landscape. From both of these we derive possible mimics which utilise aspects of the native vegetation which we hypothesise contributed to ecosystem function and resilience, but which still allow the production of a harvestable agricultural product. We conclude that situations for which we have adequate ecological knowledge from which to develop sound mimics are likely to be rare. We hence have to be prepared to make best-bet estimates and use these to develop experimental mimic systems.

1. Introduction

Studies in the *Banksia* woodlands on sandplain soils in Western Australia have provided detailed information on the functioning of these woodlands and the part that individual species are likely to play in water and nutrient cycling (Pate and Bell, this volume). From this rich information base, it is then possible to consider in detail the structure and composition required in an agricultural mimic of this system. Unfortunately, our knowledge of most Australian ecosystems is much less detailed, and the detailed physiological work described for the *Banksia* woodlands by Pate and Bell (this volume) is rarely available for most other systems. In this chapter, we take up the challenge to develop theoretical mimics for two south-western Australian ecosystems about which a complete suite of the necessary functional data is unavailable. The ecosystems in question are salmon gum (*Eucalyptus salmonophloia*) woodland and heath. We describe each in terms of what is currently known about its locale, structure and function, illustrated with one or more site-specific examples where available. There follows the design of an agroecosystem mimic of each plant community. We conclude with a discussion of the type of data that must be accessible to the agroecosystem designer.

2. Salmon gum (*Eucalyptus salmonophloia*) woodland

2.1. *Distribution*

Woodlands dominated by *Eucalyptus salmonophloia* F. Muell. (salmon gum) or a mixture of *E. salmonophloia* and other *Eucalyptus* species (Figure 1) occur throughout the semi-arid region of southwestern Australia (Beard, 1990; Yates et al., 1994b). Other major woodland types include those dominated by *Eucalyptus wandoo* or *E. capillosa* (wandoo) and *E. loxophleba* (York gum). The climate of the region where salmon gum woodlands occur is Dry Warm Mediterranean (Hobbs, 1992). Mean annual rainfall on the western edge of salmon gum's distribution is approximately 450 mm and on its eastern edge between 200 and 250 mm. Rain falls predominantly in the winter months between May and September; summer rainfall occurs but is unpredictable.

The region occupied by salmon gum includes the wheatbelt area (Avon and Stirling districts, Figure 2), which has been extensively cleared for agriculture, and the adjacent goldfields area (Coolgardie district, Figure 2), where low rainfall prevents agricultural development and extensive woodlands still remain, albeit altered in many areas by grazing and timber harvesting. In this paper, we concentrate on the woodlands occurring in the wheatbelt region. In this region, *E. salmonophloia* woodlands are most extensive on the broad valley floors on red brown sandy loam over clay (Db 1.13, Dr 2.13) and red brown clay soils (Gc 2.21, Uf 6.31, Db 1.13) but also occur to a lesser extent

Figure 1. Woodland dominated by *Eucalyptus salmonophloia* in the central wheatbelt of Western Australia (photograph by C. J. Yates).

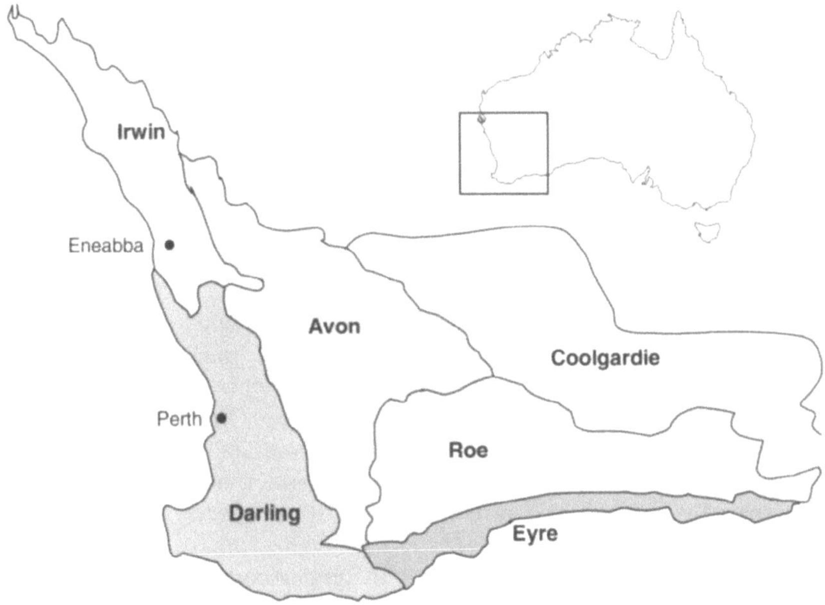

Figure 2. The districts within the south-western Australian botanical province. Salmon gum woodlands occur as remnant areas in the Avon and Stirling districts and more extensively in the Coolgardie district. Kwongan is most extensive in the Irwin and Eyre districts. The town of Eneabba is the locale of the scrub-heath sites described in detail.

on mid and upper slopes on grey, brown or dull reddish loamy sand over clay (Dy 2.13, Dy 2.23) (McArthur, 1991; Lantzke, 1992).

Salmon gum woodlands are typical of woodlands dominated by *Eucalyptus* spp. in temperate south eastern and south western Australia, many of which have been extensively cleared for agriculture (Carnahan, 1990; Hobbs et al., 1993; Prober and Thiele, 1993). In the Western Australian wheatbelt, Beard and Sprenger (1984) estimate that only 10% of the *Eucalyptus loxophleba* (York Gum) woodlands and 20% of *Eucalyptus salmonophloia/Eucalyptus salubris* (salmon gum/gimlet) woodlands remain. Furthermore, a substantial proportion of the remnant woodlands in Australia's cereal growing regions occur as isolated patches that are often degraded by livestock grazing with no understorey remaining and little or no regeneration of the tree species. In the central wheatbelt in Western Australia, approximately 40% of remnant *E. salmonophloia* woodlands had little or no understorey, 57% had no tree species in the 0–10 cm stem diameter at breast height (DBH) size class and 28% had no tree species in the 10–20 cm DBH size class (Norton et al., 1995). This indicates a failure of regeneration of the tree species, loss of understorey diversity and cover, extensive invasion by a broad spectrum of non-native annual species, loss of litter cover and deterioration of soil surface properties and soil structure.

The relatively small areas of native woodland remaining, and the relatively modified nature of this woodland, make it difficult to establish reference conditions on which to base mimic studies. Yates and Hobbs (1997) provided a conceptual framework which indicated that salmon gum woodlands can exist in a number of different states, from a relatively intact and functioning woodland system, through various states of degradation. In this section, we describe the un-degraded woodland state.

2.2. *Floristics and life form spectrum*

Salmon gum woodlands commonly consist of pure stands of canopy trees or mixtures of salmon gum and other species (e.g. *E. wandoo*, *E. capillosa*, *E. salubris*). Understoreys consist of a shrub layer which varies in composition, but often includes *Atriplex* spp. and *Acacia* spp. Often there is a sparse herbaceous layer composed of perennial herbs and annuals. Despite comprising relatively little of the total cover or biomass of the woodland system, this herbaceous component contributes greatly to the species diversity of the woodland (Figure 3).

As a comparison, a similar breakdown in another common woodland type, dominated by *Eucalyptus loxophleba*, indicates a much less diverse and abundant shrub layer and considerably greater abundance of perennial herbs and annuals (Figure 4). Studies by Yates and Hobbs (1997) suggest that salmon gum woodlands can occur in a degraded state in which the shrub layer has largely been removed and replaced with non-native herbaceous species.

Densities of trees in salmon gum woodland vary considerably, depending on the age of the stand. For instance, Norton et al. (1995) found densities varying from 650 trees ha^{-1} in a stand estimated to be about 50 years old, to 50 trees ha^{-1} in a mature stand. Tree age is difficult to determine in salmon gums, but mature individuals are thought to be over 250 years old (Rose, 1993). In intact woodlands, a mosaic of different ages is present, with either relatively large patches of even-aged trees caused by large-scale disturbances, or small patches of regeneration associated with the death of individual trees (Yates et al., 1994b), and Hobbs (unpubl. data).

2.3. *Phenology*

Few data are available on shoot growth in wheatbelt woodland species. As part of an ongoing experiment, tree and understorey shrub seedlings were planted into a degraded woodland remnant, and shoot growth has been recorded over the first year (Hobbs, Yates and Atkins, unpubl. data). These data (Figure 5) indicate that most growth occurred over the period August–December, although three species continued to grow during January. This indicates that growth commences in late winter and continues through spring and early summer. This corresponds with a period of increasing temperature, and rainfall which generally occurs only until September, with occasional

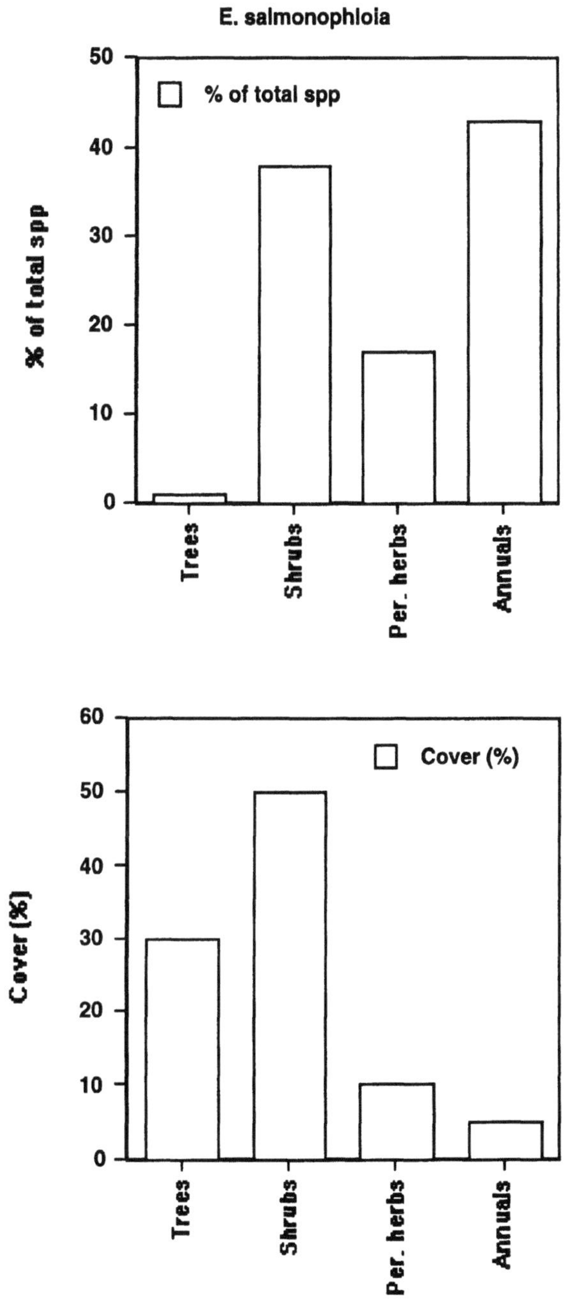

Figure 3. Relative contribution of different life forms to (a) total species richness and (b) vegetation cover in *E. salmonophloia* woodlands in the central wheatbelt. Perennial herbs comprise 19 hemicryptophytes and one chamaephyte. *Source:* Data from Yates (1995) and R. J. Hobbs (unpubl. data).

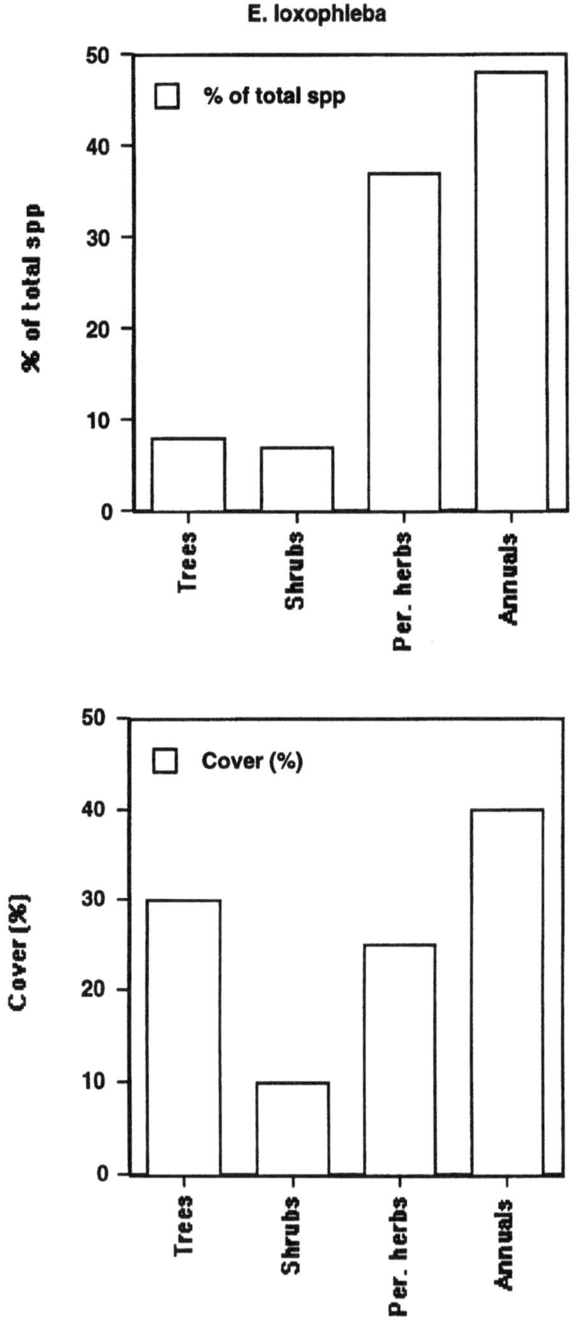

Figure 4. Relative contribution of different life forms to (a) total species richness and (b) vegetation cover in *E. loxophleba* woodlands in the central wheatbelt. Perennial herbs comprise six hemicryptophytes, three chamaephytes and one geophyte. *Source:* R. J. Hobbs (unpubl. data).

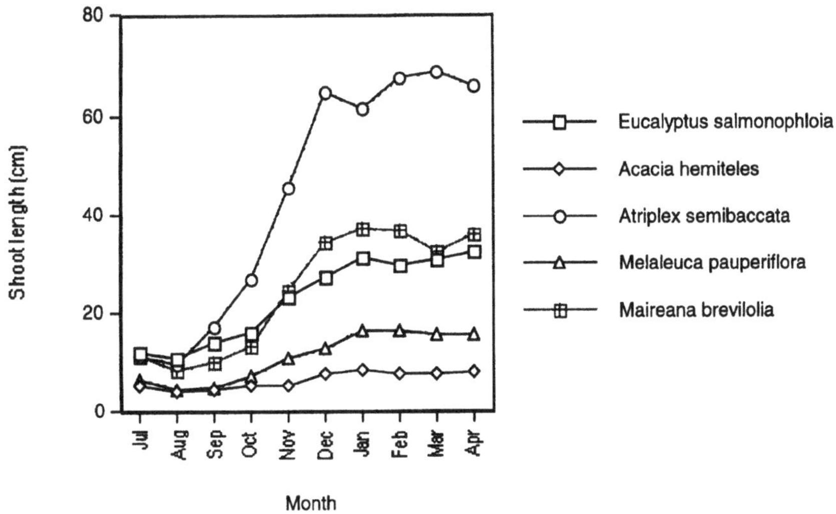

Figure 5. Shoot length (mean, n = 25) of seedlings of *Eucalyptus salmonophloia* and four understorey shrub species planted in June 1996 in soil that had been deep ripped next to a recently-dead adult tree. *Source:* C. J. Yates, R. J. Hobbs and L. Atkins (unpubl. data).

summer storms. Work by Farrington et al. (1994) indicated that transpiration rates of mature salmon gum were low during winter (10–40 L d^{-1}), rose steadily in spring to 48 L d^{-1}, and fell again in late spring, depending on the rainfall distribution. Recordings during summer indicated continued transpiration, with rapid increases in transpiration rates following a summer rain storm. This suggests both an ability to respond quickly to periodic aseasonal rainfall and to extract water from depth (Farrington et al., 1994).

Mature salmon gums have been observed to flower during the period between April and December, but flowering times vary greatly both spatially and between years (Yates et al., 1994a).

2.4. *Root architecture*

Few studies of root architecture or distribution have been conducted in wheatbelt woodlands. The only detailed study of which we are aware was carried out by Lamont (1985) in *E. wandoo* woodland. He illustrated that shrub species concentrated their roots in the top 15 cm of soil, with root mass dropping to virtually zero by 75 cm depth. Tree roots, on the other hand, were concentrated at 15–45 cm and 60–75 cm, and there was little intermixing of tree and shrub roots.

An important finding from Lamont's study was a change in root distribution with tree age. *E. wandoo* saplings had a tap-root dominated system with weak laterals and sinker roots, while mature trees retained a tap root, but had a lateral-dominated system with major horizontal roots in the 20–40 cm zone, which extended about 10 m. This is important when considering the likely water use of trees: clearly water uptake patterns will differ greatly between young and mature trees. This may help explain the difference in water use between revegetated areas and existing remnant vegetation (R. George, pers. comm.). In terms of developing a useful mimic, it may be more beneficial to maintain a vigorous component of young trees which rely mostly on a tap root system, rather than allowing the development of a lateral system which will compete with crop or pasture species.

A native woodland ecosystem is likely to consist of a spatially heterogeneous pattern of root distribution, made up of roots from mature trees, patches of juvenile trees, and understorey shrubs and herbaceous perennials (Figure 6). No details of root architecture are available for understorey species.

2.5. *Nutrition*

Of the shrub species present in salmon gum woodlands, seven species of *Acacia* possess nitrogen-fixing nodules, while one species from the Proteaceae possesses specialised (proteaceous) roots. In *E. loxophleba* woodland, only two species of *Acacia* and one Proteaceae are common.

2.6. *Fire response*

In both *E. salmonophloia* and *E. loxophleba* woodlands, most perennial species have some ability to resprout following fire, and many species combine this with effective seed regeneration. Only a few *Acacia* species are known to be strict seeders only. The dominant trees *E. salmonophloia* and *Acacia acuminata* are fire sensitive, but *E. loxophleba* will resprout. Salmon gum retains a seed store in the canopy, which is released continuously in small amounts. After a disturbance such as fire, however, a massive seed release occurs. Little seed is stored in the soil between disturbances, since ant predation removes most seed which falls (Yates et al., 1994b). Many shrub species retain a soil seed store, particularly *Acacia* spp. There is some evidence to suggest that different species require different fire treatments to germinate

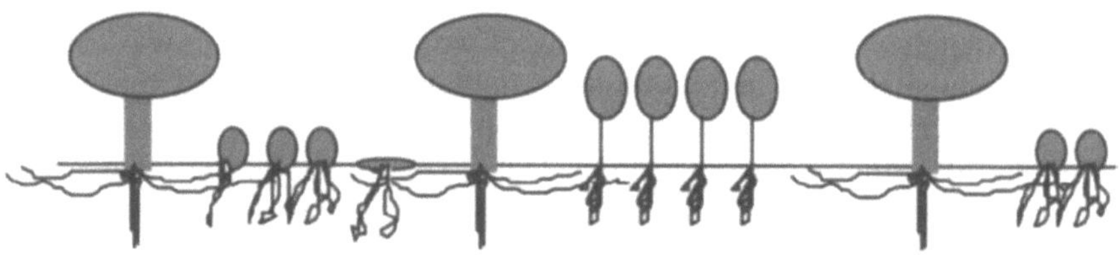

Figure 6. In an intact salmon gum woodland, roots are likely to be patchily distributed, with a strong dominance by adult trees with extensive lateral systems. Patches of juvenile trees will have mostly tap root systems, while understorey species are likely to have a variety of rooting patterns.

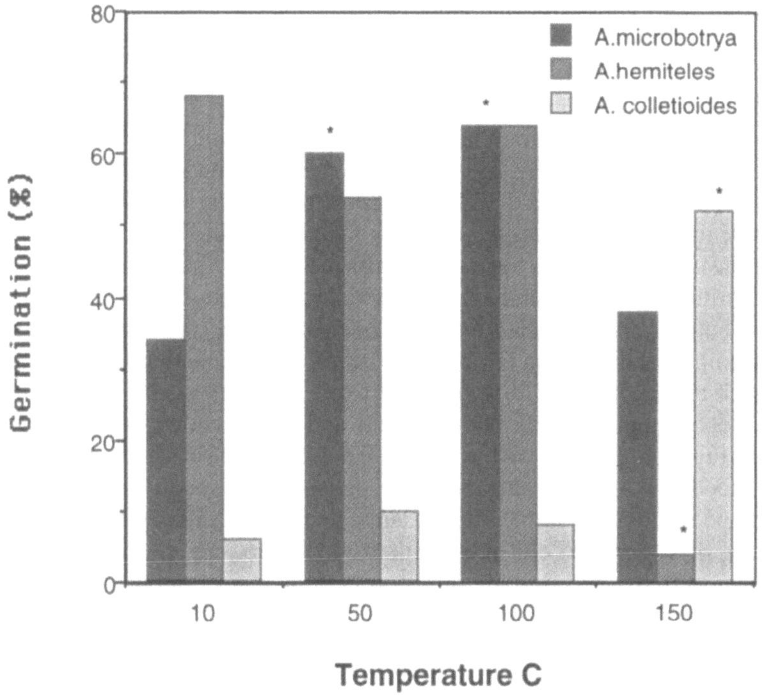

Figure 7. Germination (percent) of three *Acacia* species which are common components of the understorey in *Eucalyptus salmonophloia* woodland in the central wheatbelt, following various treatments. Seeds were scarified and subjected to different temperatures in an oven for ten minutes, placed on filter paper in petri dishes and watered regularly for thirty eight days. Five dishes were assigned to each treatment, with ten seeds per dish. Asterisks indicate significant differences from control, as determined by t-test on square root transformed data. *Source:* Adapted from Atkins and Hobbs (1995).

(Figure 7). The presence of a suite of species with different germination cues may contribute to the resilience of the system.

2.7. *Composition of a mimic*

2.7.1. *Current agricultural system*

The agricultural system currently predominant over much of the area formerly occupied by salmon gum woodlands is a crop-pasture rotation, with predominantly wheat and pastures composed of mixed annual grasses and forbs. This current system is under threat from numerous types of land degradation, especially secondary salinisation, waterlogging, acidification and water erosion (Nulsen, 1993). Responses to these threats have, until recently, been predominantly mechanical, and involved inserting structures such as interceptor banks to modify water flows. It is generally considered that an adequate response to problems such as salinity have to embrace a broader landscape approach involving large amounts of revegetation with perennial species (Agriculture Western Australia et al., 1996a, b).

2.7.2. *Mimics based on native Eucalypt woodland*

The development of a mimic based on native woodland is problematic for a number of reasons. Firstly, as noted earlier, representatives of woodland in good condition are rare in the current agricultural landscape. Hence it is quite difficult to determine what the native woodland 'should' look like or how it functions. Secondly, we have little information on the impact on system function that results from changes in woodland composition due to degradation. We know, for example, that high cover of annual species inhibits perennial seedling establishment (Hobbs and Atkins, 1991), and that soil compaction by stock results in reduced water infiltration (C. J. Yates, D. N. Norton and R. J. Hobbs, unpubl. data), but otherwise understand little else about overall system function. Finally, we have noted that woodland structure varies greatly with woodland age. For instance, tree densities are much higher in young woodland stands than in old stands, and extensive natural woodlands will be mosaics of young and old trees (Figure 8). Hence, any calculations on the numbers of trees needed in the mimic system must take into account this age difference.

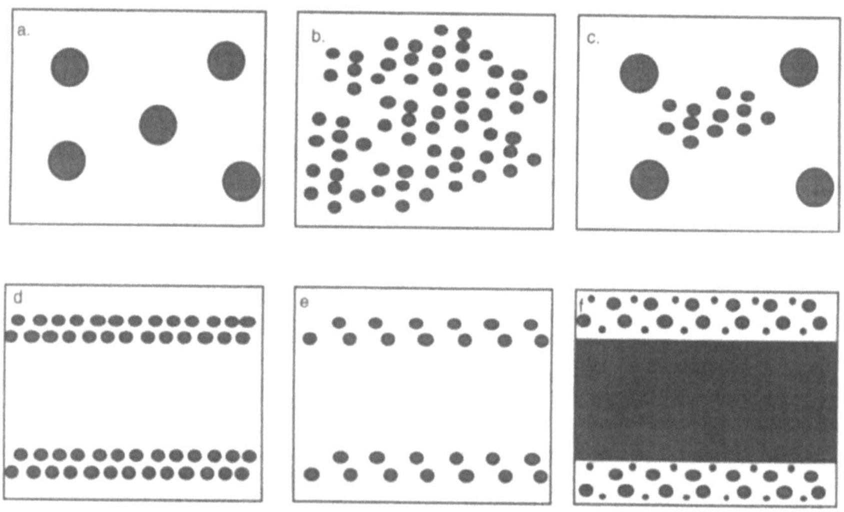

Figure 8. Construction of a mimic based on salmon gum woodland. Individual boxes represent areas approximately 40 × 50 m in dimension. Natural woodland varies in tree density from about 50 trees ha^{-1} in mature woodland (a) to 650 trees ha^{-1} in regenerating woodland (b). Natural woodland is also heterogeneous, with patches of regeneration interspersed with mature trees (c). Density in (c) is about 160 trees ha^{-1}. Arranging young trees in strips across the landscape allows high densities to be achieved; e.g. 520 trees ha^{-1} in (d) and 260 trees ha^{-1} in (e). The mimic uses the trees to form the basic template, around which other elements are introduced; for instance, understorey species and annual or perennial crop and pasture species (f).

In developing a mimic, we want to capture the essential components which contribute to function and resilience in the native system and then transpose these into a system from which a sustainable and profitable harvest can be taken. It is clear from the above descriptions of the woodland system that we have only a sketchy appreciation of the relationships between woodland composition and function. In order to initiate discussion on the likely form of a mimic, we have taken the natural densities of the dominant tree species as a starting point (Figure 8). We assume that native woodlands utilised most of the incoming rainfall, hence minimising runoff and deep drainage to the water table, and that we wish to push the agricultural system back towards this state. Hence, a density of trees not dissimilar to that in the native woodland is required. In Figure 8, we indicate ways of doing this which do not end up with the whole landscape covered in trees, but instead leave space for agricultural production. As discussed previously, it seems likely that maintaining the system with young trees may be beneficial since young trees may rely mostly on a tap root system. This is useful both because it is likely to increase the use of deep rather than surface water and because it minimises interference between trees and crops/pastures.

The example given in Figure 8 places the trees in rows about 30 m apart. Two scenarios are given, one with trees growing at a density similar to that found in a young woodland, and another with densities similar to that in a mixed mature woodland. We do not yet know the preferred density, which needs to be assessed on the basis of balancing the requirements for effective site water use and leaving sufficient 'ecological space' for the productive component of the system. If trees effectively occupy the whole site, that then leaves little scope for agricultural production.

While Figure 8 presents an option based on strips of trees (alone or mixed with other species) across the landscape, with crop or pasture between, this is one of many possible configurations. Indeed, an important question to be asked concerns the relative benefits of integration of different biotic elements *versus* segregation into different parcels (Figure 9). The scenario in Figure 8 presupposes that it is important to configure the biotic elements so that they occur throughout the landscape. How this compares with alternative arrangements needs to be assessed. For instance, is it more beneficial to have perennial species in larger blocks rather than in strips? There are certainly management benefits to be gained from this, and it is likely that there may be other benefits, for instance to nature conservation. As yet, however, there is no clear guidance from hydrological models as to which configurations are likely to be most beneficial, or indeed whether any of these alternatives are likely to address the land degradation problems fully.

In this example, we have taken a fairly simplistic approach to the problem of developing a mimic from the native system, basing the mimic on the major structural component of the woodland, i.e. the trees. Clearly, to address questions of long-term resilience and to ensure more effective water use, other elements would have to be included.

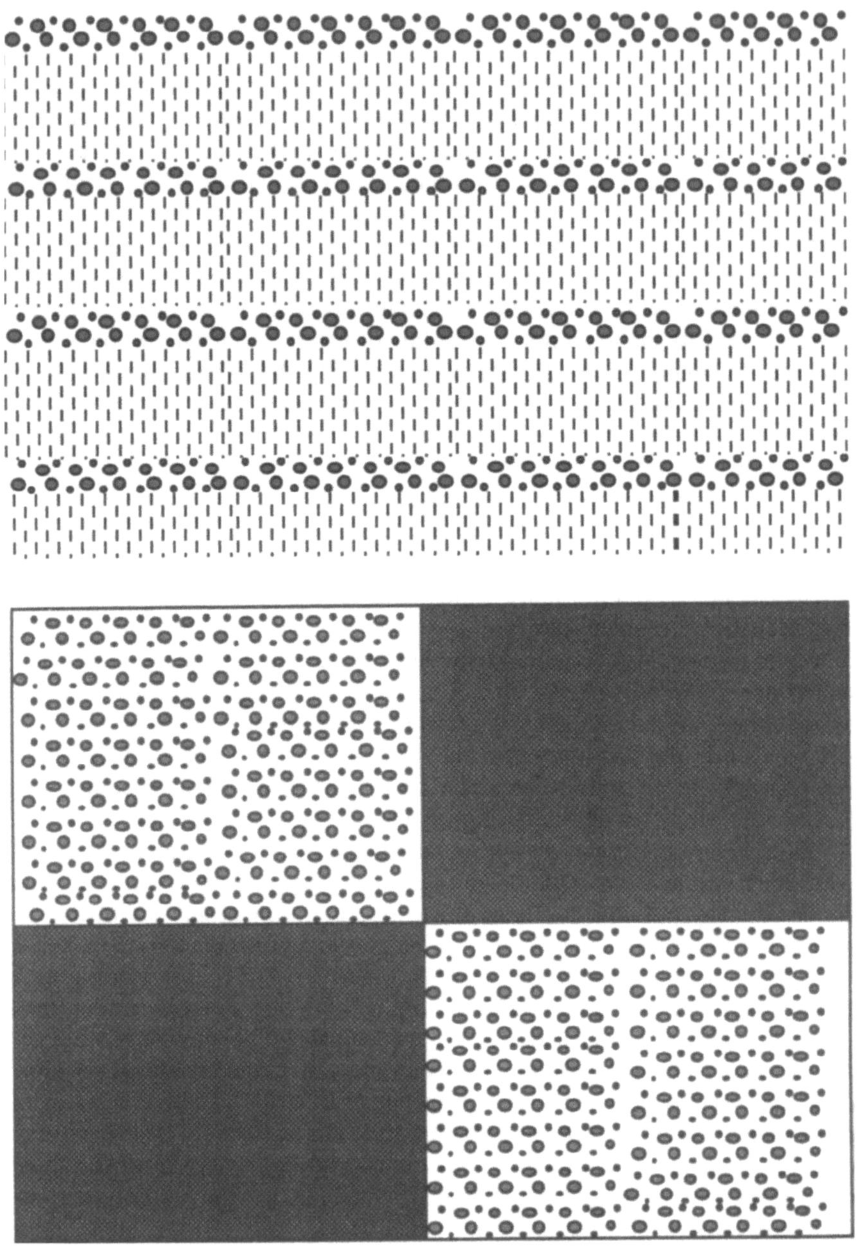

Figure 9. There are numerous ways of incorporating different elements from the native ecosystem into the agricultural system. For instance, diverse elements of the native system can be arranged within individual land units, hence increasing the diversity *within* patches (a), or different elements can be introduced to different patches, thus increasing the overall diversity *between* patches (b).

3. Heath

Unravelling the structure and function of the south-western Australian heath vegetation and its many subtypes has been the subject of much research over the last 30 years. The floristic and physiological richness that delights the botanist confounds the agroecosystem designer intent on developing a mimic. Much progress has been made in understanding heath physiological processes but there are still many blanks. In this section we describe the relevant details of locale, structure and function common across the structural subtypes of south western Australian heath.

3.1. *Definition and distribution*

Heath communities in south western Australia are known locally as *kwongan*, a term from the Nyungar aboriginal language relating to scrub vegetation over sand. The kwongan is analogous in various degrees to other mediterranean heath communities such as the maquis community of the Mediterranean, the chaparral of California, the fynbos of South Africa and the matorral of Chile (Beard and Pate, 1984). At the time of European settlement in south western Australia (1826) some 118,000 km^2 was extant, of which less than 1% was not associated with sandy soils (Beard, 1984).

In physiognomic terms kwongan is defined as (Beard and Pate, 1984):

> . . . any community of sclerophyll shrubland in south-western Australia which has a stratum ± 1 m tall or less of leptophyllous and nanophyllous shrubs. It may also contain either taller shrubs, which may be dominant – so long as the dominants are of genera other than *Eucalyptus* – or scattered trees of any kind which are not dominant.

In Beard's schema, heaths growing on sandy soils are shown to occur on two general soil types, grading into *Banksia* low woodland above 600 mm annual rainfall and into various thicket communities at the lower end of their rainfall range (Beard, 1984). The structure and function of *Banksia* woodland are discussed by Pate and Bell (this volume). The sandy soils on which they occur are very low in nutrients and highly leached, within an environment subject to fire and drought. Adaptations to this harsh environment are reflected in the wide variety of mechanisms and strategies that kwongan plants utilise to cope with these stresses.

Kwongan physiognomy encompasses five sub-types distributed over the botanical districts in south western Australia (Figure 2). The sub-types – heath, mallee-heath, scrub-heath, thicket and tree heath – vary in aspects in their gross physiognomy, particularly in respect of proportional representations (Figure 10). Kwongan was the most extensive vegetation of the Irwin and Eyre botanical provinces, and to a lesser extent the Avon (Beard, 1984). Thus, it is not surprising that many of the floristic and physiological studies originate in these three regions.

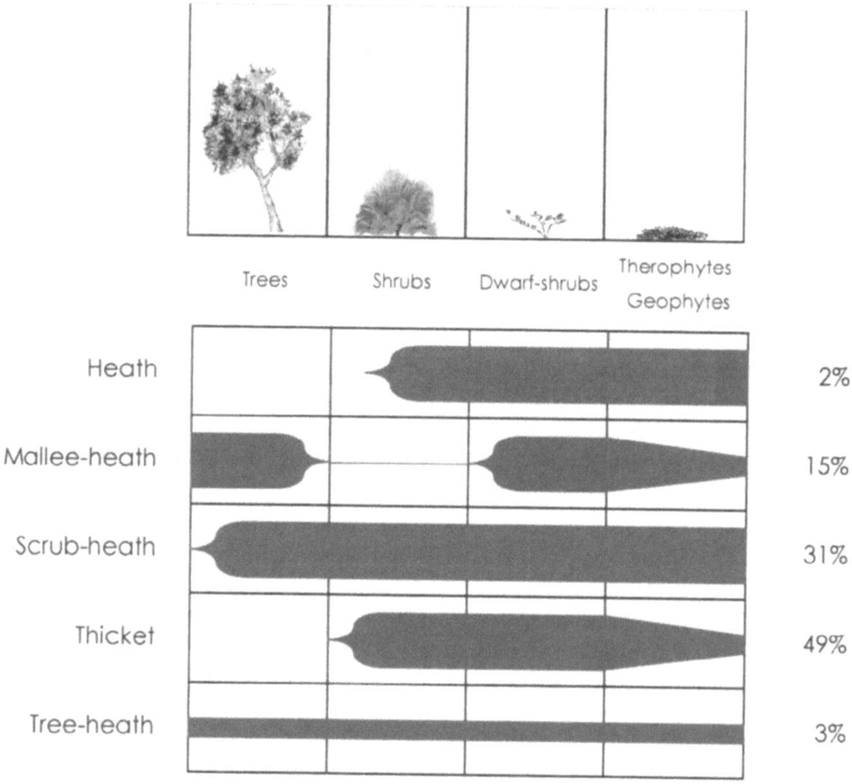

Figure 10. The gross physiognomy of the five subtypes of kwongan vegetation, with the relative proportions of life form components. The percentage figures approximate the portion of south western Australia thought to be covered by each sub-type at the time of European settlement in 1826. *Source:* Adapted from Beard (1984).

3.2. *Floristics*

Kwongan is also characterised by its floristic richness, both within and between sites. Over south-western Australia some 2500, or about 29%, of all known species occur within the kwongan. The majority of these species belong to one of six families: Myrtaceae, Proteaceae, Mimosaceae, Papillionaceae, Epacridaceae and Asteraceae (Lamont et al., 1984). Remarkable genera include *Banksia*, orchids (*Caladenia*, *Thelymitra*, *Diuris*), sundews (*Drosera*), stilt plants (*Laxmannia*, *Stylidium*) and the kangaroo paws (*Anigozanthos*). The Christmas tree (*Nuytsia floribunda*) is often seen as a lone remnant on farmland formerly under some types of kwongan, retained for its stunning orange blooms late in the year rather than for any ecological benefits it may provide.

The diversity of south western Australian heath can match that of tropical rainforests (Lamont et al., 1977), but only on a small plot area basis (Lamont et al., 1984). Thus it is species-rich within plots (33–133 species ha^{-1}), while contiguous plots are usually extremely heterogeneous in composition (Lamont et al., 1984; Specht, 1994). Species that make up the various physiognomically distinct strata of the heath accordingly vary considerably between and within the general kwongan subtypes. A recent integration of species-site data from the northern sandplains (all of Irwin and western part of Avon botanical provinces, Figure 2) drew upon 19 published and unpublished studies plus fresh site examinations to produce a highly comprehensive list of species (Griffin, 1994). Some 1964 native species were recorded, which in subsequent analysis were assigned to 20 major and 500 minor floristic groups. Despite the challenges of comprehending such variation, models of the kwongan ecosystem's structure and function are now appearing in the literature. This is not surprising given the similarities in the environments in which it is found. It would however be useful to confirm this by building and comparing functional and structural models of each kwongan sub-type and its variants (based on species present at each differing site). This is discussed further in the section below on information gaps.

3.3. *Above- and below-ground structure*

A general schema for kwongan above- and below-ground structure is illustrated in Figure 11 and described in detail in this section. Analysed by growth form, shrubs comprise the majority of species present, with perennial herbs as the other major component (Figure 12) (Pate et al., 1984). Trees are generally only present as emergents and as such contribute only 4% of total species.

A large component of the species diversity is contributed by small summer-dormant plants (e.g. perennial herbs from the Liliaceae and Haemodoraceae families). Analysis by species only would overstate the functional contribution of some forms. It is therefore pertinent to the design of mimics to correlate growth-form to measurements of the total presence of each species per unit area.

For south western Australian heaths, cover data are usually reported at the community level, and are often in the range of 65–90% depending on the nature of the emergent tree stratum. In the most extensive regional vegetation maps of WA (e.g. Beard and Parker, 1976), heaths are mapped with a mean cover in the range 30–70%. There are two published studies of cover correlated to life form (Lamont and Bell, 1990; Newbey, 1979). A floristic study of kwongan sites in the Irwin botanical province Griffin et al. (1983) noted that individual species rarely exceeded 5% cover in single quadrats, with less than 1% being the most common extent. Several larger shrubs and trees (*Calothamnus*, *Eucalyptus* and *Melaleuca*) reached cover levels of 25–33% in several quadrats.

Almost all kwongan species are perennial. An analysis of 429 species occurring across a range of kwongan sites revealed that 95% had life spans

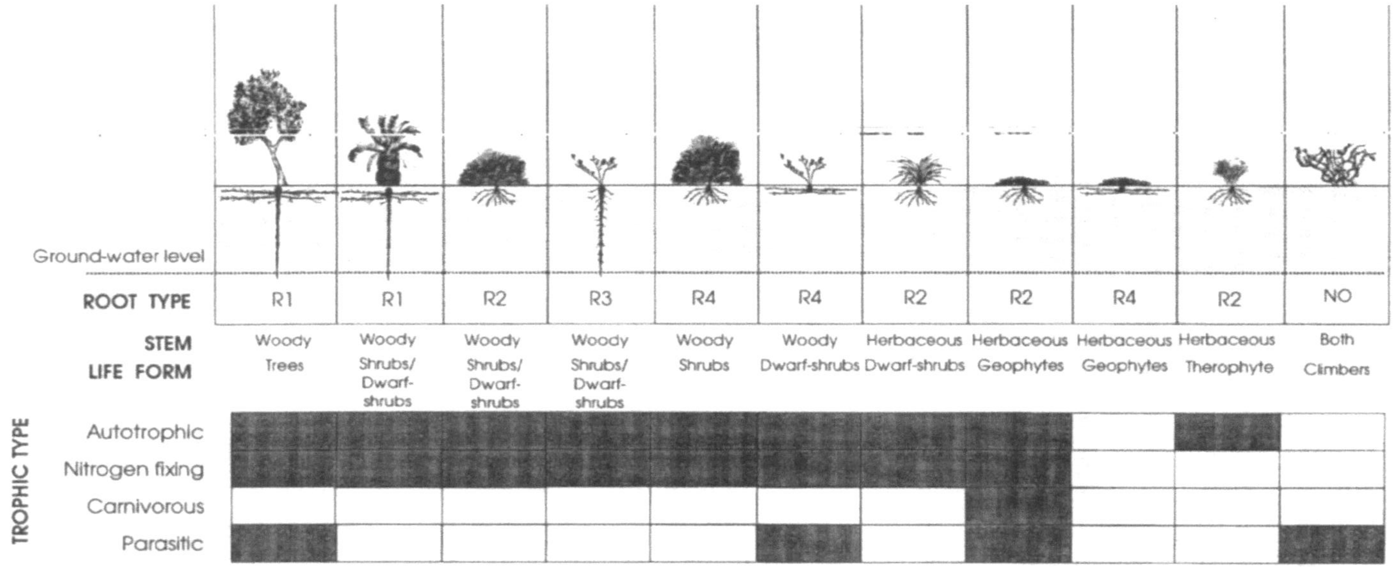

Figure 11. The above- and below-ground structure of kwongan, as a correlation of life-form, root morphology type, stem type and tropic type. Shaded boxes in the grid indicate that at least one species exhibits that combination of trophic type, life form and root morphology. *Source:* Data from Pate et al. (1984), Pate (1994).

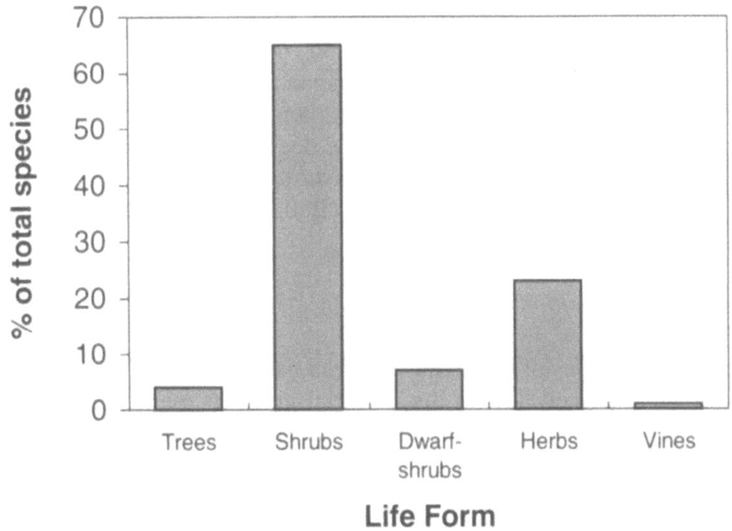

Figure 12. Relative contribution of different growth forms to total species richness in kwongan heaths based upon characteristics of 429 species found across a range of kwongan sites in south western Australia. *Source:* Data from Pate et al. (1984).

of at least five years (Pate et al., 1984). Indeed, Bell et al. (1984) suggested that many species may persist for up to 50 years, but this remains to be tested through long-term ecological site study or growth-ring analysis.

The literature does not contain comprehensive correlations of root morphology to species or life form, however generalities are becoming evident (Figure 11). The scheme for kwongan root morphologies was originally devised to have five types (Pate et al., 1984; Dodd et al., 1984) but has been revised to four by Pate (1994). A comparison with the root morphology scheme of Cannon (1949) has been published in Dodd et al. (1984), and yet another specialised scheme is suggested by Pate and Bell (this volume) specifically for *Banksia* woodland.

The trees and larger shrubs generally have type R1 – dimorphic roots incorporating near-surface laterals for nutrient acquisition and deep tap roots for accessing groundwater during summer months. Smaller shrubs, perennial herbs and annuals have a variety of root morphologies: R2 – fibrous roots confined to the upper soil layer (usually 30–50 cm) and therefore not in contact with a water table; R3 – single tap root with rootlets closer to the surface; R4 – surface laterals only. Plants without access to a water table during summer therefore lie dormant, make use of internal storage or in certain specialised cases acquire water from the xylem of various woody legumes (e.g. root hemi-parasites such as *Santalum acuminatum*; Tennakoon et al., 1997).

3.4. *Phenology*

Virtually all trees and larger shrubs of kwongan are spring- and summer-active. As shown elsewhere in this volume, such phenology is invariably associated with a dimorphic root system including a deep 'sinker' root that reaches a groundwater table. Conversely the smaller shrubs, perennial herbs and annuals are mostly winter-spring or spring-active (Bell and Stephens, 1984; Low and Lamont, 1990; Pate, 1994).

3.5. *Nutrient acquisition*

The various sub-types of kwongan are often associated with heavily leached soils that are extremely low in the essential nutrients, particularly P and N, and also low in organic matter (Table 1). In these austere environments, the heath ecosystems utilise a range of nutrient-acquisition specialisations and nutrient-retention mechanisms. As ecosystems they efficiently capture and retain available nutrients. In contrast to the agricultural systems that have replaced them, they are essentially not 'leaky'.

A total of six nutrient-acquisition specialisations are utilised by heath plants to enhance uptake (Figure 11). These include root clusters (bunches of hairy rootlets), mycorrhizae (either VA-type or host-specific), nitrogen-fixing symbioses, haustoria and carnivorous glands, with VA mycorrhizae the most common (Lamont, 1984; Pate, 1994). In the study of 429 kwongan species previously mentioned the bulk of species were of only two trophic types – 81% were autotrophic and 14% utilised an N-fixing symbiosis. An extensive study of the relative contributions of mycorrhizal and non-mycorrhizal specialisations to the movement of nutrients through the heath ecosystem is yet to be undertaken. The work of Brundrett and Abbott (1991) in the jarrah (*Eucalyptus marginata*) forest of south western Australia provides a useful approach, and indeed there are some genera and species common to both.

A few general features of particular life forms will illustrate these nutrient-acquisition specialisations. Trees and shrubs, notably the Proteaceae family, seasonally develop proteoid roots (a form of root cluster) on their surface laterals as a means of acquiring nutrients during the winter. The perennial

Table 1. Typical Eneabba scrub-heath soil profile for a deep sand in a relic dune crest.

Horizon	Depth (cm)	Description	Clay:Sand (%)	Total N ($\mu g\ g^{-1}$)	Total P ($\mu g\ g^{-1}$)	OM (%)	pH
A1	0–10	Grey brown sand	1:99	207.7	6.5	1.2	5.1
B1	10–70	Pale grey-white leached sand	1:99	70.5	3.7	0.27	5.2
B2	70–100+	Pale yellow sand	2:98	66.1	4.8	0.37	5.2

Source: Froend (1987).

herbs and some of the smaller shrubs are limited by their near-surface root systems to the uptake of nutrients during the winter months (Pate, 1994; Jeschke and Pate, 1995). Nitrogen fixing nodules are present on most members of the Mimosaceae and the Papilionaceae. Amongst seeder species the supply of N by symbiotic fixation has been shown to increase in the period two to four years after fire once seed and ash nutrient reserves are depleted (Pate and Dell, 1984).

Several nutrient-retention strategies are utilised by heath plants. The loss of aerial plant parts may occur often through disturbances such as drought and fire. Nutrients may be redeployed from leaves into stems as they age so that upon senescence fewer nutrients are lost by the plant (Pate and Dell, 1984; Low and Lamont, 1990). Concentrating essential nutrients into seeds and organs that are resistant to fire or drought also prevents the loss of nutrients after disturbance (Pate and Dell, 1984).

The nutrient-poor soils underlying the scrub-heath of south western Australia provide an interesting challenge to designers of a would-be agricultural mimic. In addition, many areas in which heath mimics may be implemented are those on which kwongan has not grown for some time. This soil has been altered by applications of fertilisers over long periods of time thereby raising the possibility of some intermediate ecosystem being necessary to ameliorate these nutrient changes.

3.6. *Effect of fire*

Fire is a regular disturbance agent of kwongan vegetation. One of the dominant elements of scrub-heath, the tall shrub *Banksia hookeriana*, was found to have an optimum fire regime of 15 to 18 years (Enright et al., 1996). Some data on species' fire response are available (e.g. Bell et al., 1984; Bell and Pate, 1996; Pate et al., 1991). The categories of response include fire ephemeral (short-lived fast growing), obligate seeder, obligate sprouter, facultative sprouter-reseeder and autoregenerating long-lived sprouter. The available data have not been correlated with growth form. Several measurements of biomass increase post-fire show the heath to have regrown to more than 95% of pre-fire levels after five years, and with very little increase thereafter (Bell et al., 1984). Species numbers appear to be relatively unchanged by fire, although as frequency increases the proportion of seeders reduces.

The heath mimic is likely to require a substantial woody component that will take considerable expertise, time and expense to establish. In some areas of south western Australia, it may be required to establish patches of heath mimic within a matrix of native heath vegetation. This retained heath is currently maintained by government agencies through periodic patch burning. This raises the possibility of fire spreading from the kwongan patches to the mimic patches. If the mimic perennials are burnt, particularly in a hot summer or autumn fire, what effect will their temporary or long-term removal have in terms of system function? Thus consideration of kwongan's adaptation to fire is also necessary to determine which species components should be selected for the mimic.

3.7. *Site example: scrub-heath*

Having discussed the generalities of the kwongan as a whole, in this section we describe the structure and function of the dominant species of the scrub-heath kwongan sub-type. The data presented here are from studies in the west Midlands agricultural region around Eneabba, WA (29°52′ S, 115°15′ E); a study of six scrub-heath sites on deep sands (Froend, 1987) and a tabulation of ecomorphological characteristics for kwongan on deep sands (Lamont and Struthers, 1988). These are higher-rainfall heath sites, and thus close to grading into open *Banksia* woodland, as described by Pate and Bell (this volume).

Prior to the arrival of European settlement scrub-heath covered some 37000 km^2 or 8.5% of the south-western Australian agricultural region. Thus, a scrub-heath functional mimic could be applied to an extensive area of south western Australia. By 1984, approximately 58% of this scrub-heath had been cleared, primarily for agriculture.

The Eneabba locale experiences a warm wet climate with long hot dry summers, which equates to Köppen classification *Csa* (Griffin et al., 1983). The mean summer temperature at Eneabba is 21 °C, with maximum summer temperature over 40 °C, while over the winter months the mean temperature is 13.5 °C. The growing season, by Prescott's formula, is about 4.5 months (Low and Lamont, 1990). Mean monthly rainfall at Eneabba in the period 1964–1978 is presented in Table 1 (no measurements 1970/71).

Scrub-heath is confined to two major soil types; sands overlying ironstone gravel (annual rainfall from 400–600 mm), and deep sands (annual rainfall from 300–600 mm). The deep sandplain soils of the Eneabba region have low water availability resulting from low rainfall, soils with poor water-holding capacity and seasonal drought associated with the pronounced Mediterranean climate (Hnatiuk and Hopkins, 1980). The typical soil profile beneath the Eneabba scrub-heaths is leached deep sand with very low clay:sand ratios (Table 1).

Structurally, scrub-heath differs from other kwongan sub-types by its very tall open upper layer (up to 6 m) and its closed heath layer (up to 1 m) (Figure 10). Grouping scrub-heath species by life forms shows that shrubs and perennial herbs again make the greatest contribution (Figure 12). Shrubs

Table 2. Average monthly rainfall at Eneabba, Western Australia. Annual average for the same period is 529 mm.

Month	F	M	A	M	J	J	A	S	O	N	D	J
Rainfall (mm)	10	9	34	67	135	104	77	42	28	12	7	4

Source: Hnatiuk and Hopkins (1980).

comprise the majority of species. The remainder of the flora is almost entirely made up of perennial herbs, and tree and annual species tend to be few in number. Thus consideration by species/life form only indicates that perennial herb species should be incorporated into the mimic.

In our specific example at Eneabba, over 98% of the above-ground biomass was concentrated in seven shrub species (Figure 13). About 80% of this comprised a mix of *Banksia hookeriana*, *B. attenuata*, *B. menziesii* and almost all of the remainder a mix of *Beaufortia elegans*, *Eremaea beaufortioides*, *Conospermum triplinervium* and *Petrophile* spp. (principally *P. drummondii*) (Froend, 1987). As might be expected the vast majority of percentage projective foliage cover is also contributed by the shrubs. On the other hand, perennial herbs, while having the second-largest proportion of species compared with other growth-forms, contribute relatively small fractions of the biomass, albeit playing a relatively greater role in specifically providing

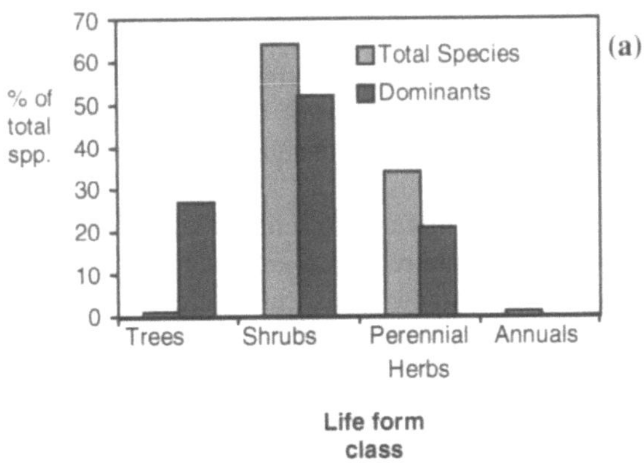

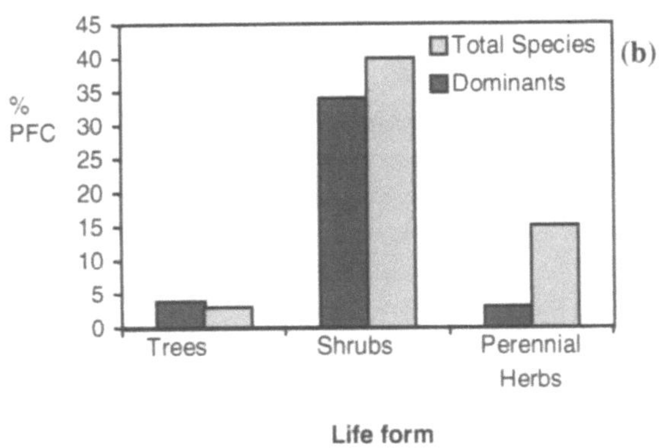

Figure 13. Relative contribution of different life forms to (a) total species richness and (b) percent projective foliage cover (PFC) for scrub-heath sites at Eneabba, WA. *Source:* Total species based on data in Froend (1987); Dominants based on data in Lamont and Struthers (1988).

ground cover and thereby possibly combating soil erosion after disturbance events such as fire.

A generalised hypothetical profile of root distribution beneath a typical scrub-heath (Figure 14) highlights the marked concentration of roots in a thick surface mat. Approximately 60% of root biomass is contained in the surface 30 cm and comprises fine roots, laterals and upper sections of tap roots. The remaining root biomass almost entirely comprises tap roots occupying the soil zone from 30 cm down to a depth of 3.5 m or more (Lamont and Bell, 1990).

This analysis shows that there are, not surprisingly, qualitative similarities between the structure and function of kwongan in general and our case study for the scrub-heath. It would be useful to investigate specific examples of the four other kwongan sub-types to develop a better picture of variation between sites, and determine to what extent the mimic described in the next section is applicable to heath sites generally.

3.8. *Composition of mimic*

In this section we develop a mimic of the Eneabba scrub-heath discussed above. The current agricultural system of this area is described, a design is proposed and the compatibility of this design with the current system is briefly considered.

It should be kept in mind that determining what to mimic is not always an easy task. Establishing the nature of pre-European vegetation in a particular locale is problematic. In some areas there are few extensive remnants available from which to interpolate, while in other areas the historical records are either scanty or await further research (O'Connor, unpubl. data). The Vegetation Survey of Western Australia map series and accompanying explanatory memoirs (e.g. Beard and Parker, 1976) are a useful starting point for determining what to mimic in any particular locale.

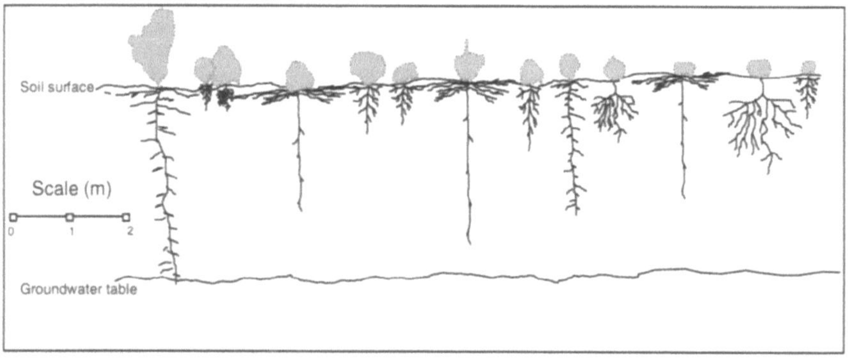

Figure 14. Typical profile of roots and aerial parts for a scrub-heath on deep sands in south western Australia. Dimorphic root systems that can reach the groundwater table are confined to the larger shrubs. *Source:* Based on data from Dodd et al. (1984) and Heddle (1980).

3.8.1. *Current land-use systems*
The west Midlands agricultural region of south western Australia was extensively alienated to agriculture in the period 1931–1977 (Anon., 1979). Today production focuses on sheep and beef cattle as well as ley farming (wheat and sheep) in the northern parts of the region (Anon., 1979). The fodder tree tagasaste (*Chamaecytisus proliferus*) has been introduced into the region for animal production in both plantation and alley farming configurations over the last decade (see Lefroy and Stirzaker, this volume). A survey in 1992 estimated that 50,000 hectares had been established to this exotic species (T. Wiley, pers. comm.) The high proportion of native vegetation relative to other areas of south western Australia provides opportunities for honey production, wildflower harvesting and tourism. Of note is the Beekeepers Reserve which is predominantly scrub-heath on deep sandy flat, extending over some 42,000 ha in a long coastal section west of Eneabba (Beard and Parker, 1976).

Limitations to the current agricultural system's sustainability include wind erosion, rising water tables and associated salinity, water repellence of soil surfaces, further clearing of native vegetation and the effect of *Phytophthora*-dieback on remnant heath (Agriculture Western Australia et al., 1996a and 1996b; Anon. undated). The development plan for this region to the year 2006 highlights increased mining, population increases and subdivision of lands into non-viable parcels as being non-ecological threats to agricultural sustainability (Anon., 1996).

3.8.2. *Mimic design*
The design presented here is a very theoretical functional prescription because the structural and functional data on which it is based does not fully describe the scrub-heath ecosystem. To be fully effective, an agricultural system attempting to mimic scrub-heath would obviously require an appropriate mix of spring- and summer-active small tree and shrub perennials distributed across the landscape. We suggest that the total projective foliage cover of the mimic should achieve an equilibrium value of at least 75% and maintain that throughout the year. These perennials should ideally have deep root systems that allow them to draw on ground water during the drier months. At least one leguminous woody species should be included to reduce the requirement for nutrient inputs. Species adapted to fire would be beneficial, except where these woody perennials are the source of saleable products (e.g. plants such as *Santalum* for sandalwood, and *Eucalyptus* which are grown as oil mallees) rather than simply utilitarian elements in counteracting rising water tables and reducing nutrient eutrophication. In that scenario it is likely that the continual development of improved cultivars of exotic and native species might reduce the risk of destruction by fire.

Considerations that apply to the arrangement of the dominant woody component in the heath mimic are similar for the salmon gum woodland mimic (Figures 8 and 9). Further hydrological modelling is required to guide the placement of the elements across the landscape. In light of the work by Lefroy and Stirzaker (this volume) the proportion of the landscape (in terms of area from which soil water is captured) that must be given over to woody perennials may not leave sufficient land for agricultural plants. An alley format would achieve two important objectives for the scrub-heath mimic. Firstly it allows wide distribution of deep-rooted woody perennials across the landscape. Secondly the alleys are wide enough to allow grain production to occur.

3.8.3. *Compatibility of mimic with existing agriculture*
The agricultural mimic system will not match the scrub-heath in its array of tightly integrated nutrient-uptake mechanisms that prevent nutrient leaching. Nor indeed may this be necessary, since the land in question will have been heavily P (and trace element) treated and be richer in N than the heath soils. These are the very circumstances in which specialised nutrient-acquiring mechanisms would be redundant. As most of this nutrient leaching potentially occurs under the alleys, it is only through timely and suitably restrained application of nutrients as fertiliser or animal excreta that leaching can be avoided. Precision farming technology may assist in achieving the required level of control.

The existing agricultural system requires an increase in the number of woody perennials distributed across the landscape. This reduces the area of land previously available to pasture or cropping which would need to be offset by the increased sustainability afforded by the mimic design and, most likely, sources of income from the introduced species.

4. Conclusions

A key precursor to designing an ecosystem mimic is the elucidation of the functional characteristics of the natural vegetation being mimicked. This is a challenging task for several reasons. Our understanding of many south-western Australian vegetation communities is still incomplete, although this is not surprising given their functional complexity. Acquiring this knowledge is difficult in some locales, often because there are not sufficiently large remnants to investigate. Previous studies may be few in number, contain superficial summaries or were conducted for other purposes (e.g. biogeography, land management, minesite rehabilitation, community analysis or plant physiology). For instance there is an extensive literature on seed production and storage in kwongan communities, which is largely irrelevant to an agricultural system which is not designed to be self-replacing. Clearly such issues could be important if the system were to be refined in the future.

As such no single set of data includes all attributes necessary for the purposes of this paper except where we focus on a single well-studied site. Thus we rely on an amalgam of several sets of data to complete the picture.

This exercise has suggested that the establishment of a centralised database of south-western Australian plant ecomorphological and functional features would be of benefit to future research in this field.

The knowledge base for plant ecophysiological characteristics (e.g. growth form, life form, root morphology, size, leaf-area, longevity, nutrient acquisition specialisation) lies strewn across a multitude of publications and unpublished data. Ideally this should be entered into the database for each species, together with details of the site and community from which the data are found. Family-level analysis is too coarse for our purpose because many families have the full range of expression of life form and root morphology (e.g. Fabaceae, Proteaceae, Myrtaceae) (Pate et al., 1984). Refining the database to the genus level would still leave some generalities that may confound (e.g. *Acacia* includes a variety of root morphologies and fire response types, and occurs in both upper- and lower-stratum shrub layers of kwongan; Dodd et al., 1984). A database would serve to highlight gaps in the structural and functional knowledge, thereby directing further field research. Looking further into the future the database could be the core of a computer-based system for modelling system function, and in turn for developing, managing and refining agricultural mimics.

Such a database would, however, perhaps more than anything highlight the lack of data on most vegetation types. The acquisition of such data is time consuming and often costly. We thus need a clear rationale for prioritising where this work should be done. Our studies have suggested that the eucalypt woodlands in the agricultural areas of Australia are perhaps the systems with the most pressing need for such information. While some quite detailed physiological and ecosystem work has been carried out in the eucalypt-dominated forest systems in higher rainfall areas (Bell and Williams, 1997; Keith, 1997), relatively little has been conducted in the lower rainfall areas.

However, even if a considerable effort is put into ecosystem studies in eucalypt woodlands, there will continue to be information gaps for all ecosystems, and we will continue to be forced to make best-bet assessments based on current knowledge and a set of assumptions about how the systems work. We cannot afford the luxury of waiting for the perfect set of data before trying to utilise what we have got already. It may be a great leap of faith (or sticking one's neck out to its limit) to take existing data and design a mimic, but at least that leap will provide us with experimental systems which we can monitor and use to test our hypotheses and assumptions. Given the experience of Jackson and Jackson and Ewel (this volume) who have done just that, it may help solve the huge problems facing agriculture in the next century.

5. References

Agriculture Western Australia, Department of Conservation and Land Management, Department of Environmental Protection, and Water and Rivers Commission (1996a) Salinity: A Situation Statement for Western Australia. Government of Western Australia, Perth

Agriculture Western Australia, Department of Conservation and Land Management, Department of Environmental Protection, and Water and Rivers Commission (1996b) Western Australian Salinity Action Plan. Government of Western Australia, Perth

Anon (1979) Western Australia: An Atlas of Human Endeavour 1829–1979. Departments of Education and Lands and Surveys, Western Australia

Anon (1996) Central Coast Regional Strategy. Western Australian Planning Commission, Perth

Anon (undated) Decade of Land Care Plan: Western Australia. Department of Agriculture, Perth

Atkins L and Hobbs RJ (1995) Measurement and effects of fire heterogeneity in south-west Australian wheatbelt vegetation. CALMScience Supplement 4: 67–76

Beard JS (1984) Biogeography of the kwongan. In: Pate JS and Beard JS (eds) Kwongan: Plant Life of the Sandplain, pp 1–26. University of Western Australia Press, Nedlands

Beard JS (1990) Plant Life in Western Australia. Kangaroo Press, Kenthurst, NSW

Beard JS and Pate JS (1984) Foreword. In: Pate JS and Beard JS (eds) Kwongan: Plant Life of the Sandplain, pp xvii–xxi. University of Western Australia Press, Nedlands

Beard JS and Sprenger BS (1984) Geographical Data from the Vegetation Survey of Western Australia. Vegmap Publications, Perth

Beard JS and Parker AZ (1976) The Vegetation of the Dongara Area, Western Australia: Map and Explanatory Memoir, 1:250000 series. Vegmap Publications, Perth

Bell DT and Stephens LJ (1984) Seasonality and phenology of kwongan species. In: Pate JS and Beard JS (eds) Kwongan: Plant Life of the Sandplain, pp 205–226. University of Western Australia Press, Nedlands

Bell DT, Hopkins AJM and Pate JS (1984) Fire in the kwongan. In: Pate JS and Beard JS (eds) Kwongan: Plant Life of the Sandplain, pp 178–204. University of Western Australia Press, Nedlands

Bell TL and Pate JS (1996) Relationships between fire response, morphology, root anatomy and starch distribution in south-west Australian *Epacridaceae*. Annals of Botany 77: 357–364

Bell DT and Williams JE (1997) Eucalypt ecophysiology. In: Williams J and Woinarski J (eds) Eucalypt Ecology: Individuals to Ecosystems, pp 168–196. Cambridge University Press, Cambridge

Brundrett MC and Abbott LK (1992) Roots of jarrah forest plants. I. Mycorrhizal associations of shrubs and herbaceous plants. Australian Journal of Botany 39: 445–457

Cannon WA (1949) A tentative classification of root systems. Ecology 30: 542–548

Carnahan JA (1990) Vegetation. Australian Surveying and Land Information Group, Australian Government Publishing Services, Canberra

Dodd J, Heddle EM, Pate JS and Dixon KW (1984) Rooting patterns of sandplain plants and their functional significance. In: Pate JS and Beard JS (eds) Kwongan: Plant Life of the Sandplain, pp 146–177. University of Western Australia Press, Nedlands

Enright NJ, Lamont BB and Marsula R (1996) Canopy seed bank dynamics and optimum fire regime for the highly serotinous shrub, *Banksia hookeriana*. Journal of Ecology 84: 9–17

Farrington P, Bartle GA, Watson GD and Salama RB (1994) Long term transpiration from two Eucalypt species in a native woodland estimated by the heat-pulse technique. Australian Journal of Ecology 19: 17–25

Froend RH (1987) Investigations into species richness patterns in the northern sandplain region of Western Australia. Unpublished PhD thesis, The University of Western Australia, WA

Griffin EA (1994) Floristic survey of northern sandplains between Perth and Geraldton. Unpublished report to the Heritage Council of WA. Department of Agriculture, South Perth

Griffin EA, Hopkins AJM and Hnatiuk RJ (1983). Regional variation in mediterranean-type shrublands near Eneabba, south-western Australia. Vegetatio 52: 103–127

Heddle EM (1980) Effects of Changes in Soil Moisture on the Native Vegetation of the Northern Swan Coastal Plain, Western Australia. Forestry Department of Western Australia, Bulletin No. 92

Hnatiuk RJ and Hopkins AJM (1980) Western Australian species-rich kwongan (sclerophyllous shrubland) affected by drought. Australian Journal of Botany 28: 573–585

Hobbs RJ (1992) Function of biodiversity in mediterranean ecosystems in Australia: definitions and background. In: Hobbs RJ (ed) Biodiversity in Mediterranean Ecosystems in Australia, pp 1–25. Surrey Beatty and Sons, Chipping Norton, NSW

Hobbs RJ and Atkins L (1991) Interactions between annuals and woody perennials in a Western Australian wheatbelt reserve. Journal of Vegetation Science 2: 643–654

Hobbs RJ, Saunders DA, Lobry de Bruyn L and Main AR (1993) Changes in the biota. In: RJ Hobbs and DA Saunders (eds) Re-integrating Fragmented Landscapes. Towards Sustainable Production and Nature Conservation, pp 65–106. Springer-Verlag: New York

Jeschke WD and Pate JS (1995) Mineral nutrition and transport in xylem and phloem of *Banksia* prionotes and proteaceous trees of dimorphic root morphology. Journal of Experimental Botany 46: 895–905

Keith H (1997) Nutrient cycling in eucalypt ecosystems. In: Williams J and Woinarski J (eds) Eucalypt Ecology: Individuals to Ecosystems, pp 197–226. Cambridge University Press, Cambridge

Lamont BB (1984) Specialised modes of nutrition. In: Pate JS and Beard JS (eds) Kwongan: Plant Life of the Sandplain, pp 126–145. University of Western Australia Press, Nedlands

Lamont BB (1985). Gradient and zonal analysis of understorey suppression by *Eucalyptus wandoo*. Vegetatio 63: 49–66

Lamont BB and Bell DT (1990) Plant and soil ecology near Eneabba. In: Bell DT and Lamont BB (eds) Plant and Soil Ecology of Natural Areas and Rehabilitated Minesites Near Eneabba. Minerals and Energy Research Institute of Western Australia, Perth. Report No. 45: 95–112

Lamont BB and Struthers R (1988) Ecomorphological characters, Western Australia: Eneabba. In: Specht RL (ed) Mediterranean-type Ecosystems: A Data Source Book, p 31. Kluwer Academic Publishers, Dordrecht

Lamont BB, Downes S and Fox JED (1977) Importance-value curves and diversity indices applied to a species-rich heathland in Western Australia. Nature 265: 438–441

Lamont BB, Hopkins AJM and Hnatiuk RJ (1984) The flora – composition, diversity and origins. In: Pate JS and Beard JS (eds) Kwongan: Plant Life of the Sandplain. University of Western Australia Press, Nedlands

Lantzke N (1992) Soils of the Northam Advisory District. Volume I The Zone of Ancient Drainage. Department of Agriculture, Perth, Western Australia

Low AB and Lamont BB (1990) Aerial and below-ground phytomass of *Banksia* scrub heath at Eneabba, southwestern Australia. In: Bell DT and Lamont BB (1990) Plant and Soil Ecology of Natural Areas and Rehabilitated Minesites near Eneabba, pp 95–112. Minerals and Energy Research Institute of Western Australia, Report No. 45

McArthur WM (1991) Reference Soils of South-western Australia. Department of Agriculture, Western Australia. Perth

Newbey K (1979) The vegetation of central south-coastal Western Australia. Unpublished MPhil thesis, Murdoch University, WA

Norton DN, Hobbs RJ and Atkins L (1995) Fragmentation, disturbance and plant distribution: mistletoes in woodland remnants in the Western Australian wheatbelt. Conservation Biology 9: 426–438

Nulsen RA (1993) Changes in soil properties. In: Hobbs Rl and Saunders DA (eds) Reintegrating Fragmented Landscapes: Towards Sustainable Production and Nature Conservation, pp 107–145. Springer-Verlag, New York

O'Connor MH (1995) Ecosystematic Agriculture: Theory and Application for South-western Australia. Report for Landcare Project 592, Curtin University of Technology

Pate JS (1994) The mycorrhizal association: just one of many nutrient acquiring specializations in natural ecosystems. Plant and Soil 159: 1–10

Pate JS, Dixon KW and Orshan G (1984) Growth and life form characteristics of kwongan species. In: Pate JS and Beard JS (eds) Kwongan: Plant Life of the Sand plain, pp 84–100. University of Western Australia Press, Nedlands, WA

Pate JS and Dell B (1984). Economy of mineral nutrients in sandplain species. In: Pate JS and Beard JS (eds) Kwongan: Plant Life of the Sand Plain, pp 227–252. University of Western Australia Press, Nedlands, WA

Pate JS, Meney KA and Dixon KW (1991) Contrasting growth and morphological characteristics of fire-sensitive (obligate seeder) and fire resistant (resprouter) species of Restionaceae (southern hemisphere restiads) from south-western Western Australia. Australian Journal of Botany 39: 505–525

Prober SM and Thiele KR (1993) The ecology and genetics of remnant grassy white box woodlands in relation to their conservation. The Victorian Naturalist 110: 30–36

Rose PW (1993) Production of Habitat Hollows by Wheatbelt Eucalyptus. Report prepared for Department of Conservation and Land Management, Perth

Specht RL (1994) Heathlands. In: Groves RH (ed) Australian Vegetation, pp 321–344. University Press, Cambridge, UK

Tennakoon KU, Pate JS and Arthur D (1997) Ecophysiological aspects of the woody root hemiparasite *Santalum acuminatum* (R.Br.) A. DC and its common hosts in south western Australia. Annals of Botany 80: 245–267

True D (1995) A Review of Grassy Woodlands in the Western Australian Wheatbelt. A Report to the Australian Nature Conservation Agency, National Biodiversity Conservation Program. Mattiske Consulting Pty Ltd, Perth

Yates CJ (1995) Factors limiting the recruitment of *Eucalyptus salmonophloia* F. Muell. (Salmon Gum). Unpublished PhD thesis, Murdoch University, WA

Yates CJ and Hobbs RJ (1997) Woodland restoration in the Western Australian wheatbelt: a conceptual framework using a state and transition model. Restoration Ecology 5: 28–35

Yates CJ, Hobbs RJ and Bell RW (1994a) Factors limiting the recruitment of *Eucalyptus salmonophloia* in remnant woodlands. I. Pattern of flowering, seed production and seed fall. Australian Journal of Botany 42: 531–542

Yates CJ, Hobbs RJ and Bell RW (1994b) Landscape-scale disturbances and regeneration of semi-arid woodlands of south-western Australia. Pacific Conservation Biology 1: 214–221

Agroforestry Systems **45**: 393–409, 1999.

Social and economic challenges in the development of complex farming systems

D. J. PANNELL
Faculty of Agriculture, The University of Western Australia, Nedlands 6907, Australia; E-mail: david.pannell@uwa.edu.au

Key words: adoption of innovations, sustainability, uncertainty

Abstract. The fundamental challenge in developing a new farming system is to have it adopted and maintained by farmers. The difficulty of achieving widespread adoption is increased if the new farming system is complex and/or radically different to current farming practice. This paper is a review of these issues with a focus on farming systems based on mimicry of natural ecosystems. It is proposed that there are four conditions which are necessary for an individual farmer to adopt an innovative farming-system: awareness of the innovation, perception that it is feasible to trial the innovation, perception that the innovation is worth trialing, and perception that the innovation promotes the farmer's objectives. Challenges involved in meeting each of these conditions are discussed. It is concluded that the most important challenges in developed countries are: (a) developing a farming system that is in fact more profitable than current practice; (b) assessing whether a system is in fact more profitable than current practice; and (c) overcoming the problem of deep uncertainty about the technology. In developing countries one must add the additional challenges of (d) high interest rates/high discount rates; and (e) insecure or inequitable land tenure.

1. Introduction

'A radical change is necessary if we are to develop a sustainable agriculture' (Jackson, 1984, p. xiii). The particular radical change being considered here is the development of farming systems which mimic natural ecosystems (e.g. Jackson and Bender, 1984). Pursuit of this goal will certainly challenge our scientific/agronomic capacities, but it should also be recognised that the social and economic challenges to development of any new farming system are formidable. The aim of this paper is to review these social and economic challenges. A recurring issue will be the complexity of the systems being considered. They are complex not only in their biology, but also in their management, in their economic impacts and in the social attitudes and perceptions which they generate. The term complex is used here to mean that there are more elements than in a simple system, there is more to learn before the system can be competently managed, and therefore there is more chance of problems occurring.

The fundamental challenge is to develop a farming system that will be adopted and maintained by farmers. Very rarely are farming systems transformed. The available examples where it has occurred are for extreme cases

– extreme opportunities to exploit or extreme problems to overcome. In practice, in a democracy it is not in society's power to directly select a sustainable integrated farming system, because it is not possible to simply order farmers to adopt the chosen system. Rather we are constrained by the effectiveness of the various tools that can be used to encourage adoption. The farming system that comes into existence will be that which results from farmers' reactions to the government policies and institutions in place (Hollick, 1990; Pannell, 1997).

There is a wealth of empirical evidence on the factors that influence farmers' adoption of innovations (e.g. Feder and Umali, 1993; Feder et al., 1985; Lindner, 1987), and it includes some very clear-cut messages. Unfortunately, responding to these messages is often not straightforward. We can identify the conditions necessary to achieve adoption of an agricultural innovation but it remains difficult to meet the conditions.

The next section is an outline of the conditions for adoption of an agricultural innovation. Thereafter these conditions are reviewed in more detail and the specific challenges for complex farming systems are identified. An example is presented which illustrates many of the points arising in the review. In the conclusion, those challenges which are likely to be most difficult to meet are suggested.

2. The conditions for adoption of an agricultural innovation

It is important to recognise that a farming system based on mimicry of natural ecosystems would be a new innovation from the point of view of farmers. They are likely to come to it with scepticism, uncertainty, ignorance, prejudices and preconceptions and with an existing farming system that may or may not be operating as they would wish, but is at least operating. Unless they are new to farming, they will have trialed other innovations in the past and concluded that at least some of them fell far short of the claims made for them. They will be particularly wary of a system that is radically different from that with which they are familiar and comfortable. They will almost certainly hold an attitude that the scientists advocating such a radical system do not understand the realities of farming, or at least of their farm.

In getting past this initial set of attitudes and beliefs, there are several specific hurdles which must be overcome. The following sub-sections describe the states of farmer awareness or knowledge which must be achieved.

2.1. *Awareness of the innovation*

In this context, 'awareness' means not just awareness that an innovation exists, but awareness that it is *potentially* of practical relevance to the farmer. Reaching this point of awareness is a trigger which prompts the farmer to open his or her ears – to begin noting and collecting information about the innovation in order to inform their decision about whether or not to go to the next step of trialing the innovation.

2.2. *Perception that it is feasible to trial the innovation*

There is strong evidence that, the world over, most farmers are 'risk-averse' (Antle, 1987; Bardsley and Harris, 1987; Binswanger, 1980; Bond and Wonder, 1980; Myers, 1987; Pluske and Fraser, 1996). This is evident from the observation that they will not leap into large-scale adoption of a new innovation. Rather, they generally employ small-scale trials, adjusting the scale either upwards towards full adoption or downwards towards disadoption as they gain knowledge and confidence in their perceptions about its performance.

This trial phase is very important, perhaps the most important phase in determining final adoption or disadoption. If small-scale trials are not possible or not enlightening for some reason, the chances of widespread adoption are greatly diminished. This is because farmers will be very unlikely to leap to full-scale adoption due to the real risk that the innovation will prove a full-scale failure. This risk of failure should be viewed as part of the cost of gaining high quality information about the innovation. Clearly, the larger the scale of the trial that is necessary, the larger is the cost of this information, and the less likely the farmer is to make the investment in trialing.

2.3. *Perception that the innovation is worth trialing*

Conducting a trial incurs costs of time, energy, finance and land that could be used productively for other purposes. To be willing to trial an innovation, the farmer's perceptions of it must be sufficiently positive to believe that there is a reasonable chance of adopting the innovation in the long run. It is not necessary for the innovation to be thought to be better than current practice, because the farmer realises that the results of a trial may revise his or her perceptions upwards. However, it cannot be too much worse or the chance of recovering the cost of the trial through later productivity improvements will be too low.

2.4. *Perception that the innovation promotes the farmer's objectives*

Lindner (1987), in a wide-ranging review of the adoption and diffusion literature concluded that the objectives of individual farmers figure centrally in the adoption and diffusion process. He found that,

> there is compelling empirical support for this emerging consensus that the final decision to adopt or reject is consistent with the producer's self interest (p. 148).

Self interest in this context is considerably broader than merely profit. It may, for example, include objectives related to risk, leisure and environmental protection. Nevertheless, profit is a particularly important element of self-interest. Indeed, the available evidence indicates that although the speed of uptake of innovations is influenced by a range of factors (including social and demographic factors), the final level of uptake seems to depend primarily on economic factors (e.g. Marsh et al., 1995). There is also strong evidence that even for innovations oriented towards resource conservation, economic considerations are the most important determinants of actual adoption decisions (Cary and Wilkinson, 1997; Sinden and King, 1990). Much has been made of the need to promote an ethic of stewardship among farmers, but based on statistical analysis of actual farmer behaviour Sinden and King (1990) concluded that,

> While the stewardship motivation and personal factors encourage perception and recognition of a problem, economic factors promote actual adoption (p. 179).

Similarly, Cary and Wilkinson (1997) found that,

> Generally, the best way to increase the use of conservation practices to overcome land degradation . . . will be to ensure the practices are economically profitable (p. 20).

The finding that self interest (broadly defined) drives adoption decisions has strong implications for sustainability-related issues in agriculture. It is likely that some farmers will respond somewhat to perceived social pressures or community expectations, but in aggregate this tendency will be swamped by the pursuit of self-interest. If the existing technologies being promoted are not sufficiently profitable (or more generally beneficial), we must either develop new technologies, or make the existing technologies more attractive through such means as subsidies, tax concessions or, in the extreme, taxes or legal penalties for non-adoption.

The use of the terms 'awareness' and 'perceptions' in the headings above highlight that these are social and economic issues. Of course they are influenced very much by biological and physical factors, as well as other socio-economic factors, in ways which are discussed in the following sections outlining the specific challenges to meeting the conditions for adoption.

3. The challenges in raising awareness

There is little empirical evidence about the rate at which farmers become aware of new innovations but what evidence there is would be of serious concern to anyone wishing for rapid adoption. Gibbs and Lindner (1986) in a survey-based study found that the time taken for farmers in South Australia to become aware of the existence of new innovations varied markedly. For many farmers it amounted to years despite the presence of extension activities designed specifically to raise awareness.

Marsh et al. (1995) found that there was very wide variation among farmers in the time taken to commence trialing a new crop, lupins, in Western Australia. This was at least partly due to long lags until awareness of the crop by some farmers. Lupins are a legume crop which have subsequently proved to be highly profitable in appropriate niches of the Western Australian farming system, but following the first release of high-yielding varieties in 1979, it took eight years for the number of farmers growing any lupins at all to peak (Figure 1). In some shires, the number of farmers growing lupins was still increasing in 1992. If it takes so long for a clearly productive and profitable innovation to diffuse fully, we might expect diffusion of less profitable innovations to occur on a time scale of decades rather than years.

4. The challenges in making a trial feasible/worthwhile

One element of 'trialability' is the size of trial that is necessary. As noted earlier, the larger this is, the less likely the farmer is to make the investment in trialing. There may be particular cause for concern about systems based on ecosystem mimicry from this point of view. Such a system clearly would require a minimum scale to be effective.

The farmer must have the resources to be able to conduct a trial. For

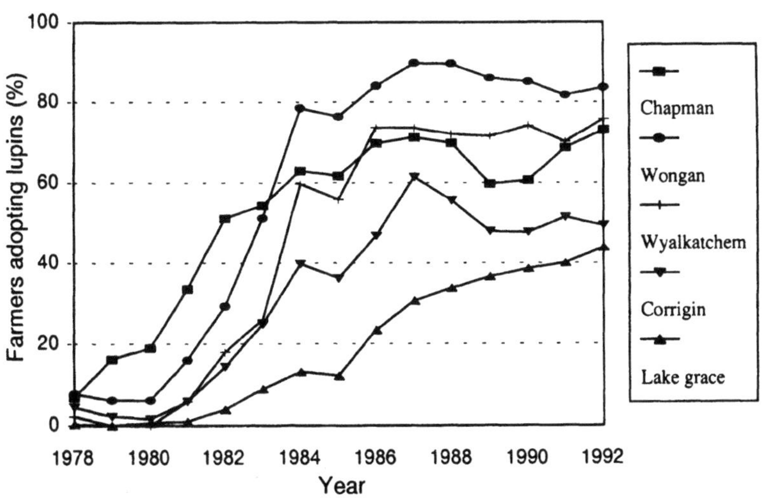

Figure 1. Adoption of lupins by farmers in different shires of Western Australia. *Source:* Marsh et al. (1995).

example, in a recent survey of Western Australian farmers, it was found that there was a positive relationship between a farmer having extra labour on the farm and the probability that they are conducting trials of new grain legumes (A. Abadi, The University of Western Australia, pers. comm.). Without access to relatively high levels of labour, the cost of the time the farmer must give up to sow the trial is too high, primarily because sowing the trial coincides with sowing of the main, money-earning crop. As a second example, upland farmers in some parts of the Philippines have been encouraged to build rock walls along contours to reduce soil erosion. However, this is extremely labour intensive, and most farmers cannot spare the time from other farming duties.

Another requirement for a trial to be worthwhile is for the results of the trial to be observable; there is no value in the trial otherwise. In terms of direct, saleable output from the system, this is usually not a problem. However if a significant part of the benefits of a system stem from reductions in resource degradation or other such indirect benefits, the issue of observability can be critical.

Many degradation processes are slow relative to the time frames used for most management decision making (e.g. dryland salinisation, soil acidification). In evaluating a trial, one requires the degradation to be continued under the old farming system for long enough for differences under the new farming system to become apparent. Historical degradation is not useful for this. Observation commences with the trial of the new system. Obviously, the slower the degradation process, the longer it will take to be convinced about differences in degradation rates. Unfortunately there are additional factors which further delay the confident recognition of any such difference. They include:

(a) variability over time and space in climatic conditions;
(b) variability over space in soil physical structure, chemical composition, and topography;
(c) variability over space and time in weeds, pests and diseases;
(d) variability over time in other management practices, due, for example, to changing economic circumstances.

These sources of variability are overlaid on the trial, and so their impacts are confounded with any effect attributable to the new farming system. Spatial variation is always an issue for interpretation of trials, but the combination of variation in space and time which affects observations of long-term trends in soil degradation makes them particularly difficult to interpret. At the very least, they increase the duration of the trial necessary to reach confident conclusions. In the extreme, they may mean that a trial could never be conclusive. In either case, the prospective benefits of conducting a trial are reduced, potentially by so much that it is not worth conducting a trial.

Can we not circumvent this problem by *telling* the farmer about the impacts of the new system? Unfortunately, this is not sufficient. As noted earlier, farmers are wary of outside experts telling them what is best for them. In general, this wariness is well-founded. It is more difficult than most realise to recognise and account correctly for the many subtle and interacting factors that determine the impact of a technology on any individual farmer's welfare. An obvious example is advice based on biological or physical considerations, without adequate attention to economics. In general, farmers will not believe what we tell them they should do. They have to see results for themselves to be convinced. Highly credible information sources, such as respected individuals, will help to promote trialing of an innovation, but their advice will almost never be accepted as a substitute for a trial.

A practical demonstration on another farmer's property can be convincing, but even then the farmer will not be convinced if the demonstration is in a situation which is too different from his or her own situation. Farmers are understandably wary that results from a remote trial may not apply to them. This could be due to factors such as soil differences, topography, labour, scale, or machinery. Indeed, bioeconomic modelling shows that the optimal farm management strategy is often highly sensitive to differences in these factors (e.g. Pannell, 1995).

Lindner et al. (1982) demonstrated the importance of distance to information source as a determinant of adoption, showing that farmer adoption of a particular innovation (trace-element fertilisers) decreased with increasing distance from the office of the department of agriculture. This is likely to be due in part to farmer perceptions about the local relevance of the information.

5. The challenges in achieving a perception of technical soundness

A fundamental challenge is to ensure that the system is, in fact, technically sound. Does it in fact deliver with sufficient reliability the biological and physical benefits being sought? A common difficulty in this regard is reliable establishment of new plants. Self-regenerating annual pasture species, shrubs and trees can all face difficulties in this regard. Usually, the problem is not whether it can be done at all, but whether it can be done on a commercial scale at an acceptable cost.

Another problem relating to actual technical soundness is that in some cases, we may wish to promote a system before we have comprehensive scientific evidence about its effectiveness. Salinity in low to medium rainfall areas of Western Australia is a case in point. We know in general that establishment of perennials are likely to help reduce the rate of rise in the saline water table. However we are unable yet to give advice on where in the landscape and at what density on a particular farm a given number of trees should be planted to maximise their impact on the water table. Nevertheless, salinity is perceived to be a major threat, so tree planting is widely endorsed and promoted. It is true that many trees have been planted, but it is widely believed that the numbers planted so far are much less than the numbers that would be needed to fully address the problem (e.g. Bartle et al., 1996; Anon., 1997).

Another example with some similarities relates to biodiversity. For a variety of reasons, we are unable to give clear advice about the farm management practices that should be recommended to enhance biodiversity. The reasons included: (a) uncertainty about what impacts different farming practices have on biodiversity, (b) ambiguity about how biodiversity should be measured, and (c) ambiguity about how much diversity is enough (e.g. Main, this volume).

Technical soundness can also be greatly affected by the quality of implementation by farmers. Unfortunately this is most likely to be true for complex farming systems, for which the risk of poor implementation is higher. For example, Garcia et al. (1995d) recorded that among upland farmers who had adopted hedgerows in the Philippine province of Palawan, almost half had used an intra-row spacing between plants of one metre or more. They commented that,

> Plants in a contour hedgerow are supposed to be very close together to effectively obstruct the flow of runoff, thereby collecting the eroded soil at the base of the hedgerows. The wide distance between plants in a hedgerow would, therefore, minimise the effectiveness of the hedgerows in controlling erosion (p. 69).

If implementation is poor, it not only means that current results from the system are poor, but it also creates the risk that farmers will wrongly draw negative conclusions about the system in general, leading to disadoption. A further possible negative spin off is non-adoption by other farmers who are standing back and watching for successful results. We know that a large proportion of farmers do wait to observe whether innovative farmers successfully apply new practices before attempting to test the practices themselves (Abadi Ghadim et al., 1996).

Finally, the farmer's perception of technical soundness is also affected by the observability of results from trials of the system, as discussed in the previous section. For technologies addressing processes of slow degradation, it may be a very long time indeed before a farmer's uncertainty about the soundness of the technology is sufficiently reduced to prompt widespread adoption.

It is important to recognise that technical soundness may not be enough to promote adoption. For example, if the technology is designed to address a public good/externality problem, the main beneficiaries may not be the farmers who we wish to have adopting the technology. Mechanisms would need to be put in place to ensure that the technology is also consistent with the farmer's own objectives (see next section).

6. The challenges in meeting farmers' objectives

6.1. *Making the system profitable*

The earlier quotes from Lindner (1987), Cary and Wilkinson (1997), and Sinden and King (1990) emphasise the importance of ensuring that the new system is profitable. In this context, profitable means that the new farming system is economically superior to the current farming system. It is not sufficient for it to generate benefits in excess of input costs (see Section 2); it must also cover opportunity costs – the profits from alternative methods of resource use which must be foregone in order to use the resources in the new way. Thus the hurdle is higher than sometimes recognised by scientists.

The discussion thus far has focused on perceptions, but the only way to create enduring perceptions of profitability is for the system to be profitable in fact. With large amounts of energy and resources devoted to persuasion, it may be possible to temporarily create an overly optimistic perception of a system, but once farmers have personal experience with the system, they will certainly put more weight on this than on any amount of persuasion or exhortation. Thus, successful trials or successful adoption are necessary for favourable perceptions in the medium to long term.

One potential threat to the actual profitability of a new, complex system is that there are likely to be substantial costs in establishing and maintaining the new system. This is particularly true of systems involving trees. Even if labour and finance availability are not absolutely constrained, they add significantly to costs which must be at least offset by the benefits.

There may be additional negative impacts of the new technology, such as,

(a) harbouring weeds, pests, or diseases;
(b) directly competing with the crop for light and nutrients (e.g. in intercropping or agroforestry systems); or
(c) physically obstructing other farm activities (e.g. preventing easy movement of agricultural machinery, or constraining grazing by livestock in order to avoid damage to seedlings).

These problems, and any others, are not necessarily fatal, but they must be set against the expected long- and short-term benefits of the new system to reach a realistic assessment of its value.

6.2. *Determining whether the system is profitable*

We have discussed the farmers' perceptions of whether a system is profitable, and some factors relating to whether the system is actually profitable. A third, related point is the capacity of scientists or extension agents to determine whether a system is profitable in order for them to be able to tell whether they should promote it or seek to modify it further. Because the process of adoption is so slow, there are great advantages in not having to wait for diffusion of the innovation to be complete before being able to assess the innovation's profitability. For a complex farming system, this is a much more difficult task than often recognised. To illustrate, Table 1 shows a list of factors which influence the profitability of legume-based farming systems (adapted from Pannell, 1995). It should be clear from this extensive list that simple economic assessments of complex farming systems are likely to be of little value, or even of negative value if they seriously mislead (as is likely).

Even if we are aware of all of these factors, uncertainty about their quan-

Table 1. Factors affecting the relative profitability of legume-based dryland farming systems in areas of Mediterranean climate.

Number	Factor
1.	*Short term profit factors*
1.1	Legume grain yield (depends on weather, soil type, weeds and so on)
1.2	Legume crop stubble (post-harvest residue) production and feed quality
1.3	Pasture production level, quality and timing
1.4	Yields of non-legume crops and pastures
1.5	Input costs
1.6	Output prices (for legume crops, livestock, livestock products and non-legume crops)
2.	*Dynamic factors (short to medium term)*
2.1	Nitrogen fixation by legumes and yield boost from other factors (e.g. disease break, improved soil structure)
2.2	Pasture density
2.3	Legume crop disease
2.4	Stubble management for crop seeding
2.5	Weed control
2.6	Tillage method
2.7	Carry over of nutrients
3.	*Sustainability factors*
3.1	Herbicide resistance
3.2	Soil degradation (acidification, organic matter decline, erosion, nutrient decline, non-wettability)
3.3	Pasture legume persistence
3.4	Pasture establishment costs
4.	*Risk factors*
4.1	Yield variability
4.2	Price variability
4.3	Yield/price covariance
4.4	Flexibility of the enterprise in response to changed conditions
4.5	The farmer's attitude to risk
5.	*Whole-farm factors*
5.1	Total crop area
5.2	Machinery capacity
5.3	Total feed supply (timing and quality)
5.4	Feed requirements of livestock on hand (timing and quality)
5.5	Finance availability and cost
5.6	Labour availability, quality and cost
5.7	The farmer's objectives (profit, risk reduction, sustainability, leisure)
5.8	The farmer's knowledge and experience

titative magnitudes is a substantial problem for new complex systems. As noted earlier, because some of the processes are very slow and are overlaid with high levels of variability, it can be difficult to attribute causality to a new farming system, even if benefits are actually being generated. In other words, the rate of reduction of uncertainty over time is low. This is true for scientists/agronomists evaluating the system as well as for farmers.

6.3. *Heterogeneity of farm situations*

It should not be thought that there is a single farming system that would apply across all farming regions, or even on all farms within a region. The reality is that optimal farm management practices depend on a wide range of factors (e.g. Table 1) which can vary markedly from farm to farm. Thus farmers' responses to a new innovation will vary, not just because the farmers have different attitudes and beliefs, but also because their farms vary so much in factors such as total area, soil types, soil fertility, machinery availability, financial resources, climate, weeds present and labour. Thus, the scientist's aim can only be to influence farming practices in the aggregate, recognising that the biophysical and human responses will vary markedly between farms.

6.4. *The timing of benefits and costs*

Farming systems based on trees or shrubs are usually characterised by high up-front costs and benefits that occur some time in the future. If farmers have to borrow money to pay the up-front costs, it is obvious that any direct comparison of the up-front costs with the eventual benefits will not be valid without allowing for the cost of interest. Even if a farmer has savings available to be invested in the system, this implies that interest that could be earned on the savings will be sacrificed. Apart from possible differences in the relevant interest rates, there is no conceptual difference between having to pay interest and having to give up earning of interest. Both would have the same impact on the decision of whether to invest in the new farming system. This is a simple version of the rationale which economists use for discounting future benefits in order to make them comparable to current costs.

Unfortunately, discounting serves to further raise the hurdle for scientists seeking to promote farming systems with high up-front costs. This is particularly a problem in developing countries where private discount rates are often extremely high due to poverty (causing future benefits to be less significant than current survival) combined with exploitive interest rates charged by small local money lenders.

6.5. *Social or institutional issues*

There are many examples of social or institutional issues which influence adoption decisions (Rogers, 1995). Examples which may be relevant to complex farming systems designed to enhance sustainability include the following;

(a) inability to enforce property rights (e.g. to prevent other farmers from causing salinity on your property),
(b) social stigmas associated with the technology (e.g. a perception that it is linked to an undesirable social group) (Fujisaka, 1994),
(c) taxation arrangements which disadvantage commercial planting of trees as a private investment relative to other shorter term investments (Bhati et al., 1992),
(d) distortions in pricing of products from private forestry. For example, Bhati et al. (1992) argue that the practice of some state forest agencies in Australia of underpricing public-sector logs might have flow on effects to small private plantations,
(e) insecure or inequitable land tenure, reducing the likely benefits from adoption because benefits will cease if tenure is lost (e.g. rangeland properties in Western Australia are leasehold rather than freehold) (Feder and Onchan, 1987; Riddell, 1987),
(f) institutional or professional inertia. For example, the professionals in a related agency may have a vested interest in maintenance of the current farming system, and so may obstruct or delay changes,
(g) directives from landlords not to adopt.

7. Example: hedgerow intercropping in the Philippine uplands

In several places in the foregoing discussion, examples have been drawn from the adoption of hedgerow intercropping in the Philippine uplands. Because this relates to a farming technology very consistent with the idea of mimicking natural ecosystems, and because it is such an extreme and classic case, it bears further examination here.

The main motivation for promoting hedgerow intercropping is the problem of soil erosion, which occurs at high rates throughout the Philippine uplands (Nelson, 1994; Fujisaka, 1994). Agriculture in the uplands has maize and upland rice as its major products. It occurs almost entirely on unprotected, steeply sloping land. This approach to farming, combined with very high rainfall, has predictable consequences for rates of soil erosion, with substantial on-site and off-site costs.

Various technologies have been put forward as potential solutions to the problem, but by far the most prominent has been hedgerow intercropping. This has been the subject of research and extension programs by the Philippine government (Gerritts, 1996), foreign non-government organisations (Garcia et al., 1995b), and foreign aid organisations. It involves planting of leguminous shrubs or other suitable plants along the contour to act as a barrier to soil movement. Alleys are planted to crops and hedgerows are trimmed, with trimmings used for mulch or animal feed. Terrace formation occurs rapidly when the alleys are ploughed. Despite a major investment in extension and promotion of versions of this farming system throughout the Philippine uplands, farmer adoption has been minimal (Fujisaka, 1994).

Garcia and co-workers conducted a series of farmer surveys in different parts of the Philippines to attempt to understand the reasons for this failure (Garcia et al., 1995a, 1995b, 1995c, 1995d, 1996a, 1996b). They found that the reasons are complex and multifaceted, with different combinations of factors being important in the different sites. Indeed it appears that in this social and physical environment, sustained adoption of hedgerow intercropping is a fragile goal. There are many possible reasons for failure, and the results of Garcia and co-workers seem to indicate that if even a small number of them are present, the system is unlikely to be adopted.

Among the reasons identified as being contributors to the failure of one or more of the extension programs were the following;

(a) farmers do not perceive that erosion is occurring (e.g. land with slopes of less than about 18% is considered 'flat'),
(b) farmers in some regions had difficulty obtaining cuttings of the recommended species for hedgerows,
(c) the quality of implementation is low (see earlier example);
(d) insecure land tenure (Fujisaka, 1994),
(e) directives from landlords not to adopt because they fear that tenants will then use the improvements they had made as the basis for a claim on ownership of the land (Sajise, 1987),
(f) inability to enforce property rights – in one region adopters could not prevent other farmers from tethering their animals to hedgerows for grazing, resulting in hedgerows being grazed to the ground,
(g) long delays after establishment before benefits are observed,
(h) high interest rates/high discount rates (due to poverty) so that current costs weigh more heavily than future benefits,
(i) large costs of establishing and maintaining hedgerows (labour and finance),
(j) although benefits are observed, they are often not sufficient to outperform the traditional farming system,
(k) in some cases, the benefits are sufficient, but this occurs in much narrower domains than recognised in the extension programs,
(l) resource constraints, especially limited availability of labour and finance for hedgerow establishment,
(m) concerns by farmers that the hedgerows harbour rats or insect pests;
(n) concern that the hedgerows occupy a significant proportion of the land which, for such small farmers, is a scarce resource,
(o) a perception that labour teams formed in some cases to facilitate hedgerow establishment are linked to communism,
(p) top-down extension, not resulting in improved understanding,

(q) institutional problems in policing and enforcing the conditions associated with land tenure or in communicating the need for a participatory approach to those conducting the extension.

8. Meeting the challenges

The discussion here has focused on farmers own motivations and behaviour. This should not be taken as advocacy for an unfettered, free-market approach to policy for agricultural resource management. There are many possible policy options, including;

(a) extension
(b) community involvement
(c) research
(d) development
(e) subsidies for adoption
(f) taxes
(g) safe minimum standards/penalties
(h) cross compliance requirements
(i) altering property rights
(j) public acquisition and management
(k) initiatives to reduce transaction costs.

All of these may help to promote changes in farm management but, apart from option j) the extent of change is determined by the issues reviewed here. The focus on farmer behaviour is essential for accurate assessment of almost any policy initiative to change farm management.

Consider, then, the key points of the review. Developing a system which is technically effective is necessary but not sufficient. The system must be economically profitable at least in the long run to be seriously considered by most farmers. Scientists need to work closely with farmers and economists to meet this challenge.

Increasing the physical productivity of a new technology is not the only way to affect its profitability relative to current practices. There are a range of institutional changes which can affect profitability, including subsidies, taxes, schemes to share costs and benefits differently, information programs, legal standards backed with penalties and schemes to alter or enforce individual rights. Policy makers need to consider implementing this type of change in order to make a new technology sufficiently attractive to farmers. For example, in developing countries, adoption may be particularly enhanced by policies that reduce interest rates, or that address insecure or inequitable land tenure.

In some cases, a key problem is deep uncertainty about the technology – demonstrating its value quickly enough and convincingly enough. This is especially true for slow, indirect effects, such as impacts on the water table.

Scientists need to consider ways to increase the observability of results from trials of such technologies, ways to help farmers recognise that the technology is what is causing the results, and ways to allow results to be observed sooner.

Moral suasion and peer pressure can be very useful in raising farmer awareness, but should not be relied upon to achieve adoption of innovations which are not clearly beneficial to the individual farmer. Any success they achieve is likely to be temporary unless the new technology is actually more profitable.

Even a simple new farming technology is subject to many of the difficulties that have been discussed in this review. For complex new farming systems, it is probably necessary to combine several of the above policy options to achieve widespread adoption.

9. Acknowledgements

The author is grateful to Don McFarlane, Dan Carter, Steven Schilizzi, Nicole Glenn, Sally Marsh, John Bartle, Warren Mason, Tony Chisholm and Simone Blennerhassett for helpful comments on an earlier draft, to Ted Lefroy for his helpful input to the paper (including the paper's title), to Bob Lindner for help and inspiration and to the Grains Research and Development Corporation and the Rural Industries Research and Development Corporation for funding of related research projects.

10. References

Abadi Ghadim AK, Pannell DJ, Bennett A and Stewart V (1996) Farmers' risk perspective on adoption of legume crops. In: Asghar M (ed) Proceedings of the 8th Australian Agronomy Conference 1996, Toowoomba, Queensland, 30 January–2 February, 1996, pp 56–59. Australian Society of Agronomy, Carlton, Victoria

Anonymous (1997) Salinity: A Situation Statement for Western Australia. Government of Western Australia

Antle JM (1987) Econometric estimation of producers' risk attitudes. American Journal of Agricultural Economics 69: 509–522

Bardsley P and Harris M (1987) An approach to the econometric estimation of attitudes to risk in agriculture. Australian Journal of Agricultural Economics 31: 112–126

Bartle JR, Campbell C and White G (1996) Can trees reverse land degradation? Australian Forest Growers Conference, Mt Gambier, SA

Bhati UN, Klijn N, Curtotti R, Dean M and Stephens M (1992) Impediments to the development of commercial forest plantations in Australia. In: Wallace N (ed) Natural Resource Management: An Economic Perspective, pp 195–205. ABARE, Canberra

Binswanger HP (1980) Attitudes toward risk: experimental measurement in rural India. American Journal of Agricultural Economics 62: 395–407

Bond G and Wonder B (1980) Risk attitudes amongst Australian farmers. Australian Journal of Agricultural Economics 24: 16–34

Cary JW and Wilkinson RL (1997) Perceived profitability and farmers' conservation behaviour. Journal of Agricultural Economics 48: 13–21

Feder G and Onchan T (1987) Land ownership security and farm investment in Thailand. American Journal of Agricultural Economics 69: 311–320
Feder G and Umali D (1993) The adoption of agricultural innovations: a review. Technological Forecasting and Social Change 43: 215–239
Feder G, Just R and Zilberman D (1985) Adoption of agricultural innovations in developing countries: A survey. Economic Development and Cultural Change 33: 255–298
Fujisaka S (1994) Learning from six reasons why farmers do not adopt innovations intended to improve sustainability of upland agriculture. Agricultural Systems 46: 409–425
Garcia JNM et al. (1995a). Soil Conservation in an Upland Farming System in Cebu: A Socio-Economic Survey. SEARCA-UQ Uplands Research Project, Survey Report No. 1. Los Baños
Garcia JNM et al. (1995b). Soil Conservation in an Upland Farming System in Davao del Sur: A Socio-Economic Survey. SEARCA-UQ Uplands Research Project, Survey Report No. 2. Los Baños
Garcia JNM et al. (1995c) Soil Conservation in an Upland Farming System in Bukidnon: A Socio-Economic Survey. SEARCA-UQ Uplands Research Project, Survey Report No. 3. Los Baños
Garcia JNM et al. (1995d) Soil Conservation in an Upland Farming System in Palawan: A Socio-Economic Survey. SEARCA-UQ Uplands Research Project, Survey Report No. 4. Los Baños
Garcia JNM et al. (1996a) Soil Conservation in an Upland Farming System in Iloilo: A Socio-Economic Survey. SEARCA-UQ Uplands Research Project, Survey Report No. 5. Los Baños
Garcia JNM et al. (1996b) Soil Conservation in an Upland Farming System in Nueva Vizcaya: A Socio-Economic Survey. SEARCA-UQ Uplands Research Project, Survey Report No. 6. Los Baños
Gerrits RV (1996) The Philippine Government Approach to Upland Development: The Integrated Social Forestry Program. SEARCA-UQ Uplands Research Project, Working Paper No. 16. Los Baños
Gibbs M and Lindner R (1986) The discovery of innovations by farmers. Paper presented at the 30th Annual Conference of the Australian Agricultural Economics Society, Canberra, 3–5 February 1986
Hollick M (1990) Land conservation policies and farmer decision making. Australian Journal of Soil and Water Conservation 3: 6–13
Jackson W (1984) Introduction. In: Jackson W, Berry W and Coleman B (eds) Meeting the Expectations of the Land: Essays in Sustainable Agriculture and Stewardship. North Point Press, San Francisco
Jackson W and Bender M (1984) Investigations into perennial polyculture. In: Jackson W, BerryW and Coleman B (eds) Meeting the Expectations of the Land: Essays in Sustainable Agriculture and Stewardship. North Point Press, San Francisco
Lindner RK (1987) Adoption and diffusion of technology: an overview. In: Champ BR, Highley E and Remenyi JV (eds) Technological Change in Postharvest Handling and Transportation of Grains in the Humid Tropics, pp 144–151. ACIAR Proceedings No. 19
Lindner RK, Pardey PG and Jarrett FG (1982) Distance to innovation source and time lag to early adoption of trace element fertilisers. Australian Journal of Agricultural Economics 26: 98–113
Main AR (1997) How much biodiversity is enough? Workshop on Agriculture as a Mimic of Natural Ecosystems, Perth, Western Australia, September 2–6, 1997
Marsh S, Pannell DJ and Lindner RK (1995) Impact of extension on adoption of lupins in Western Australia. Paper presented at the 39th Annual Conference of the Australian Agricultural Economics Society, Perth, February 14–16, 1995
Myers RJ (1989) Econometric testing for risk averse behaviour in agriculture. Applied Economics 21: 541–552
Nelson RA (1994) Soil Erosion and Conservation in the Philippine Uplands: A Review of the Literature. SEARCA-UQ Uplands Research Project, Working Paper No. 3. Los Baños
Pannell DJ (1995) Economic aspects of legume management and legume research in dryland farming systems of southern Australia. Agricultural Systems 49: 217–236

Pannell DJ (1997) Balancing economics and sustainability in building an integrated agricultural system. In: Hobbs R, Saunders D and Lambeck R (eds) An Integrated Approach to a Sustainable Future (in press)

Pearce DW and Turner RK (1990) Economics of Natural Resources and the Environment Harvester Wheatsheaf, New York

Pluske J and Fraser R (1996) Can producers place valid and reliable valuations of wool price-risk information? Review of Marketing and Agricultural Economics 63: 284–291

Riddell JC (1987) Land tenure and agroforestry: a regional overview. In: Raintree J (ed) Land, Trees and Tenure. ICRAF, Nairobi

Rogers EM (1995). Diffusion of Innovations. Free Press, New York

Sajise PE (1987) Agroforestry and land tenure in the Philippines. In: Raintree J (ed) Land, Trees and Tenure. ICRAF, Nairobi

Sinden JA and King DA (1990) Adoption of soil conservation measures in Manilla Shire, New South Wales. Review of Marketing and Agricultural Economics 58: 179–192

Agroforestry Systems **45**: 411–421, 1999.

Can we bring about a perennially peopled and productive countryside?

J. B. PASSIOURA
CSIRO Plant Industry, G.P.O. Box 1600, Canberra 2601, Australia;
E-mail: j.passioura@pi.csiro.au

Key words: agriculture, environmental stability, sustainability

Abstract. This paper is a reflection on the workshop *Agriculture as a Mimic of Natural Systems*. It analyses what is essential for protecting agricultural land from degradation, including enabling features of the social context. It looks at how natural systems maintain a stable landscape, and explores ways in which pertinent features of natural systems can be copied in agriculture to help solve particular problems of land degradation, such as: erosion; hydrologic instability and the often associated secondary salinity; and chemical imbalances, such as those leading to excessive acidity. The conservative practices that many farmers have recently adopted are in several respects based on attributes of natural systems, but the mosaic nature of land is one important aspect that has not yet strongly influenced practices, which are often too strongly tied to cadastral boundaries.

1. Introduction

Wes Jackson has argued eloquently that we need to rekindle the agrarian spirit, a close and caring connection with the land, if we are to save the agricultural countryside from continuing and eventually irretrievable degradation. There is, however, a serious problem in trying to rekindle that spirit. Most people like to live in cities, and city dwellers exert strong pressure for cheap food. 'Agrarian' agriculture has little chance at the moment of competing economically with industrialised agriculture. The pressure to produce cheap food will remain immense and with it comes pressure on the land. Most farmers will not embrace the agrarian spirit if they go broke in so doing. Indeed, the influential position of the city dweller may have been established as early as the very first towns and cities. It has sometimes been argued (e.g. Mumford, 1961) that the invention of agriculture provided enough food to enable towns and cities to develop. The reverse is as likely to be true (Allen and Starr, 1982): the development of towns and cities may have stimulated, perhaps forced, the development of agriculture.

What then can we do? The situation has about it the air of a Greek tragedy: the essence of the coming disaster is its inevitability – at least, in the minds of people concerned about land degradation.

Yet there are many well-informed people who are unconcerned. Their view is that of an optimistic accountant: if we look after the bottom line (using a discount rate that renders unimportant any returns or liabilities that accrue after a few years), all else will look after itself. In the arena of resource economics there are historical grounds for such optimism. For example, the pessimists of the 1960s and the 1970s have, at least to date, been wrong. Chisholm (1992) puts this orthodox economic view as follows:

> The lesson to be learned from the [pessimistic] predictions, taken together with statements made by the Club of Rome . . . , is that physical indicators of the supply of productive land and estimated reserves of energy and mineral resources, are by themselves inadequate and misleading measures of resource scarcity. For those natural resources for which markets exist, it is now widely accepted that the price system is the most effective indicator we have of both absolute and relative resource scarcity.

To ecologists and agriculturalists who worry about the long term degradative processes that are now in train and that will be largely irreversible (at least within the time scale of generations), this optimistic view is irresponsible. A market that is uninterested in slowly developing but eventually severe degradation, especially that producing offsite effects such as salinity damage to soil, roads, towns and rivers, does not impress the conservationist. It was such a market that no doubt led to the complete loss of soil from many parts of Greece and southern Italy (Thirgood, 1981).

Despite the current ascendancy of economic fundamentalism, with its faith in the putatively overwhelming benefits of competition and of giving the market unencumbered reign, man cannot live by the bottom line alone. Much of the wider environmental debate is about what sort of countryside we would like to create. Do we want to treat agriculture as a narrowly focused accountant might – essentially as a mining operation – with the implicit optimistic assumption that new technology and market forces will solve the problem of forever maintaining adequate supplies of cheap food. Or do we want to foster a stable agriculture, with a buoyant rural population, with negligible environmental disruption? There is no doubt that the participants in this workshop resoundingly favour the latter. But this antinomy is a little overdone. Soil forms at a rate that is alarmingly less than that at which it is being eroded by even the best current farming practices. Removing products for sale results in the removal of nutrients which must eventually be replaced from mines elsewhere. Agriculture as we know it, even perennial polyculture (Jackson and Jackson, Ewel, this volume), is everywhere, to some extent, a mining operation.

The debate, then, is in large part about what are acceptable rates of change. The rapid development of deep erosion gullies, as was common during the height of the rabbit plague in Australia, is not acceptable to anybody. An annual loss of half a millimetre of topsoil from cropped land is invisible to the market but deeply concerns many conservationists ('we lose three (five?, ten?) tonnes of soil for every tonne of grain harvested'). The market looks but a few years ahead; the conservationist looks many decades ahead.

Our interest here *is* in looking several decades ahead, and in devising agricultural systems that do not perceptibly degrade the land at this time scale. I am pessimistic about rekindling the agrarian spirit, but, although the city dwellers may have no more than a 'bumper sticker' concern (as Wes Jackson put it) for the state of the countryside, this concern is not without bite; the emergence and no doubt eventual widespread implementation of international environmental standards, such as the ISO14000 series, will help the cause. Simulating the features of natural systems that stabilise the land will also help, and that has been the theme of this meeting. Perennial polyculture is the ideal, but an economically viable system may be more than half a century away. Meanwhile what should be done to conserve the land? What follows is a collection of thoughts on this theme stimulated by the meeting.

2. The essentials of sustainability

There are several fundamental biological and physicochemical requirements for arresting degradation of agricultural land:

- negligible erosion, whether by wind or by water
- balanced leakage of water and solutes (salt in equals salt out)
- no persistent toxicity (acidity, heavy metals, agrochemicals)
- control of pests, diseases and weeds

to which should be added the enabling social and economic requirements:

- profitability of the enterprise, without which there is little chance of farmers' adopting conservation measures, and
- appropriate political and social pressures, especially global ones that are embodied in international environmental standards such as ISO14000.

This list comprises the essentials. A substantial academic industry has developed that has been exploring a plethora of 'indicators of sustainability' (e.g. Walker and Reuter, 1996), but most of these indicators are secondary. Indeed, those that do not evidently contribute to the essentials may be more related to attaining an emotional ideal about what is desirable in a landscape, rather than to directly addressing problems of degradation. There is no evidence, for example, that the cotton-strip-bioassay, which purports to measure biological activity in the soil (Walker and Reuter, 1996), has any bearing on the essential requirements listed above, yet many feel that vigorous biological activity in the soil is desirable in itself.

This is not to say that our emotional attitudes to land are unimportant – on the contrary. Schama (1995) vividly illustrates how important they are in influencing national behaviour. His arguments, although primarily addressing attitudes to wilderness and other non-agricultural land, can be extended to colonists, who typically try to impose on their new environment images of the landscapes of their homeland. Had Australia been colonised by the Spanish or Portuguese we may have ended up with stable agricultural landscapes like the dehesa that Joffre (this volume) describes.

Considering then the fundamental requirements of sustainability, what are the features of natural ecosystems that bear on them? Several emerged from discussions during the workshop, some of which are well known, and all of which are worth thinking about. Before discussing these it is first worth considering how the study of ecology, like any human activity, both reflects and influences the social conditions in which it is pursued. Two main ecological ideas that stimulated discussion were *succession* and the *impact of environmental variability* on the behaviour of both natural and agricultural communities.

Americans invented the idea of succession. They are used to one group of organisms muscling in on another. This process eventually leads to climax vegetation, in which the upperstorey makes, to the understorey plants, an offer which they can't refuse, to protect them from excess sunlight. The view is that the difficult part of an organism's environment are the other organisms, rather than the soil and climate. Competition rules.

The original Clementsian idea of succession has now been substantially revised, especially through the realisation that substantially different forms of 'climax' vegetation can develop in a given environment of soil, terrain, and climate (Holling et al., 1995). Nevertheless, at the level of environmental policy the original view still prevails, as illustrated by the poignant account by Pollan (1996) of the debate surrounding the management of a patch of remnant pine forest in Connecticut that had been destroyed by a tornado. The successionists won the debate, but their victory left the patch in a tangled mess that might take many decades before it once again becomes attractive, with no guarantee even that it would ever become a pine forest again.

Australian ecologists, in accord with the national temperament, typically see chance as the major determinant of ecological interaction. Australians are strongly imbued with a gambling ethos, perhaps because of their very variable environmental experiences – the 'droughts and flooding rains' of Mackellar's popular phrase, and the rags to occasional riches associated with mining, especially during the gold rushes. The prevailing attitude is that variability in soil and weather are the major shaping forces, rather than competing organisms.

Both views contain much truth, and in large part reflect the relative importance of biotic and abiotic influences (Ewel, this volume). The trick is in gauging their relevance in given circumstances.

3. Lessons from natural systems

Returning to the fundamental requirements for arresting land degradation, the main guiding features that emerged from the discussion of natural systems and their features that could beneficially be applied to agricultural systems are outlined in Table 1.

3.1. *Persistent ground cover and minimal soil disturbance*

The adoption of techniques of minimal tillage and stubble retention by many farmers over the last two decades has substantially lessened the risk of erosion, except in the subtropics where intense rain is still very damaging. With these changes have come new problems which require great skill to manage. They necessitate much greater reliance on herbicides, with attendant problems of the development of herbicide-resistant weeds and of toxicity to organisms other than weeds. Uncontrolled grass weeds harbour damaging root diseases such as take-all and rhizoctonia root rot. Heavy stubbles harbour leaf diseases and clog sowing equipment thereby leading to poor establishment of following crops. Many successful farmers compromise by burning heavy stubbles as late as possible before sowing, thus reducing periods of bare soil to a few weeks a year.

Despite the retention of stubble, areas with low rainfall, especially where the soils are sandy, are still at risk from wind erosion, for the amount of stubble

Table 1. Features of natural systems that are most likely to benefit agricultural systems if they can be incorporated in them.

Feature of natural systems	Benefits in agricultural systems
persistent ground cover and minimal disturbance of the soil surface which allows binding by lichens and fungal mats	minimises erosion
presence of deep-rooted perennials, which allows complementary use of water through different seasonality of uptake	enables the capture of water that has escaped the roots of annual plants during periods in which rainfall exceeds evaporation
seasonal releases of nutrients, mediated by microbial activity, that match the nutritional demands of the plants, thus promoting tight cycling with minimal losses	ensures that leaching of especially mobile nutrients such as nitrates is minimised, which not only reduces off-site effects but also reduces the acidifying effect of leaching nitrate away from where it was formed
adequate biodiversity, both above and below ground	ensures that the functional requirements of the turnover of nutrients in the soil are met; biological control of pests and diseases; reduction of risk from extremes of weather
mosaic nature of land	recognition that patches of land should be treated according to their particular properties, and not according to cadastral boundaries

produced is meagre and may in any case be eaten by grazing animals whose sharp hoofs also disturb the soil surface. Shelter belts of trees may help reduce such erosion.

3.2. *Use of deep-rooted perennials*

In southern Australia especially, winter rain substantially exceeds evaporation during the growing season of annual crops or pastures thereby resulting in considerable flows of water beyond the reach of the annuals' roots. Such recharge typically ranges from 10 to 100 times that occurring under the original native vegetation, from an average of about 30 mm in semi-arid or dry subhumid areas (with annual rainfall less than 600 mm) to 200 to 300 mm in higher rainfall areas (800 to 900 mm). The result is rising water tables, possibly locally, where the recharge occurs, or more commonly some distance (maybe tens of kilometres) away, depending on the hydrogeology of the area (Hatton and Nulsen, this volume). These rising water tables often mobilise salt previously present in the deep subsoil, bringing it to the surface where it generates dryland salinity.

Only deep-rooted perennial vegetation can use this winter-excess of water that penetrates well below the rooting depth of annuals (Pate and Bell, this volume). Where the rainfall is less than 600 mm, the deep subsoil may be able to hold up to three years' worth, on average, of this lost water. This creates the option of using perennial pasture plants, such as lucerne, to mop up most of this water episodically, thereby regenerating the storage capacity of the deep subsoil – that is, a few years of annual crops can be followed by a few years of perennial pasture to stabilise the hydrology (Dunin et al., this volume).

The usefulness of trees in the lower rainfall areas is limited, unless they can access a fresh water table, because the horizontal flow of water to roots in unsaturated soil is typically only about one or two metres (Stirzaker et al., 1998), and thus the trees would have to cover an agriculturally uneconomic proportion of the land were they to capture most of the escaped water. Where there is an accessible fresh water table in a winter-rainfall environment, quite widely spaced alleys of woody perennials such as tagasaste can stabilise the water table by continuing to transpire during the summer while simultaneously providing valuable fodder (Lefroy and Stirzaker, this volume). If the water table is saline, though, salt is transported in the water flowing to the tree root zone, where it may build up to toxic levels as the roots extract the water leaving the salt behind.

In the higher rainfall areas, the water lost by annuals each year exceeds the storage capacity of the subsoil and thus the return of substantial permanent, very deep rooted, perennial vegetation to such areas is the only way of reducing recharge to the levels that prevailed before clearing. Management of the understorey may also be important to ensure that enough of the winter excess in rainfall is shed to maintain stream flow (Dunin and Downes, 1962), rather than stored deep in the soil for later transpiration.

3.3. *Matching the rate of release of nutrients to the demand*

In natural vegetation, a large proportion of the inorganic nutrient supply is sequestered within the plants themselves. Substantial amounts of nutrients are also contained in soil organic matter, but typically in fractions that cycle slowly unless the soil is disturbed (Grierson and Adams, this volume). Leguminous pastures produce large amounts of microbially accessible soil organic matter that can mineralise so rapidly during the opening rains in winter rainfall areas that the availability of soluble nitrogen greatly exceeds the demand by the establishing annual plants, so that substantial leaching results (Peoples et al., 1998). Furthermore, in grazed pastures, the nitrogen-rich urine of sheep and cattle may also be prone to rapid nitrification, leaching, and consequent acidification (Black, 1993). Deep roots of perennials can, in some circumstances, capture the leached nitrate and accompanying cations, and return them to the surface, but generally losses seem irretrievable. Management of this problem requires great skill.

3.4. *Adequate biodiversity*

Main's penetrating discussion of biodiversity (this volume) leaves little to add, other than to emphasise three points:

The first is the extraordinary complexity of below ground ecology (see also Neher, this volume). An illustration of this complexity is given by Roget's (1995) work on soils suppressive to rhizoctonia root rot, a troublesome disease of wheat. Roget found that after about ten years of his field experiment, rhizoctonia damage, which was initially severe, essentially disappeared. Presumably the population of rhizoctonia was eventually drastically reduced by the ascendancy of other microorganisms, benign to the crop but inimical to rhizoctonia. We are a long way from learning how to manage such behaviour.

The second is that biodiversity has a crucially important dimension in time as well as in space. Farmers, even when they use monocultures, have for millennia used rotations, repeating fixed sequences of crops, as part of their husbandry. Nowadays, the sequence of crops, often including a phase of pasture, is more variable. Farmers recognise the contingent nature of their operations, the great importance of the recent history of their farms, including the impact of the weather, and choose crops depending on their reading of the state of their land – including nitrogen levels, the store of water in the soil, and the extent of the risk of damaging levels of weeds or disease. The complementary activities of different species, so important in natural systems or perennial polyculture, can thus be mimicked to an important extent (Van Noordwijk, this volume).

The third is the rather misty issue of whether or not we should maintain diversity of species to prevent 'ecosystems from collapsing', as newspapers sometimes put it. Main's treatment focused on functional biodiversity, and thus emphasised the importance, backed up with palpable arguments, of having an appropriate range of life forms available when trying to minimise land degradation (although what is 'appropriate' is often moot). Others have argued that maintaining diversity of species is, of itself, of great value. Such arguments are often based on the precautionary principle (e.g. Myers, 1993), essentially one of insurance, that we do not know what functionality we are losing if we do not protect a wide range of species. No doubt this argument will often prove to be correct in particular circumstances, although I find it hard to see its force in relation to the protection of patches of remnant natural vegetation in agricultural areas. I do think there are overwhelmingly strong arguments for protecting such patches, but they are predominantly cultural and aesthetic rather than functional. In terms of the precautionary principle, it may well be that biodiversity among the soil flora and fauna is especially important, a case that Robert May, Chief Scientist in the UK, argued during a recent lecture tour of Australia.

3.5. *The mosaic nature of land*

Land is often highly variable spatially, especially in Australia. The mosaic nature of natural vegetation, reflecting the combination of soil and terrain on which it is growing, is readily seen in aerial photographs, often at length scales smaller than many paddocks (Grierson and Adams, this volume). Yet paddock boundaries are cadastral, having generally been laid out by surveyors who liked grids and straight lines.

Aerial photographs of farms, and yield maps generated by geo-referenced yield monitors, are becoming more common and are highlighting to farmers, much more arrestingly than before, the mosaic nature of their own land. Yield maps can be especially arresting, for when converted into maps of economic gross margins they often show substantial negative areas within a paddock, despite the overall gross margins being positive. Such areas may turn out to be a strong stimulus for action, if they persist across seasons and if they are in patches large enough to be taken out of cultivation without disrupting the operation of sowing and harvesting equipment. The opportunity of converting these patches into copses of possibly native deep-rooted vegetation may bring about substantial economic and environmental benefits.

Patches of land amenable to this sort of management would have to have diameters of at least tens of metres. There are also spatial variabilities in soil, generated by native vegetation at scales of a few centimetres to a metre, that play a large part in the interaction of that vegetation with its environment. These are biopores, generally vertical pores which may penetrate for metres into the subsoil and which are colonised continuously by roots, or repeatedly recolonised by successive generations. Nulsen et al. (1986) illustrated the impact of these pores on the flow of water in the soil under mallee in WA, and illustrated further how the function of the pores was disrupted by clearing and conversion of the land to pasture.

The role of biopores in agricultural soil is unclear, but it is clear that they are disrupted by cultivation and that they can be formed by perennial agricultural plants like lucerne (Stirzaker, this volume) or by weeds with perennial roots like skeleton weed. My guess is that careful husbandry will enable the reestablishment of an array of deeply penetrating biopores in agricultural soils, which can be colonised and reinforced afresh each year by roots of annual crop plants. The crops may thus be able to grow their roots more deeply than at present and thereby capture more of the water and leached nutrients that are currently causing offsite degradation. The work of Torquebiau and Kwesiga (1996), showing that maize roots penetrate very deeply into the soil when grown after sesbania, nicely illustrates the possibilities.

3.6. *Systems: The whole is less than the sum of its parts*

Ecologists talk a lot about 'systems', by which they often mean simply collections of interacting organisms in their particular physical environment. But 'system' has a deeper sense than that, as Hobbs and Morton (this volume) and Jackson (this volume) reminded us. It implies that there is a whole that has integral behaviour that it is not discernible *a priori* from the behaviour of its parts.

The anti-aphorism in my heading sums this notion up in a more instructive way than its cliched counterpart. It not only affirms that a system has properties that transcend those of its constituent parts. It also alerts us to the fact that much of the information about the parts may be trivial and can be ignored, because the parts are constrained towards producing the transcending integral behaviour; if I ask someone how to drive from Williams to Perth, the advice I get will have to do with what roads to take, not with the detailed movements of my hands on the steering wheel or my feet on the pedals. Weiss (1969, p. 12) sums this point up mathematically:

> the complex is a system if the variance of the features of the whole collective is significantly less than the sum of variances of its constituents . . . in short, the basic characteristic of a system is its invariance beyond the much more variant flux and fluctuations of its elements.

Weiss's (1969) discussion was mostly about cells and tissues, but the principles apply at all levels of biological organisation (e.g. Passioura, 1979; Allen and Starr, 1982). Examples at the ecosystem level are given by Holling et al. (1995) in their analysis of biodiversity in relation to ecosystem function, such as work in the Serengeti which showed large seasonal effects on plant species composition while overall primary production changed little. Van Noordwijk (this volume) provides a further discussion along this line.

The behaviour of the system does of course depend on the behaviour of its parts, but which are the important parts and which are the important behaviours? Hobbs and Morton (this volume) and the other references in this section explore guide-lines for answering these questions.

Agronomists and ecologists tend to see plants, animals, land, and weather, as the constituents of their system of interest. Sociologists and economists move up a level of organisation by adding people and markets. Whether or not farmers choose to adopt practices that conserve their land depends on many considerations, including their perceptions of financial risk and of desirable life style. It is not easy to put the biological and environmental considerations of the agronomist and ecologist into a social context. That the management techniques they propose are not as well adopted as they would like is often testimony to how difficult it is for them to understand the wide range of constraints experienced by farmers (Pannell, this volume).

4. What is happening now, and where to from here?

Many of the horror stories of land degradation in Australia told passionately by White (1997) are now historical in cropped lands, with the possible exception of the subtropics. Rates of erosion are much less than they used to be, with a substantial, although still too small, proportion of farmers adopting conservative practices of land management. Long fallows in the low rainfall areas, which left the soil severely exposed to wind erosion, once the norm, are now uncommon. Techniques of minimal tillage and stubble retention greatly reduce risks of water erosion, although bring with them the risk of weed infestations, because the herbicides which make these techniques viable eventually lose their effectiveness as the weeds develop resistance to them. The rapidly increasing use of lime is helping to counter excessive acidity in many of the soils at risk.

Much of the recent progress has involved substantially increased skill in management by farmers. The physical fragility of Australian soils, coupled with their widely spread underground salt burden and propensity to hydrologic imbalance, acidification, and invasion by weeds and pathogens, means that great expertise and astute management are needed to ensure that the soil is protected against the wide range of forms of degradation that endanger it. In some types of agriculture, and especially in the system of shifting agriculture in the tropics, the more intense is the management the greater is the danger of environmental damage. In Australia the reverse is usually true, at least in the cropping areas, for conservation farming practices require meticulous preparation and timing if they are to be successful.

These improved practices generally accord with the observations outlined in Table 1, but much remains to be done, especially in countering the problem of hydrologic imbalance, a problem whose solution is very likely to be based on insights gained from studying natural ecosystems. Perennial polyculture, in effect a managed climax vegetation, is a stimulating ideal, but we cannot wait for it to become economically viable. We are obliged to continue making improvements, and these improvements must not only reduce the dangers of land degradation; they must also be profitable if farmers are to adopt them and to stay in business. If we fail, we will end up with a badly degraded and poorly populated countryside.

5. References

Allen TFH and Starr TB (1982) Hierarchy: Perspectives for Ecological Complexity. University of Chicago Press, Chicago

Black AS (1993) Soil acidification in urine- and urea-affected soil. Australian Journal of Soil Research 31: 89–99

Chisholm AH (1992) Australian agriculture: a sustainability story. Australian Journal of Agricultural Economics 36: 1–29

Dunin FX and Downes RG (1962) The effect of subterranean clover and Wimmera ryegrass in controlling surface run-off from four-acre catchments near Bacchus Marsh, Victoria. Australian Journal of Experimental Agriculture and Animal Husbandry 2: 148–152

Holling CS, Schindler DW, Walker B and Roughgarden J (1995) Biodiversity in the functioning of ecosystems: as ecological synthesis. In: Perrings C, Maler KG, Folke C, Holling CS and Jansson BO (eds) Biodiversity Loss: Economic and Ecological Issues. CUP, Cambridge

Mumford L (1961) The City in History: Its Origins, Its Transformations, and Its Prospects. Secker and Warburg, London

Myers N (1993) Biodiversity and the precautionary principle. Ambio 22: 74–79

Nulsen RA, Bligh KJ, Baxter IN, Solin EJ and Imrie DH (1986) The fate of rainfall in a mallee and heath vegetated catchment in southern Western Australia. Australian Journal of Ecology 11: 361–371

Passioura JB (1979) Accountability, philosophy, and plant physiology. Search (Sydney) 10: 347–350

Peoples MB, Gault RR, Scammell GJ, Dear BS, Virgona J, Sandral GA, Paul J, Wolfe EC and Angus JF (1998) Effect of pasture management on the contributions of fixed N to the economy of ley-farming systems. Australian Journal of Agricultural Research 49: 459–474

Pollan M (1996) Second Nature. Bloomsbury Publishing, London

Roget DK (1995) Decline in root rot (*Rhizoctonia solani* AG-8) in wheat in a tillage and rotation experiment at Avon, South Australia. Australian Journal of Experimental Agriculture 35: 1009–1013

Schama S (1995) Landscape and Memory. Fontana, London

Smith CJ, Dunin FX, Zegelin SJ and Poss R (1998) Nitrate leaching from a Riverine clay soil under cereal rotation. Australian Journal of Agricultural. Research 49: 379–389

Stirzaker RJ, Cook FJ and Knight JH (1999) Agricultural Water Management 39: 115–133

Thirgood JV (1981) Man and the Mediterranean Forest. A History of Resource Depletion. Academic Press, London

Torquebiau EF and Kwesiga F (1996) Root development in a *Sesbania sesban* fallow-maize system in Eastern Zambia. Agroforestry Systems 34: 193–211

Walker J and Reuter DJ (1996) Indicators of Catchment Health: A Technical Perspective. CSIRO Publishing, Collingwood

Weiss PW (1969) The living system: determinism stratified. In: Koestler A and Smythies WR (eds) Beyond Reductionism. Hutchinson, London

White ME (1997) Listen – Our Land is Crying. Kangaroo Press, Kenthurst

Agroforestry Systems **45**: 423–436, 1999.

What can agriculture learn from natural ecosystems?

E. C. LEFROY[1], R. J. HOBBS[2], M. H. O'CONNOR[3] and J. S. PATE[4]
[1]*Centre for Legumes in Mediterranean Agriculture, The University of Western Australia, Nedlands WA 6907, Australia;* [2]*CSIRO Wildlife and Ecology, Private Bag, PO Wembley, WA 6014, Australia;* [3]*Muresk Institute of Agriculture, Curtin University of Technology, Bentley WA 6102, Australia;* [4]*Botany Department, The University of Western Australia, Nedlands WA 6907, Australia*

In the preceding papers, the authors evaluate the usefulness of the proposition that mimicking characteristics of natural ecosystems, particularly patterns of resource capture and distribution, can lead to improved sustainability of agricultural systems. They approach this question from a diverse array of disciplines over a broad range of geographical settings and ecosystems and produce an equally broad range of views and conclusions. The aim of this paper is to draw out the major themes that have emerged concerning the usefulness of the mimic concept in generating radical solutions to the problem of agriculture. First, a brief summary of their scope and conclusions.

1. Summaries

Ewel sets the scene by classing the biomes of the world in terms of their suitability for agriculture, based on a combination of (1) the responses of living things to biotic and abiotic stress and (2) life zone maps based on temperature, humidity and altitude. The four classes he proposes are: (1) forgiving environments, where rainfall is similar to evapotranspiration, and where agriculture has historically been most successful, (2) environments which are too cold, where temperature limits plant and animal growth, (3) conditions which are too dry and where precipitation limits production, and (4) environments that are too wet, where rainfall in excess of evapotranspiration results in high levels of competition, predation and leaching. He then reports on the productivity, nutrient retention and pest response of mimics of successional ecosystems that he and co-workers have developed in the humid tropics. His conclusion is that it pays to imitate natural systems, especially the use of perennial plants, to maintain soil fertility, protect against erosion and make full use of light, water and nutrients. He notes however that it may be necessary to accept lower but more biologically sustainable-yields as (1) natural systems intrinsically feature high investments in structure and (2) high levels of fertiliser tend to switch off specialised mechanisms of nutrient input and cycling such as performed by nitrogen-fixing nodules, mycorrhizae, and proteoid roots.

way to species or life forms absorbing the shock experienced as new material is introduced. At both scales, these systems display characteristics of resilience upon invasion. To move the ecosystem concept upward to the level of our human community allows us to see the dynamics of a complex adaptive system and appreciate how the removal of large numbers of people has created disequilibria. Early invaders on the landscape were the agricultural technologies of the early Neolithic. From that era to the recent introductions of fossil fuels and agri-chemicals, our agrarian landscapes featuring interventions of small scale and effect have had to absorb one assault after another. How long this can continue very much depends on the characteristics of each member of the mosaic. He concludes that the forces at work which disrupt settings at the landscape level find striking parallels in those influences which have disrupted small human communities and now threaten to invade the nano-ecosystems of our major crops and livestock through the introduction of exotic large effect genes. Increasingly the reward for the disruption of all three integrities is oligarchic power.

Ewel sets the scene by classing the biomes of the world in terms of their suitability for agriculture, based on a combination of (1) the responses of living things to biotic and abiotic stress and (2) life zone maps based on temperature, humidity and altitude. The four classes he proposes are: (1) forgiving environments, where rainfall is similar to evapotranspiration, and where agriculture has historically been most successful, (2) environments which are too cold, where temperature limits plant and animal growth, (3) conditions which are too dry and where precipitation limits production, and (4) environments that are too wet, where rainfall in excess of evapotranspiration results in high levels of competition, predation and leaching. He then reports on the productivity, nutrient retention and pest response of mimics of successional ecosystems that he and co-workers have developed in the humid tropics. His conclusion is that it pays to imitate natural systems, especially the use of perennial plants, to maintain soil fertility, protect against erosion and make full use of light, water and nutrients. He notes however that it may be necessary to accept lower but more biologically sustainable-yields as (1) natural systems intrinsically feature high investments in structure and (2) high levels of fertiliser tend to switch off specialised mechanisms of nutrient input and cycling such as performed by nitrogen-fixing nodules, mycorrhizae, and proteoid roots.

Main poses the question 'How much biodiversity is enough?' in the context of sustainable agriculture, and provides the answer that this can only be addressed meaningfully in terms of specific criteria such as yield and the restoration of particular ecosystem services. A major problem is that we have not agreed on criteria or goals which are acceptable and desirable. Ecosystem function may be restored or retained in one area at one time, but systems are so dynamic and spatially heterogeneous that solutions must be equally dynamic and adaptable to each situation being addressed. He emphasises that there must be built-in redundancy to ensure that agriculture can adapt to changing conditions and that such risk reduction should be a primary motivation for increasing the diversity in agricultural systems. His conclusion is that if we want to decide how diverse our agricultural systems have to be, we need to distinguish what functions are paramount and the time frames over which such functions need to be considered. In particular, risk aversion calls for greater diversity.

Hobbs and Morton review the current state of ecological science as it moves from the descriptive to the prescriptive. Several new relevant ecological paradigms are examined; nature's flux, multiple stable states, patchiness and landscape ecology, prediction and the value of ecology in guiding environmental management. Next the role of biodiversity in ecological function is discussed in detail with important lessons to be learned for agricultural systems with similar function but with probably considerably less structure than displayed by comparable natural vegetation. This is followed by an evaluation of how one might apply such ecological concepts when designing new and more sustainable agricultural systems. They conclude with a prescription for implementing such an approach, observing that with ecology at the crossroads, any future path that it takes must be incorporated into our methods for prescribing agroecosystem mimics.

Joffre, Rambal and Ratte provide a fascinating account of one of the oldest continuous farming systems in Europe, the cork oak dehesa of southern Spain and Portugal, and evaluate the system from the perspective of the mimic concept. They trace its history over the last 800 years and distinguish those characteristics that have been dependent on human intervention from those that have been shaped primarily by the edaphic and climatic factors of the region. Their assessment is that the persistence of the dehesa system has been determined economically by the diversity of its plant and animal products while being continuously shaped ecologically by conservative resource use resulting from sub-optimal tree density.

Trenbath considers several traditional Indian cropping systems as examples of economically viable and environmentally selective multi-species mimics. He points out that trade-offs are always apparent in multi-species systems, while they may have distinct advantages in terms of production and stability, they are more difficult to manage effectively than monocultures. Two important messages from the paper are that there is a trade-off between short-term production and long-term sustainability and that existence of synergistic or complementary relationships between species can be important in the development of effective species mixtures. He concludes that complex systems are more difficult to manage and often defer short-term production gains for longer-term stability; and that both factors mitigate against the implementation of more complex systems.

Ong and Leakey examine resource sharing between trees and crops in the semi-arid tropics. It attempts to reconcile differences between recent agroforestry research and interactions between savannah trees and understorey vegetation by examining these from the perspective of succession. While productivity of natural vegetation under savannah trees generally increases as rainfall decreases, it appears the opposite may be true for agroforestry. One explanation is that mature savannah trees have a high proportion of woody above-ground structure compared to foliage, so that water 'saved' by reduction in soil evaporation may more than offset water 'lost' through tree transpiration. By contrast, in alley cropping and other agroforestry practices with high tree density, the contribution of trees in providing a more beneficial microclimate may be negated by reductions in soil moisture through tree transpiration and interception (evaporation directly from the tree canopy). A basic problem is that investment in woody plant structure requires time and effort, thereby reducing returns to farmers. They conclude that the greatest opportunity for agroforestry practices would appear to exist in filling niches in landscapes where resources are currently under-utilised by crops. In this way one would hope to mimic the large scale patch dynamics and successional progression of natural savannah ecosystems. Mimicking natural savannah systems at small scale comes with the penalty of a long wait before trees mature and benefits in understorey productivity can eventuate.

Van Noordwijk and Ong identify two distinct aspects of the mimic hypothesis: (1) that the resource use of agroecosystems might be improved by mimicking the structure and function of natural systems and (2) that mimic systems should benefit from reduced risk through incorporating more of the diversity of natural ecosystems. They assess these two aspects for African savannah ecosystems against the biotic and abiotic characteristics of those environments, specific constraints to resource use, and the potential for risk reduction through increased diversity. Their conclusion is that the resource use of agroecosystems can be improved by mimicking the structure and function of natural ecosystems but the related question of whether increased diversity confers stability becomes meaningful only when spatial scales of operation are made explicit (i.e. whether in relation to single plant, field, farm, industry, region). In other words, stability on a large spatial scale does not necessarily confer stability on constituent parts at lower levels, and *vice versa*.

Neher draws attention to the structure and function of highly complex soil food webs beneath agricultural and non-agricultural soils. The synergistic and competitive roles of various trophic groups of soil organisms in root growth, water movement, nutrient cycles and buffering are described, with an emphasis on the role of nematodes in resource cycling. This information is then used as basis for discussing agroecosystem management practices that might be directed at altering soil food webs to resemble more closely those found in the soils of native vegetation. She concludes that agricultural and non-agricultural soils typically show radical difference in their respective soil biota but much more research is needed to determine the role of such organisms in natural systems mimics.

Hatton and Nulsen provide a sobering contrast between the hydrological cycles of native vegetation and the mostly annual crop-based agroecosystems which have replaced it across vast tracts of southern Australia. The extent and nature of the associated impacts on agricultural landscapes are discussed in terms of lowered leaf-area index, decreased evapotranspiration and increased groundwater recharge. Functional mimics that address these adverse impacts are likely to require extensive tree planting at catchment scales. Economically viable agricultural systems that incorporate the proportionately large tree component have yet to be developed, and it is unlikely that pre-European hydrological cycles can be fully restored unless better incentives are provided. They conclude that co-ordinated intervention on a massive scale would be needed to restore the leaf area indices to levels necessary for recharge control and that this would involve revegetation of most or all of the cleared landscape.

Grierson and Adams compare the key attributes of nutrient cycling and growth of the forests of south-western Australian with those of agroecosystems in the same region in order to consider how such knowledge might be applied towards improved agroecosystem management. Key attributes for mimics include root structure, root specialisations, growth rates, fire response, ability to cope with variations in spatial and temporal heterogeneity and species diversity. They conclude that given the low productivity of natural systems in southern Australia, the best approach would appear to be to mimic those functional characteristics that confer resilience upon the agroecosystem while at the same time achieving high productivity through nutrient import and management.

Stirzaker views the problems of high input irrigated horticulture as analogous to the challenges facing life in the humid tropics where management of excess water is vital in limiting nutrient leaching and soil degradation. He asks the question is it possible to optimise management to solve these problems, or are entirely new designs required. His conclusion is that improved management is not enough, and that we need to address the way these agro-ecosystems function in the areas of water and nutrient capture and soil management. He suggests that improving the management of irrigated systems to the highest levels of precision possible will still not reduce or reverse degrading processes. New designs for production are called for where sources and sinks of water and nutrients are better matched in time and space.

Pate and Dawson describe novel and well tried experimental approaches for monitoring current and past performances of woody plant species in the utilisation of key resources within ecosystems and suggest how such assessment procedures might be useful in monitoring newly designed agricultural systems that attempt to mimic natural systems. Through case studies, the authors demonstrate the use and interpretation of stable isotope techniques coupled to collection and analysis of substances transported in the sap of xylem and phloem of an array of species. They discuss prospective uses of these techniques in studying managed ecosystems in which woody species have been introduced with the aim of maintaining functional stability. They conclude that these techniques help meet the basic requirement of understanding which species perform which functions in natural and production systems, and how effectively they operate in that regard.

Lefroy and Stirzaker examine the prospects for agroforestry as a means of managing the salinity problems which resulted following the introduction of grain- and annual pasture-based agriculture into the semi-arid woodland and heath landscapes of southern Australia. The relationships between trees, crops and their below ground environment that are likely to influence the water use of agroforestry systems are discussed. A water balance study is described for an apparently ideal situation on sandplain soils where alley and plantation plantings of the tree legume tagasaste have access to a shallow fresh water table. Factors that are likely to influence the proportion of a landscape required under trees are discussed, including recharge rates, size of the saturated zone within the reach of trees, water quality and tree crop interactions. They conclude that factors such as relief, hydraulic conductivity of soil and salt storage, all legacies of the geologic age of the study region, mean that the use of trees for salinity management would necessitate returning this landscape to its original open woodland structure, involving significant displacement of grain- and pasture-based agriculture.

Pate and Bell develop an extensive morphological and functional description of a Banksia woodland growing on deep, nutrient-impoverished sands in south western Australia. Special emphasis is placed on complementarity between species of differing structure, function and phenology. The form and function of mimic systems are then prescribed for two scenarios by integrating data on major growth forms, root morphologies, phenologies, nutrient acquisition strategies and fire responses. The two scenarios are: (1) restoration of just-cleared woodland and (2) land needing rehabilitation after long periods under agriculture. Their conclusion is that extensive structural and functional knowledge of an original ecosystem is required before one can ever hope to generate a mimic prescription which will effectively replace a complex native ecosystem.

Dunin, Williams, Verberg and Keating compare the uptake and deep drainage components of water balance data for natural ecosystems, annual species-based agricultural systems, and a cropping systems that incorporates phases of the perennial pasture legume lucerne in south eastern Australia. They observe that farming systems based on annual plants are restricted in their ability to reduce deep drainage and increase water uptake by rooting depth, which limits access to soil water reserves, and phenology, which limits transpiration to winter and spring. They conclude that lucerne grown in rotation with annual crops provides a buffer of dry soil which can halve deep drainage. They also note that this would need to be halved again to emulate natural ecosystems and achieve complete recharge control.

Hobbs and O'Connor begin with the observation that situations where we have adequate ecological knowledge from which to develop sound mimics are likely to be rare in an Australian context. They suggest therefore that we have to be prepared to make best-bet estimates and use these to develop experimental mimic systems. They assess the potential for predicting likely mimics for situations where detailed data on the ecophysiology and ecosystem function of the native ecosystem are not yet available by examining two such ecosystems in south western Australia, and draw together available information on these. Possible mimics are suggested utilising aspects of the native vegetation most likely to contribute to ecosystem function and resilience, but which still allow the production of a harvestable agricultural product. They conclude that even where we do not know much about the ecology of a native ecosystem, we can draw together whatever information is available to generate a first approximation of an agricultural mimic.

Pannell provides a comprehensive review of the issues involved in the adoption of new farming systems, particularly the complex innovations of the sort likely to be involved in mimic systems. He identifies the key problems as: (1) creating awareness of the innovation, (2) the feasibility of testing it on a small scale, (3) the perceived worthiness of testing it, and (4) the ease of demonstrating that it is more profitable than current practice. His conclusion is that the major challenges facing the idea of mimic systems is to develop farming systems which are more profitable, to find ways of demonstrating that they are indeed more profitable and to overcome the uncertainty associated with such significant change.

Passioura reviews the papers presented at the workshop from the perspective of agriculture in southern Australia and asks the questions what short and long term lessons can be learned from the study of natural ecosystems and how they might be applied. He observes that recent conservation farming techniques have been based on attributes of natural systems and comments on their efficacy and general suitability. He concludes that an aspect of natural systems that is missing from the agricultural landscape at present is a reflection of the mosaic nature of land. Were this to be incorporated, use would be more closely aligned to inherent characteristics. The problem is that lessons from natural systems must be profitable for farmers to adopt them

2. Emerging themes

In debating the consequences and value of incorporating characteristics of natural ecosystems into agricultural systems, several themes have emerged.

2.1. *What to mimic and why*

Approaching such a broad topic, it is understandable that there is potential for ambiguity in interpreting just what is meant by the proposal to mimic a natural ecosystem. It is worth remembering that our starting point was the work of Wes Jackson and The Land Institute (Jackson, 1985; Jackson and Piper, 1989), and Jack Ewel and colleagues (Ewel et al., 1991). In those cases, the structure and composition of natural ecosystems were studied as potential clues to, or surrogates for, underlying process. Their objective has been the very pragmatic one of incorporating diversity of function for improved persistence, resilience and efficiency of resource capture. Their interest, and ours in this volume, has not been the incorporation of compositional or taxonomic diversity for its intrinsic value, or in recognition of the many ways humans value natural ecosystems as aesthetic, life supporting, or more tangible resources (Fox, 1992). Nor is our interest in natural ecosystems a reflection of what Schama (1995) refers to as the penitential style of environmental history, where humans are seen as the villains in the landscape, and all that is natural is inherently good – an issue which is touched on by Van Noordwijk and Ong. Rather our interest has been the pragmatic one of improving the sustainability of agriculture. In that sense, the preceding papers can be seen as lying along a continuum that ranges from weak to strong acceptance of the mimic concept and the likely benefits which might result from its adoption. This range of views can be seen as recognition that understanding the processes occurring in natural ecosystems can be a valuable guide to:

1. Rehabilitating degraded natural ecosystems (e.g. Pate and Bell)
2. Understanding the performance of agricultural systems (e.g. Ong and Leakey; Van Noordwijk and Ong)
3. Modifying existing agricultural systems to incorporate more attributes of natural systems (e.g. Dunin et al., Passioura)
4. Developing new agricultural systems by consciously mimicking the structure and function of relevant natural ecosystems (e.g. Ewel; Jackson 1985)

2.2. *Synchronising sources and sinks of water and nutrients in time and space*

Stirzaker points out that the most sophisticated management in the world can not overcome the fundamental structural problems inherent in many high input agricultural systems, which he describes as classic mis-matches between sources and sinks of water and nutrients in space and time. Such imbalances in the farming systems of the humid tropics were the motivation for the work of Ewel and colleagues. Attempts to correct it are implicit in the farming systems described by Hatton and Nulsen, Dunin et al., Van Noordwijk and Ong, Ong and Leakey, Lefroy and Stirzaker and Hobbs and O'Connor. Taking on this characteristic of natural ecosystems represents an imperative and an opportunity. The imperative is to internalise the externalities characteristic of modern high production agriculture, i.e. to account for water and nutrients on-site rather than exporting them and leaving others to deal with the pollution and eutrophication that commonly result. The opportunity is that by so doing, resources might be more efficiently channeled into harvestable products. As Ong and Leakey point out however, this is difficult to achieve in practice. In their examination of agroforestry, they trace the historical shift from the initial expectation that complementarity of resource use between trees and crops would be the norm, to the current approach of carefully targeted opportunism based on an understanding of phenology, morphology and successional processes.

Effective balancing of sources and sinks of water and nutrients also requires an understanding of the subtleties of resource capture and transfer at a level that does not exist for many ecosystems. This makes Pate and Dawson's contribution a timely and valuable introduction to a range of monitoring techniques, some well tested, some novel.

2.3. *Perennial plants in polyculture*

Achieving the objective of matching sources and sinks of water and nutrients in space and time will almost inevitably require land use systems composed of a mixture of species of various life forms, a point particularly illustrated by the work of Ewel, Hatton and Nulsen, Pate and Bell and Dunin et al. Given that each species in a mixture is likely to have to justify its inclusion on the basis of its direct rather than its indirect values, as argued by Van Noordwijk and Ong, mimic systems are likely to be based on the multiple products of perennial polycultures. It is worth noting that there are probably more land use systems based on polycultures than on monocultures (Table 1), although the latter are more extensive and responsible for a higher proportion of the worlds food production. The point is that polycultures are not new, and as well as studying natural ecosystems, there is likely to be much to learn from traditional polycultures such as the first five types in Table 1 which are found

Table 1. Land use types by life form, product and cultural system.

Cultural type	Life forms		Product	Land use system
1. Polyculture	woody	perennial	fruit and seed	mixed orchards
2. Polyculture	woody	perennial	vegetative	mixed woodlot
3. Polyculture	woody and herbaceous	annual and perennial	fruit, seed and vegetative	alley cropping
4. Polyculture	woody and herbaceous	annual and perennial	fruit, seed and vegetative	forest garden
5. Polyculture	herbaceous	annual	fruit and seed	mixed cropping
6. Polyculture	herbaceous	annual	vegetative	pasture
7. Polyculture	herbaceous	perennial	fruit and seed	perennial grain crops
8. Polyculture	herbaceous	perennial	vegetative	pasture
9. Monoculture	woody	perennial	fruit and seed	orchard
10. Monoculture	woody	perennial	vegetative	timber plantation
11. Monoculture	herbaceous	annual	fruit and seed	cropping
12. Monculture	herbaceous	annual	vegetative	silage
13. Monoculture	herbaceous	perennial	fruit and seed	pasture seed crop
14. Monoculture	herbaceous	perennial	vegetative	hay crops and grazing

Source: Modified from Jackson (1985).

in the tropics (Ong and Leakey; Van Noordwijk and Ong; Altieri, 1992). The only land use system in Table 1 that does not exist yet is type 7, the subject of the research described by Jackson (1985), which is being developed as an alternative to type 11.

2.4. *Trade-offs between persistence and productivity*

Several papers argue that a trade off between persistence and productivity seems almost inevitable with mimic systems. Ewel, Grierson and Adams, Van Noordwijk and Ong and Pate and Bell point out that mimicking natural ecosystems is likely to result in a greater investment in infrastructure, particularly woody above- and below-ground biomass, with a consequent decrease in harvestable product. Ewel, Neher, and Grierson and Adams also point out the high investment in nutrient acquisition and cycling strategies mediated by soil biota that have an important role in persistence of natural ecosystems, but which are likely to be switched off by additions of fertiliser in more productive systems. Joffre et al., Hatton and Nulsen, Dunin et al., and Lefroy and Stirzaker refer to the investment in perennial structure in terms of the greater buffering capacity conferred by perennial species and the benefits that result from tighter cycling of water. Jackson and Dewald (1994) are dissenting voices on the subject of trade-offs, and therefore the most encouraging. Their work demonstrating that perenniality and high seed yield are not mutually exclusive trade-offs in grasses is, as Ewel observes, an exciting finding that opens the door to the prospect of perennial grass agriculture. Whether their example has general applicability to other species remains to be tested.

2.5. *Diversity and risk reduction*

Main, Hobbs and Morton, Trenbath and Van Noordwijk and Ong all tackle the complex question of the relationship between diversity and stability and how that relates to risk reduction from the farmers perspective. In terms of multiple cropping, Trenbath observes that this often means deferring short term production for longer term stability. Van Noordwijk and Ong however point out that the answer to the question of whether diversity improves stability in food production varies depending on the level at which the question is posed. Thus, food security at the regional level may be improved by low diversity at individual farm level. This question of scale emerges as a critical decision point for the application of the mimic concept, particularly as maintenance of ecosystems functions such as water and nutrient cycling and biodiversity conservation operate at larger scales and are often in conflict with the economic decisions crucial to the survival of individual farmers and foresters. Ewel makes the point that mimicry of form or function makes good ecological sense at the level of the individual species or process, for morphologies and processes are shaped by evolution and as such are proven survivors in the environment that shaped them. Mimicry of the entire complexity of natural ecosystems however is more problematic, and while it may be ill-suited to some situations, it is certain to be well-suited to others. Main similarly points out that the appropriate degree of diversity required depends entirely on the specific goals of mimic systems, which include restoration of ecosystem services as well as yield, and until those goals are made explicit, including the boundaries of consideration, the question of diversity cannot be addressed. Highlighting the complexity of this issue, Ewel points out from the experience of his constructed mimics that diversity cannot be counted on to afford protection from herbivores, and can sometimes have the opposite effect.

2.6. *The cost of persistence and the limits of theory*

If the long term persistence and yield stability of agricultural mimic systems are likely to come at the cost of short term productivity, how can they ever be sufficiently attractive to be adopted by farmers? That question may well highlight the limits of theory in resolving the problem of agriculture. Many authors conclude their analysis of the mimic concept with fairly pessimistic prognoses, particularly in terms of the trade-offs that are likely (e.g. Trenbath, Grierson and Adams, Hatton and Nulsen, Van Noordwijk and Ong, Lefroy and Stirzaker) and the complexities of management required (Pannell and Trenbath). And yet work based on the experience of constructing and testing mimic systems offer a more optimistic prognosis. Jackson and Dewald (1994), in demonstrating that high seed yield and persistence are not mutually exclusive, offer the specific hope for perennial grain agriculture despite theory to the contrary. Ewel and colleagues demonstrate that with carefully chosen but taxonomically very simple mimics, ecological targets based on the water and nutrient cycling of a complex rainforest community can be successfully met. This opens the way for land use systems based on species of high commercial value that meet the same criteria of structure and function.

The caution of many of the other contributors comes largely from their observation that there will be great difficulty in developing such systems to the point that they are manageable and can profitably compete with current practice. This difficulty is represented by the last step in the sequence below, adapted from Hobbs and Morton (this volume) and Dawson and Fry (1998), which attempts to summarise the application of the mimic concept:

1. Perception of the problems and the need for fundamental change, specifically identification of the ecosystem functions sub-optimal in the current managed systems.
2. Analysis of the form and function of natural ecosystems, particularly information on below ground plant architecture and activity, in order to identify key functions and functional groups in natural and managed ecosystems (e.g., Jackson (1985); Ewel; Pate and Bell).

3. Developing commercial crops from functional analogues – this means extending the range of conventional bio-prospecting to include consideration of the functional role of economic species in managed landscapes (e.g. Jackson (1985); Van Noordwijk and Ong). As mimicry at the landscape scale inevitably means a mosaic in time and space, there can by definition be no 'silver bullet' solutions in the form of one or two new crops as the dominant species.
4. Identify whether it is most appropriate to integrate or segregate these functions with production at field and landscape scales. This may rule out ecosystem mimicry at the field scale due to the costs of competition for resources between species in polycultures. Where the cost of that competition is greater than the benefits of added ecosystem services or biodiversity conservation, the segregation of roles (production versus resource and/or biodiversity conservation) would be the better option (Van Noordwijk and Ong).
5. Acknowledging people as part of the landscape and developing strategies to overcoming the obstacles facing adoption (see Pannell), particularly the profitability of alternatives relative to current practice and the feasibility of testing them.

2.6.1. *Adaptive management*

Despite the amount of effort put into steps 1 to 4, if the last condition is not met, little is achieved in practice, a point strongly made by Pannell. And yet to be overwhelmed by the last step may well be to give up before starting. An example of changing land use systems in south-western Australia may serve to illustrate this point. Over the last decade, the Tasmanian blue gum (*Eucalyptus globulus*) has been introduced into the higher rainfall areas (> 600 mm y^{-1}) as a alternative to meat, wool and milk production from farming systems based on annual plants. Harvested on rotation for fibre and sawn timber, it represents an alternative source of income. As a summer active perennial, it represents an opportunity to address the imbalance in the hydrological cycle that is a feature of annual based farming systems in this region. Catchments that now include plantings of blue gums, either segregated or integrated with conventional farming, cannot be considered as mimicking the original vegetation but could be regarded as a first step in that direction. In hindsight, this process can be seen as having fast tracked the above steps, with 1, 3 and 5 considered and steps 2, and 4 overlooked. The key to this alternative landuse becoming a viable option for farmers came with step 5; specifically a share farming scheme that now sees government, and increasingly private enterprise, in partnership with landowners. The landowner is paid an annuity based on the projected value of the trees at harvest while the partner bears the cost of establishment and harvest and shares the risks. There is also the prospect of longer term benefits through carbon trading.

This is not to suggest that blue gum agroforestry represents a mimic system. There are many issues to be addressed such as the appropriateness of site and species selection in the face of tree deaths, vulnerability to disease, the minimal level of functional diversity and the social protest that is occurring over the ecological and aesthetic impacts of a short term panacea based on monoculture. The point of this example however is that out of this may well evolve a more sustainable and commercially viable form of land use which may not have happened by simply progressing through the above steps in sequence, and certainly would not have happened without the intervention of step 5, the share farming scheme. In other words, solutions to complex problems at landscape scales are not going to be developed in the form of new, complete technological packages, but are more likely to evolve through a process of adaptive management (Holling, 1978; Walters, 1986) – learning by experiment and attending to the finer details once some momentum has been established.

The primary reason for interest in the ecosystem mimic concept in southern Australia is to develop crop and pasture production systems more closely tailored to the unique characteristics of Australia's climate and soils. The fact that very few new food plants have been domesticated in the last few centuries emphasises the challenge involved in identifying and domesticating new species. Finding alternatives to the wheat that occupies 15 million hectares of Australia is not a realistic option. Rather than find a substitute for annual crops, the approach suggested by the mimic concept would be to return deep rooted summer active species to the agricultural landscape in roughly the same proportions that existed prior to clearing. In some cases this will preclude broadacre farming, requiring forestry, horticulture or other land use systems based on woody perennials as shown in Table 1. Assuming the natural ecosystem is a good indicator of what is required in order to restore ecosystem processes, it is then up to the ingenuity of the designers, farmers foresters, ecologists and agronomists, to find commercially attractive products within those life form and functional constraints. The problem facing the drier areas of the continent, and in fact all biomes closer to the periphery of Ewel's ease-of-agriculture curve, is that lower growth rates and/or higher risks will limit options.

3. Voices from the periphery

A uniting theme that emerges from this collection of papers is that the physically and biologically very different environments studied by the contributors share positions at ecological and historical peripheries. In the sense used by Ewel, the semi-arid south west of Australia and the humid tropics of Central America both lie at an ecological periphery in terms of the difficulty with which the agricultural technology of their respective colonizers could be practiced. In the sense used by Jackson, many environments described here have low ecological amplitude.

In the historical sense, Robin (1997) has suggested that many significant insights in the field of ecology have come from the periphery of empire, from regions where European experience and the application of its technology have been found inappropriate. Settlers throughout history who have arrived 'with vision but not with sight', as Berry (1977) put it, have been culturally blinded to the ecological necessities and opportunities of new landscapes. Passioura alludes to the different contributions that American, British and Australian ecologists have made to that discipline, reflecting the different ways in which their respective physical and cultural environments have shaped their thinking. The problems experienced when agriculture has been introduced to environments at the periphery of Ewel's ease-of-agriculture curve is a theme that unites the contributors of this book. The resulting tensions have resulted in new insights and brought about a better appreciation of the processes that underwrite ecological sustainability.

Science at the frontier of settlement has often been directed in the first instance at the problems of adapting to new environments. Problem solving in fields such as entomology, tropical medicine, plant nutrition and veterinary science have led to new perspectives of these environments and in many cases, led to the local emergence of ecology as a discipline (Robin, 1997). The subject of this volume is a new way of looking at agriculture that is peripheral to both classical ecology and mainstream agronomy, a perspective that has resulted from the necessary marriage between these two fields (Jackson and Piper, 1989). It is hoped that the ideas presented here may encourage others to adopt this perspective and in so doing, advance both the discipline of ecology and the practice of agriculture.

4. References

Altieri MA (1995) Agroecology, 2nd ed. Westview, Boulder, 433 pp

Berry W (1977) The Unsettling of America: Culture and Agriculture. Sierra Club Books, 228 pp

Dawson TE and Fry R (1998) Agriculture in Nature's Image. Trends in Ecology and Evolution, 13(2): 50–51

Ewel JJ, Mazzarino MJ and Berish CW (1991) Tropical soil fertility changes under monocultures and successional communities of different structure. Ecological Applications 1: 289–302

Fox, W (1992) Towards Transpersonal Ecology; Developing New Foundations for Environmentalism. Shambhala Books, Boston

Holling CS (1978) Adaptive Environmental Assessment and Management. John Wiley and Sons, Chichester

Jackson W (1985) New Roots for Agriculture. University of Nebraska Press. Nebraska

Jackson W and Piper J (1989) The necessary marriage between ecology and agriculture. Ecology 70: 1591–1593

Jackson L and Dewald C (1994) Predicting evolutionary consequences of greater reproductive effort in *Tripsacum dactyloides*. Ecology 75: 627–641

Robin L (1997) Ecology: a science of empire? In: Griffiths T and Robin L (eds) Ecology and Empire; Environmental history and Settler 'ocieties, pp 63–75. Melbourne University Press, Melbourne

Schama S (1995) Landscape and Memory. Fontana, London

Walters C (1986) Adaptive Management of Renewable Resources, Macmillan, New York

Current Plant Science and Biotechnology in Agriculture

1. H.J. Evans, P.J. Bottomley and W.E. Newton (eds.): *Nitrogen Fixation Research Progress.* Proceedings of the 6th International Symposium on Nitrogen Fixation (Corvallis, Oregon, 1985). 1985 ISBN 90-247-3255-7
2. R.H. Zimmerman, R.J. Griesbach, F.A. Hammerschlag and R.H. Lawson (eds.): *Tissue Culture as a Plant Production System for Horticultural Crops.* Proceedings of a Conference (Beltsville, Maryland, 1985). 1986 ISBN 90-247-3378-2
3. D.P.S. Verma and N. Brisson (eds.): *Molecular Genetics of Plant-microbe Interactions.* Proceedings of the 3rd International Symposium on this subject (Montréal, Québec, 1986). 1987 ISBN 90-247-3426-6
4. E.L. Civerolo, A. Collmer, R.E. Davis and A.G. Gillaspie (eds.): *Plant Pathogenic Bacteria.* Proceedings of the 6th International Conference on this subject (College Park, Maryland, 1985). 1987 ISBN 90-247-3476-2
5. R.J. Summerfield (ed.): *World Crops: Cool Season Food Legumes.* A Global Perspective of the Problems and Prospects for Crop Improvement in Pea, Lentil, Faba Bean and Chickpea. Proceedings of the International Food Legume Research Conference (Spokane, Washington, 1986). 1988 ISBN 90-247-3641-2
6. P. Gepts (ed.): *Genetic Resources of* Phaseolus *Beans.* Their Maintenance, Domestication, Evolution, and Utilization. 1988 ISBN 90-247-3685-4
7. K.J. Puite, J.J.M. Dons, H.J. Huizing, A.J. Kool, M. Koorneef and F.A. Krens (eds.): *Progress in Plant Protoplast Research.* Proceedings of the 7th International Protoplast Symposium (Wageningen, The Netherlands, 1987). 1988 ISBN 90-247-3688-9
8. R.S. Sangwan and B.S. Sangwan-Norreel (eds.): *The Impact of Biotechnology in Agriculture.* Proceedings of the International Conference The Meeting Point between Fundamental and Applied in vitro Culture Research (Amiens, France, 1989). 1990. ISBN 0-7923-0741-0
9. H.J.J. Nijkamp, L.H.W. van der Plas and J. van Aartrijk (eds.): *Progress in Plant Cellular and Molecular Biology.* Proceedings of the 8th International Congress on Plant Tissue and Cell Culture (Amsterdam, The Netherlands, 1990). 1990 ISBN 0-7923-0873-5
10. H. Hennecke and D.P.S. Verma (eds.): *Advances in Molecular Genetics of Plant–Microbe Interactions.* Volume 1. 1991 ISBN 0-7923-1082-9
11. J. Harding, F. Singh and J.N.M. Mol (eds.): *Genetics and Breeding of Ornamental Species.* 1991 ISBN 0-7923-1094-2
12. J. Prakash and R.L.M. Pierik (eds.): *Horticulture – New Technologies and Applications.* Proceedings of the International Seminar on New Frontiers in Horticulture (Bangalore, India, 1990). 1991 ISBN 0-7923-1279-1
13. C.M. Karssen, L.C. van Loon and D. Vreugdenhil (eds.): *Progress in Plant Growth Regulation.* Proceedings of the 14th International Conference on Plant Growth Substances (Amsterdam, The Netherlands, 1991). 1992 ISBN 0-7923-1617-7
14. E.W. Nester and D.P.S. Verma (eds.): *Advances in Molecular Genetics of Plant–Microbe Interactions.* Volume 2. 1993 ISBN 0-7923-2045-X

Current Plant Science and Biotechnology in Agriculture

15. C.B. You, Z.L. Chen and Y. Ding (eds.): *Biotechnology in Agriculture.* Proceedings of the First Asia-Pacific Conference on Agricultural Biotechnology (Beijing, China, 1992). 1993 ISBN 0-7923-2168-5
16. J.C. Pech, A. Latché and C. Balagué (eds.): *Cellular and Molecular Aspects of the Plant Hormone Ethylene.* 1993 ISBN 0-7923-2169-3
17. R. Palacios, J. Mora and W.E. Newton (eds.): *New Horizons in Nitrogen Fixation.* Proceedings of the 9th International Congress on Nitrogen Fixation (Cancún, Mexico, 1992). 1993 ISBN 0-7923-2207-X
18. Th. Jacobs and J.E. Parlevliet (eds.): *Durability of Disease Resistance.* 1993 ISBN 0-7923-2314-9
19. F.J. Muehlbauer and W.J. Kaiser (eds.): *Expanding the Production and Use of Cool Season Food Legumes.* A Global Perspective of Peristent Constraints and of Opportunities and Strategies for Further Increasing the Productivity and Use of Pea, Lentil, Faba Bean, Chickpea, and Grasspea in Different Farming Systems. Proceedings of the Second International Food Legume Research Conference (Cairo, Egypt, 1992). 1994 ISBN 0-7923-2535-4
20. T.A. Thorpe (ed.): In Vitro *Embryogenesis in Plants.* 1995 ISBN 0-7923-3149-4
21. M.J. Daniels, J.A. Downie and A.E. Osbourn (eds.): *Advances in Molecular Genetics of Plant-Microbe Interactions.* Volume 3. 1994 ISBN 0-7923-3207-5
22. M. Terzi, R. Cella and A. Falavigna (eds.): *Current Issues in Plant Molecular and Cellular Biology.* Proceedings of the VIIIth International Congress on Plant Tissue and Cell Culture (Florence, Italy, 1994). 1995 ISBN 0-7923-3322-5
23. S.M. Jain, S.K. Sopory and R.E. Veilleux (eds.): In Vitro *Haploid Production in Higher Plants.* Volume 1: Fundamental Aspects and Methods. 1996 ISBN 0-7923-3577-5
24. S.M. Jain, S.K. Sopory and R.E. Veilleux (eds.): In Vitro *Haploid Production in Higher Plants.* Volume 2: Applications. 1996 ISBN 0-7923-3578-3
25. S.M. Jain, S.K. Sopory and R.E. Veilleux (eds.): In Vitro *Haploid Production in Higher Plants.* Volume 3: Important Selected Plants. 1996 ISBN 0-7923-3579-1
26. S.M. Jain, S.K. Sopory and R.E. Veilleux (eds.): In Vitro *Haploid Production in Higher Plants.* Volume 4: Cereals. 1996 ISBN 0-7923-3978-9
27. I.A. Tikhonovich, N.A. Provorov, V.I. Romanov and W.E. Newton (eds.): *Nitrogen Fixation: Fundamentals and Applications.* 1995 ISBN 0-7923-3707-7
28. N.L. Taylor and K.H. Quesenberry: *Red Clover Science.* 1996 ISBN 0-7923-3887-1
29. S.M. Jain, S.K. Sopory and R.E. Veilleux (eds.): In Vitro *Haploid Production in Higher Plants.* Volume 5: Oil, Ornamental and Miscellaneous Plants. 1996 ISBN 0-7923-3979-7
30. R.H. Ellis, M. Black, A.J. Murdoch and T.D. Hong (eds.): *Basic and Applied Aspects of Seed Biology.* 1997 ISBN 0-7923-4363-8
31. C. Elmerich, A. Kondorosi and W.E. Newton (eds.): *Biological Nitrogen Fixation for the 21st Century.* 1998 ISBN 0-7923-4834-6

Current Plant Science and Biotechnology in Agriculture

32. S.M. Jain, D.S. Brar and B.S. Ahloowlia (eds.): *Somaclonal Variation and Induced Mutations in Crop Improvement.* 1998 ISBN 0-7923-4862-1
33. S.J. Bennett and P.S. Cocks (eds.): *Genetic Resources of Mediterranean Pasture and Forage Legumes.* 1999 ISBN 0-7923-5522-9
34. R. Knight (ed.): *Linking Research and Marketing Opportunities for Pulses in the 21st Century.* Proceedings of the Third International Food Legumes Research Conference. 1999 ISBN 0-7923-5565-2
35. C. Oropeza, J.L. Verdeil, G.R. Ashburner, R. Cardeña and J.M. Santamaria (eds.): *Current Advances in Coconut Biotechnology.* 1999 ISBN 0-7923-5823-6
36. A. Altman, M. Ziv and S. Izhar (eds.): *Plant Biotechnology and In Vitro Biology in the 21st Century.* 1999 ISBN 0-7923-5826-0

KLUWER ACADEMIC PUBLISHERS – DORDRECHT / BOSTON / LONDON

MIX
Papier aus verantwortungsvollen Quellen
Paper from responsible sources
FSC® C105338

If you have any concerns about our products,
you can contact us on
ProductSafety@springernature.com

In case Publisher is established outside the EU,
the EU authorized representative is:
Springer Nature Customer Service Center GmbH
Europaplatz 3, 69115 Heidelberg, Germany

Printed by Libri Plureos GmbH
in Hamburg, Germany